Principles of Communication Systems

mcgraw-hill electrical and electronic engineering series

Frederick Emmons Terman, Consulting Editor
W. W. Harman, J. G. Truxal, and R. A. Rohrer, Associate Consulting Editors

HARMAN · *Principles of the Statistical Theory of Communication*
HARMAN AND LYTLE · *Electrical and Mechanical Networks*
HAYASHI · *Nonlinear Oscillations in Physical Systems*
HAYT · *Engineering Electromagnetics*
HAYT AND KEMMERLY · *Engineering Circuit Analysis*
HILL · *Electronics in Engineering*
JAVID AND BROWN · *Field Analysis and Electromagnetics*
JOHNSON · *Transmission Lines and Networks*
KOENIG, TOKAD, AND KESAVAN · *Analysis of Discrete Physical Systems*
KRAUS · *Antennas*
KRAUS · *Electromagnetics*
KUH AND PEDERSON · *Principles of Circuit Synthesis*
KUO · *Linear Networks and Systems*
LEDLEY · *Digital Computer and Control Engineering*
LEPAGE · *Complex Variables and the Laplace Transform for Engineering*
LEPAGE AND SEELY · *General Network Analysis*
LEVI AND PANZER · *Electromechanical Power Conversion*
LEY, LUTZ, AND REHBERG · *Linear Circuit Analysis*
LINVILL AND GIBBONS · *Transistors and Active Circuits*
LITTAUER · *Pulse Electronics*
LYNCH AND TRUXAL · *Introductory System Analysis*
LYNCH AND TRUXAL · *Principles of Electronic Instrumentation*
LYNCH AND TRUXAL · *Signals and Systems in Electrical Engineering*
McCLUSKEY · *Introduction to the Theory of Switching Circuits*
MANNING · *Electrical Circuits*
MEISEL · *Principles of Electromechanical-energy Conversion*
MILLMAN AND HALKIAS · *Electronic Devices and Circuits*
MILLMAN AND TAUB · *Pulse, Digital, and Switching Waveforms*
MINORSKY · *Theory of Nonlinear Control Systems*
MISHKIN AND BRAUN · *Adaptive Control Systems*
MOORE · *Traveling-wave Engineering*
MURDOCH · *Network Theory*
NANAVATI · *An Introduction to Semiconductor Electronics*
OBERMAN · *Disciplines in Combinational and Sequential Circuit Design*
PETTIT AND McWHORTER · *Electronic Switching, Timing, and Pulse Circuits*
PETTIT AND McWHORTER · *Electronic Amplifier Circuits*
PFEIFFER · *Concepts of Probability Theory*
REZA · *An Introduction to Information Theory*
REZA AND SEELY · *Modern Network Analysis*
RUSTON AND BORDOGNA · *Electric Networks: Functions, Filters, Analysis*
RYDER · *Engineering Electronics*
SCHILLING AND BELOVE · *Electronic Circuits: Discrete and Integrated*
SCHWARTZ · *Information Transmission, Modulation, and Noise*
SCHWARTZ AND FRIEDLAND · *Linear Systems*
SEELY · *Electromechanical Energy Conversion*
SEIFERT AND STEEG · *Control Systems Engineering*
SHOOMAN · *Probabilistic Reliability: An Engineering Approach*
SISKIND · *Direct-current Machinery*
SKILLING · *Electric Transmission Lines*

Principles of Communication Systems

HERBERT TAUB
Professor of Electrical Engineering
The City College of the City University of New York

DONALD L. SCHILLING
Associate Professor of Electrical Engineering
The City College of the City University of New York

McGraw-Hill Book Company
NEW YORK   ST. LOUIS   SAN FRANCISCO
DÜSSELDORF   JOHANNESBURG   KUALA LUMPUR
LONDON   MEXICO   MONTREAL   NEW DELHI
PANAMA   RIO DE JANEIRO   SINGAPORE
SYDNEY   TORONTO

PRINCIPLES OF COMMUNICATION SYSTEMS

Library of Congress Catalog Card Number 72-109255

07-062923-4

67890 KPKP 79876543

To our wives
ESTHER and ANNETTE

Preface

This book is intended to serve as a one-semester text for a senior-level undergraduate or a first-year graduate course in communication systems. Consistent with the level of presentation, every effort has been made to ensure that the material included represents the present *state of the art* and current expectations of the direction of future developments. Accordingly, although analog communication systems are accorded a full and complete treatment, the emphasis is placed on digital systems.

This book is an outgrowth of courses, both undergraduate and graduate, given at The City College of New York, at the Polytechnic Institute of Brooklyn, and also to engineering personnel at the Communications Satellite Corporation, at the National Aeronautics and Space Administration (MSC-Houston), and at the Lockheed Electronics Corporation. For the undergraduate student and beginning graduate student the book provides the background required for advanced study in communication systems. The practicing engineer will find it of service to update his knowledge in the field.

Considerable thought and effort were devoted to the pedagogy of presentation and to the clarity of presentation. Great care has been exercised in the development of approximately 400 homework problems. These problems serve to elucidate the text, in some cases to extend the discussion, and are an integral and important part of the book.

The introductory chapters cover the mathematical background required for the remainder of the text. It is assumed that the reader has had some previous exposure to the elementary notions of spectral analysis including such topics as Fourier series and the Fourier transform. Spectral analysis is a mathematical tool of such wide applicability that we have judged it rather unlikely that the senior engineering student, employing the text, will not have had some experience with the subject. Hence Chap. 1, which deals with spectral analysis, is intended for the most part as a review and to refresh the students' knowledge.

Certain more advanced topics, such as the concepts of power spectral density and of the correlation between waveforms, are dealt with more fully.

The importance of probabilistic concepts in the analysis of communication systems cannot be overemphasized. A discussion of communications that does not take into account the presence of noise omits a basic and essential feature. On the other hand, it is all too easy to devote so much time to an introduction to probabilistic concepts that, within the constraint of a single semester, hardly any time remains to cover the subject of communications itself. We believe that we have steered a reasonable and effective middle course. All of the background in random variables and processes required in this text is included in a single chapter (Chap. 2).

The discussion of communication systems begins in Chap. 3. Frequency-division multiplexing systems are covered in Chap. 3, where amplitude-modulation systems are covered, and in Chap. 4, which presents angle-modulation systems. Time-division multiplexing systems are discussed in Chaps. 5 and 6. Chapter 5 discusses the sampling theorem and analyzes pulse-amplitude-modulation systems. The concept of quantization is introduced in Chap. 6, which is devoted to pulse-code-modulation systems. This chapter also considers the transmission of quantized and coded messages by means of phase-shift keying, differential phase-shift keying, and frequency-shift keying. Delta modulation is also discussed. It is, of course, well recognized that, in the absence of noise, there is little basis for comparison between systems. Nonetheless, in these four chapters (3 through 6), communication systems are discussed without reference to the manner in which noise affects their performance. We have found that such an initial presentation, excluding considerations of noise, is pedagogically very effective.

A mathematical representation of noise, using the concepts introduced in Chap. 2, is presented in Chap. 7. Both gaussian noise and shot noise are considered. Gaussian noise is represented as an expansion in terms of spectral components. This formulation, due to Rice, more than makes up in intuitive appeal what it lacks in rigor. In Chaps. 8 and 9 the discussion of frequency-division multiplexing systems (AM and FM) is extended to the analysis of the influence of noise on system performance.

Chapter 10 discusses the subject of threshold and threshold extension in frequency modulation. Since this chapter is one of the lengthier chapters in the text and is concerned with a subject which, in other introductory texts in communications, is disposed of in a few paragraphs, a word of explanation is in order. It is true enough that a very large number of methods are available for transmitting a message over a communications channel. The fact remains, however, that when a message must be transmitted through free space, from one antenna to another, we have no alternative but to superimpose the message on a carrier either by modulation of the carrier amplitude or by some form of frequency modulation of the carrier. When noise is a problem (practically always), the advantage generally lies with frequency modulation since such modulation allows us to sacrifice bandwidth for improved performance in the presence of a noise background. Frequency modulation, however, exhibits a phenomenon referred to as the *threshold effect* which limits the usefulness of frequency modulation in the presence of high background-noise levels. In recent years a very great effort has been made to develop methods (such as phase-locked loop systems and FM systems involving feedback) to remove this limitation to the greatest extent possible. It is our judgment, therefore, that the extended discussion of threshold in frequency modulation is well justified. The presentation will acquaint the student with the basic limitation of frequency modulation and with some of the schemes devised to extend the threshold. Furthermore, it will allow the student to approach the vast literature which has developed in this area. When a shorter and more qualitative account is required, such an account may be had through a reading of Secs. 10.1 through 10.3, 10.7 through 10.10, 10.13, and 10.14.

Chapters 11 and 12 discuss the performance of digital communication systems in the presence of noise. The matched-filter concept is studied, and the notion of probability of error is introduced as a criterion for system comparison. In connection with pulse-code modulation systems the effects of both quantization noise and background thermal noise are considered, leading again to a threshold effect.

The underlying unity of the theory of communication systems is presented in Chap. 13 which includes a discussion of *information theory*. Although the presentation is introductory, we believe it is more extensive and complete than comparable discussions in other introductory texts. Because of its own intrinsic interest and because of its special relevance to information theory, error-correcting coding is also discussed in this chapter. Block codes are considered as well as convolutional coding and sequential decoding. The chapter concludes with a comparison of the various systems of communication on the basis of the insight provided by information theory.

Finally, Chap. 14 serves to provide an overall view of a communication system. Sources of noise are discussed and concepts such as *noise figure*, *noise temperature*, and *path loss* are introduced.

We are pleased to acknowledge our indebtedness to Professor Jack K. Wolf of the Polytechnic Institute of Brooklyn, who reviewed the entire manuscript. His most gracious encouragement and very valued constructive criticism are most sincerely appreciated. The section in Chap. 13 on block coding is based on material written by Professor Wolf. We are grateful to Mr. Arnold Newton, Mr. A. Steven Rosenbaum, and Mr. Tamotsu Inukai who made many valuable suggestions. We express our appreciation to Miss Sadie Silverstein, administrative assistant of the Electrical Engineering Department at The City College, for her most skillful service in the preparation of the manuscript. We also thank Mr. L. J. Taub, Miss S. M. Taub, and Miss S. L. Schilling for their assistance.

HERBERT TAUB
DONALD L. SCHILLING

Contents

Principles of Communication Systems

1

Spectral Analysis

INTRODUCTION

Suppose that two people, separated by a considerable distance, wish to communicate with one another. If there is a pair of conducting wires extending from one location to another, and if each place is equipped with a microphone and earpiece, the communication problem may be solved. The microphone, at one end of the wire communications channel, impresses an electric signal voltage on the line, which voltage is then received at the other end. The received signal, however, will have associated with it an erratic, random, unpredictable voltage waveform which is described by the term *noise*. The origin of this noise will be discussed more fully in Chaps. 7 and 14. Here we need but to note that at the atomic level the universe is in a constant state of agitation, and that this agitation is the source of a very great deal of this noise. Because of the length of the wire link, the received message signal voltage will be greatly attenuated in comparison with its level at the transmitting end of the link. As a result, the message signal voltage may not be very large in comparison with the noise voltage, and the message will be perceived with difficulty or possibly not at all. An amplifier at the receiving end will not solve the problem, since the amplifier will amplify signal and noise alike. As a matter of fact, as we shall see, the amplifier itself may well be a source of additional noise.

A principal concern of communication theory and a matter which we discuss extensively in this book is precisely the study of methods to suppress, as far as possible, the effect of noise. We shall see that, for this purpose, it may be better not to transmit directly the original signal (the microphone output in our example). Instead, the original signal is used to generate a different signal waveform, which new signal waveform is then impressed on the line. This processing of the original signal to generate the transmitted signal is called *encoding* or *modulation*. At the receiving end an inverse process called *decoding* or *demodulation* is required to recover the original signal.

It may well be that there is a considerable expense in providing the wire communication link. We are, therefore, naturally led to inquire whether we may use the link more effectively by arranging for the simultaneous transmission over the link of more than just a single waveform. It turns out that such multiple transmission is indeed possible and may be accomplished in a number of ways. Such simultaneous multiple transmission is called *multiplexing* and is again a principal area of concern of communication theory and of this book. It is to be noted that when wire communications links are employed, then, at least in principle, separate links may be used for individual messages. When, however, the communications medium is free space, as in radio communication from antenna to antenna, multiplexing is essential.

In summary, then, *communication theory* addresses itself to the following questions. Given a communication channel, how do we arrange to transmit as many simultaneous signals as possible, and how do we devise to suppress the effect of noise to the maximum extent possible? In this book, after a few mathematical preliminaries, we shall address ourselves precisely to these questions, first to the matter of multiplexing, and thereafter to the discussion of noise in communications systems.

A branch of mathematics which is of inestimable value in the study of communications systems is *spectral analysis*. Spectral analysis concerns itself with the description of waveforms in the *frequency domain* and with the correspondence between the frequency-domain description and the time-domain description. It is assumed that the reader has some familiarity with spectral analysis. The presentation in this chapter is intended as a review, and will serve further to allow a compilation of results which we shall have occasion to use throughout the remainder of this text.

1.1 FOURIER SERIES[1]

A periodic function of time $v(t)$ having a fundamental period T_0 can be represented as an infinite sum of sinusoidal waveforms. This summation, called a *Fourier series*, may be written in several forms. One such form is the following:

$$v(t) = A_0 + \sum_{n=1}^{\infty} A_n \cos \frac{2\pi nt}{T_0} + \sum_{n=1}^{\infty} B_n \sin \frac{2\pi nt}{T_0} \tag{1.1-1}$$

The constant A_0 is the average value of $v(t)$ given by

$$A_0 = \frac{1}{T_0} \int_{-T_0/2}^{T_0/2} v(t) \, dt \tag{1.1-2}$$

while the coefficients A_n and B_n are given by

$$A_n = \frac{2}{T_0} \int_{-T_0/2}^{T_0/2} v(t) \cos \frac{2\pi nt}{T_0} \, dt \tag{1.1-3}$$

and

$$B_n = \frac{2}{T_0} \int_{-T_0/2}^{T_0/2} v(t) \sin \frac{2\pi nt}{T_0} \, dt \tag{1.1-4}$$

An alternative form for the Fourier series is

$$v(t) = C_0 + \sum_{n=1}^{\infty} C_n \cos \left(\frac{2\pi nt}{T_0} - \phi_n \right) \tag{1.1-5}$$

where C_0, C_n, and ϕ_n are related to A_0, A_n, and B_n by the equations

$$C_0 = A_0 \tag{1.1-6a}$$
$$C_n = \sqrt{A_n^2 + B_n^2} \tag{1.1-6b}$$

and

$$\phi_n = \tan^{-1} \frac{B_n}{A_n} \tag{1.1-6c}$$

The Fourier series of a periodic function is thus seen to consist of a summation of harmonics of a fundamental frequency $f_0 = 1/T_0$. The coefficients C_n are called *spectral amplitudes;* that is, C_n is the amplitude of the *spectral component* $C_n \cos (2\pi n f_0 t - \phi_n)$ at frequency $n f_0$. A typical *amplitude spectrum* of a periodic waveform is shown in Fig. 1.1-1a. Here, at each harmonic frequency, a vertical line has been drawn having a length equal to the spectral amplitude associated with each harmonic frequency. Of course, such an amplitude spectrum, lacking the phase information, does not specify the waveform $v(t)$.

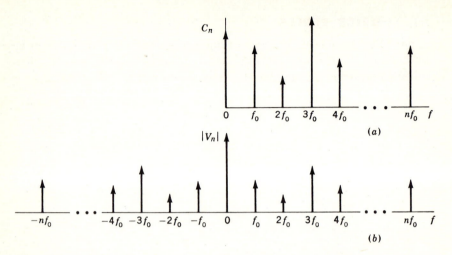

Fig. 1.1-1 (*a*) A one-sided plot of spectral amplitude of a periodic waveform. (*b*) The corresponding two-sided plot.

1.2 EXPONENTIAL FORM OF THE FOURIER SERIES

The exponential form of the Fourier series finds extensive application in communication theory. This form is given by

$$v(t) = \sum_{n=-\infty}^{\infty} V_n e^{j2\pi nt/T_0} \tag{1:2-1}$$

where V_n is given by

$$V_n = \frac{1}{T_0} \int_{-T_0/2}^{T_0/2} v(t) e^{-j2\pi nt/T_0} \, dt \tag{1.2-2}$$

The coefficients V_n have the property that V_n and V_{-n} are complex conjugates of one another, that is, $V_n = V_{-n}^*$. These coefficients are related to the C_n's in Eq. (1.1-5) by

$$V_0 = C_0 \tag{1.2-3a}$$

$$V_n = \frac{C_n}{2} e^{-j\phi_n} \tag{1.2-3b}$$

The V_n's are the *spectral amplitudes* of the *spectral components* $V_n e^{j2\pi nf_0 t}$. The amplitude spectrum of the V_n's shown in Fig. 1.1-1b corresponds to the amplitude spectrum of the C_n's shown in Fig. 1.1-1a. Observe that while $V_0 = C_0$, otherwise each spectral line in 1.1-1a at frequency f is replaced by the 2 spectral lines in 1.1-1b, each of half amplitude, one at frequency f and one at frequency $-f$. The amplitude spectrum in 1.1-1a is called a *single-sided* spectrum, while the spectrum in 1.1-1b is called a *two-sided* spectrum. We shall find it more convenient to use the two-sided amplitude spectrum and shall consistently do so from this point on.

1.3 EXAMPLES OF FOURIER SERIES

A waveform in which we shall have occasion to have some special interest is shown in Fig. 1.3-1a. The waveform consists of a periodic sequence of impulses of strength I. As a matter of convenience we have selected the time scale so that an impulse occurs at $t = 0$. The impulse at $t = 0$ is written as $I\,\delta(t)$. Here $\delta(t)$ is the delta function which has the property that $\delta(t) = 0$ except when $t = 0$ and further

$$\int_{-\infty}^{\infty} \delta(t)\,dt = 1 \tag{1.3-1}$$

The strength of an impulse is equal to the area under the impulse. Thus the strength of $\delta(t)$ is 1, and the strength of $I\,\delta(t)$ is I.

The periodic impulse train is written

$$v(t) = I \sum_{k=-\infty}^{\infty} \delta(t - kT_0) \tag{1.3-2}$$

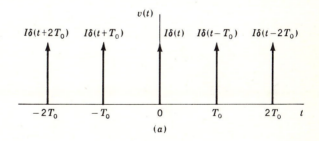

(a)

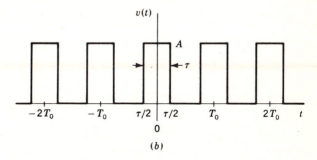

(b)

Fig. 1.3-1 Examples of periodic functions. (a) A periodic train of impulses. (b) A periodic train of pulses of duration τ.

We then find, using Eqs. (1.1-2) and (1.1-3),

$$A_0 = \frac{I}{T_0} \int_{-T_0/2}^{T_0/2} \delta(t)\, dt = \frac{I}{T_0} \tag{1.3-3}$$

$$A_n = \frac{2I}{T_0} \int_{-T_0/2}^{T_0/2} \delta(t) \cos \frac{2\pi n t}{T_0}\, dt = \frac{2I}{T_0} \tag{1.3-4}$$

and, using Eq. (1.1-4),

$$B_n = \frac{2I}{T_0} \int_{-T_0/2}^{T_0/2} \delta(t) \sin \frac{2\pi n t}{T_0}\, dt = 0 \tag{1.3-5}$$

Further we have, using Eq. (1.1-6),

$$C_0 = \frac{I}{T_0} \qquad C_n = \frac{2I}{T_0} \qquad \phi_n = 0 \tag{1.3-6}$$

and from Eq. (1.2-3)

$$V_0 = V_n = \frac{I}{T_0} \tag{1.3-7}$$

Hence $v(t)$ may be written in the forms

$$v(t) = I \sum_{k=-\infty}^{\infty} \delta(t - kT_0) = \frac{I}{T_0} + \frac{2I}{T_0} \sum_{n=1}^{\infty} \cos \frac{2\pi n t}{T_0}$$

$$= \frac{I}{T_0} \sum_{n=-\infty}^{\infty} e^{j2\pi n t/T_0} \tag{1.3-8}$$

As a second example, let us find the Fourier series for the periodic train of pulses of amplitude A and duration τ as shown in Fig. 1.3-1b.   We find

$$A_0 = C_0 = V_0 = \frac{1}{T_0} \int_{-T_0/2}^{T_0/2} v(t)\, dt = \frac{A\tau}{T_0} \tag{1.3-9}$$

$$A_n = C_n = 2V_n = \frac{2}{T_0} \int_{-T_0/2}^{T_0/2} v(t) \cos \frac{2\pi n t}{T_0}\, dt$$

$$= \frac{2A\tau}{T_0} \frac{\sin (n\pi\tau/T_0)}{n\pi\tau/T_0} \tag{1.3-10}$$

and

$$B_n = 0 \qquad \phi_n = 0 \tag{1.3-11}$$

Thus

$$v(t) = \frac{A\tau}{T_0} + \frac{2A\tau}{T_0} \sum_{n=1}^{\infty} \frac{\sin (n\pi\tau/T_0)}{n\pi\tau/T_0} \cos \frac{2\pi n t}{T_0} \tag{1.3-12a}$$

$$= \frac{A\tau}{T_0} \sum_{n=-\infty}^{\infty} \frac{\sin (n\pi\tau/T_0)}{n\pi\tau/T_0} e^{j2\pi n t/T_0} \tag{1.3-12b}$$

Suppose that in the waveform of Fig. 1.3-1*b* we reduce τ while adjusting A so that $A\tau$ is a constant, say $A\tau = I$. We would expect that in the limit, as $\tau \to 0$, the Fourier series for the pulse train in Eq. (1.3-12) should reduce to the series for the impulse train in Eq. (1.3-8). It is readily verified that such is indeed the case since as $\tau \to 0$

$$\frac{\sin (n\pi\tau/T_0)}{n\pi\tau/T_0} \to 1 \tag{1.3-13}$$

1.4 THE SAMPLING FUNCTION

A function frequently encountered in spectral analysis is the sampling function $Sa(x)$ defined by

$$Sa(x) \equiv \frac{\sin x}{x} \tag{1.4-1}$$

[A closely related function is sinc x defined by sinc $x = (\sin \pi x)/\pi x$.] The function $Sa(x)$ is plotted in Fig. 1.4-1. It is symmetrical about $x = 0$, and at $x = 0$ has the value $Sa(0) = 1$. It oscillates with an amplitude that decreases with increasing x. The function passes through zero at equally spaced intervals at values of $x = \pm n\pi$, where n is an integer other than zero. Aside from the peak at $x = 0$, the maxima and minima occur *approximately* midway between the zeros, i.e., at $x = \pm(n + \frac{1}{2})\pi$, where $|\sin x| = 1$. This approximation is poorest for the minima closest to $x = 0$ but improves as x becomes larger. Correspondingly, the approximate value of $Sa(x)$ at these extremal points is

$$Sa\left[\pm (n + \tfrac{1}{2})\pi\right] = \frac{2(-1)^n}{(2n + 1)\pi} \tag{1.4-2}$$

We encountered the sampling function in the preceding section in Eq. (1.3-12) which expresses the spectrum of a periodic sequence of rectangular pulses. In that equation we have $x = n\pi\tau/T_0$. The spectral amplitudes of Eq. (1.3-12*b*) are plotted in Fig. 1.4-1*b* for the case $A = 4$ and $\tau/T_0 = \frac{1}{4}$. The spectral components appear at frequencies which are multiples of the fundamental frequency $f_0 = 1/T_0$, that is, at frequencies $f = nf_0 = n/T_0$. The envelope of the spectral components of $Sa(\pi\tau f)$ is also shown in the figure. Here we have replaced x by

$$x = n\pi\tau/T_0 = \pi\tau nf_0 = \pi\tau f \tag{1.4-3}$$

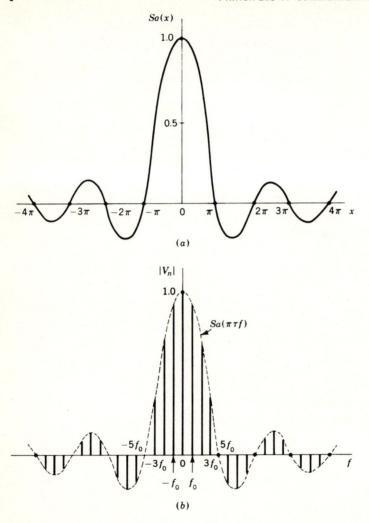

Fig. 1.4-1   (a) The function $Sa(x)$.   (b) The spectral amplitudes V_n of the two-sided Fourier representation of the pulse train of Fig. 1.3-1b for $A = 4$ and $\tau/T_0 = \frac{1}{4}$.

1.5 RESPONSE OF A LINEAR SYSTEM

The Fourier trigonometric series is by no means the only way in which a periodic function may be expanded in terms of other functions.[2]   As a matter of fact, the number of such possible alternative expansions is limitless.   However, what makes the Fourier trigonometric expansion especially useful is the distinctive and unique characteristic of the sinusoidal waveform, this characteristic being that when a sinusoidal excita-

Fig. 1.5-1 A sinusoidal waveform $v_i(t,\omega_n)$ of angular frequency ω_n is applied at the input to a network (filter) whose transfer characteristic at ω_n is $H(\omega_n)$. The output $v_o(t,\omega_n)$ differs from the input only in amplitude and phase.

tion is applied to a linear system, the response everyplace in the system is similarly sinusoidal and has the same frequency as the excitation. That is, the sinusoidal waveform preserves its waveshape. And since the waveshape is preserved, then, in order to characterize the relationship of the response to the excitation, we need but to specify how the response amplitude is related to the excitation amplitude and how the response phase is related to the excitation phase. Therefore with sinusoidal excitation, two numbers (amplitude ratio and phase difference) are all that are required to deduce a response. It turns out to be possible and very convenient to incorporate these two numbers into a single complex number.

Let the input to a linear system be the spectral component

$$v_i(t,\omega_n) = V_n e^{j2\pi nt/T_0} = V_n e^{j\omega_n t} \tag{1.5-1}$$

The waveform $v_i(t,\omega_n)$ may be, say, the voltage applied to the input of an electrical filter as in Fig. 1.5-1. Then the filter output $v_o(t,\omega_n)$ is related to the input by a complex transfer function

$$H(\omega_n) = |H(\omega_n)|e^{-j\theta(\omega_n)} \tag{1.5-2}$$

that is, the output is

$$
\begin{aligned}
v_o(t,\omega_n) = H(\omega_n)v_i(t,\omega_n) &= |H(\omega_n)|e^{-j\theta(\omega_n)}V_n e^{j\omega_n t} \\
&= |H(\omega_n)|V_n e^{j[\omega_n t - \theta(\omega_n)]}
\end{aligned} \tag{1.5-3}
$$

Actually, the spectral component in Eq. (1.5-1) is not a physical voltage. Rather, the physical input voltage $v_{ip}(t)$ which gives rise to this spectral component is the sum of this spectral component and its complex conjugate, that is,

$$v_{ip}(t,\omega_n) = V_n e^{j\omega_n t} + V_{-n}e^{-j\omega_n t} = V_n e^{j\omega_n t} + V_n^* e^{-j\omega_n t} = 2Re(V_n e^{j\omega_n t}) \tag{1.5-4}$$

The corresponding *physical* output voltage is $v_{op}(t,\omega_n)$ given by

$$v_{op}(t,\omega_n) = H(\omega_n)V_n e^{j\omega_n t} + H(-\omega_n)V_n^* e^{-j\omega_n t} \tag{1.5-5}$$

Since $v_{op}(t,\omega_n)$ must be real, the two terms in Eq. (1.5-5) must be complex conjugates, and hence we must have that $H(\omega_n) = H^*(-\omega_n)$. Therefore, since $H(\omega_n) = |H(\omega_n)|e^{-j\theta(\omega_n)}$, we must have that

$$|H(\omega_n)| = |H(-\omega_n)| \tag{1.5-6}$$

and

$$\theta(\omega_n) = -\theta(-\omega_n) \tag{1.5-7}$$

that is, $|H(\omega_n)|$ must be an even function and $\theta(\omega_n)$ an odd function of ω_n.

If, then, an excitation is expressed as a Fourier series in exponential form as in Eq. (1.2-1), the response is

$$v_o(t) = \sum_{n=-\infty}^{\infty} H(\omega_n) V_n e^{j2\pi nt/T_0} \tag{1.5-8}$$

If the form of Eq. (1.1-5) is used, the response is

$$v_o(t) = H(0)C_0 + \sum_{n=1}^{\infty} |H(\omega_n)|C_n \cos\left[\frac{2\pi nt}{T_0} - \phi_n - \theta(\omega_n)\right] \tag{1.5-9}$$

Given a periodic waveform, the coefficients in the Fourier series may be evaluated. Thereafter, if the transfer function $H(\omega_n)$ of a system is known, the response may be written out formally as in, say, Eq. (1.5-8) or (1.5-9). Actually these equations are generally of small value from a *computational* point of view. For, except in rather special and infrequent cases, we should be hard-pressed indeed to recognize the waveform of a response which is expressed as the sum of an infinite (or even a large) number of sinusoidal terms. On the other hand, the *concept* that a response may be written in the form of a linear superposition of responses to individual spectral components, as, say, in Eq. (1.5-8), is of inestimable value.

1.6 NORMALIZED POWER

In the analysis of communication systems, we shall often find that, given a waveform $v(t)$, we shall be interested in the quantity $\overline{v^2(t)}$ where the bar indicates the time-average value. In the case of periodic waveforms, the time averaging is done over one cycle. If, in our mind's eye, we were to imagine that the waveform $v(t)$ appear across a 1-ohm resistor, then the power dissipated in that resistor would be $\overline{v^2(t)}$ volts²/1 ohm $= W$ watts, where the number W would be numerically equal to the numerical value of $\overline{v^2(t)}$, the mean-square value of $v(t)$. For this reason $\overline{v^2(t)}$ is generally referred to as the *normalized power* of $v(t)$. It is to be kept in mind, however, that the dimension of normalized power is volts² and not watts.

When, however, no confusion results from so doing, we shall often follow the generally accepted practice of dropping the word "normalized" and refer instead simply to "power." We shall often have occasion also to calculate *ratios* of normalized powers. In such cases, even if the dimension "watts" is applied to normalized power, no harm will have been done, for the dimensional error in the numerator and the denominator of the ratio will cancel out.

Suppose that in some system we encounter at one point or another normalized powers S_1 and S_2. If the ratio of these powers is of interest, we need but to evaluate, say, S_2/S_1. It frequently turns out to be more convenient not to specify this ratio directly but instead to specify the quantity K defined by

$$K \equiv 10 \log \frac{S_2}{S_1} \qquad (1.6\text{-}1)$$

Like the ratio S_2/S_1, the quantity K is dimensionless. However, in order that one may know whether, in specifying a ratio, we are stating the number S_2/S_1 or the number K, the term *decibel* (abbreviated dB) is attached to the number K. Thus, for example, suppose $S_2/S_1 = 100$, then $\log S_2/S_1 = 2$ and $K = 20$ dB. The advantages of the use of the decibel are twofold. First, a very large power ratio may be expressed in decibels by a much smaller and therefore often more convenient number. Second, if power ratios are to be multiplied, such multiplication may be accomplished by the simpler arithmetic operation of addition if the ratios are first expressed in decibels. Suppose that S_2 and S_1 are, respectively, the normalized power associated with sinusoidal signals of amplitudes V_2 and V_1. Then $S_2 = V_2^2/2$, $S_1 = V_1^2/2$, and

$$K = 10 \log \frac{V_2^2/2}{V_1^2/2} = 20 \log \frac{V_2}{V_1} \qquad (1.6\text{-}2)$$

The use of the decibel was introduced in the early days of communications systems in connection with the transmission of signals over telephone lines. (The "bel" in decibel comes from the name Alexander Graham Bell.) In those days the decibel was used for the purpose of specifying ratios of *real* powers, not normalized powers. Because of this early history, occasionally some confusion occurs in the meaning of a ratio expressed in decibels. To point out the source of this confusion and, we hope, thereby to avoid it, let us consider the situation represented in Fig. 1.6-1. Here a waveform $v_i(t) = V_i \cos \omega t$ is applied to the input of a linear amplifier of input impedance R_i. An output signal $v_o(t) = V_o \cos (\omega t + \theta)$ then appears across the load resistor R_o. A real power $P_i = V_i^2/2R_i$ is supplied to the input, and the real power delivered

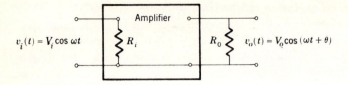

Fig. 1.6-1 An amplifier of input impedance R_i with load R_o.

to the load is $P_o = V_o^2/2R_o$. The real power gain P_o/P_i of the amplifier expressed in decibels is

$$K_{\text{real}} = 10 \log \frac{V_o^2/2R_o}{V_i^2/2R_i} \qquad (1.6\text{-}3)$$

If it should happen that $R_i = R_o$, then K_{real} may be written as in Eq. (1.6-2).

$$K_{\text{real}} = 20 \log \frac{V_o}{V_i} \qquad (1.6\text{-}4)$$

But if $R_i \neq R_o$, then Eq. (1.6-4) does not apply. On the other hand, if we calculate the normalized power gain, then we have

$$K_{\text{norm}} = 10 \log \frac{V_o^2/2}{V_i^2/2} = 20 \log \frac{V_o}{V_i} \qquad (1.6\text{-}5)$$

So far as the normalized power gain is concerned, the impedances R_i and R_o in Fig. 1.6-1 are *absolutely irrelevant*. If it should happen that $R_i = R_o$, then $K_{\text{real}} = K_{\text{norm}}$, but otherwise they would be different.

1.7 NORMALIZED POWER IN A FOURIER EXPANSION

Let us consider two typical terms of the Fourier expansion of Eq. (1.1-5). If we take, say, the fundamental and first harmonic, we have

$$v'(t) = C_1 \cos \left(\frac{2\pi t}{T_0} - \phi_1 \right) + C_2 \cos \left(\frac{4\pi t}{T_0} - \phi_2 \right) \qquad (1.7\text{-}1)$$

To calculate the normalized power S' of $v'(t)$, we must square $v'(t)$ and evaluate

$$S' = \frac{1}{T_0} \int_{-T_0/2}^{T_0/2} [v'(t)]^2 \, dt \qquad (1.7\text{-}2)$$

When we square $v'(t)$, we get the square of the first term, the square of the second term, and then the cross-product term. However, the two cosine functions in Eq. (1.7-1) are *orthogonal*. That is, when their product is integrated over a complete period, the result is zero. Hence in evaluating the normalized power, we find no term corresponding to this cross product. We find actually that S' is given by

$$S' = \frac{C_1^2}{2} + \frac{C_2^2}{2} \tag{1.7-3}$$

By extension it is apparent that the normalized power associated with the entire Fourier series is

$$S = C_0^2 + \sum_{n=1}^{\infty} \frac{C_n^2}{2} \tag{1.7-4}$$

Hence we observe that because of the orthogonality of the sinusoids used in a Fourier expansion, the total normalized power is the sum of the normalized power due to each term in the series separately. If we write a waveform as a sum of terms which are not orthogonal, this very simple and useful result will not apply. We may note here also that, in terms of the A's and B's of the Fourier representation of Eq. (1.1-1), the normalized power is

$$S = A_0^2 + \sum_{n=1}^{\infty} \frac{A_n^2}{2} + \sum_{n=1}^{\infty} \frac{B_n^2}{2} \tag{1.7-5}$$

It is to be observed that power and normalized power are to be associated with a *real* waveform and not with a *complex* waveform. Thus suppose we have a term $A_n \cos (2\pi nt/T_0)$ in a Fourier series. Then the normalized power contributed by this term is $A_n^2/2$ quite independently of all other terms. And this normalized power comes from averaging, over time, the product of the term $A_n \cos (2\pi nt/T)$ by *itself*. On the other hand, in the complex Fourier representation of Eq. (1.2-1) we have terms of the form $V_n e^{j2\pi nt/T_0}$. The average value of the square of such a term is zero. We find as a matter of fact that the contributions to normalized power come from product terms

$$V_n e^{j2\pi nt/T_0} V_{-n} e^{-j2\pi nt/T_0} = V_n V_{-n} = V_n V_n^* \tag{1.7-6}$$

The total normalized power is

$$S = \sum_{n=-\infty}^{n=+\infty} V_n V_n^* \tag{1.7-7}$$

Thus, in the complex representation, the power associated with a particular *real* frequency $n/T_0 = nf_0$ (f_0 is the fundamental frequency) is associated neither with the spectral component at nf_0 nor with the component at $-nf_0$, but rather with the *combination* of spectral components, one in the

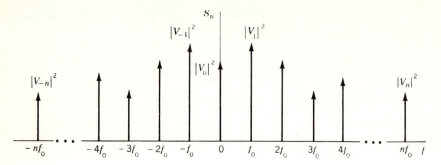

Fig. 1.7-1 A two-sided power spectrum.

positive-frequency range and one in the negative-frequency range. This power is

$$V_n V_n^* + V_{-n} V_{-n}^* = 2 V_n V_n^* \tag{1.7-8}$$

It is nonetheless a procedure of great convenience to associate one-half of the power in this combination of spectral components (that is, $V_n V_n^*$) with the frequency $n f_0$ and the other half with the frequency $-n f_0$. Such a procedure will always be valid provided that we are careful to use the procedure to calculate only the *total* power associated with frequencies $n f_0$ and $-n f_0$. Thus we may say that the power associated with the spectral component at $n f_0$ is $V_n V_n^*$ and the power associated with the spectral component at $-n f_0$ is similarly $V_n V_n^*$ ($= V_{-n} V_{-n}^*$). If we use these associations only to arrive at the result that the total power is $2 V_n V_n^*$, we shall make no error.

In correspondence with the one-sided and two-sided spectral amplitude pattern of Fig. 1.1-1 we may construct one-sided and two-sided spectral (normalized) power diagrams. A two-sided power spectral diagram is shown in Fig. 1.7-1. The vertical axis is labeled S_n, the power associated with each spectral component. The height of each vertical

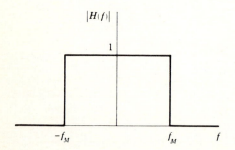

Fig. 1.7-2 The transfer characteristic of an idealized low-pass filter.

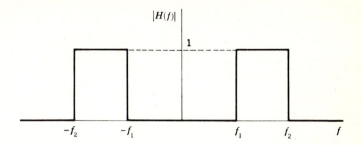

Fig. 1.7-3 The transfer characteristic of an idealized bandpass filter with passband from f_1 to f_2.

line is $|V_n|^2$. Because of its greater convenience and because it lends a measure of systemization to very many calculations, we shall use the two-sided amplitude and power spectral pattern exclusively throughout this text.

We shall similarly use a two-sided representation to specify the transmission characteristics of filters. Thus, suppose we have a low-pass filter which transmits without attenuation all spectral components up to a frequency f_M and transmits nothing at a higher frequency. Then the magnitude of the transfer function will be given as in Fig. 1.7-2. The transfer characteristic of a bandpass filter will be given as in Fig. 1.7-3.

1.8 POWER SPECTRAL DENSITY

Suppose that, in Fig. 1.7-1 where S_n is given for each spectral component, we start at $f = -\infty$ and then, moving in the positive-frequency direction, we add the normalized powers contributed by each power spectral line up to the frequency f. This sum is $S(f)$, a function of frequency. $S(f)$ typically will have the appearance shown in Fig. 1.8-1. It does not change as f goes from one spectral line to another, but jumps abruptly as the normalized power of each spectral line is added. Now let us inquire about the normalized power at the frequency f in a range df. This quantity of normalized power $dS(f)$ would be written

$$dS(f) = \frac{dS(f)}{df} \, df \qquad (1.8\text{-}1)$$

The quantity $dS(f)/df$ is called the (normalized) *power spectral density* $G(f)$; thus

$$G(f) \equiv \frac{dS(f)}{df} \qquad (1.8\text{-}2)$$

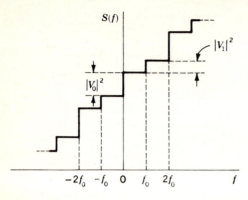

Fig. 1.8-1 The sum $S(f)$ of the normalized power in all spectral components from $f = -\infty$ to f.

The power in the range df at f is $G(f)\, df$. The power in the positive-frequency range f_1 to f_2 is

$$S(f_1 \leq f \leq f_2) = \int_{f_1}^{f_2} G(f)\, df \tag{1.8-3}$$

The power in the negative-frequency range $-f_2$ to $-f_1$ is

$$S(-f_2 \leq f \leq -f_1) = \int_{-f_2}^{-f_1} G(f)\, df \tag{1.8-4}$$

The quantities in Eqs. (1.8-3) and (1.8-4) do not have physical significance. However, the total power in the *real* frequency range f_1 to f_2 does have physical significance, and this power $S(f_1 \leq |f| \leq f_2)$ is given by

$$S(f_1 \leq |f| \leq f_2) = \int_{-f_2}^{-f_1} G(f)\, df + \int_{f_1}^{f_2} G(f)\, df \tag{1.8-5}$$

To find the power spectral density, we must differentiate $S(f)$ in Fig. 1.8-1. Between harmonic frequencies we would have $G(f) = 0$. At a harmonic frequency, $G(f)$ would yield an impulse of strength equal to the size of the jump in $S(f)$. Thus we would find

$$G(f) = \sum_{n=-\infty}^{\infty} |V_n|^2\, \delta(f - nf_0) \tag{1.8-6}$$

If, in plotting $G(f)$, we were to represent an impulse by a vertical arrow of height proportional to the impulse strength, then a plot of $G(f)$ versus f as given by Eq. (1.8-6) would have exactly the same appearance as the plot of S_n shown in Fig. 1.7-1.

1.9 EFFECT OF TRANSFER FUNCTION ON POWER SPECTRAL DENSITY

Let the input signal $v_i(t)$ to a filter have a power spectral density $G_i(f)$. If V_{in} are the spectral amplitudes of this input signal, then, using

Eq. (1.8-6)

$$G_i(f) = \sum_{n=-\infty}^{\infty} |V_{in}|^2 \, \delta(f - nf_0) \tag{1.9-1}$$

where, from Eq. (1.2-2),

$$V_{in} = \frac{1}{T_0} \int_{-T_0/2}^{T_0/2} v_i(t) e^{-j2\pi nt/T_0} \, dt \tag{1.9-2}$$

Let the output signal of the filter be $v_o(t)$. If V_{on} are the spectral amplitudes of this output signal, then the corresponding power spectral density is

$$G_o(f) = \sum_{n=-\infty}^{\infty} |V_{on}|^2 \, \delta(f - nf_0) \tag{1.9-3}$$

where

$$V_{on} = \frac{1}{T_0} \int_{-T_0/2}^{T_0/2} v_o(t) e^{-j2\pi nt/T_0} \, dt \tag{1.9-4}$$

As discussed in Sec. 1.5, if the transfer function of the filter is $H(f)$, then the output coefficient V_{on} is related to the input coefficient by

$$V_{on} = V_{in}H(f = nf_0) \tag{1.9-5}$$

Hence,

$$|V_{on}|^2 = |V_{in}|^2|H(f = nf_0)|^2 \tag{1.9-6}$$

Substituting Eq. (1.9-6) into Eq. (1.9-3) and comparing the result with Eq. (1.9-1) yields the important result

$$G_o(f) = G_i(f)|H(f)|^2 \tag{1.9-7}$$

Equation (1.9-7) is of the greatest importance since it relates the power spectral density, and hence the power, at one point in a system to the power spectral density at another point in the system. Equation (1.9-7) was derived for the special case of periodic signals; however, it applies to nonperiodic signals and signals represented by random processes (Sec. 2.19) as well.

As a special application of interest of the result given in Eq. (1.9-7), assume that an input signal $v_i(t)$ with power spectral density $G_i(f)$ is passed through a differentiator. The differentiator output $v_o(t)$ is related to the input by

$$v_o(t) = \tau \frac{d}{dt} v_i(t) \tag{1.9-8}$$

where τ is a constant. The operation indicated in Eq. (1.9-8) multiplies each spectral component of $v_i(t)$ by $j2\pi f\tau = j\omega\tau$. Hence $H(f) = j\omega\tau$, and $|H(f)|^2 = \omega^2\tau^2$. Thus from Eq. (1.9-7) the spectral density of the output is

$$G_0(f) = \omega^2\tau^2 G_i(f) \tag{1.9-9}$$

1.10 THE FOURIER TRANSFORM

A periodic waveform may be expressed, as we have seen, as a sum of spectral components. These components have finite amplitudes and are separated by finite frequency intervals $f_0 = 1/T_0$. The normalized power of the waveform is finite, as is also the normalized energy of the signal in an interval T_0. Now suppose we increase without limit the period T_0 of the waveform. Thus, say, in Fig. 1.3-1b the pulse centered around $t = 0$ remains in place, but all other pulses move outward away from $t = 0$ as $T_0 \rightarrow \infty$. Then eventually we would be left with a single-pulse nonperiodic waveform.

As $T_0 \rightarrow \infty$, the spacing of spectral components becomes infinitesimal. The frequency of the spectral components, which in the Fourier series was a discontinuous variable with a one-to-one correspondence with the integers, becomes instead a continuous variable. The normalized energy of the nonperiodic waveform remains finite, but, since the waveform is not repeated, its normalized power becomes infinitesimal. The spectral amplitudes similarly become infinitesimal. The Fourier series for the periodic waveform

$$v(t) = \sum_{n=-\infty}^{\infty} V_n e^{j2\pi n f_0 t} \tag{1.10-1}$$

becomes (see Prob. 1.10-1)

$$v(t) = \int_{-\infty}^{\infty} V(f) e^{j2\pi f t} \, df \tag{1.10-2}$$

The finite spectral amplitudes V_n are analogous to the infinitesimal spectral amplitudes $V(f) \, df$. The quantity $V(f)$ is called the *amplitude spectral density* or more generally the *Fourier transform* of $v(t)$. The Fourier transform is given by

$$V(f) = \int_{-\infty}^{\infty} v(t) e^{-j2\pi f t} \, dt \tag{1.10-3}$$

in correspondence with V_n, which is given by

$$V_n = \frac{1}{T_0} \int_{-T_0/2}^{T_0/2} v(t) e^{-j2\pi n f_0 t} \, dt \tag{1.10-4}$$

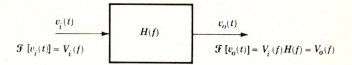

Fig. 1.10-1 A waveform $v_i(t)$ of transform $V_i(f)$ is transmitted through a network of transfer function $H(f)$. The output waveform $v_o(t)$ has a transform $V_o(f) = V_i(f)H(f)$.

Again, in correspondence with Eq. (1.5-8), let $H(f)$ be the transfer function of a network. If the input signal is $v_i(t)$, then the output signal will be $v_o(t)$ given by

$$v_o(t) = \int_{-\infty}^{\infty} H(f)V(f)e^{j2\pi ft}\,df \qquad (1.10\text{-}5)$$

Comparing Eq. (1.10-5) with Eq. (1.10-2), we see that the Fourier transform $V_o(f) \equiv \mathcal{F}[v_o(t)]$ is related to the transform $V_i(f)$ of $v_i(t)$ by

$$\mathcal{F}[v_o(t)] = H(f)\mathcal{F}[v_i(t)] \qquad (1.10\text{-}6)$$

or

$$V_o(f) = H(f)V_i(f) \qquad (1.10\text{-}7)$$

as indicated in Fig. 1.10-1.

In the following examples we shall evaluate the transforms of a number of functions both for the sake of review and also because we shall have occasion to refer to the results.

Example 1.10-1 If $v(t) = \cos \omega_0 t$, find $V(f)$.

Solution   The function $v(t) = \cos \omega_0 t$ is periodic, and therefore has a Fourier series representation as well as a Fourier transform.

The exponential Fourier series representation of $v(t)$ is

$$v(t) = \tfrac{1}{2}e^{+j\omega_0 t} + \tfrac{1}{2}e^{-j\omega_0 t} \qquad \omega_0 = \frac{2\pi}{T_0} \qquad (1.10\text{-}8)$$

Thus

$$V_1 = V_{-1} = \tfrac{1}{2} \qquad (1.10\text{-}9)$$

and

$$V_n = 0 \qquad n \neq \pm 1 \qquad (1.10\text{-}10)$$

The Fourier transform $V(f)$ is found using Eq. (1.10-3):

$$V(f) = \int_{-\infty}^{\infty} \cos \omega_0 t e^{-j2\pi ft} \, dt = \frac{1}{2} \int_{-\infty}^{\infty} e^{-j2\pi(f-f_0)t} \, dt$$
$$+ \frac{1}{2} \int_{-\infty}^{\infty} e^{-j2\pi(f+f_0)t} \, dt \quad (1.10\text{-}11a)$$

$$= \frac{1}{2}\delta(f - f_0) + \frac{1}{2}\delta(f + f_0) \quad\quad\quad (1.10\text{-}11b)$$

[See Eq. (1.10-29) below.]

From Eqs. (1.10-8) and (1.10-11) we draw the following conclusion: The Fourier transform of a sinusoidal signal (or other periodic signal) consists of impulses located at each harmonic frequency of the signal, i.e., at $f_n = n/T_0 = nf_0$. The strength of each impulse is equal to the amplitude of the Fourier coefficient of the exponential series.

Example 1.10-2 A signal $m(t)$ is multiplied by a sinusoidal waveform of frequency f_c. The product signal is

$$v(t) = m(t) \cos 2\pi f_c t \quad\quad\quad (1.10\text{-}12)$$

If the Fourier transform of $m(t)$ is $M(f)$, that is,

$$M(f) = \int_{-\infty}^{\infty} m(t)e^{-j2\pi ft} \, dt \quad\quad\quad (1.10\text{-}13)$$

find the Fourier transform of $v(t)$.

Solution   Since

$$m(t) \cos 2\pi f_c t = \frac{1}{2}m(t)e^{j2\pi f_c t} + \frac{1}{2}m(t)e^{-j2\pi f_c t} \quad\quad\quad (1.10\text{-}14)$$

then the Fourier transform $V(f)$ is given by

$$V(f) = \frac{1}{2} \int_{-\infty}^{\infty} m(t)e^{-j2\pi(f+f_c)t} \, dt + \frac{1}{2} \int_{-\infty}^{\infty} m(t)e^{-j2\pi(f-f_c)t} \, dt$$
$$(1.10\text{-}15)$$

Comparing Eq. (1.10-15) with Eq. (1.10-13), we have the result that

$$V(f) = \frac{1}{2}M(f + f_c) + \frac{1}{2}M(f - f_c) \quad\quad\quad (1.10\text{-}16)$$

The relationship of the transform $M(f)$ of $m(t)$ to the transform $V(f)$ of $m(t) \cos 2\pi f_c t$ is illustrated in Fig. 1.10-2. In Fig. 1.10-2b we see the spectral pattern of $M(f)$ replaced by two patterns of the same form. One is shifted to the right and one to the left, each by amount f_c. Further, the amplitudes of each of these two spectral patterns is one-half the amplitude of the spectral pattern $M(f)$.

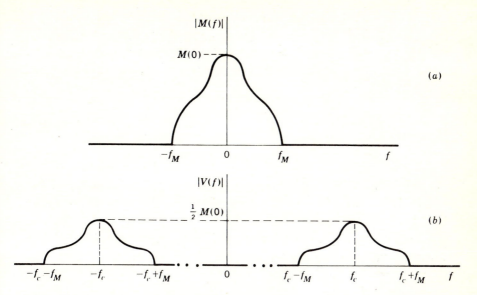

Fig. 1.10-2 (*a*) The amplitude spectrum of a waveform with no spectral component beyond f_M. (*b*) The amplitude spectrum of the waveform in (*a*) multiplied by $\cos 2\pi f_c t$.

A case of special interest arises when the waveform $m(t)$ is itself sinusoidal. Thus assume

$$m(t) = m \cos 2\pi f_m t \qquad (1.10\text{-}17)$$

where m is a constant. We then find that $V(f)$ is given by

$$V(f) = \frac{m}{4}\,\delta(f + f_c + f_m) + \frac{m}{4}\,\delta(f + f_c - f_m)$$

$$+ \frac{m}{4}\,\delta(f - f_c + f_m) + \frac{m}{4}\,\delta(f - f_c - f_m) \qquad (1.10\text{-}18)$$

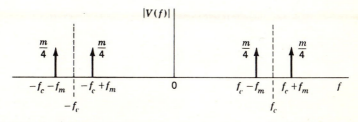

Fig. 1.10-3 The two-sided amplitude spectrum of the product waveform $v(t) = m \cos 2\pi f_m t \cos 2\pi f_c t$.

This spectral pattern is shown in Fig. 1.10-3. Observe that the pattern has four spectral lines corresponding to two real frequencies $f_c + f_m$ and $f_c - f_m$. The waveform itself is given by

$$v(t) = \frac{m}{4} \left[e^{j2\pi(f_c+f_m)t} + e^{-j2\pi(f_c+f_m)t} \right] + \frac{m}{4} \left[e^{j2\pi(f_c-f_m)t} + e^{-j2\pi(f_c-f_m)t} \right]$$

$$(1.10\text{-}19a)$$

$$= \frac{m}{4} \left[\cos 2\pi(f_c + f_m)t + \cos 2\pi(f_c - f_m)t \right] \qquad (1.10\text{-}19b)$$

Example 1.10-3 A pulse of amplitude A extends from $t = -\tau/2$ to $t = +\tau/2$. Find its Fourier transform $V(f)$. Consider also the Fourier series for a periodic sequence of such pulses separated by intervals T_0. Compare the Fourier series coefficients V_n with the transform in the limit as $T_0 \to \infty$.

Solution We have directly that

$$V(f) = \int_{-\tau/2}^{\tau/2} A e^{-j2\pi ft} \, dt = A\tau \frac{\sin \pi f\tau}{\pi f\tau} \qquad (1.10\text{-}20)$$

The Fourier series coefficients of the periodic pulse train are given by Eq. (1.3-12b) as

$$V_n = \frac{A\tau}{T_0} \frac{\sin (n\pi\tau/T_0)}{n\pi\tau/T_0} \qquad (1.10\text{-}21)$$

The fundamental frequency in the Fourier series is $f_0 = 1/T_0$. We shall set $f_0 \equiv \Delta f$ in order to emphasize that $f_0 \equiv \Delta f$ is the frequency interval between spectral lines in the Fourier series. Hence, since $1/T_0 = \Delta f$, we may rewrite Eq. (1.10-21) as

$$V_n = A\tau \frac{\sin (\pi n \, \Delta f\tau)}{\pi n \, \Delta f\tau} \Delta f \qquad (1.10\text{-}22)$$

In the limit, as $T_0 \to \infty$, $\Delta f \to 0$. We then may replace Δf by df and replace $n \, \Delta f$ by a continuous variable f. Equation (1.10-22) then becomes

$$\lim_{\Delta f \to 0} V_n = A\tau \frac{\sin \pi ft}{\pi ft} \, df \qquad (1.10\text{-}23)$$

Comparing this result with Eq. (1.10-20), we do indeed note that

$$V(f) = \lim_{\Delta f \to 0} \frac{V_n}{\Delta f} \qquad (1.10\text{-}24)$$

Thus we confirm our earlier interpretation of $V(f)$ as an *amplitude spectral density*.

Example 1.10-4 (a) Find the Fourier transform of $\delta(t)$, an impulse of unit strength.

(b) Given a network whose transfer function is $H(f)$. An impulse $\delta(t)$ is applied at the input. Show that the response $v_o(t) \equiv h(t)$ at the output is the inverse transform of $H(f)$, that is, show that $h(t) = \mathfrak{F}^{-1}[H(f)]$.

Solution (a) The impulse $\delta(t) = 0$ except at $t = 0$ and, further, has the property that

$$\int_{-\infty}^{\infty} \delta(t)\, dt = 1 \tag{1.10-25}$$

Hence

$$V(f) = \int_{-\infty}^{\infty} \delta(t) e^{-j2\pi ft}\, dt = 1 \tag{1.10-26}$$

Thus the spectral components of $\delta(t)$ extend with uniform amplitude and phase over the entire frequency domain.

(b) Using the result given in Eq. (1.10-7), we find that the transform of the output $v_o(t) \equiv h(t)$ is $V_o(f)$ given by

$$V_o(f) = 1 \times H(f) \tag{1.10-27}$$

since the transform of $\delta(t)$, $\mathfrak{F}[\delta(t)] = 1$. Hence the inverse transform of $V_o(f)$, which is the function $h(t)$, is also the inverse transform of $H(f)$. Specifically, for an impulse input, the output is

$$h(t) = \int_{-\infty}^{\infty} H(f) e^{j2\pi ft}\, df \tag{1.10-28}$$

We may use the result given in Eq. (1.10-28) to arrive at a useful representation of $\delta(t)$ itself. If $H(f) = 1$, then the response $h(t)$ to an impulse $\delta(t)$ is the impulse itself. Hence, setting $H(f) = 1$ in Eq. (1.10-28), we find

$$\delta(t) = \int_{-\infty}^{\infty} e^{j2\pi ft}\, df = \int_{-\infty}^{\infty} e^{-j2\pi ft}\, df \tag{1.10-29}$$

1.11 CONVOLUTION

Suppose that $v_1(t)$ has the Fourier transform $V_1(f)$ and $v_2(t)$ has the transform $V_2(f)$. What then is the waveform $v(t)$ whose transform is the product $V_1(f)V_2(f)$? This question arises frequently in spectral analysis and is answered by the *convolution theorem*, which says that

$$v(t) = \int_{-\infty}^{\infty} v_1(\tau) v_2(t - \tau)\, d\tau \tag{1.11-1}$$

or equivalently

$$v(t) = \int_{-\infty}^{\infty} v_2(\tau)v_1(t - \tau) \, d\tau \tag{1.11-2}$$

The integrals in Eq. (1.11-1) or (1.11-2) are called *convolution integrals*, and the process of evaluating $v(t)$ through these integrals is referred to as *taking the convolution* of the functions $v_1(t)$ and $v_2(t)$.

To prove the theorem, we begin by writing

$$v(t) = \mathcal{F}^{-1}[V_1(f)V_2(f)] \tag{1.11-3a}$$

$$= \frac{1}{2\pi} \int_{-\infty}^{\infty} V_1(f)V_2(f)e^{j\omega t} \, d\omega \tag{1.11-3b}$$

By definition we have

$$V_1(f) = \int_{-\infty}^{\infty} v_1(\tau)e^{-j\omega\tau} \, d\tau \tag{1.11-4}$$

Substituting $V_1(f)$ as given by Eq. (1.11-4) into the integrand of Eq. (1.11-3b), we have

$$v(t) = \frac{1}{2\pi} \int_{-\infty}^{\infty} \int_{-\infty}^{\infty} v_1(\tau)e^{-j\omega\tau} \, d\tau \, V_2(f)e^{j\omega t} \, d\omega \tag{1.11-5}$$

Interchanging the order of integration, we find

$$v(t) = \int_{-\infty}^{\infty} v_1(\tau) \left[\frac{1}{2\pi} \int_{-\infty}^{\infty} V_2(f)e^{j\omega(t-\tau)} \, d\omega \right] d\tau \tag{1.11-6}$$

We recognize that the expression in brackets in Eq. (1.11-6) is $v_2(t - \tau)$, so that finally

$$v(t) = \int_{-\infty}^{\infty} v_1(\tau)v_2(t - \tau) \, d\tau \tag{1.11-7}$$

Examples of the use of the convolution integral are given in Probs. 1.11-2 and 1.11-3. These problems will also serve to recall to the reader the relevance of the term *convolution*.

A special case of the convolution theorem and one of very great utility is arrived at through the following considerations. Suppose a waveform $v_i(t)$ whose transform is $V_i(f)$ is applied to a linear network with transfer function $H(f)$. The transform of the output waveform is $V_i(f)H(f)$. What then is the waveform of $v_o(t)$?

In Eq. (1.11-7) we identify $v_1(\tau)$ with $v_i(\tau)$ and $v_2(t)$ with the inverse transform of $H(f)$. But we have seen [in Eq. (1.10-28)] that the inverse transform of $H(f)$ is $h(t)$, the impulse response of the network. Hence Eq. (1.11-7) becomes

$$v_o(t) = \int_{-\infty}^{\infty} v_i(\tau)h(t - \tau) \, d\tau \tag{1.11-8}$$

in which the output $v_o(t)$ is expressed in terms of the input $v_i(t)$ and the impulse response of the network.

1.12 PARSEVAL'S THEOREM

We saw that for periodic waveforms we may express the normalized power as a summation of powers due to individual spectral components. We also found for periodic signals that it was appropriate to introduce the concept of *power spectral density*. Let us keep in mind that we may make the transition from the periodic to the nonperiodic waveform by allowing the period of the periodic waveform to approach infinity. A nonperiodic waveform, so generated, has a finite *normalized energy*, while the normalized power approaches zero. We may therefore expect that for a nonperiodic waveform the energy may be written as a continuous summation (integral) of energies due to individual spectral components in a continuous distribution. Similarly we should expect that with such nonperiodic waveforms it should be possible to introduce an *energy spectral density*.

The normalized energy of a periodic waveform $v(t)$ in a period T_0 is

$$E = \int_{-T_0/2}^{T_0/2} [v(t)]^2 \, dt \tag{1.12-1}$$

From Eq. (1.7-7) we may write E as

$$E = T_0 S = T_0 \sum_{n=-\infty}^{n=+\infty} V_n V_n^* \tag{1.12-2}$$

Again, as in the illustrative Example 1.10-3, we let $\Delta f \equiv 1/T_0 = f_0$, where f_0 is the fundamental frequency, that is, Δf is the spacing between harmonics. Then we have

$$V_n V_n^* = \frac{V_n}{\Delta f} \frac{V_n^*}{\Delta f} (\Delta f)^2 \tag{1.12-3}$$

and Eq. (1.12-2) may be written

$$E = \sum_{n=-\infty}^{\infty} \frac{V_n}{\Delta f} \frac{V_n^*}{\Delta f} \Delta f \tag{1.12-4}$$

In the limit, as $\Delta f \to 0$, we replace Δf by df, we replace $V_n/\Delta f$ by the transform $V(f)$ [see Eq. (1.10-24)], and the summation by an integral. Equation (1.12-4) then becomes

$$E = \int_{-\infty}^{\infty} V(f) V^*(f) \, df = \int_{-\infty}^{\infty} |V(f)|^2 \, df = \int_{-\infty}^{\infty} [v(t)]^2 \, dt \tag{1.12-5}$$

This equation expresses *Parseval's theorem*. Parseval's theorem is the extension to the nonperiodic case of Eq. (1.7-7) which applies for periodic waveforms. Both results simply express the fact that the power (periodic case) or the energy (nonperiodic case) may be written as the superposition of power or energy due to individual spectral components separately. The validity of these results depends on the fact that the spectral components are orthogonal. In the periodic case the spectral components are orthogonal over the interval T_0. In the nonperiodic case this interval of orthogonality extends over the entire time axis, i.e., from $-\infty$ to $+\infty$.

From Eq. (1.12-5), we find that the *energy density* is $G_E(f)$ given by

$$G_E(f) \equiv \frac{dE}{df} = |V(f)|^2 \tag{1.12-6}$$

in correspondence with Eq. (1.8-6), which gives the power spectral density for a periodic waveform.

Parseval's theorem may be derived more formally by a direct application of the convolution theorem and by other methods (see Prob. 1.12-1).

1.13 POWER AND ENERGY TRANSFER THROUGH A NETWORK

Suppose that a periodic signal $v_i(t)$ is applied to a network of transfer function $H(f)$ which yields an output $v_o(t)$. If $v_i(t)$ is written

$$v_i(t) = \sum_{n=-\infty}^{\infty} V_n e^{j2\pi n f_0 t} \tag{1.13-1}$$

then

$$v_o(t) = \sum_{n=-\infty}^{\infty} H(nf_0) V_n e^{j2\pi n f_0 t} \tag{1.13-2}$$

The power of $v_i(t)$ is

$$S_i = \sum_{n=-\infty}^{\infty} |V_n|^2 \tag{1.13-3}$$

while the power of $v_o(t)$ is

$$S_0 = \sum_{n=-\infty}^{\infty} |H(nf_0)|^2 |V_n|^2 \tag{1.13-4}$$

Thus, it is seen from Eq. (1.13-4) that, depending on $H(nf_0)$, the power associated with particular spectral components may be increased or decreased. Suppose for example that $H(nf_0) = h$ (a constant) for values of n between $n = i$ and $n = j$, and that $H(nf_0) = 0$ otherwise. Then the power associated with spectral components outside the range $i \leq |n| \leq j$

will be *lost,* and the network output waveform will be

$$v_o(t) = \sum_{n=-j}^{-i} h V_n e^{j2\pi n f_0 t} + \sum_{n=i}^{j} h V_n e^{j2\pi n f_0 t} \tag{1.13-5}$$

while the power output will be

$$S_0 = \sum_{n=-j}^{-i} h^2 |V_n|^2 + \sum_{n=i}^{j} h^2 |V_n|^2 = 2 \sum_{n=i}^{j} h^2 |V_n|^2 \tag{1.13-6}$$

Alternatively, as may be verified, S_0 may be written in terms of the output-power spectral density as

$$S_0 = \int_{-jf_0}^{-if_0} G(f) \, df + \int_{if_0}^{jf_0} G(f) \, df = 2 \int_{if_0}^{jf_0} G(f) \, df \tag{1.13-7}$$

Similarly, suppose the signal is nonperiodic and is transmitted through a network for which $H(f) = 1$ between the real frequencies f_1 and f_2 and $H(f) = 0$ otherwise. Then in correspondence with Eq. (1.13-7) the output energy is

$$E_0 = \int_{-f_2}^{-f_1} G_E(f) \, df + \int_{f_1}^{f_2} G_E(f) \, df = 2 \int_{f_1}^{f_2} G_E(f) \, df \tag{1.13-8}$$

where G_E is the energy spectral density.

1.14 BANDLIMITING OF WAVEFORMS

Rather generally, waveforms, periodic or nonperiodic, have spectral components which extend, at least in principle, to infinite frequency. Periodic waveforms may or may not have a dc component. Nonperiodic waveforms usually have spectral components extending to zero or near-zero frequency. Therefore, again, at least in principle, if an arbitrary waveform is to pass through a network without changing its shape, the transfer function of the network must not discriminate among spectral components. All spectral amplitudes must be increased or decreased by the same amount, and each spectral component must be delayed equally. A network which introduces no distortion, therefore, has the transfer function

$$H(f) = h_0 e^{-j\omega T_D} \tag{1.14-1}$$

in which h_0 is a constant and T_D is the delay time introduced by the network. An example of a network which produces no distortion is a length of transmission line of uniform cross section, having no losses, and properly terminated.

A signal may have spectral components which extend in the upper-frequency direction only up to a maximum frequency f_M. Such a signal is described as being *bandlimited* (at the high-frequency end) to f_M. If such a signal is passed through a network (filter) for which $H(f) = 1$ for $f \leq f_M$, the signal will be transmitted without distortion. Suppose, however, the signal is not precisely bandlimited to f_M but nonetheless, a very large part of its power or energy lies in spectral components below f_M. For example, suppose 99 percent of the power or energy lies below f_M. Then we may reasonably expect that the filter will introduce no serious distortion. Similarly, a signal may be bandlimited at the low-frequency end to a frequency f_L. And, again, if the signal is not precisely so bandlimited, but if a negligible fraction of the signal energy or power lies below f_L, a filter with low-frequency cutoff at f_L will produce negligible distortion.

We shall frequently have occasion to deal with waveforms which are not of themselves bandlimited but are passed through filters, introducing bandlimiting and hence producing distortion. The effect of such bandlimiting may be seen in a general sort of way by considering the response of the networks shown in Fig. 1.14-1 to a step function. We select the step function because it represents a combination of the fastest possible rate of change of voltage (rise time equal to zero) and the slowest possible (zero) rate of change of voltage after the abrupt rise. We select the circuit of Fig. 1.14-1a because it produces bandlimiting (albeit not abrupt) at the high-frequency end, while the circuit in Fig. 1.14-1b produces bandlimiting at the low-frequency end.

The low-pass RC circuit of Fig. 1.14-1a has a transfer function

$$H(f) = \frac{1}{1 + jf/f_2} \qquad (1.14\text{-}2)$$

where $f_2 = 1/(2\pi RC)$. The magnitude of $H(f)$ is plotted in Fig. 1.14-1c. At the frequency $f = f_2$, $|H(f)|$ has fallen to the value $1/\sqrt{2} = 0.707$, corresponding to a reduction of 3 dB from its value at $f = 0$. The frequency f_2 is called the 3-dB frequency of the network and is sometimes referred to as the *passband* of the network. For a step input of amplitude V the output is

$$v_o(t) = V(1 - e^{-t/RC}) = V(1 - e^{-2\pi f_2 t}) \qquad (1.14\text{-}3)$$

The essential distortion which has been introduced by the frequency discrimination of the network is that the output rises gradually rather than abruptly as does the input. We thus associate with the output waveform a *rise time*. This rise time may be defined in a variety of ways. One definition commonly employed is that the rise time is the time required for the output $v_o(t)$ to change from 0.1 to 0.9 volt. It may be verified from Eq. (1.14-3) that this rise time t_r is related to f_2 by

$$t_r f_2 = 0.35 \qquad (1.14\text{-}4)$$

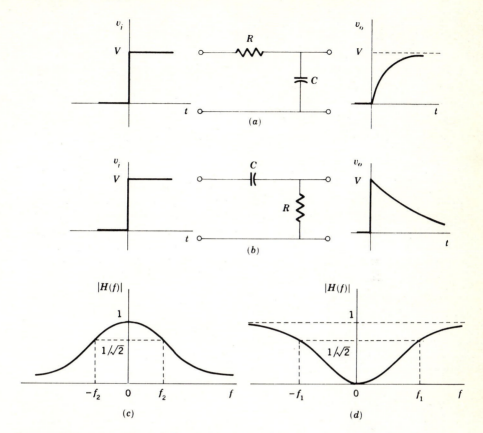

Fig. 1.14-1 (a) A low-pass RC circuit. (b) A high-pass RC circuit. (c) $|H(f)|$ for the circuit in (a). (d) $|H(f)|$ for the circuit in (b).

This rise time is a measure of the promptness with which the output responds to a voltage change at the input. A long rise time indicates a sluggish response. We note then from Eq. (1.14-4) that the more the upper-frequency range of the transfer function is restricted, the longer the rise time becomes.

The principle enunciated in Eq. (1.14-4) is of general validity. Even for filter circuits of much greater complexity and sharper cutoff than the simple RC circuit of Fig. 1.14-1a, the product $t_r f_2$ remains approximately constant. By way of example, suppose we postulate an ideal (unrealizable) filter. Such a filter has an arbitrarily sharp cutoff. That is, $H(f) = 1$ for $0 \leq f \leq f_M$, and $H(f) = 0$ for $f > f_M$. Then in this case it turns out that $t_r f_2 = 0.44$. On the other hand, Eq. (1.14-4) will often be encountered in texts and in literature with a rather different constant. For example, the result for the ideal filter is often given as

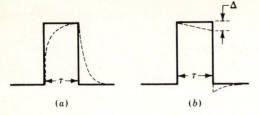

Fig. 1.14-2 (*a*) A rectangular pulse (solid) and the response (dashed) of a low-pass RC circuit. The rise time $t_r \approx 0.35\tau$. (*b*) The response of a high-pass RC circuit for $f_1\tau = 0.02$.

$t_r f_2 = 1$. These differences result from different definitions of the rise time.

If the input to the RC low-pass filter is a pulse, then there will be a rise time associated with the leading and trailing edges of the pulse. Let the pulse duration be τ. Then as a rule of thumb it is generally considered that to preserve the waveshape of the pulse with reasonable fidelity, it is necessary that the bandwidth f_2 be at least large enough to satisfy the condition $f_2\tau = 1$. For combining this condition with Eq. (1.14-4), we find $t_r = 0.35\tau$. In this case, with the rise time about one-third the pulse duration, the output pulse waveform has the form shown in Fig. 1.14-2*a*.

The high-pass RC circuit of Fig. 1.14-1*b* has a transfer function

$$H(f) = \frac{1}{1 - jf_1/f} \tag{1.14-5}$$

where

$$f_1 = \frac{1}{2\pi RC} \tag{1.14-6}$$

The magnitude of $H(f)$ is plotted in Fig. 1.14-1*d*. The frequency f_1 is the 3-dB frequency of the network and is generally described as the low-frequency cutoff of the network. For a step input of amplitude V, the output is

$$v_o(t) = Ve^{-t/RC} = Ve^{-2\pi f_1 t} \tag{1.14-7}$$

The essential distortion introduced by the frequency discrimination of this network is that the output does not sustain a constant voltage level when the input is constant. Instead, the output begins immediately to decay toward zero, which is its asymptotic limit. From Eq. (1.14-7) we have

$$\frac{1}{v_o}\frac{dv_o}{dt} = -\frac{1}{RC} = -2\pi f_1 \tag{1.14-8}$$

Thus the low-frequency cutoff f_1 alone determines the percentage drop in voltage per unit time. Again the importance of Eq. (1.14-8) is that, at least as a reasonable approximation, it applies to high-pass networks quite generally, even when the network is very much more complicated than the simple RC circuit.

If the input to the RC high-pass circuit is a pulse of duration τ, then the output has the waveshape shown in Fig. 1.14-2b. The output exhibits a tilt and an undershoot. As a rule of thumb, we may assume that the pulse is reasonably faithfully reproduced if the tilt Δ is no more than $0.1V$. Correspondingly, this condition requires that f_1 be no higher than given by the condition

$$f_1\tau \approx 0.02 \tag{1.14-9}$$

1.15 CORRELATION BETWEEN WAVEFORMS

The correlation between waveforms is a measure of the similarity or relatedness between the waveforms. Suppose that we have waveforms $v_1(t)$ and $v_2(t)$, not necessarily periodic nor confined to a finite time interval. Then the correlation between them, or more precisely the *average cross correlation* between $v_1(t)$ and $v_2(t)$, is $R_{12}(\tau)$ defined as

$$R_{12}(\tau) \equiv \lim_{T \to \infty} \frac{1}{T} \int_{-T/2}^{T/2} v_1(t)v_2(t + \tau)\, dt \tag{1.15-1}$$

If $v_1(t)$ and $v_2(t)$ are periodic with the same fundamental period T_0, then the average cross correlation is

$$R_{12}(\tau) = \frac{1}{T_0} \int_{-T_0/2}^{T_0/2} v_1(t)v_2(t + \tau)\, dt \tag{1.15-2}$$

If $v_1(t)$ and $v_2(t)$ are waveforms of finite energy (for example, nonperiodic pulse-type waveforms), then the cross correlation is defined as

$$R_{12}(\tau) = \int_{-\infty}^{\infty} v_1(t)v_2(t + \tau)\, dt \tag{1.15-3}$$

The need for introducing the parameter τ in the definition of cross correlation may be seen by the example illustrated in Fig. 1.15-1. Here the two waveforms, while different, are obviously related. They have the same period and nearly the same form. However, the integral of the product $v_1(t)v_2(t)$ is zero since at all times one or the other function is zero. The function $v_2(t + \tau)$ is the function $v_2(t)$ shifted to the left by amount τ. It is clear from the figure that, while $R_{12}(0) = 0$, $R_{12}(\tau)$ will increase as τ increases from zero, becoming a maximum when $\tau = \tau_0$. Thus τ is a "searching" or "scanning" parameter which may be adjusted to a proper time shift to reveal, to the maximum extent possible, the relatedness or correlation between the functions. The term *coherence* is sometimes used as a synonym for correlation. Functions for which $R_{12}(\tau) = 0$ for all τ are described as being *uncorrelated* or *incoherent*.

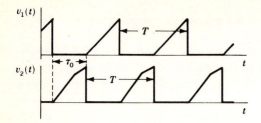

Fig. 1.15-1 Two related waveforms. The timing is such that the product $v_1(t)v_2(t) = 0$.

In scanning to see the extent of the correlation between functions, it is necessary to specify which function is being shifted. In general, $R_{12}(\tau)$ is not equal to $R_{21}(\tau)$. It is readily verified (Prob. 1.15-2) that

$$R_{21}(\tau) \equiv \lim_{T \to \infty} \frac{1}{T} \int_{-T/2}^{T/2} v_1(t + \tau)v_2(t)\, dt = R_{12}(-\tau) \qquad (1.15\text{-}4)$$

with identical results for periodic waveforms or waveforms of finite energy.

1.16 POWER AND CROSS CORRELATION

Let $v_1(t)$ and $v_2(t)$ be waveforms which are not periodic nor confined to a finite time interval. Suppose that the normalized power of $v_1(t)$ is S_1 and the normalized power of $v_2(t)$ is S_2. What, then, is the normalized power of $v_1(t) + v_2(t)$? Or, more generally, what is the normalized power S_{12} of $v_1(t) + v_2(t + \tau)$? We have

$$S_{12} = \lim_{T \to \infty} \frac{1}{T} \int_{-T/2}^{T/2} [v_1(t) + v_2(t + \tau)]^2\, dt \qquad (1.16\text{-}1a)$$

$$= \lim_{T \to \infty} \frac{1}{T} \left\{ \int_{-T/2}^{T/2} v_1^2(t)\, dt + \int_{-T/2}^{T/2} [v_2(t + \tau)]^2\, dt \right.$$
$$\left. + 2 \int_{-T/2}^{T/2} v_1(t)v_2(t + \tau)\, dt \right\} \qquad (1.16\text{-}1b)$$

$$= S_1 + S_2 + 2R_{12}(\tau) \qquad (1.16\text{-}1c)$$

In writing Eq. (1.16-1c), we have taken account of the fact that the normalized power of $v_2(t + \tau)$ is the same as the normalized power of $v_2(t)$. For, since the integration in Eq. (1.16-1b) extends eventually over the entire time axis, a time shift in v_2 will clearly not affect the value of the integral.

From Eq. (1.16-1c) we have the important result that if two waveforms are uncorrelated, that is, $R_{12}(\tau) = 0$ for all τ, then no matter how these waveforms are time-shifted with respect to one another, the normalized power due to the superposition of the waveforms is the sum of the powers due to the waveforms individually. Similarly if a waveform is the sum of any number of mutually uncorrelated waveforms, the normalized power is the sum of the individual powers. It is readily verified that the same result applies for periodic waveforms. For finite energy waveforms, the result applies to the normalized energy.

Suppose that two waveforms $v_1'(t)$ and $v_2'(t)$ are uncorrelated. If dc components V_1 and V_2 are added to the waveforms, then the waveforms $v_1(t) = v_1'(t) + V_1$ and $v_2(t) = v_2'(t) + V_2$ will be correlated with correlation $R_{12}(\tau) = V_1 V_2$. In most applications where the correlation between waveforms is of concern, there is rarely any interest in the dc component. It is customary, then, to continue to refer to waveforms as being uncorrelated if the only source of the correlation is the dc components.

1.17 AUTOCORRELATION

The correlation of a function with itself is called the *autocorrelation*. Thus with $v_1(t) = v_2(t)$, $R_{12}(\tau)$ becomes $R(\tau)$ given, in the general case, by

$$R(\tau) = \lim_{T \to \infty} \frac{1}{T} \int_{-T/2}^{T/2} v(t)v(t + \tau) \, dt \qquad (1.17\text{-}1)$$

A number of the properties of $R(\tau)$ are listed in the following:

(a) $$R(0) = \lim_{T \to \infty} \frac{1}{T} \int_{-T/2}^{T/2} [v(t)]^2 \, dt = S \qquad (1.17\text{-}2)$$

That is, the autocorrelation for $\tau = 0$ is the average power S of the waveform.

(b) $$R(0) \geq R(\tau) \qquad (1.17\text{-}3)$$

This result is rather intuitively obvious since we would surely expect that similarity between $v(t)$ and $v(t + \tau)$ be a maximum when $\tau = 0$. The student is guided through a more formal proof in Prob. 1.17-1.

(c) $$R(\tau) = R(-\tau) \qquad (1.17\text{-}4)$$

Thus the autocorrelation function is an even function of τ. To prove Eq. (1.17-4), assume that the axis $t = 0$ is moved in the negative t direction by amount τ. Then the integrand in Eq. (1.17-1) would become $v(t - \tau)v(t)$, and $R(\tau)$ would become $R(-\tau)$. Since, however, the integration eventually extends from $-\infty$ to ∞, such a shift in time axis can have no effect on the value of the integral. Thus $R(\tau) = R(-\tau)$.

The three characteristics given in Eqs. (1.17-2) to (1.17-4) are features not only of $R(\tau)$ defined by Eq. (1.17-1) but also for $R(\tau)$ as defined by Eqs. (1.15-2) and (1.15-3) for the periodic case and the nonperiodic case of finite energy. In the latter case, of course, $R(0) = E$, the energy rather than the power.

1.18   AUTOCORRELATION OF A PERIODIC WAVEFORM

When the function is periodic, we may write

$$v(t) = \sum_{n=-\infty}^{\infty} V_n e^{j2\pi nt/T_0} \qquad (1.18\text{-}1)$$

and, using the correlation integral in the form of Eq. (1.15-2), we have

$$R(\tau) = \frac{1}{T_0} \int_{-T_0/2}^{T_0/2} \left(\sum_{m=-\infty}^{\infty} V_m e^{j2\pi mt/T_0} \right) \left(\sum_{n=-\infty}^{\infty} V_n e^{j2\pi n(t+\tau)/T_0} \right) dt \quad (1.18\text{-}2)$$

The order of integration and summation may be interchanged in Eq. (1.18-2). If we do so, we shall be left with a double summation over m and n of terms $I_{m,n}$ given by

$$I_{m,n} = \frac{1}{T_0} e^{j2\pi n\tau/T_0} \int_{-T_0/2}^{T_0/2} V_m V_n e^{j2\pi(m+n)t/T_0} \, dt \qquad (1.18\text{-}3a)$$

$$= V_m V_n e^{j2\pi n\tau/T_0} \frac{[\sin \pi(m+n)]}{\pi(m+n)} \qquad (1.18\text{-}3b)$$

Since m and n are integers, we see from Eq. (1.18-3b) that $I_{m,n} = 0$ except when $m = -n$ or $m + n = 0$. To evaluate $I_{m,n}$ in this latter case, we return to Eq. (1.18-3a) and find

$$I_{m,n} = I_{-n,n} = V_n V_{-n} e^{j2\pi n\tau/T_0} \qquad (1.18\text{-}4)$$

Finally, then,

$$R(\tau) = \sum_{n=-\infty}^{\infty} V_n V_{-n} e^{j2\pi n\tau/T_0} = \sum_{n=-\infty}^{\infty} |V_n|^2 e^{j2\pi n\tau/T_0} \qquad (1.18\text{-}5a)$$

$$= |V_0|^2 + 2 \sum_{n=1}^{\infty} |V_n|^2 \cos 2\pi n \frac{\tau}{T_0} \qquad (1.18\text{-}5b)$$

We note from Eq. (1.18-5b) that $R(\tau) = R(-\tau)$ as anticipated, and we note as well that for this case of a periodic waveform the correlation $R(\tau)$ is also periodic with the same fundamental period T_0.

We shall now relate the correlation function $R(\tau)$ of a periodic waveform to its power spectral density. For this purpose we compute the Fourier transform of $R(\tau)$. We find, using $R(\tau)$ as in Eq. (1.18-5a), that

$$\mathcal{F}[R(\tau)] = \int_{-\infty}^{\infty} \left(\sum_{n=-\infty}^{\infty} |V_n|^2 e^{j2\pi n\tau/T_0} \right) e^{-j2\pi f\tau} \, d\tau \qquad (1.18\text{-}6)$$

Interchanging the order of integration and summation yields

$$\mathfrak{F}[R(\tau)] = \sum_{n=-\infty}^{\infty} |V_n|^2 \int_{-\infty}^{\infty} e^{-j2\pi(f-n/T_0)\tau}\, d\tau \qquad (1.18\text{-}7)$$

Using Eq. (1.10-29), we may write Eq. (1.18-7) as

$$\mathfrak{F}[R(\tau)] = \sum_{n=-\infty}^{\infty} |V_n|^2 \delta\left(f - \frac{n}{T_0}\right) \qquad (1.18\text{-}8)$$

Comparing Eq. (1.18-8) with Eq. (1.8-6), we have the interesting result that for a periodic waveform

$$G(f) = \mathfrak{F}[R(\tau)] \qquad (1.18\text{-}9)$$

and, of course, conversely

$$R(\tau) = \mathfrak{F}^{-1}[G(f)] \qquad (1.18\text{-}10)$$

Expressed in words, we have the following: *The power spectral density and the correlation function of a periodic waveform are a Fourier transform pair.*

1.19 AUTOCORRELATION OF NONPERIODIC WAVEFORM OF FINITE ENERGY

For pulse-type waveforms of finite energy there is a relationship between the correlation function of Eq. (1.15-3) and the energy spectral density which corresponds to the relationship given in Eq. (1.18-9) for the periodic waveform. This relationship is that the correlation function $R(\tau)$ and the *energy* spectral density are a Fourier transform pair. This result is established as follows.

We use the convolution theorem. We combine Eqs. (1.11-1) and (1.11-3) for the case where the waveforms $v_1(t)$ and $v_2(t)$ are the same waveforms, that is, $v_1(t) = v_2(t) = v(t)$, and get

$$\mathfrak{F}^{-1}[V(f)V(f)] = \int_{-\infty}^{\infty} v(\tau)v(t - \tau)\, d\tau \qquad (1.19\text{-}1)$$

Since $V(-f) = V^*(f) = \mathfrak{F}[v(-t)]$, Eq. (1.19-1) may be written

$$\mathfrak{F}^{-1}[V(f)V^*(f)] = \mathfrak{F}^{-1}[|V(f)|^2] = \int_{-\infty}^{\infty} v(\tau)v(\tau - t)\, d\tau \qquad (1.19\text{-}2)$$

The integral in Eq. (1.19-2) is a function of t, and hence this equation expresses $\mathfrak{F}^{-1}[V(f)V^*(f)]$ as a function of t. If we want to express $\mathfrak{F}^{-1}[V(f)V^*(f)]$ as a function of τ without changing the form of the function, we need but to interchange t and τ. We then have

$$\mathfrak{F}^{-1}[V(f)V^*(f)] = \int_{-\infty}^{\infty} v(t)v(t - \tau)\, dt \qquad (1.19\text{-}3)$$

The integral in Eq. (1.19-3) is precisely $R(\tau)$, and thus

$$\mathfrak{F}[R(\tau)] = V(f)V^*(f) = |V(f)|^2 \qquad (1.19\text{-}4)$$

which verifies that $R(\tau)$ and the energy spectral density $|V(f)|^2$ are Fourier transform pairs.

1.20 AUTOCORRELATION OF OTHER WAVEFORMS

In the preceding sections we discussed the relationship between the auto-correlation function and power or energy spectral density of *deterministic* waveforms. We use the term "deterministic" to indicate that at least, in principle, it is possible to write a function which specifies the value of the function at all times. For such deterministic waveforms, the availability of an autocorrelation function is of no particularly great value. The autocorrelation function does not include within itself complete information about the function. Thus we note that the autocorrelation function is related only to the amplitudes and not to the phases of the spectral components of the waveform. The waveform cannot be reconstructed from a knowledge of the autocorrelation functions. Any characteristic of a deterministic waveform which may be calculated with the aid of the autocorrelation function may be calculated by direct means at least as conveniently.

On the other hand, in the study of communication systems we encounter waveforms which are not deterministic but are instead random and unpredictable in nature. Such waveforms are discussed in Chap. 2. There we shall find that for such random waveforms no explicit function of time can be written. The waveforms must be described in statistical and probabilistic terms. It is in connection with such waveforms that the concepts of correlation and autocorrelation find their true usefulness. Specifically, it turns out that even for such random waveforms, the auto-correlation function and power spectral density are a Fourier transform pair. The proof[3] that such is the case is formidable and will not be undertaken here.

PROBLEMS

1.1-1. Verify the relationship of C_n and ϕ_n to A_n and B_n as given by Eq. (1.1-6).

1.1-2. Calculate A_n, B_n, C_n, and ϕ_n for a waveform $v(t)$ which is a symmetrical square wave and which makes peak excursions to $+\frac{3}{4}$ volt and $-\frac{1}{4}$ volt, and has a period $T = 1$ sec. A positive going transition occurs at $t = 0$.

1.2-1. Verify the relationship of the complex number V_n to C_n and ϕ_n as given by Eq. (1.2-3).

1.2-2. The function

$$p(t) = \begin{cases} e^{-t} & 0 \le t \le 1 \\ 0 & \text{elsewhere} \end{cases}$$

is repeated every $T = 1$ sec. Thus, with $u(t)$ the unit step function

$$v(t) = \sum_{n=-\infty}^{\infty} p(t-n)u(t-n)$$

Find V_n and the exponential Fourier series for $v(t)$.

1.3-1. The impulse function can be defined as:

$$\delta(t) = \lim_{a \to \infty} \frac{a}{2} e^{-a|t|}$$

Discuss.

1.3-2. In Eq. (1.3-8) we see that

$$v(t) = I \sum_{k=-\infty}^{\infty} \delta(t-kT_0) = \frac{I}{T_0} + \frac{2I}{T_0} \sum_{n=1}^{\infty} \cos \frac{2\pi nt}{T_0}$$

Set $I = 1$, $T_0 = 1$. Show by plotting

$$v(t) = 1 + 2 \sum_{n=1}^{3} \cos 2\pi nt$$

that the Fourier series does approximate the train of impulses.

1.4-1. $Sa(x) \equiv (\sin x)/x$. Determine the maxima and minima of $Sa(x)$ and compare your result with the approximate maxima and minima obtained by letting $x = (2n+1)\pi/2$, $n = 1, 2, \ldots$.

1.4-2. A train of rectangular pulses, making excursions from zero to 1 volt, have a duration of 2 μsec and are separated by intervals of 10 μsec.

(a) Assume that the center of one pulse is located at $t = 0$. Write the exponential Fourier series for this pulse train and plot the spectral amplitude as a function of frequency. Include at least 10 spectral components on each side of $f = 0$, and draw also the envelope of these spectral amplitudes.

(b) Assume that the left edge of a pulse rather than the center is located at $t = 0$. Correct the Fourier expansion accordingly. Does this change affect the plot of spectral amplitudes? Why not?

1.5-1. (a) A periodic waveform $v_i(t)$ is applied to the input of a network. The output $v_o(t)$ of the network is $v_o(t) = \tau[dv_i(t)/dt]$, where τ is a constant. What is the transfer function $H(\omega)$ of this network?

(b) A periodic waveform $v_i(t)$ is applied to the RC network shown, whose time constant is $\tau \equiv RC$. Assume that the highest frequency spectral component of $v_i(t)$ is of a frequency $f \ll 1/\tau$. Show that, under these circumstances, the output $v_o(t)$ is approximately $v_o(t) \simeq \tau[dv_i(t)/dt]$.

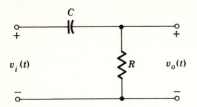

Fig. P 1.5-1

1.5-2. A voltage represented by an impulse train of strength I and period T is filtered by a low-pass RC filter having a 3-dB frequency f_c.

    (a) Find the Fourier series of the output voltage across the capacitor.

    (b) If the third harmonic of the output is to be attenuated by 1000, find $f_c T$.

1.6-1. Measurements on a voltage amplifier indicate a gain of 20 dB.

    (a) If the input voltage is 1 volt, calculate the output voltage.

    (b) If the input power is 1 mw, calculate the output power.

1.6-2. A voltage gain of 0.1 is produced by an attenuator.

    (a) What is the gain in decibels?

    (b) What is the power gain (not in decibels)?

1.7-1. A periodic triangular waveform $v(t)$ is defined by

$$v(t) = \frac{2t}{T} \quad \text{for} \ -\frac{T}{2} < t < \frac{T}{2}$$

and

$$v(t \pm T) = v(t)$$

and has the Fourier expansion

$$v(t) = \frac{2}{\pi} \sum_{n=1}^{\infty} \frac{(-1)^{n+1}}{n} \sin 2\pi n \frac{t}{T}$$

Calculate the fraction of the normalized power of this waveform which is contained in its first three harmonics.

1.7-2. The complex spectral amplitudes of a periodic waveform are given by

$$V_n = \frac{1}{|n|} e^{-j \arctan(n/2)} \qquad n = \pm 1, \pm 2, \ldots$$

Find the ratio of the normalized power in the second harmonic to the normalized power in the first harmonic.

1.8-1. Find $G(f)$ for the following voltages:

    (a) An impulse train of strength I and period T.

    (b) A pulse train of amplitude A, duration $\tau = I/A$, and period T.

1.8-2. Plot $G(f)$ for a voltage source represented by an impulse train of strength I and period nT for $n = 1, 2, 10$, infinity. Comment on this limiting result.

1.9-1. $G_i(f)$ is the power spectral density of a *square-wave* voltage of peak-to-peak amplitude 1 and period 1. The square wave is filtered by a low-pass RC filter with a 3-dB frequency 1. The output is taken across the capacitor.

    (a) Calculate $G_i(f)$.

    (b) Find $G_0(f)$.

1.9-2. (a) A symmetrical square wave of zero mean value, peak-to-peak voltage 1 volt, and period 1 sec is applied to an ideal low-pass filter. The filter has a transfer function $|H(f)| = \frac{1}{2}$ in the frequency range $-3.5 \le f \le 3.5$ Hz, and $H(f) = 0$ elsewhere. Plot the power spectral density of the filter output.

    (b) What is the normalized power of the input square wave? What is the normalized power of the filter output?

1-10-1. In Eqs. (1.2-1) and (1.2-2) write $f_0 \equiv 1/T_0$. Replace f_0 by Δf, i.e., Δf is the frequency interval between harmonics. Replace $n\,\Delta f$ by f, i.e., as $\Delta f \to 0$, f becomes a continuous variable ranging from 0 to ∞ as n ranges from 0 to ∞. Show that in the limit as $\Delta f \to 0$, so that Δf may be replaced by the differential df, Eq. (1.2-1) becomes

$$v(t) = \int_{-\infty}^{\infty} V(f) e^{j2\pi ft}\, df$$

in which $V(f)$ is

$$V(f) = \lim_{\substack{f_0 = \Delta f \to 0 \\ nf_0 \to f}} \int_{-1/2f_0}^{1/2f_0} v(t) e^{-j2\pi nf_0 t}\, dt = \int_{-\infty}^{\infty} v(t) e^{-j2\pi ft}\, dt$$

1.10-2. Find the Fourier transform of $\sin \omega_0 t$. Compare with the transform of $\cos \omega_0 t$. Plot and compare the power spectral densities of $\cos \omega_0 t$ and $\sin \omega_0 t$.

1.10-3. The waveform $v(t)$ has the Fourier transform $V(f)$. Show that the waveform delayed by time t_d, i.e., $v(t - t_d)$ has the transform $V(f) e^{-j\omega t_d}$.

1.10-4. (a) The waveform $v(t)$ has the Fourier transform $V(f)$. Show that the time derivative $(d/dt)v(t)$ has the transform $(j2\pi f)V(f)$.

    (b) Show that the transform of the integral of $v(t)$ is given by

$$\mathfrak{F}\left[\int_{-\infty}^{t} v(\lambda)\, d\lambda\right] = \frac{V(f)}{j2\pi f}$$

1.11-1. Derive the convolution formula in the frequency domain. That is, let $V_1(f) = \mathfrak{F}[v_1(t)]$ and $V_2(f) = \mathfrak{F}[v_2(t)]$. Show that if $V(f) = \mathfrak{F}[v_1(t)v_2(t)]$, then

$$V(f) = \frac{1}{2\pi} \int_{-\infty}^{\infty} V_1(\lambda)V_2(f - \lambda)\, d\lambda$$

or

$$V(f) = \frac{1}{2\pi} \int_{-\infty}^{\infty} V_2(\lambda)V_1(f - \lambda)\, d\lambda$$

1.11-2. (a) A waveform $v(t)$ has a Fourier transform which extends over the range from $-f_M$ to $+f_M$. Show that the waveform $v^2(t)$ has a Fourier transform which extends over the range from $-2f_M$ to $+2f_M$. (*Hint:* Use the result of Prob. 1.11-1).

(*b*) A waveform $v(t)$ has a Fourier transform $V(f) = 1$ in the range $-f_M$ to $+f_M$ and $V(f) = 0$ elsewhere. Make a plot of the transform of $v^2(t)$.

1.11-3. A filter has an impulse response $h(t)$ as shown. The input to the network is a pulse of unit amplitude extending from $t = 0$ to $t = 2$. By graphical means, determine the output of the filter.

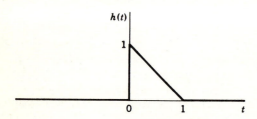

Fig. P 1.11-3

1.12-1. The energy of a nonperiodic waveform $v(t)$ is

$$E = \int_{-\infty}^{\infty} v^2(t) \, dt$$

(*a*) Show that this can be written as

$$E = \int_{-\infty}^{\infty} dt \, v(t) \int_{-\infty}^{\infty} V(f) e^{j2\pi ft} \, df$$

(*b*) Show that by interchanging the order of integration we have

$$E = \int_{-\infty}^{\infty} V(f) V^*(f) \, df = \int_{-\infty}^{\infty} |V(f)|^2 \, df$$

which proves Eq. (1.12-5). This is an alternate proof of Parseval's theorem.

1.12-2. If $V(f) = AT \sin 2\pi fT / 2\pi fT$, find the energy E contained in $v(t)$.

1.12-3. A waveform $m(t)$ has a Fourier transform $M(f)$ whose magnitude is as shown.

(*a*) Find the normalized energy content of the waveform.

(*b*) Calculate the frequency f_1 such that one-half of the normalized energy is in the range $-f_1$ to f_1.

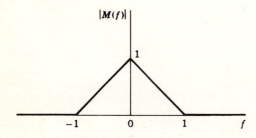

Fig. P 1.12-3

1.13-1. The signal $v(t) = \cos \omega_0 t + 2 \sin 3\omega_0 t + 0.5 \sin 4\omega_0 t$ is filtered by an RC low-pass filter with a 3-dB frequency $f_c = 2f_0$.

    (a) Find $G_i(f)$.

    (b) Find $G_0(f)$.

    (c) Find S_0.

1.13-2. The waveform $v(t) = e^{-t/\tau}u(t)$ is passed through a high-pass RC circuit having a time constant τ.

    (a) Find the energy spectral density at the output of the circuit.

    (b) Show that the total output energy is one-half the input energy.

1.14-1. (a) An impulse of strength I is applied to a low-pass RC circuit of 3-dB frequency f_2. Calculate the output waveform.

    (b) A pulse of amplitude A and duration τ is applied to the low-pass RC circuit. Show that, if $\tau \ll 1/f_2$, the response at the output is approximately the response that would result from the application to the circuit of an impulse of strength $I' = A\tau$. Generalize this result by considering any voltage waveform of area I' and duration τ.

1.14-2. A pulse extending from 0 to A volts and having a duration τ is applied to a high-pass RC circuit. Show that the area under the response waveform is zero.

1.15-1. Find the cross correlation of the functions $\sin \omega t$ and $\cos \omega t$.

1.15-2. Prove that $R_{21}(\tau) = R_{12}(-\tau)$.

1.16-1. Find the cross-correlation function $R_{12}(\tau)$ of the two periodic waveforms shown.

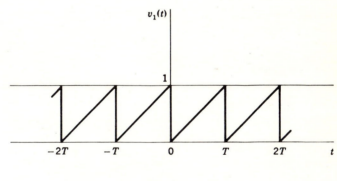

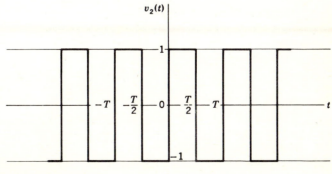

Fig. P 1.16-1

1.17-1. $R(\tau) = \lim_{T \to \infty} \frac{1}{T} \int_{-T/2}^{T/2} v(t)v(t + \tau)\, dt$

Prove that $R(0) \geq R(\tau)$. *Hint:* Consider

$$I = \lim_{T \to \infty} \frac{1}{T} \int_{-T/2}^{T/2} [v(t) - v(t + \tau)]^2\, dt$$

Is $I \geq 0$? Expand and integrate term by term. Show that

$$I = 2[R(0) - R(\tau)] \geq 0$$

1.17-2. Determine an expression for the correlation function of a square wave having the values 1 or 0 and a period T.

1.18-1. (*a*) Find the power spectral density of a square-wave voltage by Fourier transforming the correlation function. Use the results of Prob. 1.17-2.

(*b*) Compare the answer to (*a*) with the spectral density obtained from the Fourier series (Prob. 1.9-1*a*) itself.

1.18-2. If $v(t) = \sin \omega_0 t$,

(*a*) Find $R(\tau)$.

(*b*) If $G(f) = \mathfrak{F}[R(\tau)]$, find $G(f)$ directly and compare.

1.19-1. A waveform consists of a single pulse of amplitude A extending from $t = -\tau/2$ to $t = \tau/2$.

(*a*) Find the autocorrelation function $R(\tau)$ of this waveform.

(*b*) Calculate the energy spectral density of this pulse by evaluating $G_E(f) = \mathfrak{F}[R(\tau)]$.

(*c*) Calculate $G_E(f)$ directly by Parseval's theorem and compare.

REFERENCES

1. Javid, M., and E. Brenner: "Analysis of Electric Circuits," 2d ed., McGraw-Hill Book Company, New York, 1967.
   Papoulis, A.: "The Fourier Integral and Applications," McGraw-Hill Book Company, New York, 1963.
2. Churchill, R. V.: "Fourier Series," McGraw-Hill Book Company, New York, 1941.
3. Papoulis, A.: "Probability, Random Variables, and Stochastic Processes," McGraw-Hill Book Company, New York, 1965.

2
Random Variables and Processes

A waveform which can be expressed, at least in principle, as an explicit function of time $v(t)$ is called a *deterministic* waveform. Such a waveform is determined for all times in that, if we select an arbitrary time $t = t_1$, there need be no *uncertainty* about the value of $v(t)$ at that time. The waveforms encountered in communication systems, on the other hand, are in many instances unpredictable. Consider, say, the very waveform itself which is transmitted for the purpose of communication. This waveform, which is called the *signal*, must, at least in part, be unpredictable. If such were not the case, i.e., if the signal were predictable, then its transmission would be unnecessary, and the entire communications system would serve no purpose. This point concerning the unpredictability of the signal is explored further in Chap. 13. Further, as noted briefly in Chap. 1, transmitted signals are invariably accompanied by *noise* which results from the ever-present agitation of the universe at the atomic level. These noise waveforms are also not predictable. Unpredictable waveforms such as a signal voltage $s(t)$ or a noise voltage $n(t)$ are examples of *random processes*. [Note that in writing symbols like $s(t)$

and $n(t)$ we do not imply that we can write explicit functions for these time functions.]

While random processes are not predictable, neither are they completely unpredictable. It is generally possible to predict the future performance of a random process with a certain *probability* of being correct. Accordingly, in this chapter we shall present some elemental ideas of probability theory and apply them to the description of random processes. We shall rather generally limit our discussion to the development of only those aspects of the subject which we shall have occasion to employ in this text.

2.1 PROBABILITY[1]

The concept of probability occurs naturally when we contemplate the possible outcomes of an experiment whose outcome is not always the same. Suppose that one of the possible outcomes is called A and that when the experiment is repeated N times the outcome A occurs N_A times. The relative frequency of occurrence of A is N_A/N, and this ratio N_A/N is not predictable unless N is very large. For example, let the experiment consist of the tossing of a die and let the outcome A correspond to the appearance of, say, the number 3 on the die. Then in 6 tosses the number 3 may not appear at all, or it may appear 6 times, or any number of times in between. Thus with $N = 6$, N_A/N may be 0 or 1/6, etc., up to $N_A/N = 1$. On the other hand, we know from experience that when an experiment, whose outcomes are determined by chance, is repeated *very many times*, the relative frequency of a particular outcome approaches a fixed limit. Thus, if we were to toss a die very many times we would expect that N_A/N would turn out to be very close to 1/6. This limiting value of the relative frequency of occurrence is called the probability of outcome A, written $P(A)$, so that

$$P(A) = \lim_{N \to \infty} \frac{N_A}{N} \tag{2.1-1}$$

In many cases the experiments needed to determine the probability of an event are done more in thought than in practice. Suppose that we have 10 balls in a container, the balls being identical in every respect except that 8 are white and 2 are black. Let us ask about the probability that, in a single draw, we shall select a black ball. If we draw blindly, so that the color has no influence on the outcome, we would surely judge that the probability of drawing the black ball is 2/10. We arrive at this conclusion on the basis that we have postulated that there is absolutely nothing which favors one ball over another. There are 10 possible outcomes of the experiment, that is, any of the 10 balls may be drawn; of these 10 outcomes, 2 are favorable to our interest. The only reasonable outcome we can imagine is that, in very many drawings, 2 out of 10 will be black. Any other outcome would immediately suggest that either the black or the white balls had been favored. These considerations lead to an alternative definition of the probability of occurrence of an event A, that is

$$P(A) = \frac{\text{number of possible favorable outcomes}}{\text{total number of possible equally likely outcomes}} \qquad (2.1\text{-}2)$$

It is apparent from either definition, Eq. (2.1-1) or (2.1-2), that the probability of occurrence of an event P is a positive number and that $0 \leq P \leq 1$. If an event is not possible, then $P = 0$, while if an event is certain, $P = 1$.

2.2 MUTUALLY EXCLUSIVE EVENTS

Two possible outcomes of an experiment are defined as being *mutually exclusive* if the occurrence of one outcome precludes the occurrence of the other. In this case, if the events are A_1 and A_2 with probabilities $P(A_1)$ and $P(A_2)$, then the probability of occurrence of *either* A_1 or A_2 s written $P(A_1 \text{ or } A_2)$ and given by

$$P(A_1 \text{ or } A_2) = P(A_1) + P(A_2) \qquad (2.2\text{-}1)$$

This result follows directly from Eq. (2.1-1). For suppose that in a very large number N of repetitions of the experiment, outcome A_1 had occurred N_1 times and outcome A_2 had occurred N_2 times. Then A_1 or A_2 will have occurred $N_1 + N_2$ times and

$$P(A_1 \text{ or } A_2) = \frac{N_1 + N_2}{N} = \frac{N_1}{N} + \frac{N_2}{N} = P(A_1) + P(A_2) \qquad (2.2\text{-}2)$$

Equation (2.2-2) may be extended, of course, to more than two mutually exclusive outcomes. Thus

$$P(A_1 \text{ or } A_2 \text{ or } \cdots \text{ or } A_L) = \sum_{j=1}^{L} P(A_j) \qquad (2.2\text{-}3)$$

and if it should happen that there are only L possible events, then, of course,

$$\sum_{j=1}^{L} P(A_j) = 1 \qquad (2.2\text{-}4)$$

As an example of the calculation of the probability of mutually exclusive events, we ask, in connection with the tossing of a die, about the probability that either a 1 or a 2 will appear. Since the probability of either event is $\frac{1}{6}$ and since, one having occurred the other cannot take place, the probability of one or the other is $\frac{1}{6} + \frac{1}{6} = \frac{1}{3}$.

2.3 JOINT PROBABILITY OF RELATED AND INDEPENDENT EVENTS

Suppose that we contemplate two experiments A and B with outcomes $A_1, A_2, \ldots$ and $B_1, B_2, \ldots$. The probability of the joint occurrence of, say, A_j and B_k is written $P(A_j \text{ and } B_k)$ or more simply $P(A_j, B_k)$.

It may be that the probability of event B_k depends on whether A_j does indeed occur. For example, imagine 4 balls in a box, 2 white and 2 black. The probability of drawing a white ball is $1/2$. If, having drawn a ball, we replace it and draw again, then we are repeating the same experiment and the probability of drawing a white ball is again $1/2$. Suppose, however, that the first ball drawn is not replaced. Then if we draw a second ball, we shall be performing a new experiment. If the first ball drawn was white, the probability of a white draw in the second experiment is $1/3$. If the first ball drawn was black, the probability of a white draw in the second experiment is $2/3$. Here, then, we have a situation in which the outcome of the second experiment is *conditional* on the outcome of the first experiment. The probability of the outcome B_k, given that A_j is known to have occurred, is called the *conditional probability* and written $P(B_k|A_j)$.

Suppose that we perform N times (N a very large number) the experiment of determining which pairs of outcomes of experiment A and B occur jointly. Let N_j be the number of times A_j occurs with or without B_k, N_k the number of times B_k occurs with or without A_j, and N_{jk} the number of times of joint occurrence. Then

$$P(B_k|A_j) = \frac{N_{jk}}{N_j} = \frac{N_{jk}/N}{N_j/N} = \frac{P(A_j, B_k)}{P(A_j)} \tag{2.3-1}$$

Similarly we have, since $N_{jk} = N_{kj}$,

$$P(A_j|B_k) = \frac{N_{kj}}{N_k} = \frac{N_{kj}/N}{N_k/N} = \frac{P(A_j, B_k)}{P(B_k)} \tag{2.3-2}$$

From Eqs. (2.3-1) and (2.3-2) we have

$$P(A_j, B_k) = P(A_j)P(B_k|A_j) = P(B_k)P(A_j|B_k) \tag{2.3-3}$$

so that

$$P(A_j|B_k) = \frac{P(A_j)}{P(B_k)} P(B_k|A_j) \tag{2.3-4}$$

which result is known as *Bayes' theorem*.

It is apparent that Eq. (2.3-4) applies equally well to the case of a *single* experiment whose outcome is characterized by two events. This result follows from a simple restatement of the considerations leading up to Bayes' theorem. We need only view the successive performance of experiment A followed by experiment B as a single joint experiment. If we consider the example of the 2 white and 2 black balls in a box, it is obvious that the probability of picking a black ball (second experiment) after having picked a white ball (first experiment) is the same as the probability that, in picking 2 balls (joint experiment), we find the first white and the second black.

2.4 STATISTICAL INDEPENDENCE

Suppose, as before, that A_j and B_k are the possible outcomes of two successive experiments or the joint outcome of a single experiment. And suppose that it turns out that the probability of the occurrence of outcome B_k simply does not depend at all on which outcome A_j accompanies it. Then we say that the outcomes A_j and B_k are *independent*. In this case of complete independence

$$P(B_k|A_j) = P(B_k) \tag{2.4-1}$$

and Eq. (2.3-3) yields

$$P(A_j,B_k) = P(A_j)P(B_k) \tag{2.4-2}$$

Expressed in words, when outcomes are independent, the probability of a joint occurrence of particular outcomes is the product of the probabilities of the individual independent outcomes. This result may be extended to any arbitrary number of outcomes. Thus

$$P(A_j,B_k,C_l, \ . \ . \ .) = P(A_j)P(B_k)P(C_l) \ \cdot \ \cdot \ \cdot \tag{2.4-3}$$

2.5 RANDOM VARIABLES

The term random variable is used to signify a *rule* by which a real number is assigned to each possible outcome of an experiment. Let the possible outcomes of an experiment be identified by the symbols λ_i. These symbols λ_i need not be numbers. For example, the experiment might consist of running a horse race where the outcome of interest is the name of the winner. The symbols λ_i might then be the names of the horses. Let us now arbitrarily establish some rule by which we assign real numbers $X(\lambda_i)$ to each possible outcome. Then the *rule* or *functional relationship* represented by the symbol $X(\ \ )$ is called a *random variable*. When used in this sense, the term random variable is a misnomer, since $X(\ )$ is not a variable at all and is in no way random, being a perfectly definite rule.

By a rather easy extension of meaning, the term random variable is also used to refer to a *variable* which may assume any of the *numbers* $X(\lambda_i)$. When used in this sense, the random variable may be represented by the symbol $X(\lambda_i)$.

In many cases, the identifying symbols λ_i may turn out to be numbers. To illustrate this point, consider that we perform the experiment of measuring a random voltage V between a set of terminals and find a number of possible outcomes v_1, v_2, or v_3. Then for identifying symbols for the outcomes we would rather naturally be led to use the numbers themselves; that is, we would use $\lambda_1 = v_1$, $\lambda_2 = v_2$, and $\lambda_3 = v_3$. Further, we might very naturally use a rule in which the number assigned to an outcome is the very same number used for identifications. We would then have $X(\lambda_i) = X(v_i) = v_i$. With these considerations in mind, we might well refer to the random voltage itself as a random variable and represent it by the symbol V.

A random variable may be *discrete* or *continuous*. If in any finite interval $X(\lambda)$ assumes only a finite number of distinct values, then the random variable is *discrete*. An example of an experiment which yields such a discrete random variable is the tossing of a die. If, however, $X(\lambda)$ can assume any value within an interval, the random variable is continuous. Thus suppose we fire a bullet at a target. Because of wind currents and other unpredictable influences the bullet may miss its mark, and the magnitude of the miss will be a *continuous random variable*.

2.6 CUMULATIVE DISTRIBUTION FUNCTION

The *cumulative distribution function* associated with a random variable is defined as the probability that the outcome of an experiment will be one of the outcomes for which $X(\lambda) \leq x$, where x is any given number. This probability will depend on the number x and also on $X(\ )$, that is, on the rule by which we assign numbers to outcomes. Thus we use the symbol $F_{X(\ )}(x)$ to represent the cumulative distribution function, and we have the definition

$$F_{X(\ )}(x) \equiv P[X(\lambda) \leq x] \tag{2.6-1}$$

Observe that in Eq. (2.6-1) we used the symbol $X(\ )$ to represent a rule and the symbol $X(\lambda)$ to represent a variable which ranges over the numbers assigned to possible outcomes. If a different rule $Y(\ )$ were used to assign numbers, $F_{Y(\ )}(x)$ would differ from $F_{X(\ )}(x)$ for the same x. It is for this reason that the subscript $X(\ )$ is included in the notation

$F_{X()}(x)$. Generally, for simplicity of notation, Eq. (2.6-1) is written more simply in the form

$$F_X(x) \equiv P(X < x) \tag{2.6-2}$$

Even further, when there is only a single random variable under discussion and no ambiguity will result, we shall drop the subscript X and use the symbol $F(x)$ to represent the cumulative distribution function.

The cumulative distribution function has the properties

$$0 \le F(x) \le 1 \tag{2.6-3a}$$

$$F(-\infty) = 0 \qquad F(\infty) = 1 \tag{2.6-3b}$$

$$F(x_1) \le F(x_2) \qquad \text{if } x_1 < x_2 \tag{2.6-3c}$$

The property in Eq. (2.6-3a) follows from the fact that $F(x)$ is a probability. The property in Eq. (2.6-3b) follows from the fact that $F(-\infty)$ includes no possible events, while $F(\infty)$ includes all possible events. Finally Eq. (2.6-3c) holds since for $x_1 < x_2$, $F(x_2)$ includes as many or more of the possible outcomes as does $F(x_1)$.

The features of the cumulative distribution function, as well as the concept of a random variable, will be clarified in the course of the following illustrative discussion:—We consider the experiment which consists in the rolling of 2 dice. There are 36 possible outcomes since each die may show the numbers 1 through 6. These 36 outcomes may be represented by 36 symbols λ_{ij}, where, say, i is the number that shows on the first die, and j the number that shows on the second die. Suppose however that our interest in the outcome extends only to knowing the sum of the numbers appearing on the dice. Then we may use as our random variable the function $N(\lambda_{ij})$ defined by $N(\lambda_{ij}) = i + j$. Observe that in so doing we are assigning the same number to $N(\lambda_{ij})$ and to $N(\lambda_{ji})$. Let

$$n \equiv N(\lambda_{ij}) = i + j$$

Then each integral value of n from $n = 2$ to $n = 12$ corresponds to outcomes which we care to distinguish from one another. The probabilities $P(n)$ are readily calculated. We find $P(1) = 0$, $P(2) = P(12) = 1/36$, $P(3) = P(11) = 2/36$, $P(4) = P(10) = 3/36$, $P(5) = P(9) = 4/36$, $P(6) = P(8) = 5/36$, and $P(7) = 6/36$.

Let us calculate, as an example, the cumulative distribution function $F(n)$ for, say, $n = 3$. We have from the definition of Eq. (2.6-2)

$$F(3) = P[N(\lambda_{ij}) \le 3] = P(n \le 3) \tag{2.6-4}$$

since $N(\lambda_{ij}) \equiv n$. We must now add up the probabilities of all outcomes corresponding to $n \le 3$. We find that

$$F(3) = P(1) + P(2) + P(3) = 0 + 1/36 + 2/36 = 3/36 \tag{2.6-5}$$

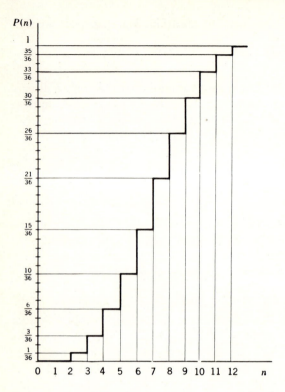

Fig. 2.6-1 Cumulative distribution function associated with rolling two dice.

In a similar way $F(n)$ for other values of n may be determined. $F(n)$ is plotted as a function of n in Fig. 2.6-1. Note that $F(n)$ satisfies all the conditions given in Eq. (2.6-3a to c) above. Observe also that there is a significance to $F(n)$ for nonintegral values of n. Note also that if the function $N(\lambda_{ij})$ had been selected in a different manner, say, $N'(\lambda_{ij})$, $F_{N'}(n)$ would be different from $F_N(n)$.

2.7 PROBABILITY DENSITY FUNCTION

The *probability density function* (p.d.f.) $f_X(x)$ is defined in terms of the cumulative distribution function $F_X(x)$ as

$$f_X(x) = \frac{d}{dx} F_X(x) \tag{2.7-1}$$

[Having made the point in Eq. (2.7-1) that $f_X(x)$, like $F_X(x)$, requires a subscript in principle, we shall again omit it where no confusion will result.]

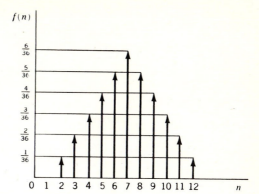

Fig. 2.7-1 The probability density function corresponding to the cumulative distribution function of Fig. 2.6-1. The heights of the arrows represent the strengths of impulses.

Thus $f(x)$ is simply the derivative of the cumulative distribution function $F(x)$. The p.d.f. has the following properties:

(a) $f(x) \geq 0$ for all x (2.7-2)

This results from the fact that $F(x)$ increases monotonically, for as x increases, more outcomes are included in the probability of occurrence represented by $F(x)$.

(b) $\int_{-\infty}^{\infty} f(x)\, dx = 1$ (2.7-3)

This result is to be seen from the fact that

$$\int_{-\infty}^{\infty} f(x)\, dx = F(\infty) - F(-\infty) = 1 - 0 = 1$$ (2.7-4)

(c) $F(x) = \int_{-\infty}^{x} f(x)\, dx$ (2.7-5)

This result follows directly from Eq. (2.7-1).

The p.d.f. corresponding to the cumulative distribution function of Fig. 2.6-1 is shown in Fig. 2.7-1. Since the derivative of a step of amplitude A is an impulse of strength $I = A$, we have $f(2) = \frac{1}{36}\delta(n - 2)$, $f(3) = \frac{2}{36}\delta(n - 3)$, etc. In Fig. 2.7-1 the height of the arrow is proportional to the strength of the impulse.

As we have already noted, a random variable may be discrete or continuous. The random variable associated, in the discussion above, with the tossing of 2 dice is a discrete random variable. The corresponding cumulative distribution function increases by steps as in Fig. 2.6-1, and the probability density displays impulses as in Fig. 2.7-1. With a continuous random variable, on the other hand, the cumulative distribution function and the probability density will generally be *smooth* as shown by the typical plots of Fig. 2.7-2.

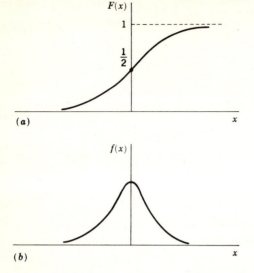

Fig. 2.7-2 (*a*) A continuous cumulative distribution function and (*b*) the corresponding probability density function.

2.8 RELATION BETWEEN PROBABILITY AND PROBABILITY DENSITY

From Eqs. (2.6-1) and (2.7-5) we find that the probability of the outcome X being less than or equal to x_1 is

$$P(X \le x_1) = F(x_1) = \int_{-\infty}^{x_1} f(x)\,dx \qquad (2.8\text{-}1)$$

Similarly, the probability that the outcome X is less than or equal to x_2 is

$$P(X \le x_2) = F(x_2) = \int_{-\infty}^{x_2} f(x)\,dx \qquad (2.8\text{-}2)$$

The probability that the outcome lies in the range $x_1 \le X \le x_2$ is

$$P(x_1 \le X \le x_2) = P(X \le x_2) - P(X < x_1) \qquad (2.8\text{-}3)$$

Note that the second term of the right-hand member of Eq. (2.8-3) is $P(X < x_1)$ and not $P(X \le x_1)$. The exclusion of the equal sign is necessary since $P(x_1 \le X \le x_2)$ includes the possibility that X is exactly $X = x_1$. Using Eqs. (2.8-1) and (2.8-2) with Eq. (2.8-3), we have

$$P(x_1 \le X \le x_2) = \int_{-\infty}^{x_2} f(x)\,dx - \int_{-\infty}^{x_1-\epsilon} f(x)\,dx = \int_{x_1-\epsilon}^{x_2} f(x)\,dx$$
$$(2.8\text{-}4)$$

where ϵ is a number which in the limit may approach zero and is introduced simply to exclude $x = x_1$ from the range of integration. If $f(x)$ is everywhere finite, then ϵ need not be included in Eq. (2.8-4). Changing the limit by an infinitesimal amount will change the integral only infinitesi-

mally. If, however, the probability density contains impulses as in Fig.
2.7-1, and if there is an impulse at $x = x_1$, then the presence of the ϵ
reminds us to include this impulse when evaluating Eq. (2.8-4). For the
case where $f(x)$ has no impulses, we have the result which expresses the
most important property of the probability density, namely,

$$P(x_1 \leq X \leq x_2) = \int_{x_1}^{x_2} f(x)\,dx \tag{2.8-5}$$

or

$$P(x \leq X \leq x + dx) = f(x)\,dx \tag{2.8-6}$$

Example 2.8-1 Consider the probability density $f(x) = ae^{-b|x|}$, where
X is a random variable whose allowable values range from $x = -\infty$
to $x = +\infty$. Find (a) the cumulative distribution function $F(x)$,
(b) the relationship between a and b, and (c) the probability that the
outcome X lies between 1 and 2.

Solution (a) The cumulative distribution function is

$$F(x) = P(X \leq x) = \int_{-\infty}^{x} f(x)\,dx = \int_{-\infty}^{x} ae^{-b|x|}\,dx \tag{2.8-7}$$

$$= \begin{cases} \dfrac{a}{b}\,e^{bx} & x \leq 0 \\[2mm] \dfrac{a}{b}\,(2 - e^{-bx}) & x \geq 0 \end{cases} \tag{2.8-8}$$

(b) In order that $f(x)$ be a probability density, it is necessary
that

$$\int_{-\infty}^{\infty} f(x)\,dx = \int_{-\infty}^{\infty} ae^{-b|x|}\,dx = \frac{2a}{b} = 1 \tag{2.8-9}$$

so that $a/b = \frac{1}{2}$.

(c) The probability that X lies in the range between 1 and 2 is

$$P(1 \leq X \leq 2) = \frac{b}{2} \int_{1}^{2} e^{-b|x|}\,dx = \frac{1}{2}(e^{-b} - e^{-2b}) \tag{2.8-10}$$

2.9 JOINT CUMULATIVE DISTRIBUTION AND PROBABILITY DENSITY

It may be necessary to identify the outcome of an experiment by two (or
more) random variables. These random variables may or may not be
independent of one another. The concepts of a cumulative distribution
and a probability density are readily extended to such cases.

For the case of a single random variable, Eq. (2.8-6) expresses, in
terms of the probability density, the probability that the random variable

X lies in the range $x \leq X \leq x + dx$. Correspondingly, for two random variables X and Y, the probability that $x \leq X \leq x + dx$ while at the same time $y \leq Y \leq y + dy$ is written

$$P(x \leq X \leq x + dx, y \leq Y \leq y + dy) = f_{XY}(x,y) \, dx \, dy \qquad (2.9\text{-}1)$$

A comparison of a probability density $f_X(x)$, involving a single random variable, and a density $f_{XY}(x,y)$, involving two variables, is indicated in Fig. 2.9-1. Part (a) shows a portion of a *curve* which is a plot of $f_X(x)$ as a function of x. Part (b) shows a portion of a *surface* which is a plot of

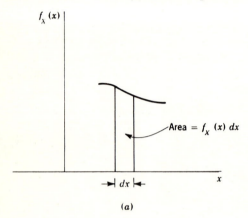

(a)

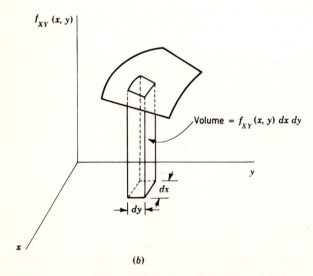

(b)

Fig. 2.9-1   A comparison between a probability density function involving a simple random variable in (a) and a density function involving two random variables as in (b).

$f_{XY}(x,y)$ as a function of the two variables x and y. The *area* indicated in (a) is the probability that $x \leq X \leq x + dx$, while the *volume* indicated in (b) is the probability that $x \leq X \leq x + dx$ and that $y \leq Y \leq y + dy$.

Extending Eq. (2.9-1) to a finite interval, we have, in correspondence with Eq. (2.8-5),

$$P(x_1 \leq X \leq x_2, y_1 \leq Y \leq y_2) = \int_{y_1}^{y_2} \int_{x_1}^{x_2} f_{XY}(x,y) \, dx \, dy \qquad (2.9\text{-}2)$$

The cumulative distribution function is

$$F_{XY}(x,y) = P(X \leq x, Y \leq y) = \int_{-\infty}^{y} \int_{-\infty}^{x} f_{XY}(x,y) \, dx \, dy \qquad (2.9\text{-}3)$$

If we should be concerned only with the cumulative probability up to, say, some value of x quite independently of y, we would write

$$F_X(x) = P(X \leq x, -\infty \leq Y \leq \infty) = \int_{-\infty}^{\infty} \int_{-\infty}^{x} f_{XY}(x,y) \, dx \, dy$$
$$(2.9\text{-}4)$$

The probability density corresponding to $F_X(x)$ in Eq. (2.9-4) is

$$f_X(x) = \frac{d}{dx} F_X(x) = \int_{-\infty}^{\infty} f_{XY}(x,y) \, dy \qquad (2.9\text{-}5)$$

If the random variables X and Y are *independent*, then by an easy extension of the considerations of Sec. 2.4, the probability given in Eq. (2.9-1) may be written as a product in which each factor involves only one random variable. That is, we may write

$$P(x \leq X \leq x + dx, y \leq Y \leq y + dy) = [f_X(x) \, dx][f_Y(y) \, dy]$$
$$(2.9\text{-}6)$$

The function $f_X(x)$, which is given by Eq. (2.9-5), depends only on x and need bear no simple relationship to $f_{XY}(x,y)$ given in Eq. (2.9-1). A similar comment applies, of course, to $f_Y(y)$. We then have, from Eqs. (2.9-1) and (2.9-6), that, if X and Y are independent,

$$f_{XY}(x,y) = f_X(x)f_Y(y) \qquad (2.9\text{-}7)$$

Further, we have, when X and Y are independent,

$$P(x_1 \leq X \leq x_2, y_1 \leq Y \leq y_2) = \left[\int_{x_1}^{x_2} f_X(x) \, dx \right] \left[\int_{y_1}^{y_2} f_Y(y) \, dy \right]$$
$$(2.9\text{-}8)$$

Henceforth, where no confusion will be caused thereby, we shall drop the subscripts from the probability density functions even when more than one such function is involved. Thus we shall write $f_X(x) = f(x)$,

$f_Y(y) = f(y)$, and $f_{XY}(x,y) = f(x,y)$. Thus the symbol $f(\ \ )$ represents not a particular probability density but rather a probability density function in general. And the particular random variable or variables referred to are to be determined by the argument of the function.

Example 2.9-1   The joint probability density of the random variables X and Y is

$$f(x,y) = \tfrac{1}{4}e^{-|x|-|y|} \qquad -\infty < x < \infty, -\infty < y < \infty$$

(a) Are X and Y statistically independent random variables?

(b) Calculate the probability that $X \le 1$ and $Y \le 0$.

Solution   (a) Since $f(x,y)$ can be written as

$$f(x,y) = \tfrac{1}{2}e^{-|x|}\tfrac{1}{2}e^{-|y|} = f(x)f(y)$$

X and Y are statistically independent.

(b) $P(X \le 1, Y \le 0) = \int_{-\infty}^{1} dx \int_{-\infty}^{0} dy\, f(x,y)$

$$= \int_{-\infty}^{1} \frac{1}{2}\, e^{-|x|}\, dx \int_{-\infty}^{0} \frac{1}{2}\, e^{-|y|}\, dy$$

$$= \left(\frac{2 - e^{-1}}{2}\right)\frac{1}{2} = \frac{1}{4}\,(2 + e^{-1})$$

2.10 AVERAGE VALUE OF A RANDOM VARIABLE

Consider now that we have the values and their associated probabilities of a discrete random variable. The possible numerical values of the random variable X are x_1, x_2, x_3, . . . , with probabilities of occurrence $P(x_1)$, $P(x_2)$, $P(x_3)$ . . . . As the number of measurements N of X becomes very large, we would expect that we would find the outcome $X = x_1$ would occur $NP(x_1)$ times, the outcome $X = x_2$ would occur $NP(x_2)$ times, etc. Hence the arithmetic sum of all the N measurements would be

$$x_1 P(x_1)N + x_2 P(x_2)N + \cdots = N \sum_i x_i P(x_i) \qquad (2.10\text{-}1)$$

The *mean* or *average value* of all these measurements and hence the average value of the random variable is calculated by dividing the sum in Eq. (2.10-1) by the number of measurements N. The mean of a random variable X is also called the *expectation* of X and is represented either by the notation $\bar{X}$ or by $E(X)$. We shall use these notations interchangeably.

Thus, using m to represent the value of the average or expectation of X, we have, from Eq. (2.10-1),

$$\bar{X} \equiv E(X) = m = \sum_i x_i P(x_i) \tag{2.10-2}$$

To calculate the average for a continuous random variable, let us divide the range of the variable into small intervals Δx. Then from Eq. (2.8-6) the probability that X lies in the range between x_i and $x_i + \Delta x$ is $P(x_i \leq X \leq x_i + \Delta x) \equiv P(x_i)$ is given approximately by

$$P(x_i) = f(x_i) \, \Delta x \tag{2.10-3}$$

Substituting Eq. (2.10-3) into Eq. (2.10-2), we have

$$m = \sum_i x_i f(x_i) \, \Delta x \tag{2.10-4}$$

In the limit, as $\Delta x \to 0$ and is replaced by dx, the summation in Eq. (2.10-4) becomes an integral, and

$$m = \int_{-\infty}^{\infty} x f(x) \, dx \tag{2.10-5}$$

In general, the average value, or expectation, of a function $g(X)$ of the random variable X is

$$\overline{g(X)} = E[g(X)] = \int_{-\infty}^{\infty} g(x) f(x) \, dx \tag{2.10-6}$$

If the function $g(X)$ is X raised to a power, that is, $g(X) = X^n$, the average value $E(X^n)$ is referred to as the *nth moment* of the random variable. For this reason, the average value $\bar{X}$ is also called the *first moment* of X.

If a random variable Z is a function of two random variables X and Y, say $Z = w(X,Y)$, then by an extension of the above discussion it may be shown that

$$\bar{Z} = \int_{-\infty}^{\infty} \int_{-\infty}^{\infty} w(x,y) f_{XY}(x,y) \, dx \, dy \tag{2.10-7}$$

2.11 VARIANCE OF A RANDOM VARIABLE

In Fig. 2.11-1 are shown two probability density functions $f(x)$ and $f'(x)$ for two random variables X and X'. As a matter of simplicity we have drawn them of the same general form and have drawn them symmetrically about a common average value m. But these features are not essential to the ensuing discussion. Rather, the important point is that $f(x)$ is *narrower* than is $f'(x)$. Suppose, then, that experimental determinations were made of X and X' yielding numerical outcomes x and x'. We would

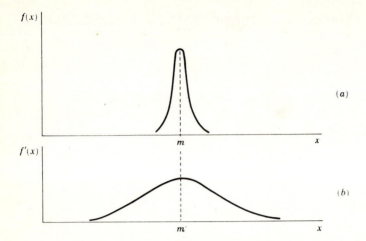

Fig. 2.11-1 Two probability density functions corresponding to random variables with different variances.

surely find that, on the average, x would be closer to m than x' would be to m'. Thus in comparing X with X', we find that the outcomes of X have a higher probability of occurring in a smaller range. In other words, if a number of determinations were made of X and X', we would expect to find that the outcomes of X would *cluster* more closely around m than would be the case for X'.

It is convenient to have a number which serves as a measure of the "width" of a probability density function. We might suggest as a candidate for such a number the average value of $(X - m)$ that is, $\overline{X - m}$. However, $\overline{X - m} = 0$, since positive and negative contributions from the portions of $f(x)$ above and below x cancel. A second possibility is $\overline{|X - m|}$, since taking the absolute value of $X - m$ would avoid the cancellation. However, a more useful measure is the square root of the average value of $(X - m)^2$, that is, the second moment of $X - m$. This second moment is represented by the symbol σ^2 and is called the *variance* of the random variable. Thus

$$\sigma^2 \equiv E[(X - m)^2] = \int_{-\infty}^{\infty} (x - m)^2 f(x) \, dx \qquad (2.11\text{-}1)$$

Writing $(x - m)^2 = x^2 - 2mx + m^2$ in the integral of Eq. (2.11-1) and integrating term by term, we find

$$\sigma^2 = E(X^2) - 2m^2 + m^2 \qquad (2.11\text{-}2a)$$

$$= E(X^2) - m^2 \qquad (2.11\text{-}2b)$$

The quantity σ itself is called the standard deviation and is the *root mean square* (rms) value of $(X - m)$. If the average value $m = 0$, then

$$\sigma^2 = E(X^2)$$

2.12 THE GAUSSIAN PROBABILITY DENSITY

The *gaussian* (also called *normal*) probability density function is of the greatest importance because many naturally occurring experiments are characterized by random variables with a gaussian density. It is of special relevance to us because the random variables of concern to us will be described by, almost exclusively, the gaussian density function. A further importance is attached to the gaussian density because it is involved in a remarkable theorem called the *central-limit theorem* which we discuss in Sec. 2.18.

The gaussian probability density function is defined as

$$f(x) = \frac{1}{\sqrt{2\pi\sigma^2}}\, e^{-(x-m)^2/2\sigma^2} \tag{2.12-1}$$

and is plotted in Fig. 2.12-1. In using the symbols m and σ^2 in Eq. (2.12-1), we have taken cognizance of the fact that m and σ^2 are indeed the average value and variance associated with $f(x)$. Thus we find that

$$\bar{X} = \int_{-\infty}^{\infty} \frac{xe^{-(x-m)^2/2\sigma^2}}{\sqrt{2\pi\sigma^2}}\, dx = m \tag{2.12-2}$$

and

$$E[(X - m)^2] = \int_{-\infty}^{\infty} \frac{(x - m)^2 e^{-(x-m)^2/2\sigma^2}}{\sqrt{2\pi\sigma^2}}\, dx = \sigma^2 \tag{2.12-3}$$

It may also be verified that

$$\int_{-\infty}^{\infty} f(x)\, dx = 1 \tag{2.12-4}$$

as is required for a probability density function.

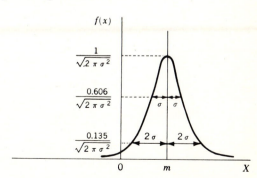

Fig. 2.12-1 The gaussian density function.

As is indicated in Fig. 2.12-1, when $x - m = \pm\sigma$, that is, at values of x separated from m by the standard deviation, $f(x)$ has fallen to 0.606 of its peak value. When $x - m = \pm 2\sigma$, $f(x)$ falls to 0.135 of the peak value, and at $x - m = 3\sigma$ (not shown in the figure), $f(x)$ has fallen to 0.01 of the peak value.

2.13 THE ERROR FUNCTION

The cumulative distribution corresponding to the gaussian probability density, for $m = 0$, is

$$P(X \leq x) = F(x) = \int_{-\infty}^{x} \frac{e^{-x^2/2\sigma^2}}{\sqrt{2\pi\sigma^2}}\, dx \tag{2.13-1}$$

The integral in Eq. (2.13-1) is not easily evaluated. It is, however, directly related to the *error function*, tabulated values of which are readily available in mathematical tables.[2] A short table is available in the appendix. The error function of u, written erf u, is defined as

$$\text{erf } u \equiv \frac{2}{\sqrt{\pi}} \int_{0}^{u} e^{-u^2}\, du \tag{2.13-2}$$

The error function has the values erf $(0) = 0$ and erf $(\infty) = 1$. The *complementary error function*, written erfc u, is defined by erfc $u \equiv 1 - \text{erf } u$ and is given by

$$\text{erfc } u \equiv 1 - \text{erf } u = \frac{2}{\sqrt{\pi}} \int_{u}^{\infty} e^{-u^2}\, du \tag{2.13-3}$$

The cumulative distribution $F(x)$ of Eq. (2.13-1) may be expressed in terms of the error function and the complementary error function. For $x \geq 0$ we write

$$F(x) = \int_{-\infty}^{x} \frac{e^{-x^2/2\sigma^2}}{\sqrt{2\pi\sigma^2}}\, dx = \int_{-\infty}^{\infty} \frac{e^{-x^2/2\sigma^2}}{\sqrt{2\pi\sigma^2}}\, dx - \int_{x}^{\infty} \frac{e^{-x^2/2\sigma^2}}{\sqrt{2\pi\sigma^2}}\, dx \tag{2.13-4}$$

The first term of the right-hand member of Eq. (2.13-4) is the integral from $-\infty$ to $+\infty$ of the probability density $f(x)$ and hence has the value 1. If we let $u \equiv x/\sqrt{2}\,\sigma$, Eq. (2.13-4) becomes

$$F(x) = 1 - \frac{1}{2}\left(\frac{2}{\sqrt{\pi}} \int_{x/\sqrt{2}\,\sigma}^{\infty} e^{-u^2}\, du\right) = 1 - \frac{1}{2}\,\text{erfc}\left(\frac{x}{\sqrt{2}\,\sigma}\right) \tag{2.13-5}$$

For $x \leq 0$, since tabulated values of erfc $(x/\sqrt{2}\,\sigma) = $ erfc u are readily available only for positive u, we proceed as follows:

$$F(x) = F(-|x|) = \int_{-\infty}^{-|x|} \frac{e^{-x^2/2\sigma^2}}{\sqrt{2\pi\sigma^2}}\, dx = \frac{1}{\sqrt{\pi}} \int_{-\infty}^{-|x|/\sqrt{2}\,\sigma} e^{-u^2}\, du \tag{2.13-6}$$

Letting $\xi = -u$ yields

$$F(x) = \frac{1}{2} \left(\frac{2}{\sqrt{\pi}} \int_{|x|/\sqrt{2}\,\sigma}^{\infty} e^{-\xi^2} \, d\xi \right) = \frac{1}{2} \operatorname{erfc} \left(\frac{|x|}{\sqrt{2}\,\sigma} \right) \qquad (2.13\text{-}7)$$

The error function and complementary error function may be used for many additional useful calculations in connection with the gaussian density. A matter of interest, for example, is the probability that a measurement will yield an outcome that falls within a certain range about the average value of the random variable. Since the "width" of the probability density depends on the standard deviation σ, we ask for the probability $P(m - k\sigma \leq X \leq m + k\sigma)$, that is, the probability that the random variable X is not further from $x = m$ than $k\sigma$, where k is a constant number. It may be shown (Prob. 2.13-3) that

$$P_{\pm k\sigma} \equiv P(m - k\sigma \leq X \leq m + k\sigma) = \operatorname{erf} \left(\frac{k}{\sqrt{2}} \right) \qquad (2.13\text{-}8)$$

For future reference some values of $P_{\pm k\sigma}$ are tabulated here.

k	$P_{\pm k\sigma}$	k	$P_{\pm h\sigma}$
0.5	0.383	2.5	0.988
1.0	0.683	3.0	0.997
1.5	0.866	3.5	0.9995
2.0	0.955	4.0	0.99994

Observe how small is the likelihood that a measured value will fall outside the range $\pm 3\sigma$.

As a further example of a probability calculation, consider the following situation. Across a set of terminals there appears a voltage V which is either $V = v_0$ volts or $V = 0$ volt. One of the two possible constant voltages is transmitted over wires from a distant point to alert us as to which of two possible situations prevails there. We are to determine the situation by performing the experiment of making an instantaneous measurement of the voltage across the terminals. A difficulty arises because superimposed on the constant 0 volt or v_0 volts is noise. We assume that the noise has a gaussian probability density and zero average value. Because of the noise our measurement will, in general, yield neither the result 0 volt nor the result v_0 volts. What procedure are we to use to make a decision? What is the probability that our decision will be in error?

When the transmitted voltage is 0 volt, the terminal voltage is $V = N$, where N is the gaussian random variable representing the noise. Hence V is also a gaussian random variable with probability density as

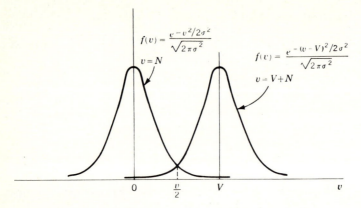

Fig. 2.13-1 Probability density function of a gaussian random variable N and of the sum $v_0 + N$.

indicated in Fig. 2.13-1. When the transmitted voltage is v_0, then $V = v_0 + N$. In this latter case, V is a gaussian random variable with mean v_0 but with the same variance as in the former case (Prob. 2.12-2). The probability density of $V = v_0 + N$ is also shown in Fig. 2.13-1. On the basis of the symmetry in Fig. 2.13-1 it is apparent that the decision level should be $V = v_0/2$. That is, when we measure V to be $V < v_0/2$, we should decide that the transmitted voltage is 0 volt, and when $V > v_0/2$, we should decide that v_0 volts was transmitted.

When v_0 volts is transmitted, the probability of making an error is the probability $P(V = v_0 + N < v_0/2)$, for, if $v_0 + N < v_0/2$, the v_0-volt transmission will be mistaken for a 0-volt transmission. This *probability of error* is

$$P_{\text{error}} = P\left(V = v_0 + N < \frac{v_0}{2}\right) = \int_{-\infty}^{v_0/2} \frac{e^{-(v-v_0)^2/2\sigma^2}}{\sqrt{2\pi\sigma^2}} \, dv \qquad (2.13\text{-}9)$$

Letting $u \equiv (v_0 - v)/\sqrt{2}\sigma$, we find

$$P_{\text{error}} = \frac{1}{\sqrt{\pi}} \int_{v_0/(2\sqrt{2}\sigma)}^{\infty} e^{-u^2} \, du = \frac{1}{2} \text{erfc} \frac{v_0}{2\sqrt{2}\sigma} \qquad (2.13\text{-}10)$$

Alternatively this P_{error} may be calculated as the probability that the noise alone is *negative* and of magnitude greater than $v_0/2$.

Similarly 0 volt will be mistaken for v_0 volts if the noise is *positive* and greater in magnitude than $v_0/2$. Since the gaussian density is symmetrical, the probability that either transmission will be mistaken for the other is the same and is given by Eq. (2.13-10). Finally, then, P_{error} in Eq. (2.13-10) gives the probability of an error *without reference* to which voltage is transmitted.

We may note that, because of the manner in which the error probabilities are related to the error function, this function is appropriately named.

2.14 THE RAYLEIGH PROBABILITY DENSITY

We consider now the Rayleigh probability density function. The Rayleigh density is of interest to us principally because of a special relationship which it holds to the gaussian density. For reasons which will be apparent shortly, we use the symbol R to represent the random variable and r to represent the value assumed by the variable. The Rayleigh density is defined by

$$f(r) = \begin{cases} \dfrac{r}{\alpha^2} e^{-r^2/2\alpha^2} & 0 \leq r \leq \infty \\ 0 & r < 0 \end{cases} \qquad (2.14\text{-}1)$$

Note particularly that $f(r)$ is nonzero only for positive values of r. A plot of $f(r)$ as a function of r is shown in Fig. 2.14-1. It attains a maximum value $1/(\alpha \sqrt{e})$ at $r = \alpha$. It may be verified that the mean value $\bar{R} = \sqrt{\pi/2}\, \alpha$, the mean-square value $\overline{R^2} = 2\alpha^2$, and the variance $\sigma^2 = (2 - \pi/2)\alpha^2$.

Now let X and Y be two independent gaussian random variables each with average value zero and each with variance σ^2. The joint density function is, using Eq. (2.9-7),

$$f(x,y) = f(x)f(y) = \frac{e^{-x^2/2\sigma^2}}{\sqrt{2\pi\sigma^2}} \frac{e^{-y^2/2\sigma^2}}{\sqrt{2\pi\sigma^2}} = \frac{e^{-(x^2+y^2)/2\sigma^2}}{2\pi\sigma^2} \qquad (2.14\text{-}2)$$

If we should now make a plot of $f(x,y)$ as a function of x and y, we should find a bell-shaped surface above the x-y plane. The probability

$$P(x \leq X \leq x + dx, y \leq Y \leq y + dy) = f(x,y)\, dx\, dy \qquad (2.14\text{-}3)$$

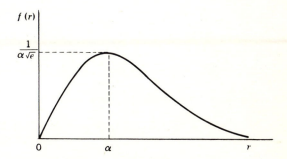

Fig. 2.14-1 The Rayleigh density.

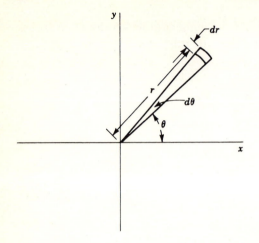

Fig. 2.14-2 Cartesian and polar representation of an area.

has the significance indicated in Fig. 2.9-1*b*. That is, to find the probability specified in Eq. (2.14-3), we mark off an area $dx\,dy$ on the x-y plane, and the probability specified is equal to the volume above the surface area $dx\,dy$ and below the surface $f(x,y)$.

Suppose, however, that to locate a point on the x-y plane we use not the x and y coordinates but rather the coordinates r and θ, where r is the length of the radius vector to the point from the coordinate system origin and θ is the angle measured from the x axis as shown in Fig. 2.14-2. A differential area in such an r, θ coordinate system is $r\,dr\,d\theta$. The volume above this area is then equal to the probability $P(r \leq R \leq r + dr,\ \theta \leq \Theta \leq \theta + d\theta)$. Since $r^2 = x^2 + y^2$, we have, from Eqs. (2.14-2) and (2.14-3), that

$$P(r \leq R \leq r + dr,\ \theta \leq \Theta \leq \theta + d\theta) = \frac{e^{-r^2/2\sigma^2}}{2\pi\sigma^2}\,r\,dr\,d\theta$$

$$= \left(\frac{re^{-r^2/2\sigma^2}}{\sigma^2}\,dr\right)\left(\frac{d\theta}{2\pi}\right)$$

$$(2.14\text{-}4)$$

Hence it appears that the probability specified in Eq. (2.14-4) can be expressed in terms of the probability density functions associated with two *independent* random variables R and Θ. The probability densities are

$$f_R(r) = \begin{cases} \dfrac{re^{-r^2/2\sigma^2}}{\sigma^2} & r \geq 0 \\ 0 & r < 0 \end{cases} \qquad (2.14\text{-}5)$$

and

$$f_\Theta(\theta) = \frac{1}{2\pi} \qquad -\pi \leq \theta \leq \pi \qquad (2.14\text{-}6)$$

The independence is apparent from the fact, as appears in Eq. (2.14-4), that the joint density $f_{R,\Theta}(r,\theta)$ appears as the product $f_R(r)f_\Theta(\theta)$.

We note that $f_R(r)$ is precisely the Rayleigh density with $\sigma^2 = \alpha^2$, while $f_\Theta(\theta)$ is a uniform density independent of the angle θ. We observe in Fig. 2.14-1 that the most probable value of r is $r = \alpha = \sigma$.

2.15 MEAN AND VARIANCE OF THE SUM OF RANDOM VARIABLES

Let X and Y be two random variables with means m_x and m_y. Let $Z = X + Y$ and have the mean m_z. Then $m_z = m_x + m_y$. This result follows directly from the definition of the mean. We have, using Eqs. (2.10-7), (2.9-5), and (2.10-5), that

$$m_z = \int_{-\infty}^{\infty} \int_{-\infty}^{\infty} (x + y)f(x,y)\, dx$$

$$= \int_{-\infty}^{\infty} \int_{-\infty}^{\infty} xf(x,y)\, dx\, dy + \int_{-\infty}^{\infty} \int_{-\infty}^{\infty} yf(x,y)\, dx\, dy \quad (2.15\text{-}1a)$$

$$= m_x \qquad\qquad\qquad + m_y \qquad\qquad\qquad (2.15\text{-}1b)$$

Expressed in words, *the mean of the sum is equal to the sum of the means.* This result holds whether the variables X and Y are independent or not.

We calculate next the second moment of $Z = X + Y$. In this calculation, however, we shall restrict ourselves to the circumstance that X and Y are *independent*. We have

$$\overline{Z^2} = \overline{(X + Y)^2} = \int_{-\infty}^{\infty} \int_{-\infty}^{\infty} (x + y)^2 f(x,y)\, dx\, dy \quad (2.15\text{-}2)$$

Because of the *independence* of X and Y, $f(x,y) = f(x)f(y)$ so that

$$\overline{Z^2} = \int_{-\infty}^{\infty} x^2 f(x)\, dx \int_{-\infty}^{\infty} f(y)\, dy + \int_{-\infty}^{\infty} y^2 f(y)\, dy \int_{-\infty}^{\infty} f(x)\, dx$$

$$+ 2 \int_{-\infty}^{\infty} xf(x)\, dx \int_{-\infty}^{\infty} yf(y)\, dy \quad (2.15\text{-}3)$$

However

$$\int_{-\infty}^{\infty} f(x)\, dx = \int_{-\infty}^{\infty} f(y)\, dy = 1 \qquad\qquad (2.15\text{-}4)$$

so that

$$\overline{Z^2} = \overline{X^2} + \overline{Y^2} + 2\bar{X}\bar{Y} \qquad\qquad (2.15\text{-}5)$$

If either $\bar{X}$ or $\bar{Y}$ or both are zero, then $\overline{Z^2} = \overline{X^2} + \overline{Y^2}$.

The variance of Z is found, using Eqs. (2.11-2) and (2.15-5), to be

$$\sigma_z^2 = \sigma_x^2 + \sigma_y^2 \qquad\qquad (2.15\text{-}6)$$

This result can be obtained directly by noting that $\sigma_z^2 = \overline{(Z - m_z)^2}$.

2.16 PROBABILITY DENSITY OF $Z = X + Y$

We now calculate the probability density $f(z)$ of $Z = X + Y$ in terms of the joint density $f(x,y)$. Assume an arbitrary value of Z and call it z. Then the region $Y \leq z - X$ is shown as the shaded region in Fig. 2.16-1. Hence the probability that $Z \leq z$ is the same as the probability that $Y \leq z - X$ independently of the value of X, that is, for $-\infty \leq X \leq +\infty$. This probability is

$$F(z) = P(Z \leq z) = P(X \leq \infty, Y \leq z - X) \qquad (2.16\text{-}1)$$

Using Eq. (2.9-3), we have

$$F(z) = \int_{-\infty}^{\infty} dx \int_{-\infty}^{z-x} f(x,y)\, dy \qquad (2.16\text{-}2)$$

The probability density of Z is found by differentiating $F(z)$ with respect to z. We then have

$$f(z) = \frac{dF(z)}{dz} = \int_{-\infty}^{\infty} f(x, z - x)\, dx \qquad (2.16\text{-}3)$$

which expresses $f_Z(z)$ in terms of $f_{XY}(x,y = z - x)$.

When X and Y are *independent*, $f(x,y) = f(x)f(y)$, and Eq. (2.16-3) may be written

$$f(z) = \int_{-\infty}^{\infty} f(x)f(z - x)\, dx \qquad (2.16\text{-}4)$$

Comparing Eq. (2.16-4) with Eq. (1.11-1), we recognize that $f(z)$ is the *convolution* of $f(x)$ and $f(y)$.

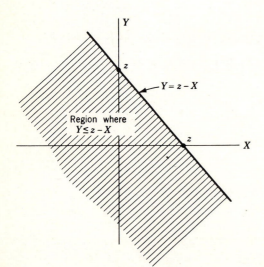

Fig. 2.16-1 Related to the calculation of the probability density of a sum of random variables.

As a most important consequence of the result given in Eq. (2.16-4), let us consider the case where $f(x)$ and $f(y)$ are both gaussian densities.

$$f(x) = \frac{e^{-x^2/2\sigma_x^2}}{\sqrt{2\pi\sigma_x^2}} \tag{2.16-5}$$

$$f(y) = \frac{e^{-y^2/2\sigma_y^2}}{\sqrt{2\pi\sigma_y^2}} \tag{2.16-6}$$

Then $f(z)$ is given by

$$f(z) = \frac{1}{2\pi\sigma_x\sigma_y} \int_{-\infty}^{\infty} e^{-x^2/2\sigma_x^2} e^{-(z-x)^2/2\sigma_y^2} \, dx \tag{2.16-7}$$

If we define a variable u by

$$u \equiv \left(\frac{1}{\sigma_x^2} + \frac{1}{\sigma_y^2}\right)^{\frac{1}{2}} x - \frac{z}{\sigma_y^2(1/\sigma_x^2 + 1/\sigma_y^2)^{\frac{1}{2}}} \tag{2.16-8}$$

then we find, after some algebraic manipulation, that Eq. (2.16-7) may be written

$$f(z) = \frac{e^{-z^2/2(\sigma_x^2+\sigma_y^2)}}{\sqrt{2\pi(\sigma_x^2 + \sigma_y^2)}} \int_{-\infty}^{\infty} \frac{e^{-u^2/2}}{\sqrt{2\pi}} \, du \tag{2.16-9}$$

The definite integral in Eq. (2.16-9) has the value 1. Hence if $\sigma^2 \equiv \sigma_x^2 + \sigma_y^2$, Eq. (2.16-9) becomes

$$f(z) = \frac{e^{-z^2/2\sigma^2}}{\sqrt{2\pi\sigma^2}} \tag{2.16-10}$$

which is a gaussian density function. Thus we have the extremely interesting and important result: Given two independent gaussian random variables, the sum of these variables is itself a gaussian random variable. We note, of course, that the variance σ^2 of the sum variable is the sum $\sigma_x^2 + \sigma_y^2$ of the individual variables. This result, $\sigma^2 = \sigma_x^2 + \sigma_y^2$, applies for independent random variables generally, not just gaussian variables. Further, in the discussion above we assumed, for simplicity, that the variables X and Y had zero mean. However, this assumption does not make the result above any less general. For if we had assumed average values m_x and m_y for X and Y, then we would have found that Z had an average value $m_z = m_x + m_y$ and Eq. (2.16-10) would become

$$f(z) = \frac{e^{-(z-m_z)^2/2\sigma^2}}{\sqrt{2\pi\sigma^2}} \tag{2.16-11}$$

This result, concerning the sum of independent gaussian random variables, may be extended somewhat. If X_1 is a gaussian random variable, then, as may be easily verified, if c_1 is a constant, c_1X_1 is also a gaussian variable. It follows, by extension of the result above for two independent gaussian random variables, that a linear combination of independent gaussian variables is also gaussian. Explicitly, if X_1, X_2, X_3 . . . are independent gaussian variables and c_1, c_2, c_3 . . . are constants, then

$$X = c_1 X_1 + c_2 X_2 + c_3 X_3 \cdots \tag{2.16-12}$$

is also a gaussian random variable. This result is rather distinctive of the gaussian form. In general, the probability density function of a sum of variables of like probability density does not preserve the form of the probability density of the individual variables.

Our proof does not extend to the case of *dependent* gaussian variables. However, it is interesting to note that, even in the case where the gaussian variables are dependent, a linear combination of such variables is still gaussian.

2.17 CORRELATION BETWEEN RANDOM VARIABLES

The *covariance* μ of two random variables X and Y is defined as

$$\mu \equiv E\{(X - m_x)(Y - m_y)\} \tag{2.17-1}$$

If X and Y are independent random variables, we find, using Eq. (2.9-7) and (2.10-7), that

$$\mu = E\{(X - m_x)(Y - m_y)\}$$
$$= \int_{-\infty}^{\infty} (x - m_x)f(x) \, dx \int_{-\infty}^{\infty} (y - m_y)f(y) \, dy$$
$$= (m_x - m_x)(m_y - m_y) = 0 \tag{2.17-2}$$

This result is rather to have been expected on intuitive grounds. For, as assumed, when an experiment is performed to determine a joint outcome specified by the set of numbers x and y, the outcome y is in no way conditional on the outcome x. Hence we would find that a particular outcome x_1 would occur jointly sometimes with a value y which was positive with respect to its average m_y, and sometimes with a value y negative with respect to m_y. In the course of many experiments with the outcome x_1, the sum of the numbers $x_1(y - m_y)$ would add up to zero.

On the other hand, suppose X and Y were dependent. Suppose, for example, that outcome y was conditioned on the outcome x in such manner that there was an enhanced probability that $(y - m_y)$ was of the same sign as $(x - m_x)$. In such a case we would anticipate that the expected value $E\{(X - m_x)(Y - m_y)\} > 0$. Similarly, if there were an enhanced probability that $(x - m_x)$ and $(y - m_y)$ were of opposite sign, we would expect $E\{(X - m_x)(Y - m_y)\} < 0$.

As an extreme case, let us assume the maximum possible dependency between X and Y. Let us assume that $X = Y$ or $X = -Y$. In these cases we would find (with $m_x = m_y = 0$)

$$E\{XY\} = E\{X^2\} = E\{Y^2\} = \sigma_x^2 = \sigma_y^2 = \sigma_x \sigma_y \tag{2.17-3}$$

or

$$E\{XY\} = E\{-X^2\} = E\{-Y^2\} = -\sigma_x^2 = -\sigma_y^2 = -\sigma_x\sigma_y \quad (2.17\text{-}4)$$

Consider then the quantity ρ defined by

$$\rho \equiv \frac{\mu}{\sigma_x\sigma_y} = \frac{E\{XY\}}{\sigma_x\sigma_y} \quad (2.17\text{-}5)$$

This number ρ is called the *correlation coefficient* between the variables X and Y and serves as a measure of the extent to which X and Y are dependent. From Eqs. (2.17-3) to (2.17-5) we have that ρ falls in the range

$$-1 \le \rho \le +1 \quad (2.17\text{-}6)$$

When X and Y are independent, $\rho = 0$. When $X = Y$, $\rho = 1$; when $X = -Y$, $\rho = -1$. If X and Y are neither identical nor independent, then ρ will have a magnitude between 0 and 1. When $\rho = 0$, the random variables X and Y are said to be *uncorrelated*.

When random variables are independent, they are uncorrelated. However, the fact that they are uncorrelated does not ensure that they are independent. A simple illustrative example will establish this point.

Example 2.17-1 Let Z be a random variable with probability density $f(z) = \frac{1}{2}$ in the range $-1 \le z \le 1$. Let the random variable $X = Z$ and the random variable $Y = Z^2$. Obviously X and Y are not independent since $X^2 = Y$. Show, nonetheless, that X and Y are uncorrelated.

Solution   We have

$$E\{Z\} = \int_{-1}^{1} \frac{1}{2}\, dz = 0 \quad (2.17\text{-}7)$$

Since $X = Z$, $E\{X\} = E\{Z\} = 0$. Since $Y = Z^2$, $E\{Y\} = E\{Z^2\}$, so that

$$E\{Y\} = \int_{-1}^{1} \frac{1}{2} z^2\, dz = \frac{1}{3} \quad (2.17\text{-}8)$$

The covariance μ is

$$\mu = E\{(X - m_x)(Y - m_y)\} = E\left\{(X)\left(Y - \frac{1}{3}\right)\right\}$$

$$= E\left\{XY - \frac{1}{3}X\right\} \quad (2.17\text{-}9a)$$

$$= E\left\{Z^3 - \frac{Z}{3}\right\} = \int_{-1}^{1} \frac{1}{2}\left(z^3 - \frac{z}{3}\right) dz = 0 \quad (2.17\text{-}9b)$$

Hence $\rho = 0$, and the variables are uncorrelated.

As we have seen, it is easy enough to invent random variables which are uncorrelated but are nevertheless dependent. It is even possible to do so when the random variables are gaussian. On the other hand, the random variables of interest to us will be variables which occur in the description of such natural physical processes as noise. As already noted, for the most part, such variables are gaussian. It does indeed turn out that, with such gaussian random variables, an absence of correlation does imply independence. We shall therefore generally assume that two gaussian random variables, for which the covariance $\mu = 0$, are independent variables.

2.18 THE CENTRAL-LIMIT THEOREM[3]

The result of the previous section concerning the probability density of the sum of gaussian random variables is a special case of the *central-limit theorem*. The so-called central-limit theorem is actually a group of related theorems which are collectively grouped under a single name. We shall not even undertake to state these theorems precisely, let alone undertake to prove them. For our purposes it will be adequate to note that the central-limit theorem indicates that the probability density of a sum of N independent random variables tends to approach a gaussian density as the number N increases. The mean and variance of this gaussian density are respectively the sum of the means and the sum of the variances of the N independent random variables. The theorem applies even when (with a few special exceptions) the individual random variables are not gaussian. In addition, the theorem applies in certain special cases even when the individual random variables are not independent.

As an illustration of the central-limit theorem consider the case where the individual random variable has a uniform (constant) probability density as shown in Fig. 2.18-1a. Note here that the area under the density has been adjusted to unity. Then, as indicated by Eq. (2.16-4), if there were two terms in the sum, the density of the sum would be determined as the convolution of the density in Fig. 2.18-1a with itself. The result of such a convolution is indicated in Fig. 2.18-1b. Similarly, the density of a sum of the three random variables is the convolution of the density in Fig. 2.18-1b with the density in Fig. 2.18-1a. The result (Prob. 2.18-1) is shown in Fig. 2.18-1c. Note that even for this sum of only three terms the result suggests a gaussian density. In the limit, as more and more terms are added, the density would indeed become gaussian.

2.19 RANDOM PROCESSES

To determine the probabilities of the various possible outcomes of an experiment, it is necessary to repeat the experiment many times. Sup-

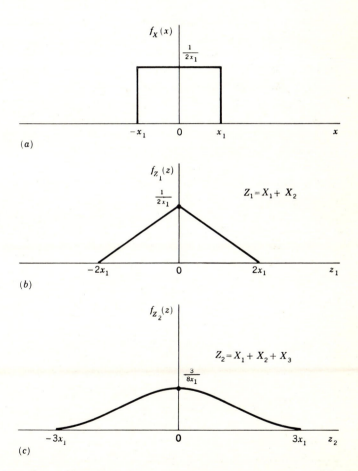

Fig. 2.18-1 (*a*) A random variable X has a uniform probability density. (*b*) The probability density of the random variable $X_1 + X_2$. (*c*) The density of the random variable $X_1 + X_2 + X_3$.

pose then that we are interested in establishing the statistics associated with the tossing of a die. We might proceed in either of two ways. On one hand, we might use a single die and toss it repeatedly. Alternatively, we might toss simultaneously a very large number of dice. Intuitively, we would expect that both methods would give the same results. Thus, we would expect that a single die would yield a particular outcome, on the average, of 1 time out of 6. Similarly, with many dice we would expect that 1/6 of the dice tossed would yield a particular outcome.

Analogously, let us consider a random process such as a noise waveform $n(t)$ mentioned at the beginning of this chapter. To determine the statistics of the noise, we might make repeated measurements of the noise voltage output of a single noise source, or we might, at least conceptually, make simultaneous measurements of the output of a very large collection of statistically identical noise sources. Such a collection of sources is called an *ensemble*, and the individual noise waveforms are called *sample functions*. A statistical average may be determined from measurements made at some fixed time $t = t_1$ on all the sample functions of the ensemble. Thus to determine, say, $\overline{n^2(t)}$, we would, at $t = t_1$, measure the voltages $n(t_1)$ of each noise source, square and add the voltages, and divide by the (large) number of sources in the ensemble. The average so determined is the *ensemble average* of $n^2(t)$.

Now $n(t_1)$ is a random variable and will have associated with it a probability density function. The ensemble averages will be identical with the statistical averages computed earlier in Secs. 2.10 and 2.11 and may be represented by the same symbols. Thus the statistical or ensemble average of $n^2(t_1)$ may be written $E[n^2(t_1)] = \overline{n^2(t_1)}$. The averages determined by measurements on a single sample function at successive times will yield a *time average*, which we represent as $\langle n^2(t) \rangle$.

In general, ensemble averages and time averages are not the same. Suppose, for example, that the statistical characteristics of the sample functions in the ensemble were changing with time. Such a variation could not be reflected in measurements made at a fixed time, and the ensemble averages would be different at different times. When the statistical characteristics of the sample functions do not change with time, the random process is described as being *stationary*. However, even the property of being stationary does not ensure that ensemble and time averages are the same. For it may happen that while each sample function is stationary the individual sample functions may differ statistically from one another. In this case, the time average will depend on the particular sample function which is used to form the average. When the nature of a random process is such that ensemble and time averages are identical, the process is referred to as *ergodic*. An ergodic process is stationary, but, of course, a stationary process is not necessarily ergodic.

Throughout this text we shall assume that the random processes with which we shall have occasion to deal are ergodic. Hence the ensemble average $E\{n(t)\}$ is the same as the time average $\langle n(t) \rangle$, the ensemble average $E\{n^2(t)\}$ is the same as the time average $\langle n^2(t) \rangle$, etc.

Example 2.19-1 Consider the random process

$$V(t) = \cos(\omega_0 t + \Theta) \tag{2.19-1}$$

where Θ is a random variable with a probability density

$$f(\theta) = \frac{1}{2\pi} \qquad \pi \le \theta \le \pi \tag{2.19-2}$$

$$= 0 \qquad \text{elsewhere}$$

(a) Show that the first and second moments of $V(t)$ are independent of time.

(b) If the random variable Θ in Eq. (2.19-1) is replaced by a fixed angle θ_0, will the ensemble mean of $V(t)$ be time independent?

Solution (a) Choose a fixed time $t = t_1$. Then

$$E\{V(t_1)\} = \int_{-\pi}^{\pi} \frac{1}{2\pi} \cos(\omega_0 t_1 + \theta) \, d\theta = 0 \tag{2.19-3}$$

and

$$E\{V^2(t_1)\} = \int_{-\pi}^{\pi} \frac{1}{2\pi} \cos^2(\omega_0 t_1 + \theta) \, d\theta = \frac{1}{2} \tag{2.19-4}$$

We note that these moments are independent of t_1 and hence independent of time.

In a similar manner it can be established that all of the moments and all other statistical characteristics of $V(t)$ are independent of time. *Hence $V(t)$ is a stationary process.*

(b) Since θ is known, $V(t)$ is deterministic.

$$E\{V(t)\} = V(t) = \cos(\omega_0 t + 30°) \ne \text{constant} \tag{2.19-5}$$

Thus, $V(t)$ is not stationary.

Example 2.19-2 A voltage $V(t)$, which is a gaussian ergodic random process with a mean of zero and a variance of 1 volt^2, is measured by a dc meter, a true rms meter, and a meter which first squares $V(t)$ and then reads its dc value.

Find the output of each meter.

Solution   (*a*) The dc meter reads

$$\langle V(t) \rangle = E\{V(t)\}$$

since $V(t)$ is ergodic. Since $E\{V(t)\} = 0$, the dc meter reads zero.
(*b*) The true rms meter reads

$$\sqrt{\langle V^2(t) \rangle} = \sqrt{E\{V^2(t)\}}$$

since $V(t)$ is ergodic. Since $V(t)$ has a zero mean, the true rms
meter reads $\sigma = 1$ volt.
(*c*) The square and average meter (a full-wave rectifier meter) yields
a deflection proportional to

$$\langle v^2(t) \rangle = E\{V^2(t)\} = \sigma^2 = 1$$

2.20 AUTOCORRELATION

A random process $n(t)$, being neither periodic nor of finite energy has an
autocorrelation function defined by Eq. (1.17-1). Thus

$$R(\tau) = \lim_{T \to \infty} \frac{1}{T} \int_{-T/2}^{T/2} n(t)n(t + \tau) \, dt \qquad (2.20\text{-}1)$$

In connection with deterministic waveforms we were able to give a
physical significance to the concept of a power spectral density $G(f)$ and
to show that $G(f)$ and $R(\tau)$ constitute a Fourier transform pair. As an
extension of that result we shall *define* the power spectral density of a
random process in the same way. Thus for a random process we take
$G(f)$ to be

$$G(f) = \mathfrak{F}[R(\tau)] = \int_{-\infty}^{\infty} R(\tau)e^{-j\omega\tau} \, d\tau \qquad (2.20\text{-}2)$$

It is of interest to inquire whether $G(f)$ defined in Eq. (2.20-2) for a random process has a physical significance which corresponds to the physical
significance of $G(f)$ for deterministic waveforms.

For this purpose consider a *deterministic* waveform $v(t)$ which extends
from $-\infty$ to ∞. Let us select a section of this waveform which extends
from $-T/2$ to $T/2$. This waveform $v_T(t) = v(t)$ in this range, and otherwise $v_T(t) = 0$. The waveform $v_T(t)$ has a Fourier transform $V_T(f)$. We
recall that $|V_T(f)|^2$ is the energy spectral density; that is, $|V_T(f)|^2 \, df$ is the
normalized energy in the spectral range df. Hence, over the interval T
the normalized power density is $|V_T(f)|^2/T$. As $T \to \infty$, $v_T(t) \to v(t)$,
and we then have the result that the physical significance of the power
spectral density $G(f)$, at least for a deterministic waveform, is that

$$G(f) = \lim_{T \to \infty} \frac{1}{T} |V_T(f)|^2 \qquad (2.20\text{-}3)$$

Correspondingly, we state, without proof, that when $G(f)$ is defined for a random process, as in Eq. (2.20-2), as the transform of $R(\tau)$, then $G(f)$ has the significance that

$$G(f) = \lim_{T \to \infty} E \left\{ \frac{1}{T} |N_T(f)|^2 \right\} \qquad (2.20\text{-}4)$$

where $E\{\ \}$ represents the ensemble average or expectation and $N_T(f)$ represents the Fourier transform of a truncated section of a sample function of the random process $n(t)$.

The autocorrelation function $R(\tau)$ is, as indicated in Eq. (2.20-1), a *time average* of the product $n(t)$ and $n(t + \tau)$. Since we have assumed an ergodic process, we are at liberty to perform the averaging over any sample function of the ensemble, since every sample function will yield the same result. However, again because the noise process is ergodic, we may replace the time average by an ensemble average and write, instead of Eq. (2.20-1),

$$R(\tau) = E\{n(t)n(t + \tau)\} \qquad (2.20\text{-}5)$$

The averaging indicated in Eq. (2.20-5) has the following significance. At some *fixed* time t, $n(t)$ is a random variable, the possible values for which are the values $n(t)$ assumed at time t by the individual sample functions of the ensemble. Similarly, at the *fixed* time $t + \tau$, $n(t + \tau)$ is also a random variable. It then appears that $R(\tau)$ as expressed in Eq. (2.20-5) is the covariance between these two random variables.

Suppose then that we should find that for some τ, $R(\tau) = 0$. Then the random variables $n(t)$ and $n(t + \tau)$ are uncorrelated, and for the gaussian process of interest to us, $n(t)$ and $n(t + \tau)$ are independent. Hence, if we should select some sample function, a knowledge of the value of $n(t)$ at time t would be of no assistance in improving our ability to predict the value attained by that same sample function at time $t + \tau$.

The physical fact about the noise, which is of principal concern in connection with communications systems, is that such noise has a power spectral density $G(f)$ which is uniform over all frequencies. Such noise is referred to as "white" noise in analogy with the consideration that white light is a combination of all colors, that is, colors of all frequencies. Actually as is pointed out in Sec. 14.5, there is an upper-frequency limit beyond which the spectral density falls off sharply. However, this upper-frequency limit is so high that we may ignore it for our purposes.

Now, since the autocorrelation $R(\tau)$ and the power spectral density $G(f)$ are a Fourier transform pair, they have the properties of such pairs. Thus when $G(f)$ extends over a wide frequency range, $R(\tau)$ is restricted to a narrow range of τ. In the limit, if $G(f) = I$ (a constant) for all frequencies from $-\infty \leq f \leq +\infty$, then $R(\tau)$ becomes $R(\tau) = I\,\delta(\tau)$, where

$\delta(\tau)$ is the delta function with $\delta(\tau) = 0$ except for $\tau = 0$. Since, then, for white noise, $R(\tau) = 0$ except for $\tau = 0$, Eq. (2.20-5) says that $n(t)$ and $n(t + \tau)$ are uncorrelated and hence independent, no matter how small τ. We shall have occasion to take account of this characteristic of white noise in our discussion in Chap. 7 of the mathematical representation of noise.

PROBLEMS

2.1-1. Six dice are thrown simultaneously. What is the probability that at least 1 die shows a 3?

2.1-2. A card is drawn from a deck of 52 cards.
    (a) What is the probability that a 2 is drawn?
    (b) What is the probability that a 2 of *clubs* is drawn?
    (c) What is the probability that a spade is drawn?

2.2-1. A card is picked from each of four 52-card decks of cards.
    (a) What is the probability of selecting at least one 6 of spades?
    (b) What is the probability of selecting at least 1 card larger than an 8?

2.3-1. A card is drawn from a 52-card deck, and without replacing the first card a second card is drawn. The first and second cards are not replaced and a third card is drawn.
    (a) If the first card is a heart, what is the probability of the second card being a heart?
    (b) If the first and second cards are hearts, what is the probability that the third card is the king of clubs?

2.3-2. Two factories produce identical clocks. The production of the first factory consists of 10,000 clocks of which 100 are defective. The second factory produces 20,000 clocks of which 300 are defective. What is the probability that a particular defective clock was produced in the first factory?

2.3-3. One box contains two black balls. A second box contains one black and one white ball. We are told that a ball was withdrawn from one of the boxes and that it turned out to be black. What is the probability that this withdrawal was made from the box that held the two black balls?

2.4-1. Two dice are tossed.
    (a) Find the probability of a 3 and a 4 appearing.
    (b) Find the probability of a 7 being rolled.

2.4-2. A card is drawn from a deck of 52 cards, then replaced, and a second card drawn.
    (a) What is the probability of drawing the same card twice?
    (b) What is the probability of first drawing a 3 of hearts and then drawing a 4 of spades?

2.6-1. A die is tossed and the number appearing is n_i. Let N be the random variable identifying the outcome of the toss defined by the specifications $N = n_i$ when n_i appears. Make a plot of the probability $P(N \leq n_i)$ as a function of n_i.

2.6-2. A coin is tossed four times. Let H be the random variable which identifies the number of heads which occur in these four tosses. It is defined by $H = h$, where h is the number of heads which appear. Make a plot of the probability $P(H \leq h)$ as a function of h.

2.6-3. A coin is tossed until a head appears. Let T be the random variable which identifies the number of tosses t required for the appearance of this first head. Make a plot of the probability $P(T \leq t)$ as a function of t up to $t = 5$.

2.7-1. An important probability density function is the Rayleigh density

$$f(x) = \begin{cases} xe^{-x^2/2} & x \geq 0 \\ 0 & x < 0 \end{cases}$$

(a) Prove that $f(x)$ satisfies Eqs. (2.7-2) and (2.7-3).
(b) Find the distribution function $F(x)$.

2.8-1. Refer to Fig. 2.6-1.
(a) Find $P(2 < n \leq 11)$.
(b) Find $P(2 \leq n < 11)$.
(c) Find $P(2 \leq n \leq 11)$.
(d) Find $F(9)$.

2.8-2. Refer to the Rayleigh density function given in Prob. 2.7-1. Find the probability $P(x_1 < x \leq x_2)$, where $x_2 - x_1 = 1$, so that $P(x_1 < x \leq x_2)$ is a maximum. *Hint:* Find $P(x_1 < x \leq x_2)$; replace x_2 by $1 + x_1$, and maximize P with respect to x_1.

2.8-3. Refer to the Rayleigh density function given in Prob. 2.7-1. Find
(a) $P(0.5 < x \leq 2)$.
(b) $P(0.5 \leq x < 2)$.

2.9-1. The joint probability density of the random variables X and Y is $f(x,y) = ke^{-(x+y)}$ in the range $0 \leq x \leq \infty$, $0 \leq y \leq \infty$, and $f(x,y) = 0$ otherwise.
(a) Find the value of the constant k.
(b) Find the probability density $f(x)$, the probability density of X independently of Y.
(c) Find the probability $P(0 \leq X \leq 2; 2 \leq Y \leq 3)$.
(d) Are the random variables dependent or independent?

2.9-2. X is a random variable having a gaussian density. $E(X) = 0, \sigma_x^2 = 1$. V is a random variable having the values 1 or -1, each with probability $\frac{1}{2}$.
(a) Find the joint density $f_{X,V}(x,v)$.
(b) Show that $f_V(v) = \int_{-\infty}^{\infty} f_{XV}(x,v)\, dx$.

2.9-3. The joint probability density of the random variables X and Y is $f(x,y) = xe^{-x(y+1)}$ in the range $0 \leq x \leq \infty$, $0 \leq y \leq \infty$, and $f(x,y) = 0$ otherwise.
(a) Find $f(x)$ and $f(y)$, the probability density of X independently of Y and Y independently of X.
(b) Are the random variables dependent or independent?

2.10-1. If $f_X(x) = \dfrac{1}{\sqrt{2\pi}} e^{-x^2/2}$ for all x, show that

(a) $E(X^{2n}) = 1 \cdot 3 \cdot 5 \cdots (n-1)$, $n = 1, 2, \ldots$.
(b) $E(X^{2n-1}) = 0$, $n = 1, 2, \ldots$.

2.10-2. Compare the most probable [$f(x)$ is a maximum] and the average value of X when

(a) $f_{X_1}(x) = \dfrac{1}{\sqrt{2\pi}} \, e^{-(x-m)^2/2}$ for all x

(b) $f_{X_2}(x) = \begin{cases} xe^{-x^2/2} & \text{for } x \geq 0 \\ 0 & \text{elsewhere} \end{cases}$

2.11-1. Calculate the variance of the random variables having densities:

(a) The gaussian density $f_{X_1}(x) = \dfrac{1}{\sqrt{2\pi}} \, e^{-(x-m)^2/2}$, all x.

(b) The Rayleigh density $f_{X_2}(x) = xe^{-x^2/2}$, $x \geq 0$.

(c) The uniform density $f_{X_3}(x) = 1/a$, $-a/2 \leq x \leq a/2$.

2.11-2. Consider the Cauchy density function

$$f(x) = \frac{K}{1 + x^2} \qquad -\infty \leq x \leq \infty$$

(a) Find K so that $f(x)$ is a density function.

(b) Find $E(X)$.

(c) Find the variance of X.   Comment on the significance of this result.

2.11-3. The random variable X has a variance σ^2 and a mean m.   The random variable Y is related to X by $Y = aX + b$, where a and b are constants.   Find the mean and variance of Y.

2.12-1. Refer to the gaussian density given in Eq. (2.12-1).

(a) Show that $E((X - m)^{2n-1}) = 0$.

(b) Show that $E((X - m)^{2n}) = 1 \cdot 3 \cdot 5 \cdots (n - 1)\sigma^2$.

2.12-2. A random variable $V = b + X$, where X is a gaussian distributed random variable with mean 0 and variance σ^2, and b is a constant.   Show that V is a gaussian distributed random variable with mean b and variance σ^2.

2.12-3. The joint density function of two dependent random variables X and Y is

$$f(x,y) = \frac{1}{\pi\sqrt{3}} \, e^{-2(x^2-xy+y^2)/3}$$

(a) Show that, when X and Y are each considered without reference to the other, each is a gaussian variable, i.e., $f(x)$ and $f(y)$ are gaussian density functions.

(b) Find σ_x^2 and σ_y^2.

2.13-1. Obtain values for and plot erf u versus u.

2.13-2. On the same set of axes used in Prob. 2.13-1 plot e^{-u^2} and erfc u versus u.   Compare your results.

2.13-3. The probability

$$P_{\pm k\sigma} \equiv P(m - k\sigma \leq X \leq m + k\sigma) = \int_{m-k\sigma}^{m+k\sigma} \frac{e^{-(x-m)^2/2\sigma^2}}{\sqrt{2\pi}\,\sigma} \, dx$$

(a) Change variables by letting $u = x - m/\sqrt{2}\,\sigma$.

(b) Show that $P_{\pm k\sigma} = $ erf $(k/\sqrt{2})$.

2.14-1. Show that a random variable with a Rayleigh density as in Eq. (2.14-1) has a mean value $R = \sqrt{(\pi/2)}\,\alpha$, a mean square value $R^2 = 2\alpha^2$, and a variance $\sigma^2 = (2 - \pi/2)\alpha^2$.

2.14-2. (a) A voltage V is a function of time t and is given by

$$V(t) = X \cos \omega t + Y \sin \omega t$$

in which ω is a constant angular frequency and X and Y are independent gaussian variables each with zero mean and variance σ^2. Show that $V(t)$ may be written

$$V(t) = R \cos (\omega t + \Theta)$$

in which R is a random variable with a Rayleigh probability density and Θ is a random variable with uniform density.

(b) If $\sigma^2 = 1$, what is the probability that $R \geq 1$?

2.15-1. Derive Eq. (2.15-6) directly from the definition $\sigma_z^2 = E\{(Z - m_z)^2\}$.

2.15-2. $Z = X_1 + X_2 + \cdots + X_N$, $E(X_i) = m$.

(a) Find $E(Z)$.

(b) If $E(X_i X_j) = \begin{cases} 1 & j = i \\ \rho & j = i \pm 1 \\ 0 & \text{otherwise} \end{cases}$

find (1) $E(Z^2)$ and (2) σ_z^2.

2.16-1. The independent random variables X and Y are added to form Z. If

$$f_X(x) = xe^{-x^2/2} \quad 0 \leq x \leq \infty \quad \text{and} \quad f_Y(y) = \tfrac{1}{2}e^{-|y|} \quad |y| < \infty$$

find $f_Z(z)$.

2.16-2. The independent random variables X and Y have the probability densities

$$f(x) = e^{-x} \quad 0 \leq x \leq \infty$$
$$f(y) = 2e^{-2y} \quad 0 \leq y \leq \infty$$

Find and plot the probability density of the variable $Z = X + Y$.

2.16-3. The random variable X has a probability density uniform in the range $0 \leq x \leq 1$ and zero elsewhere. The independent variable Y has a density uniform in the range $0 \leq x \leq 2$ and zero elsewhere. Find and plot the density of $Z = X + Y$.

2.16-4. The N independent gaussian random variables $X_1, \ldots, X_N$ are added to form Z. If the mean of X_i is 1 and its variance is 1, find $f_Z(z)$.

2.17-1. Two gaussian random variables X and Y, each with mean zero and variance σ^2, between which there is a correlation coefficient ρ, have a joint probability density given by

$$f(x,y) = \frac{1}{2\pi\sigma^2 \sqrt{1 - \rho^2}} \exp - \left[\frac{x^2 - 2\rho xy + y^2}{2\sigma^2(1 - \rho^2)} \right]$$

(a) Verify that the symbol ρ in the expression for $f(x,y)$ is indeed the correlation coefficient. That is, evaluate $E\{XY\}/\sigma^2$ and show that the result is ρ as required by Eq. (2.17-5).

(b) Show that the case $\rho = 0$ corresponds to the circumstance where X and Y are independent.

2.17-2. The random variables X and Y are related to the random variable Θ by $X = \sin \Theta$ and $Y = \cos \Theta$. The variable Θ has a uniform probability density in the range from 0 to 2π. Show that X and Y are not independent but that, nonetheless, $E(XY) = 0$ so that they are uncorrelated.

2.17-3. The random variables X_1, X_2, X_3, . . . are dependent but uncorrelated. $Z = X_1 + X_2 + X_3 + \cdots$. Show that $\sigma_Z^2 = \sigma_1^2 + \sigma_2^2 + \sigma_3^2 + \cdots$.

2.18-1. The random variables X_1, X_2, and X_3 are independent and each has a uniform probability density in the range $0 \leq x \leq 1$. Find and plot the probability density of $X_1 + X_2$ and of $X_1 + X_2 + X_3$.

2.19-1. The function of time $Z(t) = X_1 \cos \omega_0 t - X_2 \sin \omega_0 t$ is a random process. If X_1 and X_2 are independent gaussian random variables each with zero mean and variance σ^2, find

(a) $E(Z)$, $E(Z^2)$, σ_z^2, and

(b) $f_Z(z)$.

2.19-2. $Z(t) = M(t) \cos (\omega_0 t + \Theta)$. $M(t)$ is a random process, with $E(M(t)) = 0$ and $E(M^2(t)) = M_0$.

(a) If $\Theta = 0$, find $E(Z^2)$. Is $Z(t)$ stationary?

(b) If Θ is an independent random variable such that $f_\Theta(\theta) = 1/2\pi$, $-\pi \leq \theta \leq \pi$, show that $E(Z^2(t)) = E(M^2(t))E(\cos^2 (\omega_0 t + \Theta)) = M_0/2$. Is $Z(t)$ now stationary?

2.20-1. Refer to Prob. 2.19-1. Find $R_z(\tau)$.

2.20-2. A random process $n(t)$ has a power spectral density $G(f) = \eta/2$ for $-\infty \leq f \leq \infty$. The random process is passed through a low-pass filter which has a transfer function $H(f) = 2$ for $-f_M \leq f \leq f_M$ and $H(f) = 0$ otherwise. Find the power spectral density of the waveform at the output of the filter.

2.20-3. White noise $n(t)$ with $G(f) = \eta/2$ is passed through a low-pass RC network with a 3-dB frequency f_c.

(a) Find the autocorrelation $R(\tau)$ of the output noise of the network.

(b) Sketch $\rho(\tau) = R(\tau)/R(0)$.

(c) Find $\omega_c \tau$ such that $\rho(\tau) \leq 0.1$.

REFERENCES

1. Mood, A., and F. Graybill: "Introduction to the Theory of Statistics," McGraw-Hill Book Company, New York, 1963.
2. Peirce, B. O.: "A Short Table of Integrals," Ginn and Company, Boston, 1956.
3. Papoulis, A.: "Probability, Random Variables, and Stochastic Processes," McGraw-Hill Book Company, 1965.

3
Amplitude-modulation Systems

One of the basic problems of communication engineering is the design and analysis of systems which allow many individual messages to be transmitted simultaneously over a single communication channel. A method by which such multiple transmission, called *multiplexing*, may be achieved consists in translating each message to a different position in the *frequency* spectrum. Such multiplexing is called *frequency multiplexing*. The individual message can eventually be separated by filtering. Frequency multiplexing involves the use of an auxiliary waveform, usually sinusoidal, called a *carrier*. The operations performed on the signal to achieve frequency multiplexing results in the generation of a waveform which may be described as the carrier modified in that its amplitude, frequency, or phase, individually or in combination, varies with time. Such a modified carrier is called a *modulated* carrier. In some cases the modulation is related simply to the message; in other cases the relationship is quite complicated. In this chapter, we discuss the generation and characteristics of amplitude-modulated carrier waveforms.[1]

3.1 FREQUENCY TRANSLATION

It is often advantageous and convenient, in processing a signal in a communications system, to translate the signal from one region in the frequency domain to another region. Suppose that a signal is bandlimited, or nearly so, to the frequency range extending from a frequency f_1 to a frequency f_2. The process of frequency translation is one in which the original signal is replaced with a new signal whose spectral range extends from f_1' to f_2' and which *new* signal bears, in recoverable form, the same *information* as was borne by the original signal. We discuss now a number of useful purposes which may be served by frequency translation.

FREQUENCY MULTIPLEXING

Suppose that we have several different signals, all of which encompass the same spectral range. Let it be required that all these signals be transmitted along a single communications channel in such a manner that, at the receiving end, the signals be separately recoverable and distinguishable from each other. The single channel may be a single pair of wires or the free space that separates one radio antenna from another. Such multiple transmissions, i.e., multiplexing, may be achieved by translating each one of the original signals to a different frequency range. Suppose, say, that one signal is translated to the frequency range f_1' to f_2', the second to the range f_1'' to f_2'', and so on. If these new frequency ranges do not overlap, then the signal may be separated at the receiving end by appropriate bandpass filters, and the outputs of the filters processed to recover the original signals.

PRACTICABILITY OF ANTENNAS

When free space is the communications channel, antennas radiate and receive the signal. It turns out that antennas operate effectively only when their dimensions are of the order of magnitude of the wavelength of the signal being transmitted. A signal of frequency 1 kHz (an audio tone) corresponds to a wavelength of 300,000 m, an entirely impractical length. The required length may be reduced to the point of practicability by translating the audio tone to a higher frequency.

NARROWBANDING

Returning to the matter of the antenna, just discussed, suppose that we wanted to transmit an audio signal directly from the antenna, and that the inordinate length of the antenna were no problem. We would still be left with a problem of another type. Let us assume that the audio range extends from, say, 50 to 10^4 Hz. The ratio of the highest audio frequency to the lowest is 200. Therefore, an antenna suitable for use at one end of the range would be entirely too short or too long for the other end. Suppose, however, that the audio spectrum were translated so that it occupied the range, say, from $(10^6 + 50)$ to $(10^6 + 10^4)$ Hz. Then the ratio of highest to lowest frequency would be only 1.01. Thus the processes of frequency translation may be used to change a "wideband" signal into a "narrowband" signal which may well be more conveniently processed. The terms "wideband" and "narrowband" are being used here to refer not to an absolute range of frequencies but rather to the fractional change in frequency from one band edge to the other.

COMMON PROCESSING

It may happen that we may have to process, in turn, a number of signals similar in general character but occupying different spectral ranges. It will then be necessary, as we go from signal to signal, to adjust the frequency range of our processing apparatus to correspond to the frequency range of the signal to be processed. If the processing apparatus is rather elaborate, it may well be wiser to leave the processing apparatus to operate in some fixed frequency range and instead to translate the frequency range of each signal in turn to correspond to this fixed frequency.

3.2 A METHOD OF FREQUENCY TRANSLATION

A signal may be translated to a new spectral range by *multiplying* the signal with an auxiliary sinusoidal signal. To illustrate the process, let us consider initially that the signal is sinusoidal in waveform and given by

$$v_m(t) = A_m \cos \omega_m t = A_m \cos 2\pi f_m t \qquad (3.2\text{-}1a)$$

$$= \frac{A_m}{2} \left(e^{j\omega_m t} + e^{-j\omega_m t} \right) = \frac{A_m}{2} \left(e^{j2\pi f_m t} + e^{-j2\pi f_m t} \right) \qquad (3.2\text{-}1b)$$

in which A_m is the constant amplitude and $f_m = \omega_m/2\pi$ is the frequency. The two-sided spectral amplitude pattern of this signal is shown in Fig. 3.2-1a. The pattern consists of two lines, each of amplitude $A_m/2$, located at $f = f_m$ and at $f = -f_m$. Consider next the result of the multiplication of $v_m(t)$ with an auxiliary sinusoidal signal

$$v_c(t) = A_c \cos \omega_c t = A_c \cos 2\pi f_c t \qquad (3.2\text{-}2a)$$

$$= \frac{A_c}{2} \left(e^{j\omega_c t} + e^{-j\omega_c t} \right) = \frac{A_c}{2} \left(e^{j2\pi f_c t} + e^{-j2\pi f_c t} \right) \qquad (3.2\text{-}2b)$$

in which A_c is the constant amplitude and f_c is the frequency. Using the trigonometric identity $\cos \alpha \cos \beta = \frac{1}{2}\cos(\alpha + \beta) + \frac{1}{2}\cos(\alpha - \beta)$, we

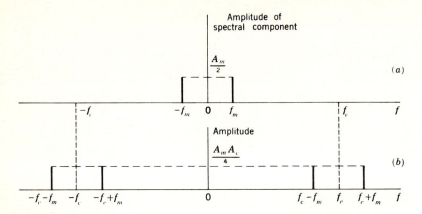

Fig. 3.2-1 (*a*) Spectral pattern of $A_m \cos \omega_m t$. (*b*) Spectral pattern of the product $A_m A_c \cos \omega_m t \cos \omega_c t$.

have for the product $v_m(t)v_c(t)$

$$v_m(t)v_c(t) = \frac{A_m A_c}{2} [\cos (\omega_c + \omega_m)t + \cos (\omega_c - \omega_m)t] \qquad (3.2\text{-}3a)$$

$$= \frac{A_m A_c}{4} (e^{j(\omega_c+\omega_m)t} + e^{-j(\omega_c+\omega_m)t}$$
$$+ e^{j(\omega_c-\omega_m)t} + e^{-j(\omega_c-\omega_m)t}) \qquad (3.2\text{-}3b)$$

The new spectral amplitude pattern is shown in Fig. 3.2-1*b*. Observe that the two original spectral lines have been *translated*, both in the positive-frequency direction by amount f_c and also in the negative-frequency direction by the same amount. There are now four spectral components resulting in two sinusoidal waveforms, one of frequency $f_c + f_m$ and the other of frequency $f_c - f_m$. Note that while the product signal has four spectral components each of amplitude $A_m A_c/4$, there are only two frequencies, and the amplitude of each sinusoidal component is $A_m A_c/2$.

A generalization of Fig. 3.2-1 is shown in Fig. 3.2-2. Here a signal is chosen which consists of a superposition of four sinusoidal signals, the highest in frequency having the frequency f_M. Before translation by multiplication, the two-sided spectral pattern displays eight components centered around zero frequency. After multiplication, we find this spectral pattern translated both in the positive- and the negative-frequency directions. The 16 spectral components in this two-sided spectral pattern give rise to eight sinusoidal waveforms. While the original signal extends in range up to a frequency f_M, the signal which results from multiplication has sinusoidal components covering a range $2f_M$, from $f_c - f_M$ to $f_c + f_M$.

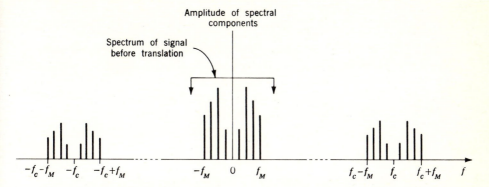

Fig. 3.2-2 An original signal consisting of four sinusoids of differing frequencies is translated through multiplication and becomes a signal containing eight frequencies symmetrically arranged about f_c.

Finally, we consider in Fig. 3.2-3 the situation in which the signal to be translated may not be represented as a superposition of a number of sinusoidal components at sharply defined frequencies. Such would be the case if the signal were of finite energy and nonperiodic. In this case the signal is represented in the frequency domain in terms of its Fourier transform, that is, in terms of its spectral density. Thus let the signal

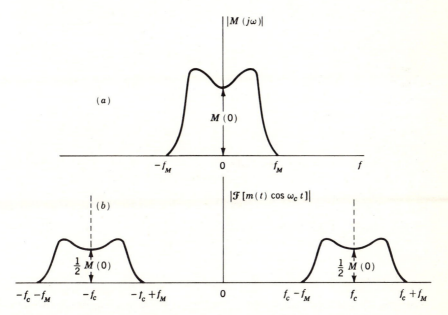

Fig. 3.2-3 (a) The spectral density $|M(j\omega)|$ of a nonperiodic signal $m(t)$. (b) The spectral density of $m(t) \cos 2\pi f_c t$.

$m(t)$ be bandlimited to the frequency range 0 to f_M. Its Fourier transform is $M(j\omega) = \mathfrak{F}[m(t)]$. The magnitude $|M(j\omega)|$ is shown in Fig. 3.2-3a. The transform $M(j\omega)$ is symmetrical about $f = 0$ since we assume that $m(t)$ is a real signal. The spectral density of the signal which results when $m(t)$ is multiplied by $\cos \omega_c t$ is shown in Fig. 3.2-3b. This spectral pattern is deduced as an extension of the results shown in Figs. 3.2-1 and 3.2-2. Alternatively, we may easily verify (Prob. 3.2-2) that if $M(j\omega) = \mathfrak{F}[m(t)]$, then

$$\mathfrak{F}[m(t) \cos \omega_c t] = \frac{1}{2}[M(j\omega + j\omega_c) + M(j\omega - j\omega_c)] \tag{3.2-4}$$

The spectral range occupied by the original signal is called the *baseband frequency range* or simply the *baseband*. On this basis, the original signal itself is referred to as the baseband signal. The operation of multiplying a signal with an auxiliary sinusoidal signal is called *mixing* or *heterodyning*. In the translated signal, the part of the signal which consists of spectral components *above* the auxiliary signal, in the range f_c to $f_c + f_M$, is called the *upper-sideband* signal. The part of the signal which consists of spectral components below the auxiliary signal, in the range $f_c - f_M$ to f_c, is called the *lower-sideband* signal. The two sideband signals are also referred to as the *sum* and the *difference* frequencies, respectively. The auxiliary signal of frequency f_c is variously referred to as the *local oscillator signal*, the *mixing signal*, the *heterodyning signal*, or as the *carrier signal*, depending on the application. The student will note, as the discussion proceeds, the various contexts in which the different terms are appropriate.

We may note that the process of translation by multiplication actually gives us something somewhat different from what was intended. Given a signal occupying a baseband, say, from zero to f_M, and an auxiliary signal f_c, it would often be entirely adequate to achieve a simple translation, giving us a signal occupying the range f_c to $f_c + f_M$, that is, the upper sideband. We note, however, that translation by multiplication results in a signal that occupies the range $f_c - f_M$ to $f_c + f_M$. This feature of the process of translation by multiplication may, depending on the application, be a nuisance, a matter of indifference, or even an advantage. Hence, this feature of the process is, of itself, neither an advantage nor a disadvantage. It is, however, to be noted that there is no other operation so simple which will accomplish translation.

3.3 RECOVERY OF THE BASEBAND SIGNAL

Suppose a signal $m(t)$ has been translated out of its baseband through multiplication with $\cos \omega_c t$. How is the signal to be recovered? The recovery may be achieved by a reverse translation, which is accomplished simply by multiplying the translated signal with cos $\omega_c t$. That such is the case may be seen by drawing spectral plots as in Fig. 3.2-2 or 3.2-3 and noting that the difference-frequency signal obtained by multiplying $m(t)$ cos $\omega_c t$ by cos $\omega_c t$ is a signal whose spectral range is back at base-band. Alternatively, we may simply note that

$$[m(t) \cos \omega_c t] \cos \omega_c t = m(t) \cos^2 \omega_c t = m(t)(\tfrac{1}{2} + \tfrac{1}{2} \cos 2\omega_c t) \tag{3.3-1a}$$

$$= \frac{m(t)}{2} + \frac{m(t)}{2} \cos 2\omega_c t \tag{3.3-1b}$$

Thus, the baseband signal $m(t)$ reappears. We note, of course, that in addition to the recovered baseband signal there is a signal whose spectral range extends from $2f_c - f_M$ to $2f_c + f_M$. As a matter of practice, this latter signal need cause no difficulty. For most commonly $f_c \gg f_M$, and consequently the spectral range of this double-frequency signal and the baseband signal are widely separated. Therefore the double-frequency signal is easily removed by a low-pass filter.

This method of signal recovery, for all its simplicity, is beset by an important inconvenience when applied in a physical communication system. Suppose that the auxiliary signal used for recovery differs in phase from the auxiliary signal used in the initial translation. If this phase angle is θ, then, as may be verified (Prob. 3.3-1), the recovered baseband waveform will be proportional to $m(t) \cos \theta$. Therefore, unless it is possible to maintain $\theta = 0$, the signal strength at recovery will suffer. If it should happen that $\theta = \pi/2$, the signal will be lost entirely. Or consider, for example, that θ drifts back and forth with time. Then in this case the signal strength will wax and wane, in addition, possibly, to disappearing entirely from time to time.

Alternatively, suppose that the recovery auxiliary signal is not precisely at frequency f_c but is instead at $f_c + \Delta f$. In this case we may verify (Prob. 3.3-2) that the recovered baseband signal will be proportional to $m(t) \cos 2\pi \Delta f t$, resulting in a signal which will wax and wane or even be entirely unacceptable if Δf is comparable to, or larger than, the frequencies present in the baseband signal. This latter contingency is a distinct possibility in many an instance, since usually $f_c \gg f_M$ so that a small percentage change in f_c will cause a Δf which may be comparable or larger than f_M. In telephone or radio systems, an offset $\Delta f \leq 30$ Hz is deemed acceptable.

We note, therefore, that signal recovery using a second multiplication requires that there be available at the recovery point a signal which is precisely *synchronous* with the corresponding auxiliary signal at the point of the first multiplication. In such a *synchronous* or *coherent* system a *fixed* initial phase discrepancy is of no consequence since a simple phase

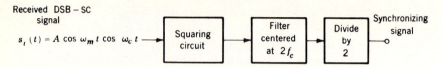

Fig. 3.3-1 A simple squaring synchronizer.

shifter will correct the matter. Similarly it is not essential that the recovery auxiliary signal be sinusoidal (see Prob. 3.3-3). What is essential is that, in any time interval, the number of cycles executed by the two auxiliary-signal sources be the same. Of course, in a physical system, where some signal distortion is tolerable, some lack of synchronism may be allowed.

When the use of a common auxiliary signal is not feasible, it is necessary to resort to rather complicated means to provide a synchronous auxiliary signal at the location of the receiver. One commonly employed scheme is indicated in Fig. 3.3-1. To illustrate the operation of the synchronizer, we assume that the baseband signal is a sinusoid $\cos \omega_m t$. The received signal is $s_i(t) = A \cos \omega_m t \cos \omega_c t$, with A a constant amplitude. This signal $s_i(t)$ does not have a spectral component at the angular frequency ω_c. The output of the squaring circuit is

$$s_i^2(t) = A^2 \cos^2 \omega_m t \cos^2 \omega_c t \qquad (3.3\text{-}2a)$$

$$= A^2(\tfrac{1}{2} + \tfrac{1}{2} \cos 2\omega_m t)(\tfrac{1}{2} + \tfrac{1}{2} \cos 2\omega_c t) \qquad (3.3\text{-}2b)$$

$$= \frac{A^2}{4} [1 + \tfrac{1}{2} \cos 2(\omega_c + \omega_m)t + \tfrac{1}{2} \cos 2(\omega_c - \omega_m)t$$
$$+ \cos 2\omega_m t + \cos 2\omega_c t] \quad (3.3\text{-}2c)$$

The filter selects the spectral component $(A^2/4) \cos 2\omega_c t$, which is then applied to a circuit which divides the frequency by a factor of 2. (See Prob. 3.3-4.) This frequency division may be accomplished by using, for example, a bistable multivibrator. The output of the divider is used to demodulate (multiply) the incoming signal and thereby recover the baseband signal $\cos \omega_m t$.

We turn our attention now to a modification of the method of frequency translation, which has the great merit of allowing recovery of the baseband signal by an extremely simple means. This technique is called *amplitude modulation.*

3.4 AMPLITUDE MODULATION

A frequency-translated signal from which the baseband signal is easily recoverable is generated by adding, to the product of baseband and carrier, the carrier signal itself. Such a signal is shown in Fig. 3.4-1. Figure

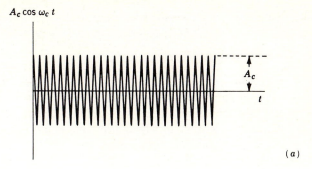

$A_c \cos \omega_c t$

(a)

$m(t)$

(b)

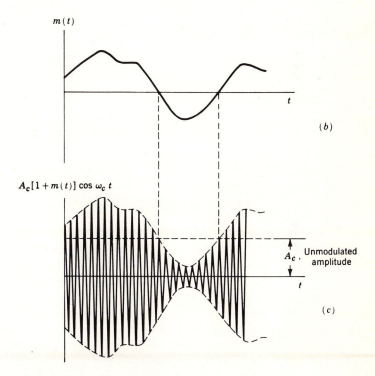

$A_c[1 + m(t)] \cos \omega_c t$

A_c, Unmodulated amplitude

(c)

Fig. 3.4-1 (a) A sinusoidal carrier. (b) A modulating waveform. (c) The sinusoidal carrier in (a) modulated by the waveform in (b).

3.4-1a shows the carrier signal with amplitude A_c; in Fig. 3.4-1b we see the baseband signal. The translated signal (Fig. 3.4-1c) is given by

$$v(t) = A_c[1 + m(t)] \cos \omega_c t \qquad (3.4\text{-}1)$$

We observe, from Eq. (3.4-1) as well as from Fig. 3.4-1c, that the resultant waveform is one in which the carrier $A_c \cos \omega_c t$ is *modulated in amplitude*. The process of generating such a waveform is called *amplitude modulation*, and a communication system which employs such a method of frequency translation is called an *amplitude-modulation* system, or *AM* for short. The designation "carrier" for the auxiliary signal $A_c \cos \omega_c t$ seems especially appropriate in the present connection since this signal now "carries" the baseband signal as its envelope. The term "carrier" probably originated, however, in the early days of radio when this relatively high-frequency signal was viewed as the *messenger* which actually "carried' the baseband signal from one antenna to another.

The very great merit of the amplitude-modulated carrier signal is the ease with which the baseband signal can be recovered. The recovery of the baseband signal, a process which is referred to as *demodulation* or *detection*, is accomplished with the simple circuit of Fig. 3.4-2a, which consists of a diode D and the resistor-capacitor RC combination. We now discuss the operation of this circuit briefly and qualitatively. For simplicitly, we assume that the amplitude-modulated carrier which is applied at the input terminals is supplied by a voltage source of zero internal impedance. We assume further that the diode is ideal, i.e., of zero or infinite resistance, depending on whether the diode current is positive or the diode voltage negative.

Let us initially assume that the input is of fixed amplitude and that the resistor R is not present. In this case, the capacitor charges to the peak positive voltage of the carrier. The capacitor holds this peak voltage, and the diode would not again conduct. Suppose now that the input-carrier amplitude is increased. The diode again conducts, and the capacitor charges to the new higher carrier peak. In order to allow the capacitor voltage to follow the carrier peaks when the carrier amplitude is decreasing, it is necessary to include the resistor R, so that the capacitor may discharge. In this case the capacitor voltage v_c has the form shown in Fig. 3.4-2b. The capacitor charges to the peak of each carrier cycle and decays slightly between cycles. The time constant RC is selected so that the change in v_c between cycles is at least equal to the decrease in carrier amplitude between cycles. This constraint on the time constant RC is explored in Probs. 3.4-1 and 3.4-2.

It is seen that the voltage v_c follows the carrier envelope except that v_c also has superimposed on it a sawtooth waveform of the carrier frequency. In Fig. 3.4-2b the discrepancy between v_c and the envelope is

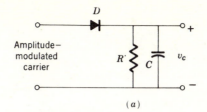

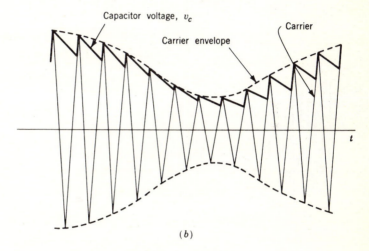

(b)

Fig. 3.4-2 (a) A demodulator for an AM signal. (b) Input waveform and output voltage v_c across capacitor.

greatly exaggerated. In practice, the normal situation is one in which the time interval between carrier cycles is extremely small in comparison with the time required for the envelope to make a sizable change. Hence v_c follows the envelope much more closely than is suggested in the figure. Further, again because the carrier frequency is ordinarily much higher than the highest frequency of the modulating signal, the sawtooth distortion of the envelope waveform is very easily removed by a filter.

3.5 MAXIMUM ALLOWABLE MODULATION

If we are to avail ourselves of the convenience of demodulation by the use of the simple diode circuit of Fig. 3.4-2a, we must limit the extent of the modulation of the carrier. That such is the case may be seen from Fig. 3.5-1. In Fig. 3.5-1a is shown a carrier modulated by a sinusoidal signal. It is apparent that the envelope of the carrier has the waveshape

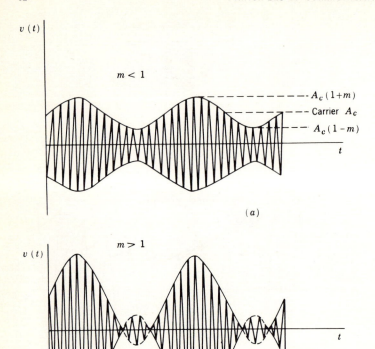

Fig. 3.5-1 (a) A sinusoidally modulated carrier ($m < 1$). (b) A carrier overmodulated ($m > 1$) by a sinusoidal modulating waveform.

of the modulating signal. The modulating signal is sinusoidal; hence $m(t) = m \cos \omega_m t$, where m is a constant. Equation (3.4-1) becomes

$$v(t) = A_c(1 + m \cos \omega_m t) \cos \omega_c t \tag{3.5-1}$$

In Fig. 3.5-1b we have shown the situation which results when, in Eq. (3.5-1), we adjust $m > 1$. Observe now that the diode demodulator which yields as an output the positive envelope (a negative envelope if the diode is reversed) will not reproduce the sinusoidal modulating waveform. In this latter case, where $m > 1$, we may recover the modulating waveform but not with the diode modulator. Recovery would require the use of a coherent demodulation scheme such as was employed in connection with the signal furnished by a multiplier.

It is therefore necessary to restrict the excursion of the modulating signal in the direction of decreasing carrier amplitude to the point where the carrier amplitude is just reduced to zero.   No such similar restriction applies when the modulation is increasing the carrier amplitude.   With sinusoidal modulation, as in Eq. (3.5-1), we require that $|m| \leq 1$.   More generally in Eq. (3.4-1) we require that the maximum negative excursion of $m(t)$ be -1.

The extent to which a carrier has been amplitude-modulated is expressed in terms of a *percentage modulation*.   Let A_c, $A_c(\max)$, and $A_c(\min)$, respectively, be the unmodulated carrier amplitude and the maximum and minimum carrier levels.   Then if the modulation is symmetrical, the percentage modulation is defined as P, given by

$$\frac{P}{100\%} = \frac{A_c(\max) - A_c}{A_c} = \frac{A_c - A_c(\min)}{A_c} = \frac{A_c(\max) - A_c(\min)}{2A_c}$$

$$(3.5\text{-}2)$$

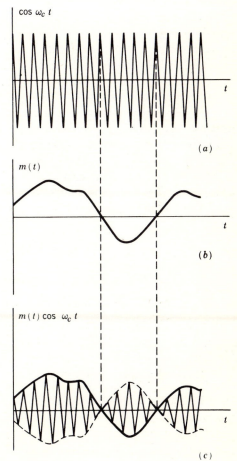

In the case of sinusoidal modulation, given by Eq. (3.5-1) and shown in Fig. 3.5-1a, $P = m \times 100$ percent.

Having observed that the signal $m(t)$ may be recovered from the waveform $A_c[1 + m(t)] \cos \omega_c t$ by the simple circuit of Fig. 3.4-2a, it is of interest to note that a similar easy recovery of $m(t)$ is not possible from the waveform $m(t) \cos \omega_c t$. That such is the case is to be seen from Fig. 3.5-2. Figure 3.5-2a shows the carrier signal. The modulation or base-band signal $m(t)$ is shown in Fig. 3.5-2b, and the product $m(t) \cos \omega_c t$ is shown in Fig. 3.5-2c. We note that the envelope in Fig. 3.5-2c has the waveform not of $m(t)$ but rather of $|m(t)|$, the *absolute value* of $m(t)$. Observe the reversal of phase of the carrier in Fig. 3.5-2c whenever $m(t)$ passes through zero.

3.6 THE SQUARE-LAW DEMODULATOR

An alternative method of recovering the baseband signal which has been superimposed as an amplitude modulation on a carrier is to pass the AM signal through a nonlinear device. Such demodulation is illustrated in Fig. 3.6-1. We assume here for simplicity that the device has a square-law relationship between input signal x (current or voltage) and output signal y (current or voltage). Thus $y = kx^2$, with k a constant. Because of the nonlinearity of the transfer characteristic of the device, the output response is different for positive and for negative excursions of the carrier away from the quiescent operating point O of the device. As a result, and as is shown in Fig. 3.6-1c, the output, when averaged over a time which encompasses many carrier cycles but only a very small part of the modulation cycle, has the waveshape of the envelope.

The applied signal is

$$x = A_o + A_c[1 + m(t)] \cos \omega_c t \qquad (3.6\text{-}1)$$

Thus the output of the squaring circuit is

$$y = k\{A_o + A_c[1 + m(t)] \cos \omega_c t\}^2 \qquad (3.6\text{-}2)$$

Squaring, and dropping dc terms as well as terms whose spectral components are located near ω_c and $2\omega_c$, we find that the output signal $s_o(t)$, that is, the signal output of a low-pass filter located after the squaring circuit, is

$$s_o(t) = kA_c^2[m(t) + \tfrac{1}{2}m^2(t)] \qquad (3.6\text{-}3)$$

Observe that the modulation $m(t)$ is indeed recovered but that $m^2(t)$ appears as well. Thus the total recovered signal is a distorted version of the original modulation. The distortion is small, however, if $\tfrac{1}{2}m^2(t) \ll |m(t)|$ or if $|m(t)| \ll 2$.

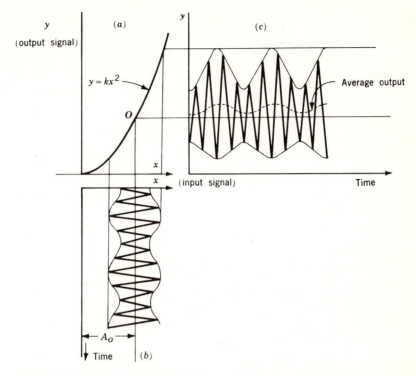

Fig. 3.6-1 Illustrating the operation of a square-law demodulator. The output is the value of y averaged over many carrier cycles.

There are two points of interest to be noted in connection with the type of demodulation described here; the first is that the demodulation does not depend on the nonlinearity being square-law. Any type of non-linearity which does not have odd-function symmetry with respect to the initial operating point will similarly accomplish demodulation. The second point is that even when demodulation is not intended, such demodulation may appear incidentally when the modulated signal is passed through a system, say, an amplifier, which exhibits some nonlinearity.

3.7 SPECTRUM OF AN AMPLITUDE-MODULATED SIGNAL

The spectrum of an amplitude-modulated signal is similar to the spectrum of a signal which results from multiplication except, of course, that in the former case a carrier of frequency f_c is present. If in Eq. (3.4-1) $m(t)$ is the superposition of three sinusoidal components $m(t) = m_1 \cos \omega_1 t + m_2 \cos \omega_2 t + m_3 \cos \omega_3 t$, then the (one-sided) spectrum of this baseband signal appears as at the left in Fig. 3.7-1a. The

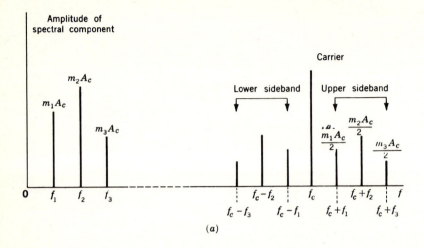

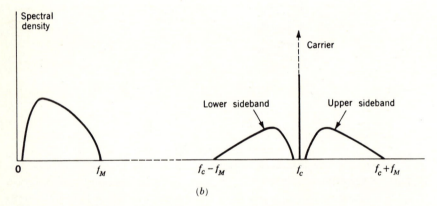

(b)

Fig. 3.7-1 (a) At left the one-sided spectrum of $m(t)A_c$, where $m(t)$ has three spectral components. At right the spectrum of $A_c[1 + m(t)] \cos 2\pi f_c t$. (b) Same as in (a) except $m(t)$ is a nonperiodic signal and the vertical axis is spectral density rather than spectral amplitude.

spectrum of the modulated carrier is shown at the right. The spectral lines at the sum frequencies $f_c + f_1$, $f_c + f_2$, and $f_c + f_3$ constitute the *upper-sideband* frequencies. The spectral lines at the difference frequencies constitute the *lower sideband*.

The spectrum of the baseband signal and modulated carrier are shown in Fig. 3.7-1b for the case of a bandlimited nonperiodic signal of finite energy. In this figure the ordinate is the spectral density, i.e., the magnitude of the Fourier transform rather than the spectral amplitude, and consequently the carrier is represented by an impulse.

3.8 MODULATORS AND BALANCED MODULATORS

We have described a "multiplier" as a device that yields as an output a signal which is the product of two input signals. Actually no simple physical device now exists which yields the product alone. On the contrary, all such devices yield, at a minimum, not only the product but the input signals themselves. Suppose, then, that such a device has as inputs a carrier $\cos \omega_c t$ and a modulating baseband signal $m(t)$. The device output will then contain the product $m(t) \cos \omega_c t$ and also the signals $m(t)$ and $\cos \omega_c t$. Ordinarily, the baseband signal will be bandlimited to a frequency range very much smaller than $f_c = \omega_c/2\pi$. Suppose, for example, that the baseband signal extends from zero frequency to 1000 Hz, while $f_c = 1$ MHz. In this case, the carrier and its sidebands extend from 999,000 to 1,001,000 Hz, and the baseband signal is easily removed by a filter.

The overall result is that the devices available for multiplication yield an output carrier as well as the lower- and upper-sideband signals. The output is therefore an amplitude-modulated signal. If we require the product signal alone, we must take steps to cancel or *suppress* the carrier. Such a suppression may be achieved by adding, to the amplitude-modulated signal, a signal of carrier frequency equal in amplitude but opposite in phase to the carrier of the amplitude-modulated signal. Under these circumstances only the sideband signals will remain. For this reason, a product signal is very commonly referred to as a *double-sideband suppressed-carrier* signal, abbreviated DSB-SC.

An alternative arrangement for carrier suppression is shown in Fig. 3.8-1. Here two *physical* multipliers are used which are labeled in the diagram as *amplitude modulators*. The carrier inputs to the two modulators are of reverse polarity, as are the modulating signals. The modulator outputs are added with consequent suppression of the carrier. We observe a cancellation not only of the carrier but of the baseband signal

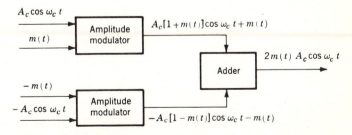

Fig. 3.8-1 Showing how the outputs of two amplitude modulators are combined to produce a double-sideband suppressed-carrier output.

$m(t)$ as well. This last feature is not of great import, since, as noted previously, the baseband signal is easily eliminated by a filter. We note that the product terms of the two modulators reinforce. The arrangement of Fig. 3.8-1 is called a *balanced modulator*.

3.9 SINGLE-SIDEBAND MODULATION

We have seen that the baseband signal may be recovered from a double-sideband suppressed-carrier signal by multiplying a second time, with the same carrier. It can also be shown that the baseband signal can be recovered in a similar manner even if only one sideband is available. For suppose a spectral component of the baseband signal is mult plied by a carrier $\cos \omega_c t$, giving rise to an upper sideband at $\omega_c + \omega$ and a lower sideband at $\omega_c - \omega$. Now let us assume that we have filtered out one sideband and are left with, say, only the upper sideband at $\omega_c + \omega$. If now this sideband signal is again multiplied by $\cos \omega_c t$, we shall generate a signal at $2\omega_c + \omega$ and the original baseband spectral component at ω. If we had used the lower sideband at $\omega_c - \omega$, the second multiplication would have yielded a signal at $2\omega_c - \omega$ and again restored the baseband spectral component. Since it is possible to recover the baseband signal from a single sideband, there is an obvious advantage in doing so, since spectral space is used more economically. In principle, two single-sideband (abbreviated SSB) communications systems can now occupy the spectral range previously occupied by a single amplitude-modulation system or a double-sideband suppressed-carrier system.

The baseband signal may *not* be recovered from a single-sideband signal by the use of a diode modulator. That such is the case is easily seen by considering, for example, that the modulating signal is a sinusoid of frequency f. In this case the single-sideband signal consists also of a single sinusoid of frequency, say, $f_c + f$, and there is no amplitude variation at all at the baseband frequency to which the diode modulator can respond.

Baseband recovery is achieved at the receiving end of the single-sideband communications channel by heterodyning the received signal with a local carrier signal which is synchronous (coherent) with the carrier used at the transmitting end to generate the sideband. As in the double-sideband case it is necessary, in principle, that the synchronism be exact and, in practice, that synchronism be maintained to a high order of precision. The effect of a lack of synchronism is different in a double-sideband system and in a single-sideband system. Suppose that the carrier received is $\cos \omega_c t$ and that the local carrier is $\cos (\omega_c t + \theta)$. Then with DSB-SC, as noted in Sec. 3.3, the spectral component $\cos \omega t$ will, upon demodulation, reappear as $\cos \omega t \cos \theta$. In SSB, on the other hand, the spectral component cos ωt will reappear (Prob. 3.9-2) in the form cos $(\omega t - \theta)$. Thus, in one case a phase offset in carriers affects the amplitude of the recovered signal and, for $\varphi = \pi/2$, may result in a total loss of the signal. In the other case the offset produces a phase change but not an amplitude change.

Alternatively, let the local oscillator carrier have an angular frequency offset $\Delta\omega$ and so be of the form cos $(\omega_c + \Delta\omega)t$. Then as already noted, in DSB-SC, the recovered signal has the form cos ωt cos $\Delta\omega t$. In SSB, however, the recovered signal will have the form cos $(\omega + \Delta\omega)t$. Thus, in one case the recovered spectral component cos ωt reappears with a "warble," that is, an amplitude fluctuation at the rate $\Delta\omega$. In the other case the amplitude remains fixed, but the frequency of the recovered signal is in error by amount $\Delta\omega$.

A phase offset between the received carrier and the local oscillator will cause distortion in the recovered baseband signal. In such a case each spectral component in the baseband signal will, upon recovery, have undergone the *same* phase shift. Fortunately, when SSB is used to transmit voice or music, such *phase distortion* does not appear to be of major consequence, because the human ear seems to be insensitive to the phase distortion.

A frequency offset between carriers in amount Δf will cause each recovered spectral component of the baseband signal to be in error by the same amount Δf. Now, if it had turned out that the frequency error were proportional to the frequency of the spectral component itself, then the recovered signal would sound like the original signal except that it would be at a higher or lower pitch. Such, however, is not the case, since the frequency error is fixed. Thus frequencies in the original signal which were harmonically related will no longer be so related after recovery. The overall result is that a frequency offset between carriers adversely affects the intelligibility of spoken communication and is not well tolerated in connection with music. As a matter of experience, it turns out that an error Δf of 30 Hz is acceptable to the ear.

The need to keep the frequency offset Δf between carriers small normally imposes severe restrictions on the frequency stabilities of the carrier signal generators at both ends of the communications system. For suppose that we require to keep Δf to 10 Hz or less, and that our system uses a carrier frequency of 10 MHz. Then the sum of the frequency drift in the two carrier generators may not exceed 1 part in 10^6. The required equality in carrier frequency may be maintained through the. use of quartz crystal oscillators using crystals cut for the same frequency at transmitter and receiver. The receiver must use as many crystals (or equivalent signals derived from crystals) as there are channels in the communications system.

It is also possible to tune an SSB receiver manually and thereby reduce the frequency offset.   To do this, the operator manually adjusts the frequency of the receiver carrier generator until the received signal sounds "normal."   Experienced operators are able to tune correctly to within 10 or 20 Hz.   However, because of oscillator drift, such tuning must be readjusted periodically.

When the carrier frequency is very high, even quartz crystal oscillators may be hard pressed to maintain adequate stability.   In such cases it is necessary to transmit the carrier itself along with the sideband signal. At the receiver the carrier may be separated by filtering and used to synchronize a local carrier generator.   When used for such synchronization, the carrier is referred to as a "pilot carrier" and may be transmitted at a substantially reduced power level.

It is interesting to note that the squaring circuit used to recover the frequency and phase information of the DSB-SC system cannot be used here.   In any event, it is clear that a principal complication in the way of more widespread use of single sideband is the need for supplying an accurate carrier frequency at the receiver.

3.10 METHODS OF GENERATING AN SSB SIGNAL

FILTER METHOD

A straightforward method of generating an SSB signal is illustrated in Fig. 3.10-1.   Here the baseband signal and a carrier are applied to a balanced modulator.   The output of the balanced modulator bears both the upper- and the lower-sideband signals.   One or the other of these signals is then selected by a filter.   The filter is a bandpass filter whose passband encompasses the frequency range of the sideband selected.   The filter must have a cutoff sharp enough to separate the selected sideband

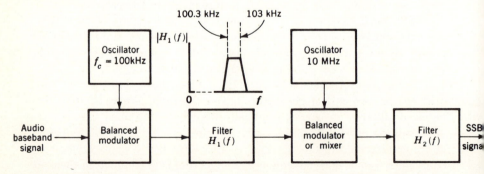

Fig. 3.10-1 Block diagram of the filter method of generating a single-sideband signal.

from the other sideband. The frequency separation of the sidebands is twice the frequency of the lowest frequency spectral components of the baseband signal. Human speech contains spectral components as low as about 70 Hz. However, to alleviate the sideband filter selectivity requirements in an SSB system, it is common to limit the lower spectral limit of speech to about 300 Hz. It is found that such restriction does not materially affect the intelligibility of speech. Similarly, it is found that no serious distortion results if the upper limit of the speech spectrum is cut off at about 3000 Hz. Such restriction is advantageous for the purpose of conserving bandwidth. Altogether, then, a typical sideband filter has a passband which, measured from f_c, extends from about 300 to 3000 Hz and in which range its response is quite flat. Outside this passband the response falls off sharply, being down about 40 dB at 4000 Hz and rejecting the unwanted sideband also by at least 40 dB. The filter may also serve, further, to suppress the carrier itself. Of course, in principle, no carrier should appear at the output of a balanced modulator. In practice, however, the modulator may not balance exactly, and the precision of its balance may be subject to some variation with time. Therefore, even if a pilot carrier is to be transmitted, it is well to suppress it at the output of the modulator and to add it to the signal at a later point in a controllable manner.

Now consider that we desire to generate an SSB signal with a carrier of, say, 10 MHz. Then we require a passband filter with a selectivity that provides 40 dB of attenuation within 600 Hz at a frequency of 10 MHz, a percentage frequency change of 0.006 percent. Filters with such sharp selectivity are very elaborate and difficult to construct. For this reason, it is customary to perform the translation of the baseband signal to the final carrier frequency in several stages. Two such stages of translation are shown in Fig. 3.10-1. Here we have selected the first carrier to be of frequency 100 kHz. The upper sideband, say, of the output of the balanced modulator ranges from 100.3 to 103 kHz. The filter following the balanced modulator which selects this upper sideband need now exhibit a selectivity of only a hundredth of the selectivity (40 dB in 0.6 percent frequency change) required in the case of a 10-MHz carrier. Now let the filter output be applied to a second balanced modulator, supplied this time with a 10-MHz carrier. Let us again select the upper sideband. Then the second filter must provide 40 dB of attenuation in a frequency range of 200.6 kHz, which is nominally 2 percent of the carrier frequency.

We have already noted that the simplest physical frequency-translating device is a multiplier or mixer, while a balanced modulator is a balanced arrangement of two mixers. A mixer, however, has the disadvantage that it presents at its output not only sum and difference frequencies but the input frequencies as well. Still, when it is feasible to discriminate against these input signals, there is a merit of simplicity in using a mixer rather than a balanced modulator. In the present case, if the second frequency-translating device in Fig. 3.10-1 were a mixer rather than a multiplier, then in addition to the upper and lower sidebands, the output would contain a component encompassing the range 100.3 to 103 kHz as well as the 10-MHz carrier. The range 100.3 to 103 kHz is well out of the range of the second filter intended to pass the range 10,100,300 to 10,103,000 Hz. And it is realistic to design a filter which will suppress the 10-MHz carrier, since the carrier frequency is separated from the lower edge of the upper sideband (10,100,300) by nominally a 1-percent frequency change.

Altogether, then, we note in summary that when a single-sideband signal is to be generated which has a carrier in the megahertz or tens-of-megahertz range, the frequency translation is to be done in more than one stage—frequently two but not uncommonly three. If the baseband signal has spectral components in the range of hundreds of hertz or lower (as in an audio signal), the first stage invariably employs a balanced modulator, while succeeding stages may use mixers.

PHASING METHOD

An alternative scheme for generating a single-sideband signal is shown in Fig. 3.10-2. Here two balanced modulators are employed. The carrier signals of angular frequency ω_c which are applied to the modulators differ in phase by 90°. Similarly the baseband signal, before application to the modulators, is passed through a 90° phase-shifting network so that there

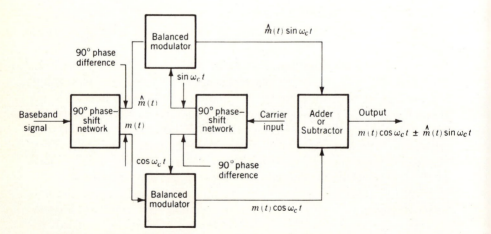

Fig. 3.10-2 A method of generating a single-sideband signal using balanced modulators and phase shifters.

is a 90° phase shift between any spectral component of the baseband signal applied to one modulator and the like-frequency component applied to the other modulator.

To see most simply how the arrangement of Fig. 3.10-2 operates, let us assume that the baseband signal is sinusoidal and appears at the input to one modulator as $\cos \omega_m t$ and hence as $\sin \omega_m t$ at the other. Also, let the carrier be $\cos \omega_c t$ at one modulator and $\sin \omega_c t$ at the other. Then the outputs of the balanced modulators (multipliers) are

$$\cos \omega_m t \cos \omega_c t = \tfrac{1}{2}[\cos (\omega_c - \omega_m)t + \cos (\omega_c + \omega_m)t] \qquad (3.10\text{-}1)$$

$$\sin \omega_m t \sin \omega_c t = \tfrac{1}{2}[\cos (\omega_c - \omega_m)t - \cos (\omega_c + \omega_m)t] \qquad (3.10\text{-}2)$$

If these waveforms are added, the lower sideband results; if substracted, the upper sideband appears at the output. In general, if the modulation $m(t)$ is given by

$$m(t) = \sum_{i=1}^{m} A_i \cos (\omega_i t + \theta_i) \qquad (3.10\text{-}3)$$

then, using Fig. 3.10-2, we see that the output of the SSB modulator is in general

$$m(t) \cos \omega_c t \pm \hat{m}(t) \sin \omega_c t \qquad (3.10\text{-}4)$$

where

$$\hat{m}(t) \equiv \sum_{i=1}^{m} A_i \sin (\omega_i t + \theta_i) \qquad (3.10\text{-}5)$$

The single-sideband generating system of Fig. 3.10-2 generally enjoys less popularity than does the filter method. The reason for this lack of favor is that the present phasing method requires, for satisfactory operation, that a number of constraints be rather precisely met if the carrier and one sideband are adequately to be suppressed. It is required that each modulator be rather carefully balanced to suppress the carrier. It requires also that the baseband signal phase-shifting network provide to the modulators signals in which equal frequency spectral components are of exactly equal amplitude and differ in phase by precisely 90°. Such a network is difficult to construct for a baseband signal which extends over many octaves. It is also required that each modulator display equal sensitivity to the baseband signal. Finally, the carrier phase-shift network must provide exactly 90° of phase shift. If any of these constraints is not satisfied, the suppression of the rejected sideband and of the carrier will suffer. The effect on carrier and sideband suppression due to a failure precisely to meet these constraints is explored in Probs. 3.10-3 and 3.10-4. Of course, in any physical system a certain level of carrier and rejected sideband is tolerable. Still, there seems to be a general inclination to achieve a single sideband by the use of passive filters rather than by a method which requires many exactly maintained balances in passive and active circuits. There is an alternative single-sideband generating scheme[2] which avoids the need for a wideband phase-shifting network but which uses four balanced modulators.

3.11 VESTIGIAL-SIDEBAND MODULATION

As a preliminary to a discussion of vestigial-sideband modulation, let us consider the situation when a single-sideband signal is accompanied by its carrier. Suppose a carrier of angular frequency ω_c is amplitude modulated by a sinusoid of angular frequency ω_m to the extent where the resultant signal displays a percentage modulation m. Then the waveform is

$$f_1(t) = A(1 + m \cos \omega_m t) \cos \omega_c t \tag{3.11-1}$$

$$= A \cos \omega_c t + \frac{mA}{2} [\cos (\omega_c + \omega_m)t + \cos (\omega_c - \omega_m)t] \tag{3.11-2}$$

If one of the sidebands is removed, leaving, however, the carrier, we have

$$f_2(t) = A \cos \omega_c t + \frac{mA}{2} \cos (\omega_c + \omega_m)t \tag{3.11-3}$$

To calculate the response of a diode demodulator to $f_2(t)$ we need to have the form of the envelope of $f_2(t)$. We have

$$f_2(t) = A \cos \omega_c t + \frac{mA}{2} \cos \omega_c t \cos \omega_m t - \frac{mA}{2} \sin \omega_c t \sin \omega_m t$$

$$= A \left(1 + \frac{m}{2} \cos \omega_m t\right) \cos \omega_c t - \frac{mA}{2} \sin \omega_m t \sin \omega_c t \tag{3.11-4}$$

The amplitude $A(t)$ of $f_2(t)$ is

$$A(t) = \sqrt{A^2 \left(1 + \frac{m}{2} \cos \omega_m t\right)^2 + \left(\frac{mA}{2} \sin \omega_m t\right)^2}$$

$$= \sqrt{A^2 \left(1 + \frac{m^2}{4}\right) + A^2 m \cos \omega_m t} \tag{3.11-5}$$

and for $m \ll 1$,

$$A(t) \cong A \left(1 + \frac{m}{2} \cos \omega_m t\right) \tag{3.11-6}$$

We note that if m is small, the diode demodulator does demodulate a signal which is lacking one sideband. Comparing the amplitude $A(t)$

given in Eq. (3.11-6) with the factor in parentheses in Eq. (3.11-1), we observe that the baseband signal output with one sideband suppressed is half as large as it would be if both sidebands had been present, a result to have been anticipated.

The diode-demodulator method for SSB application is of interest since it allows recovery of the baseband signal with a receiving system intended for double-sideband amplitude-modulation signals. Many amplitude-modulation "communications" type receivers are equipped with an adjustable oscillator which can be adjusted in frequency to serve as the local carrier and added to a single-sideband signal. Hence such AM receivers may demodulate SSB signals with, however, some distortion.

This technique should be compared with the use of the synchronous demodulator. Although the synchronous demodulator yields no distortion when the carrier phase is perfectly adjusted, the diode demodulator introduces some distortion (see Prob. 3.11-1). However, for synchronous demodulation we need, at the receiver, information about both the frequency and phase of the carrier. With a diode demodulator we need to know only the frequency.

Turning now to vestigial-sideband modulation, we shall take account of the fact that the principal application of this modulation method is to be found in commercial television broadcasting. Therefore, by way of illustration and for the sake of motivation, we refer specifically to that application.

The television picture signal, i.e., the video signal, in accordance with practice that prevails in the United States, occupies a bandwidth of nominally 4.5 MHz. A carrier, amplitude-modulated with such a signal, would give rise to a signal extending over 9 MHz. Since this bandwidth is about 9 times the frequency range which encompasses *all* standard AM radio broadcasting stations, some means of conserving bandwidth is surely needed. Single sideband is not feasible because of the complexity it must introduce into each of the millions of receivers. A workable compromise between the spectrum-conserving characteristic of single-sideband modulation and the demodulation simplicity of double-sideband amplitude modulation is found in the vestigial-sideband system which is standard in television broadcasting.

In the vestigial-sideband system, an amplitude-modulated signal, carrier plus double sideband, is passed through a filter before transmission to the receiving end. The response of the filter is plotted in Fig. 3.11-1a as a function of the deviation Δf of the frequency from the carrier frequency. (The sound which accompanies the video signal is transmitted by frequency modulation, discussed in the next chapter, on a carrier located 4.5 MHz above the picture carrier. A frequency range of 100 kHz

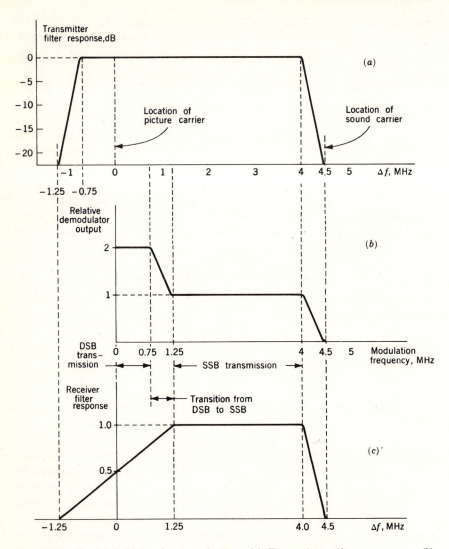

Fig. 3.11-1 Vestigial-sideband transmission. (*a*) Transmitter filter response. (*b*) Relative diode demodulator output. (*c*) Receiver filter response.

is allowed on each side of the sound carrier for the sound sidebands.) The upper sideband of the picture carrier is transmitted without attenuation up to 4 MHz. Thereafter the sideband is attenuated so that it does not interfere with the lower sideband of the sound carrier. The lower sideband of the picture carrier is transmitted without attenuation over the range 0.75 MHz and is entirely attenuated at 1.25 MHz. Thus the picture signal is transmitted double sideband over the range 0 to

0.75 MHz, single sideband over the range 1.25 MHz and above, while in the intermediate range, 0.75 to 1.25 MHz, the transition is made from one to the other. Altogether, however, the entire transmission is confined to a range of about 6 MHz, a saving of *one-third* of the bandwidth that would be required for full double-sideband transmission.

We noted above that when only a single sideband is present, the output of a diode demodulator is half the output yielded when both sidebands are present. Therefore, with a vestigial-sideband signal, the relative demodulator output, plotted against frequency for a fixed percentage modulation, has the form shown in Fig. 3.11-1b. This lack of uniformity is corrected at the receiver by passing the received signal through a filter before demodulation. The relative response of this filter is shown in Fig. 3.11-1c. Over the range 1.25 MHz on either side of the picture carrier the response varies linearly as shown. As a result, for modulating frequencies up to 1.25 MHz, the *sum* of the amplitudes of the two sidebands, and hence of the demodulator output, is the same as is yielded by the single sideband above 1.25 MHz. This result is easily verified. For example, the received signal frequency component at $\Delta f = 0$ is attenuated by a factor of 2. Referring to Fig. 3.11-1b, we see that this reduces the modulation to 1 when $\Delta f = 0$. As a second illustration, let $\Delta f = \pm 0.75$. The sideband amplitude due to $\Delta f = -0.75$ is 0.2, while the sideband amplitude due to $\Delta f = +0.75$ is 0.8. The sum is again unity, as expected.

It is, of course, to be anticipated that the vestigial-sideband system will introduce some distortion into the demodulated signal, especially at high-percentage modulation. The experimental fact is, however, that in the transmission of picture information the distortion can be kept within tolerable levels. We may even wonder why the cutoff frequency of the filter which removes the lower sideband is not adjusted to be even closer to the carrier frequency, thereby conserving additional bandwidth. It is found rather generally, in real filters, that spectral components which lie within the passband but close to the cutoff frequency suffer distortion-producing phase shifts, even when the amplitude response of the filter is maintained uniform. The nature of the picture signal is such that its waveforms suffer substantial distortion from relatively small phase shifts of its low-frequency components. Thus the decision to leave a *vestige* of the nominally suppressed sideband is dictated by an engineering compromise between bandwidth economy and faithfulness of reproduction of the picture.

3.12 COMPATIBLE SINGLE SIDEBAND

We are now rather naturally led to inquire whether it is possible to generate an amplitude-modulated signal whose bandwidth is f_M Hz, for a baseband signal whose highest spectral component is f_M Hz, and from which the baseband signal may be faithfully recovered by the simple diode demodulator, even when no carrier signal is transmitted. Such a single-sideband signal would be *compatible* for reception by an AM radio receiver and is referred to as a *compatible single-sideband* signal, abbreviated CSSB. It has been shown[3] that such a signal exists. However, the involved signal processing required for the production of such a CSSB signal makes the system presently impractical for commercial use.

3.13 MULTIPLEXING

We have explored the principles of operation of a number of amplitude-modulation *modems* (systems of modulation and demodulation). The manner in which such systems are used for multiplexing (i.e., transmitting many baseband signals over a common communications channel) is shown in Fig. 3.13-1. The individual baseband signals $m_1(t), m_2(t), \ldots,$ $m_\lambda(t)$, each bandlimited to f_M, are applied to individual modulators, each modulator being supplied as well with carrier waveforms of frequency $f_1, f_2, \ldots, f_\lambda$. The individual modulator-output signals extend over a limited range in the neighborhood of the individual carrier frequencies. Most importantly the carrier frequencies are selected so the spectral ranges of the modulator-output signals do not overlap. This separation in frequency is precisely the feature that allows the eventual recovery of

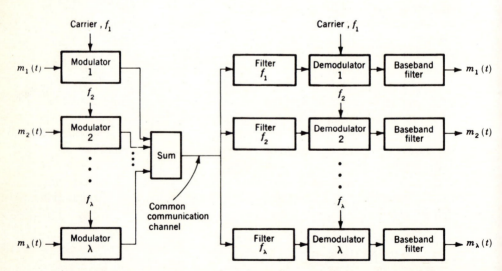

Fig. 3.13-1 Multiplexing many baseband signals over a single communications channel.

the individual signals, and for this reason this multiplexing system is referred to as *frequency multiplexing*. As a matter of fact, to facilitate this separation of the individual signals, the carrier frequencies are selected to leave a comfortable margin (guard band) between the limit of one frequency range and the beginning of the next. The combined output of all the modulators, i.e., the *composite* signal, is applied to a common communications channel. In the case of radio transmission, the channel is free space, and coupling to the channel is made by means of an antenna. In other cases wires are used.

At the receiving end the composite signal is applied to each of a group of bandpass filters whose passbands are in the neighborhood $f_1, f_2, \ldots, f_\lambda$. The filter f_1 is a bandpass filter which passes only the spectral range of the output of modulator 1 and similarly for the other bandpass filters. The signals have thus been separated. They are then applied to individual demodulators which extract the baseband signals from the carrier. The carrier inputs to the demodulators are required only for synchronous demodulation and are not used otherwise.

The final operation indicated in Fig. 3.13-1 consists in passing the demodulator output through a baseband filter. The baseband filter is a low-pass filter with cutoff at the frequency f_M to which the baseband signal is limited. This baseband filter will pass, without modification, the baseband signal output of the modulator and in this sense serves no function in the system as described up to the present point. We shall, however, see in Chaps. 8 and 9 that such baseband filters are essential to suppress the noise which invariably accompanies the signal.

PROBLEMS

3.2-1. A signal $v_m(t)$ is bandlimited to the frequency range 0 to f_M. It is frequency-translated by multiplying it by the signal $v_c(t) = \cos 2\pi f_c t$. Find f_c so that the bandwidth of the translated signal is 1 percent of the frequency f_c.

3.2-2. The Fourier transform of $m(t)$ is $\mathcal{F}[m(t)] \equiv M(f)$. Show that

$$\mathcal{F}[m(t) \cos 2\pi f_c t] = \tfrac{1}{2}[M(f + f_c) + M(f - f_c)]$$

3.2-3. The signals

$$v_1(t) = 2 \cos \omega_1 t + \cos 2\omega_1 t$$

and

$$v_2(t) = \cos \omega_2 t + 2 \cos 2\omega_2 t$$

are multiplied. Plot the resultant amplitude-frequency characteristic, assuming that $\omega_2 > 2\omega_1$, but not a harmonic of ω_1. Repeat for $\omega_2 = 2\omega_1$.

3.3-1. The baseband signal $m(t)$ in the frequency-translated signal

$$v(t) = m(t) \cos 2\pi f_c t$$

is recovered by multiplying $v(t)$ by the waveform $\cos (2\pi f_c t + \theta)$.

(a) The product waveform is transmitted through a low-pass filter which rejects the double-frequency signal.   What is the output signal of the filter?

(b) What is the maximum allowable value for the phase θ if the recovered signal is to be 90 percent of the maximum possible value?

(c) If the baseband signal $m(t)$ is bandlimited to 10 kHz, what is the minimum value of f_c for which it is possible to recover $m(t)$ by filtering the product waveform $v(t) \cos (2\pi f_c t + \theta)$?

3.3-2. The baseband signal $m(t)$ in the frequency-translated signal

$$v(t) = m(t) \cos 2\pi f_c t$$

is recovered by multiplying $v(t)$ by the waveform $\cos 2\pi (f_c + \Delta f)t$.   The product waveform is transmitted through a low-pass filter which rejects the double-frequency signal.   Find the output signal of the filter.

3.3-3. (a) The baseband signal $m(t)$ in the frequency-translated signal $v(t) = m(t) \cos 2\pi f_c t$ is to be recovered.   There is available a waveform $p(t)$ which is periodic with period $1/f_c$.   Show that $m(t)$ may be recovered by appropriately filtering the product waveform $p(t)v(t)$.

(b) Show that $m(t)$ may be recovered as well if the periodic waveform has a period n/f_c, where n is an integer.   Assume $m(t)$ bandlimited to the frequency range from 0 to 5 kHz and let $f_c = 1$ MHz.   Find the largest n which will allow $m(t)$ to be recovered.   Are all periodic waveforms acceptable?

3.3-4. The signal $m(t)$ in the DSB-SC signal $v(t) = m(t) \cos (\omega_c t + \theta)$ is to be reconstructed by multiplying $v(t)$ by a signal derived from $v^2(t)$.

(a) Show that $v^2(t)$ has a component at the frequency $2f_c$.   Find its amplitude.

(b) If $m(t)$ is bandlimited to f_M and has a probability density

$$f(m) = \frac{1}{\sqrt{2\pi}} e^{-m^2/2} \qquad -\infty \leq m \leq \infty$$

find the expected value of the amplitude of the component of $v^2(t)$ at $2f_c$.

3.4-1. The envelope detector shown in Fig. 3.4-2a is used to recover the signal $m(t)$ from the AM signal $v(t) = [1 + m(t)] \cos \omega_c t$, where $m(t)$ is a square wave taking on the values 0 and -0.5 volt and having a period $T \gg 1/f_c$.   Sketch the *recovered* signal if $RC = T/20$ and $4T$.

3.4-2. (a) The waveform $v(t) = (1 + m \cos \omega_m t) \cos \omega_c t$, with m a constant ($m \leq 1$), is applied to the diode demodulator of Fig. 3.4-2a.   Show that, if the demodulator output is to follow the envelope of $v(t)$, it is required that at any time t_0:

$$\frac{1}{RC} \geq \omega_m \left(\frac{m \sin \omega_m t_0}{1 + m \cos \omega_m t_0} \right)$$

(b) Using the result of part (a), show that if the demodulator is to follow the envelope at all times then m must be less than or equal to the value of m_0, determined from the equation

$$RC = \frac{1}{\omega_m} \frac{\sqrt{1 - m_0^2}}{m_0}$$

(c) Draw, qualitatively, the form of the demodulator output when the condition specified in part (b) is not satisfied.

3.5-1. The signal $v(t) = (1 + m \cos \omega_m t) \cos \omega_c t$ is detected using a diode envelope detector. Sketch the detector output when $m = 2$.

3.6-1. The signal $v(t) = [1 + 0.2 \cos (\omega_M/3)t] \cos \omega_c t$ is demodulated using a square-law demodulator having the characteristic $v_o = (v + 2)^2$. The output $v_o(t)$ is then filtered by an ideal low-pass filter having a cutoff frequency at f_M Hz. Sketch the amplitude-frequency characteristic of the output waveform in the frequency range $0 \le f \le f_M$.

3.6-2. Repeat Prob. 3.6-1 if the square-law demodulator is centered at the origin so that $v_o = v^2$.

3.6-3. The signal $v(t) = [1 + m(t)] \cos \omega_c t$ is square-law detected by a detector having the characteristic $v_o = v^2$. If the Fourier transform of $m(t)$ is a constant M_0 extending from $-f_M$ to $+f_M$, sketch the Fourier transform of $v_o(t)$ in the frequency range $-f_M < f < f_M$. *Hint:* Convolution in the frequency domain is needed to find the Fourier transform of $m^2(t)$. See Prob. 1.11-2.

3.6-4. The signal $v(t) = (1 + 0.1 \cos \omega_1 t + 0.1 \cos 2\omega_1 t) \cos \omega_c t$ is detected by a square-law detector, $v_o = 2v^2$. Plot the amplitude-frequency characteristic of $v_o(t)$.

3.9-1. (a) Show that the signal

$$v(t) = \sum_{i=1}^{N} [\cos \omega_c t \cos (\omega_i t + \theta_i) - \sin \omega_c t \sin (\omega_i t + \theta_i)]$$

is an SSB-SC signal ($\omega_c \gg \omega_N$). Is it the upper or lower sideband?

(b) Write an expression for the missing sideband.

(c) Obtain an expression for the total DSB-SC signal.

3.9-2. The SSB signal in Prob. 3.9-1 is multiplied by $\cos \omega_c t$ and then low-pass filtered to recover the modulation.

(a) Show that the modulation is completely recovered if the cutoff frequency of the low-pass filter f_0 is $f_M < f_0 < 2f_c$.

(b) If the multiplying signal were $\cos (\omega_c t + \theta)$, find the recovered signal.

(c) If the multiplying signal were $\cos (\omega_c + \Delta\omega)t$, find the recovered signal. Assume that $\Delta\omega \ll \omega_1$.

3.9-3. Show that the squaring circuit shown in Fig. 3.3-1 will not permit the generation of a local oscillator signal capable of demodulating an SSB-SC signal.

3.10-1. A baseband signal, bandlimited to the frequency range 300 to 3000 Hz, is to be superimposed on a carrier of frequency of 40 MHz as a single-sideband modulation using the filter method. Assume that bandpass filters are available which will provide 40 dB of attenuation in a frequency interval which is about

1 percent of the filter center frequency. Draw a block diagram of a suitable system. At each point in the system draw plots indicating the spectral range occupied by the signal present there.

3.10-2. The system shown in Fig. 3.10-2 is used to generate a single-sideband signal. However, an ideal 90° phase-shifting network which is independent of a frequency is unattainable. The 90° phase shift is approximated by a lattice network having the transfer function

$$H(f) = e^{-j \arctan(f/30)}$$

The input to this network is $m(t)$, given by Eq. (3.10-3). If $f_1 = 300$ Hz and $f_M = 3000$ Hz show that $H(f) \cong e^{-j\pi/2}e^{j30/f}$, for $f_1 \leq f \leq f_M$.

3.10-3. In the SSB generating system of Fig. 3.10-2, the carrier phase-shift network produces a phase shift which differs from 90° by a *small* angle α. Calculate the output waveform and point out the respects in which the output no longer meets the requirements for an SSB waveform. Assume that the input is a single spectral component $\cos \omega_m t$.

3.10-4. Repeat Prob. 3.10-3 except, assume instead, that the baseband phase-shift network produces a phase shift differing from 90° by a *small* angle α.

3.11-1. A received SSB signal in which the modulation is a single spectral component has a normalized power of 0.5 volt². A carrier is added to the signal, and the carrier plus signal are applied to a diode demodulator. The carrier amplitude is to be adjusted so that at the demodulator output 90 percent of the normalized power is in the recovered modulating waveform. Neglect dc components. Find the carrier amplitude required.

REFERENCES

1. Transmission Systems for Communications, Bell Telephone Laboratories, published by Western Electric Company, Tech. Pub., Winston-Salem, N.C., 1964.
2. Norgaard, D. E.: A Third Method of Generation and Detection of Single-sideband Signals, *Proc. IRE*, Dec. 1956.
3. Voelcker, H.: Demodulation of Single-sideband Signals Via Envelope Detection, *IEEE, Trans. on Communication Technology*, pp. 22–30, Feb. 1966.

4
Frequency-modulation Systems

In the amplitude-modulation systems described in Chap. 3, the modulator output consisted of a carrier which displayed variations in its amplitude. In the present chapter we discuss modulation systems in which the modulator output is of constant amplitude and in which the signal information is superimposed on the carrier through variations of the carrier frequency.

4.1 ANGLE MODULATION[1]

All the modulation schemes considered up to the present point have two principal features in common. In the first place, each spectral component of the baseband signal gives rise to one or two spectral components in the modulated signal. These components are separated from the carrier by a frequency difference equal to the frequency of the baseband component. Most importantly, the nature of the modulators is such that the spectral components which they produce depend only on the carrier frequency and and the baseband frequencies. The amplitudes of the spectral components of the modulator output may depend on the amplitude of the input signals; however, the frequencies of the spectral components do not.   In the second place, all the operations performed on the signal (addition, subtraction, and multiplication) are linear operations so that superposition applies.   Thus, if a baseband signal $m_1(t)$ introduces one spectrum of components into the modulated signal and a second signal $m_2(t)$ introduces a second spectrum, the application of the sum $m_1(t) + m_2(t)$ will introduce a spectrum which is the sum of the spectra separately introduced.   All these systems are referred to under the designation "amplitude or linear modulation."   This terminology must be taken with some reservation, for we have noted that, at least in the special case of single sideband using modulation with a single sinusoid, there is no amplitude variation at all.   And even more generally, when the amplitude of the modulated signal does vary, the carrier envelope need not have the waveform of the baseband signal.

We now turn our attention to a new type of modulation which is not characterized by the features referred to above.   The spectral components in the modulated waveform depend on the amplitude as well as the frequency of the spectral components in the baseband signal.   Furthermore, the modulation system is *not* linear and superposition does *not* apply.   Such a system results when, in connection with a carrier of constant amplitude, the phase angle is made to respond in some way to a baseband signal.   Such a signal has the form

$$v(t) = A \cos [\omega_c t + \varphi(t)] \qquad (4.1\text{-}1)$$

in which A and ω_c are constant but in which the phase angle $\varphi(t)$ is a function of the baseband signal.   Modulation of this type is called *angle modulation* for obvious reasons.   It is also referred to as *phase modulation* since $\varphi(t)$ is the phase angle of the argument of the cosine function.   Still another designation is *frequency modulation* for reasons to be discussed in the next section.

4.2 PHASE AND FREQUENCY MODULATION

To review some elementary ideas in connection with sinusoidal waveforms, let us recall that the function $A \cos \omega_c t$ can be written as

$$A \cos \omega_c t = \text{real part } (A e^{j\omega_c t}) \qquad (4.2\text{-}1)$$

The function $A e^{j\theta}$ is represented in the complex plane by a phasor of length A and an angle θ measured counterclockwise from the real axis.   If $\theta = \omega_c t$, then the phasor rotates in the counterclockwise direction with an angular velocity ω_c.   With respect to a coordinate system which also rotates in the counterclockwise direction with angular velocity ω_c, the phasor will be stationary. If in Eq. (4.1-1) φ is actually not time-dependent but is a constant, then $v(t)$ is to be represented precisely in the manner just described. But suppose $\varphi = \varphi(t)$ does change with time and makes positive and negative excursions. Then $v(t)$ would be represented by a phasor of amplitude A which runs ahead of and falls behind the phasor representing $A \cos \omega_c t$. We may, therefore, consider that the angle $\omega_c t + \varphi(t)$, of $v(t)$, undergoes a *modulation* around the angle $\theta = \omega_c t$. The waveform of $v(t)$ is, therefore, a representation of a signal which is *modulated in phase*.

If the phasor of angle $\theta + \varphi(t) = \omega_c t + \varphi(t)$ alternately runs ahead of and falls behind the phasor $\theta = \omega_c t$, then the first phasor must alternately be rotating more, or less, rapidly than the second phasor. Therefore we may consider that the angular velocity of the phasor of $v(t)$ undergoes a modulation around the nominal angular velocity ω_c. The signal $v(t)$ is, therefore, an angular-velocity-modulated waveform. The angular velocity associated with the argument of a sinusoidal function is equal to the time rate of change of the argument (i.e., the angle) of the function. Thus we have that the instantaneous radial frequency $\omega = d(\theta + \varphi)/dt$, and the corresponding frequency $f = \omega/2\pi$ is

$$f = \frac{1}{2\pi} \frac{d}{dt} [\omega_c t + \varphi(t)] = \frac{\omega_c}{2\pi} + \frac{1}{2\pi} \frac{d}{dt} \varphi(t) \tag{4.2-2}$$

The waveform $v(t)$ is, therefore, *modulated in frequency*.

In initial discussions of the sinusoidal waveform it is customary to consider such a waveform as having a fixed frequency and phase. In the present discussion we have generalized these concepts somewhat. To acknowledge this generalization, it is not uncommon to refer to the frequency f in Eq. (4.2-2) as the *instantaneous frequency* and $\varphi(t)$ as the *instantaneous phase*. If the frequency variation about the nominal frequency ω_c is small, that is, if $d\varphi(t)/dt \ll \omega_c$, then the resultant waveform will have an appearance which is readily recognizable as a "sine wave," albeit with a period which changes somewhat from cycle to cycle. Such a waveform is represented in Fig. 4.2-1. In this figure the modulating signal is a square wave. The frequency-modulated signal changes frequency whenever the modulation changes level.

Among the possibilities which suggest themselves for the design of a modulator are the following. We might arrange that the phase $\varphi(t)$ in Eq. (4.1-1) be directly proportional to the modulating signal, or we might arrange a direct proportionality between the modulating signal and the derivative, $d\varphi(t)/dt$. From Eq. (4.2-2), with $f_c = \omega_c/2\pi$

$$\frac{d\varphi(t)}{dt} = 2\pi(f - f_c) \tag{4.2-3}$$

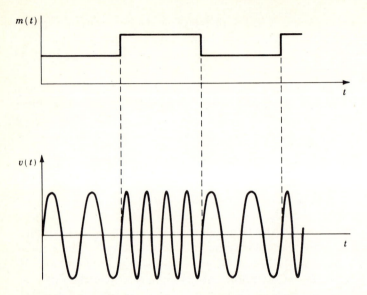

Fig. 4.2-1 An angle-modulated waveform. (*a*) Modulating signal.
(*b*) Frequency-modulated sinusoidal carrier signal.

where f is the instantaneous frequency. Hence in this latter case the
proportionality is between modulating signal and the departure of the
instantaneous frequency from the carrier frequency. Using standard
terminology, we refer to the modulation of the first type as *phase* modulation, and the term *frequency modulation* refers only to the second type.
On the basis of these definitions it is, of course, not possible to determine
which type of modulation is involved simply from a visual examination of
the waveform or from an analytical expression for the waveform. We
would also have to be given the waveform of the modulating signal. This
information is, however, provided in any practical communication system.

4.3 RELATIONSHIP BETWEEN PHASE
AND FREQUENCY MODULATION

The relationship between phase and frequency modulation may be visu-
alized further by a consideration of the diagrams of Fig. 4.3-1. In Fig.
4.3-1*a* the phase-modulator block represents a device which furnishes an
output $v(t)$ which is a carrier, phase-modulated by the input signal $m_i(t)$.
Thus

$$v(t) = A \cos [\omega_c t + k' m_i(t)] \tag{4.3-1}$$

k' being a constant. Let the waveform $m_i(t)$ be derived as the integral of the modulating signal $m(t)$ so that

$$m_i(t) = k'' \int_{-\infty}^{t} m(t) \, dt \qquad (4.3\text{-}2)$$

in which k'' is also a constant. Then with $k = k'k''$ we have

$$v(t) = A \cos \left[\omega_c t + k \int_{-\infty}^{t} m(t) \, dt\right] \qquad (4.3\text{-}3)$$

The instantaneous angular frequency is

$$\omega = \frac{d}{dt}\left[\omega_c t + k \int_{-\infty}^{t} m(t) \, dt\right] = \omega_c + km(t) \qquad (4.3\text{-}4)$$

The deviation of the instantaneous frequency from the carrier frequency $\omega_c/2\pi$ is

$$\nu \equiv f - f_c = \frac{k}{2\pi} m(t) \qquad (4.3\text{-}5)$$

Since the deviation of the instantaneous frequency is directly proportional to the modulating signal, the combination of *integrator* and *phase modulator* of Fig. 4.3-1a constitutes a device for producing a *frequency-modulated* output. Similarly, the combination in Fig. 4.3-1b of the differentiator and frequency modulator generates a *phase-modulated output*, i.e., a signal whose phase departure from the carrier is proportional to the modulating signal.

In summary, we have referred generally to the waveform given by Eq. (4.1-1) as an *angle-modulated* waveform, an appropriate designation when we have no interest in, or information about, the modulating signal. When $\varphi(t)$ is proportional to the modulating signal $m(t)$, we use the designation *phase modulation* or PM. When the time derivative of $\varphi(t)$ is proportional to $m(t)$, we use the term *frequency modulation* or FM. In an FM waveform, the form of Eq. (4.3-3) is of special interest, since here

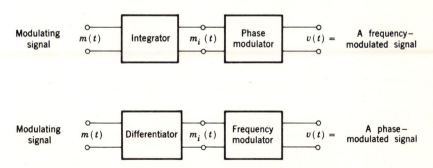

Fig. 4.3-1 Illustrating the relationship between phase and frequency modulation.

the instantaneous frequency deviation is directly proportional to the signal $m(t)$ which appears explicitly in the expression. In general usage, however, we find that such precision of language is not common. Very frequently the terms angle modulation, phase modulation, and frequency modulation are used rather interchangeably and without reference to, or even interest in, the modulating signal.

4.4 PHASE AND FREQUENCY DEVIATION

In the waveform of Eq. (4.1-1) the maximum value attained by $\varphi(t)$, that is, the maximum phase deviation of the total angle from the carrier angle $\omega_c t$, is called the *phase deviation*. Similarly the maximum departure of the instantaneous frequency from the carrier frequency is called the *frequency deviation*.

When the angular (and consequently the frequency) variation is sinusoidal with frequency f_m, we have, with $\omega_m = 2\pi f_m$

$$v(t) = A \cos(\omega_c t + \beta \sin \omega_m t) \tag{4.4-1}$$

where β is the peak amplitude of $\varphi(t)$. In this case β which is the maximum phase deviation, is usually referred to as the *modulation index*. The instantaneous frequency is

$$f = \frac{\omega_c}{2\pi} + \frac{\beta \omega_m}{2\pi} \cos \omega_m t \tag{4.4-2a}$$

$$= f_c + \beta f_m \cos \omega_m t \tag{4.4-2b}$$

The maximum frequency deviation is defined as Δf and is given by

$$\Delta f = \beta f_m \tag{4.4-3}$$

Equation (4.4-1) can, therefore, be written

$$v(t) = A \cos\left(\omega_c t + \frac{\Delta f}{f_m} \sin \omega_m t\right) \tag{4.4-4}$$

While the instantaneous frequency f lies in the range $f_c \pm \Delta f$, it should not be concluded that all spectral components of such a signal lie in this range. We consider next the spectral pattern of such an angle-modulated waveform.

4.5 SPECTRUM OF AN FM SIGNAL: SINUSOIDAL MODULATION

In this section we shall look into the frequency spectrum of the signal

$$v(t) = \cos(\omega_c t + \beta \sin \omega_m t) \tag{4.5-1}$$

which is the signal of Eq. (4.4-1) with the amplitude arbitrarily set at unity as a matter of convenience. We have

$$\cos (\omega_c t + \beta \sin \omega_m t) = \cos \omega_c t \cos (\beta \sin \omega_m t)$$
$$- \sin \omega_c t \sin (\beta \sin \omega_m t) \quad (4.5\text{-}2)$$

Consider now the expression $\cos (\beta \sin \omega_m t)$ which appears as a factor on the right-hand side of Eq. (4.5-2). It is an *even*, periodic function having an angular frequency ω_m. Therefore it is possible to expand this expression in a Fourier series in which $\omega_m/2\pi$ is the fundamental frequency. We shall not undertake the evaluation of the coefficients in the Fourier expansion of $\cos (\beta \sin \omega_m t)$ but shall instead simply write out the results. The coefficients are, of course, functions of β, and, since the function is *even*, the coefficients of the odd harmonics are zero. The result is

$$\cos (\beta \sin \omega_m t) = J_0(\beta) + 2J_2(\beta) \cos 2\omega_m t + 2J_4(\beta) \cos 4\omega_m t$$
$$+ \cdots + 2J_{2n}(\beta) \cos 2n\omega_m t + \cdots \quad (4.5\text{-}3)$$

while for $\sin (\beta \sin \omega_m t)$, which is an *odd* function, we find the expansion contains only odd harmonics and is given by

$$\sin (\beta \sin \omega_m t) = 2J_1(\beta) \sin \omega_m t + 2J_3(\beta) \sin 3\omega_m t$$
$$+ \cdots + 2J_{2n-1}(\beta) \sin (2n - 1)\omega_m t + \cdots \quad (4.5\text{-}4)$$

The functions $J_n(\beta)$ occur often in the solution of engineering problems. They are known as Bessel functions of the first kind and of order n. The numerical values of $J_n(\beta)$ are tabulated in texts of mathematical tables.[2]

Putting the results given in Eqs. (4.5-3) and (4.5-4) back into Eq. (4.5-2) and using the identities

$$\cos A \cos B = \tfrac{1}{2} \cos (A - B) + \tfrac{1}{2} \cos (A + B) \quad (4.5\text{-}5)$$

$$\sin A \sin B = \tfrac{1}{2} \cos (A - B) - \tfrac{1}{2} \cos (A + B) \quad (4.5\text{-}6)$$

we find that $v(t)$ in Eq. (4.5-1) becomes

$$v(t) = J_0(\beta) \cos \omega_c t - J_1(\beta)[\cos (\omega_c - \omega_m)t - \cos (\omega_c + \omega_m)t]$$
$$+ J_2(\beta)[\cos (\omega_c - 2\omega_m)t + \cos (\omega_c + 2\omega_m)t]$$
$$- J_3(\beta)[\cos (\omega_c - 3\omega_m)t - \cos (\omega_c + 3\omega_m)t]$$
$$+ \cdots \quad (4.5\text{-}7)$$

Observe that the spectrum is composed of a carrier with an amplitude $J_0(\beta)$ and a set of sidebands spaced symmetrically on either side of the carrier at frequency separations of ω_m, $2\omega_m$, $3\omega_m$, etc. In this respect the result is unlike that which prevails in the amplitude-modulation systems discussed earlier, since in AM a sinusoidal modulating signal gives rise to only one sideband or one pair of sidebands. A second difference, which is left for verification by the student (Prob. 4.5-1), is that the present modulation system is nonlinear, as anticipated from the discussion of Sec. 4.1.

4.6 SOME FEATURES OF THE BESSEL COEFFICIENTS

Several of the Bessel functions which determine the amplitudes of the spectral components in the Fourier expansion are plotted in Fig. 4.6-1. We note that, at $\beta = 0$, $J_0(0) = 1$, while all other J_n's are zero. Thus, as expected when there is no modulation, only the carrier, of normalized amplitude unity, is present, while all sidebands have zero amplitude. When β departs slightly from zero, $J_1(\beta)$ acquires a magnitude which is significant in comparison with unity, while all higher-order J's are negligible in comparison. That such is the case may be seen either from Fig. 4.6-1 or from the approximations[3] which apply when $\beta \ll 1$, that is,

$$J_0(\beta) \cong 1 - \left(\frac{\beta}{2}\right)^2 \tag{4.6-1}$$

$$J_n(\beta) \cong \frac{1}{n!}\left(\frac{\beta}{2}\right)^n \qquad n \neq 0 \tag{4.6-2}$$

Accordingly, for β very small, the FM signal is composed of a carrier and a single pair of sidebands with frequencies $\omega_c \pm \omega_m$. An FM signal which is so constituted, that is, a signal where β is small enough so that only a single sideband pair is of significant magnitude, is called a *narrowband* FM signal. We see further, in Fig. 4.6-1, as β becomes somewhat larger, that the amplitude J_1 of the first sideband pair increases and that also the amplitude J_2 of the second sideband pair becomes significant. Further, as β continues to increase, J_3, J_4, etc. begin to acquire significant magnitude, giving rise to sideband pairs at frequencies $\omega_c \pm 2\omega_m$, $\omega_c \pm 3\omega_m$, etc.

Another respect in which FM is unlike the linear-modulation schemes described earlier is that in an FM signal the amplitude of the spectral component at the carrier frequency is not constant independent of β. It is to be expected that such should be the case on the basis of the following considerations. The envelope of an FM signal has a constant amplitude. Therefore the power of such a signal is a constant independent of the modulation, since the power of a periodic waveform depends only on the square of its amplitude and not on its frequency. The power of a unit amplitude signal, as in Eq. (4.5-1), is $P_v = \frac{1}{2}$ and is independent of β. When the carrier is modulated to generate an FM signal, the power in the sidebands may appear only at the expense of the power originally in the carrier. Another way of arriving at the same conclusion is to make use of the identity[3] $J_0^2 + 2J_1^2 + 2J_2^2 + 2J_3^2 + \cdots = 1$. We calculate the

$J_n\ (n = 0, 1, 2)$

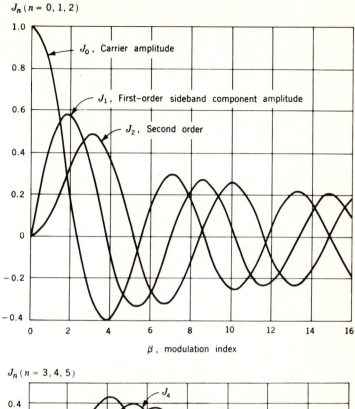

β , modulation index

$J_n\ (n = 3, 4, 5)$

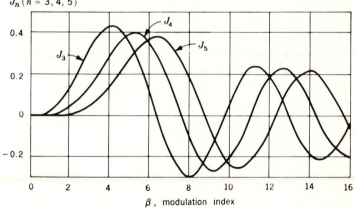

β , modulation index

Fig. 4.6-1 The Bessel functions $J_n(\beta)$ plotted as a function of β for $n = 0, 1, 2, \ldots , 5$.

power P_v by squaring $v(t)$ in Eq. (4.5-7) and then averaging $v^2(t)$. Keeping in mind that cross-product terms average to zero, we find, independently of β, that

$$P_v = \tfrac{1}{2}\left(J_0^2 + 2\sum_{n=1}^{\infty} J_n^2\right) = \tfrac{1}{2} \tag{4.6-3}$$

as expected. We observe in Fig. 4.6-1 that, at various values of β, $J_0(\beta) = 0$. At these values of β all the power is in the sidebands and none in the carrier.

4.7 BANDWIDTH OF A SINUSOIDALLY MODULATED FM SIGNAL

In principle, when an FM signal is modulated, the number of sidebands is infinite and the bandwidth required to encompass such a signal is similarly infinite in extent. As a matter of practice, it turns out that for any β, so large a fraction of the total power is confined to the sidebands which lie within some finite bandwidth that no serious distortion of the signal results if the sidebands outside this bandwidth are lost. We see in Fig. 4.6-1 that, except for $J_0(\beta)$, each $J_n(\beta)$ hugs the zero axis initially and that as n increases, the corresponding J_n remains very close to the zero axis up to a larger value of β. For any value of β only those J_n need be considered which have succeeded in making a significant departure from the zero axis. How many such sideband components need to be considered may be seen from an examination of Table 4.7-1 where $J_n(\beta)$ is tabulated for various values of n and of β.

It is found experimentally that the distortion resulting from band-limiting an FM signal is tolerable as long as 98 percent or more of the power is passed by the bandlimiting filter. This definition of the bandwidth of a filter is, admittedly, somewhat vague, especially since the term "tolerable" means different things in different applications. However, using this definition for bandwidth, one can proceed with an initial tentative design of a system. When the system is built, the bandwidth may thereafter be readjusted, if necessary. In each column of Table 4.7-1, a line has been drawn after the entries which account for at least 98 percent of the power. To illustrate this point, consider $\beta = 1$. Then the power contained in the terms $n = 0$, 1, and 2 are

$$P = \tfrac{1}{2}J_0^2(1) + J_1^2(1) + J_2^2(1)$$
$$= 0.289 + 0.193 + 0.013 = 0.495 \tag{4.7-1}$$

The sum 0.495 is 99 percent of the power in the FM signal, which is $\tfrac{1}{2}$.

Table 4.7-1 Values of the Bessel functions $J_n(\beta)$ for various orders n and integral values of β

n \ β	1	2	3	4	5	6	7	8	9	10
0	.7652	.2239	−.2601	−.3971	−.1776	.1506	.3001	.1717	−.09033	−.2459
1	.4401	.5767	.3391	−.06604	−.3276	−.2767	−.004683	.2346	.2453	.04347
2	.1149	.3528	.4861	.3641	.04657	−.2429	−.3014	−.1130	.1448	.2546
3	.01956	.1289	.3091	.4302	.3648	.1148	−.1676	−.2911	−.1809	.05838
4	.002477	.03400	.1320	.2811	.3912	.3576	.1578	−.1054	−.2655	.2196
5		.007040	.04303	.1321	.2611	.3621	.3479	.1858	−.05504	−.2341
6		.001202	.01139	.04909	.1310	.2458	.3392	.3376	.2043	−.01446
7			.002547	.01518	.05338	.1296	.2336	.3206	.3275	.2167
8				.004029	.01841	.05653	.1280	.2235	.3051	.3179
9					.005520	.02117	.05892	.1263	.2149	.2919
10					.001468	.006964	.02354	.06077	.1247	.2075
11						.002048	.008335	.02560	.06222	.1231
12							.002656	.009624	.02739	.06337
13								.003275	.01083	.02897
14								.001019	.003895	.01196
15									.001286	.004508
16										.001567

We note that the horizontal lines in Table 4.7-1, which indicate the value of n for 98 percent power transmission, always occur just after $n = \beta + 1$. Thus, for sinusoidal modulation the bandwidth required to transmit or receive the FM signal is

$$B = 2(\beta + 1)f_m \tag{4.7-2}$$

By way of example, when $\beta = 5$, the sideband components furthest from the carrier which have adequate amplitude to require consideration are those which occur at frequencies $f_c \pm 6f_m$. From the table of Bessel functions published in Jahnke and Emde[2] it may be verified on a numerical basis that the rule given in Eq. (4.7-2) holds without exception up to $\beta = 29$, which is the largest value of β for which J_m is tabulated there.

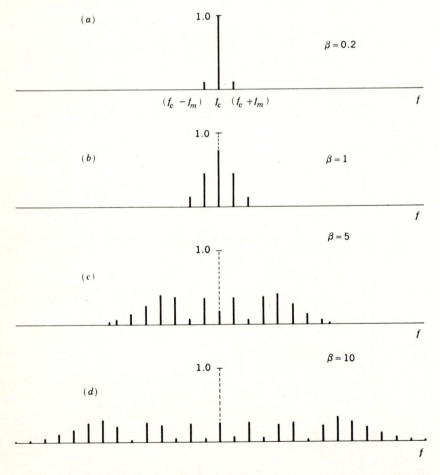

Fig. 4.7-1 The spectra of sinusoidally modulated FM signals for various values of β.

Using Eq. (4.4-3), we may put Eq. (4.7-2) in a form which is more immediately significant. We find

$$B = 2(\Delta f + f_m) \tag{4.7-3}$$

Expressed in words, *the bandwidth is twice the sum of the maximum frequency deviation and the modulating frequency.* This rule for bandwidth is called *Carson's rule.*

We deduced Eqs. (4.7-2) and (4.7-3) as a generalization from Table 4.7-1, which begins with $\beta = 1$. We may note, however, that the bandwidth approximation applies quite well even when $\beta \ll 1$. For, in that case, we find that Eq. (4.7-2) gives $B = 2f_m$, which we know to be correct from our earlier discussion of narrowband FM.

The spectra of several FM signals with sinusoidal modulation are shown in Fig. 4.7-1 for various values of β. These spectra are constructed directly from the entries in Table 4.7-1 except that the signs of the terms have been ignored. The spectral lines have, in every case, been drawn upward even when the corresponding entry is negative. Hence, the lines represent the *magnitudes* only of the spectral components. Not all spectral components have been drawn. Those, far removed from the carrier, which are too small to be drawn conveniently to scale, have been omitted.

4.8 EFFECT OF THE MODULATION INDEX β ON BANDWIDTH

The modulation index β plays a role in FM which is not unlike the role played by the parameter m in connection with AM. In the AM case, and for sinusoidal modulation, we established that to avoid distortion we must observe $m = 1$ as an upper limit. It was also apparent that when it is feasible to do so, it is advantageous to adjust m to be close to unity, that is, 100 percent modulation; by so doing, we keep the magnitude of the recovered baseband signal at a maximum. On this same basis we expect the advantage to lie with keeping β as large as possible. For, again, the larger is β, the stronger will be the recovered signal. While in AM the constraint that $m \leq 1$ is imposed by the necessity to avoid distortion, there is no similar absolute constraint on β.

There is, however, a constraint which needs to be imposed on β for a different reason. From Eq. (4.7-2) for $\beta \gg 1$ we have $B \cong 2\beta f_m$. Therefore the maximum value we may allow for β is determined by the maximum allowable bandwidth and the modulation frequency. In comparing AM with FM, we may then note, in review, that in AM the recovered modulating signal may be made progressively larger subject to the onset of distortion in a manner which keeps the occupied bandwidth constant. In FM there is no similar limit on the modulation, but increasing the magnitude of the recovered signal is achieved at the expense of bandwidth. A more complete comparison is deferred to Chaps. 8 and 9, where we shall take account of the presence of noise and also of the relative power required for transmission.

4.9 SPECTRUM OF "CONSTANT BANDWIDTH" FM

Let us consider that we are dealing with a modulating signal voltage $v_m \cos 2\pi f_m t$ with v_m the peak voltage. In a phase-modulating system the phase angle $\varphi(t)$ would be proportional to this modulating signal so that $\varphi(t) = k'v_m \cos 2\pi f_m t$, with k' a constant. The phase deviation is $\beta = k'v_m$, and, for constant v_m, the bandwidth occupied increases linearly with modulating frequency since $B \cong 2\beta f_m = 2k'v_m f_m$. We may avoid this variability of bandwidth with modulating frequency by arranging that $\varphi(t) = (k/2\pi f_m)v_m \sin 2\pi f_m t$ (k a constant). For, in this case

$$\beta = \frac{kv_m}{2\pi f_m} \qquad\qquad (4.9\text{-}1)$$

and the bandwidth is $B \cong (2k/2\pi)v_m$, independently of f_m. In this latter case, however, the instantaneous frequency is $\omega = \omega_c + kv_m \cos 2\pi f_m t$. Since the instantaneous frequency is proportional to the modulating signal, the initially angle-modulated signal has become a frequency-modulated signal. Thus a signal intended to occupy a nominally constant bandwidth is a frequency-modulated rather than an angle-modulated signal.

In Fig. 4.9-1 we have drawn the spectrum for three values of β for the condition that βf_m is kept constant. The nominal bandwidth $B \cong 2\Delta f = 2\beta f_m$ is consequently constant. The amplitude of the unmodulated carrier at f_c is shown by a dashed line. Note that the extent to which the actual bandwidth extends beyond the nominal bandwidth is greatest for small β and large f_m and is least for large β and small f_m.

In commercial FM broadcasting, the Federal Communications Commission allows a frequency deviation $\Delta f = 75$ kHz. If we assume that the highest audio frequency to be transmitted is 15 kHz, then at this frequency $\beta = \Delta f/f_m = 75/15 = 5$. For all other modulation frequencies β is larger than 5. When $\beta = 5$, there are $\beta + 1 = 6$ significant sideband pairs so that at $f_m = 15$ kHz the bandwidth required is $B = 2 \times 6 \times 15 = 180$ kHz, which is to be compared with $2\Delta f = 150$ kHz. When $\beta = 20$, there are 21 significant sideband pairs, and $B = 2 \times 21 \times 15/4 = 157.5$ kHz. In the limiting case of very large β and correspondingly very small f_m, the actual bandwidth becomes equal to the nominal bandwidth $2\Delta f$.

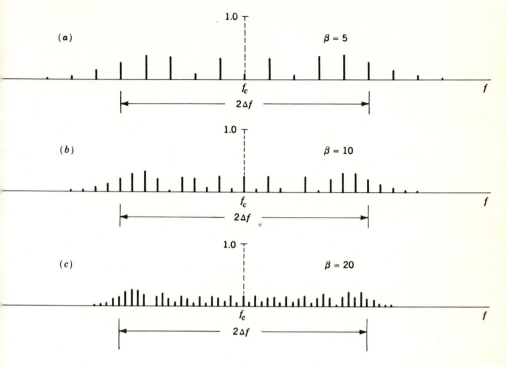

Fig. 4.9-1 Spectra of sinusoidally modulated FM signals. The nominal bandwidth $B \approx 2\beta f_m = 2\,\Delta f$ is kept fixed.

4.10 PHASOR DIAGRAM FOR FM SIGNALS

With the aid of a phasor diagram we shall be able to arrive at a rather physically intuitive understanding of how so odd an assortment of sidebands as in Eq. (4.5-7) yields an FM signal of constant amplitude. The diagram will also make clear the difference between AM and narrowband FM (NBFM). In both of these cases there is only a single pair of sideband components.

Let us consider first the case of narrowband FM. From Eqs. (4.4-1), (4.6-1), and (4.6-2) we have for $\beta \ll 1$ that

$$v(t) = \cos\left(\omega_c t + \beta \sin \omega_m t\right) \tag{4.10-1a}$$

$$\cong \cos \omega_c t - \frac{\beta}{2} \cos\left(\omega_c - \omega_m\right)t + \frac{\beta}{2} \cos\left(\omega_c + \omega_m\right)t \tag{4.10-1b}$$

Refer to Fig. 4.10-1a. Assuming a coordinate system which rotates counterclockwise at an angular velocity ω_c, the phasor for the carrier-frequency term in Eq. (4.10-1) is fixed and oriented in the horizontal direction. In the same coordinate system, the phasor for the term $(\beta/2) \cos (\omega_c + \omega_m)t$ rotates in a counterclockwise direction at an angular velocity ω_m, while the phasor for the term $-(\beta/2) \cos (\omega_c - \omega_m)t$ rotates in a clockwise direction, also at the angular velocity ω_m. At the time $t = 0$, both phasors, which represent the sideband components, have maximum projections in the horizontal direction. At this time one is parallel to, and one is antiparallel to, the phasor representing the carrier, so that the two cancel. The situation depicted in Fig. 4.10-1a corresponds to a time shortly after $t = 0$. At this time, the rotation of the sideband phasors which are in opposite directions, as indicated by the curved arrows, have given rise to a sum phasor Δ_1. In the coordinate system in which the carrier phasor is stationary, the phasor Δ_1 always stands perpendicularly to the carrier phasor and has the magnitude

$$\Delta_1 = \beta \sin \omega_m t \qquad \qquad (4.10\text{-}2)$$

The carrier, now slightly reduced in amplitude, and Δ_1 combine to give rise to a resultant R. The angular departure of R from the carrier phasor is φ. It is readily seen from Fig. 4.10-1a that since $\beta \ll 1$, the maximum value of $\varphi \simeq \tan \varphi = \beta$, as is to be expected. The small variation in the amplitude of the resultant which appears in Fig. 4.10-1a is only the result of the fact that we have neglected higher-order sidebands.

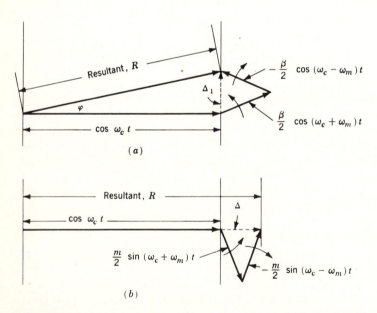

Fig. 4.10-1 (a) Phasor diagram for a narrowband FM signal. (b) Phasor diagram for an AM signal.

Now let us consider the phasor diagram for AM. The AM signal is

$$(1 + m \sin \omega_m t) \cos \omega_c t = \cos \omega_c t + \frac{m}{2} \sin (\omega_c + \omega_m)t$$

$$- \frac{m}{2} \sin (\omega_c - \omega_m)t \quad (4.10\text{-}3)$$

and the individual terms are represented as phasors in Fig. 4.10-1b. Comparing Eqs. (4.10-1) and (4.10-3), we see that there is a 90° phase shift in the phases of the sidebands between the FM and AM cases. In Fig. 4.10-1b the sum Δ of the sideband phasors is given by

$$\Delta = m \sin \omega_m t \qquad (4.10\text{-}4)$$

The important difference between the FM and AM cases is that in the former the sum Δ_1 is always perpendicular to the carrier phasor, while in the latter the sum Δ is always parallel to the carrier phasor. Hence in the AM case, the resultant R does not rotate with respect to the carrier phasor but instead varies in amplitude between $1 + m$ and $1 - m$.

Another way of looking at the difference between AM and NBFM is to note that in NBFM where $\beta \ll 1$

$$v(t) \approx \cos \omega_c t - \beta \sin \omega_m t \sin \omega_c t \qquad (4.10\text{-}5)$$

while in AM

$$v(t) = \cos \omega_c t + m \sin \omega_m t \cos \omega_c t \qquad (4.10\text{-}6)$$

Note that in NBFM the first term is $\cos \omega_c t$, while the second term involves $\sin \omega_c t$, a *quadrature* relationship. In AM both first and second terms involve $\cos \omega_c t$, an *in-phase* relationship.

To return now to the FM case and to Fig. 4.10-1a, the following point is worth noting. When the angle φ completes a full cycle, that is, Δ_1 varies from $+\beta$ to $-\beta$ and back again to $+\beta$, the magnitude of the resultant R will have executed *two* full cycles. For R is a maximum at $\Delta_1 = \beta$, a minimum at $\Delta_1 = 0$, a maximum again when $\Delta_1 = -\beta$, and so on. On this basis, it may well be expected that if an additional sideband pair is to be added to the first to make R more nearly constant, this new pair must give rise to a resultant Δ_2 which varies at the frequency $2\omega_m$. Thus, we are not surprised to find that as the phase deviation β increases, a sideband pair comes into existence at the frequencies $\omega_c \pm 2\omega_m$.

As long as we depend on the first-order sideband pair only, we see from Fig. 4.10-1a that φ cannot exceed 90°. A deviation of such magnitude is hardly adequate. For consider, as above, that $\Delta f = 75$ kHz and that $f_m = 50$ Hz. Then $\varphi_m = 75,000/50 = 1500$ rad, and the resultant R must, in this case, spin completely about $1500/2\pi$ or about 240 times. Such wild whirling is made possible through the effect of the higher-order sidebands.   As noted, the first-order sideband pair gives rise to a phasor $\Delta_1 = J_1(\beta) \sin \omega_m t$, which phasor is perpendicular to the carrier phasor. It may also be established by inspection of Eq. (4.5-7) that the second-order sideband pair gives rise to a phasor $\Delta_2 = J_2(\beta) \cos 2\omega_m t$ and that this phasor is *parallel* to the carrier phasor.   Continuing, we easily establish that all odd-numbered sideband pairs give rise to phasors

$$\Delta_n = J_n(\beta) \sin n\omega_m t \qquad n \text{ odd} \tag{4.10-7}$$

which are perpendicular to the carrier phasor, while all even-numbered sideband pairs give rise to phasors

$$\Delta_n = J_n(\beta) \cos n\omega_m t \qquad n \text{ even} \tag{4.10-8}$$

which are parallel to the carrier phasor.   Thus, phasors Δ_1, Δ_2, Δ_3, etc., alternately perpendicular and parallel to the carrier phasor, are added to carry the end point of the resultant phasor R completely around as many times as may be required, while maintaining R at constant magnitude. It is left as an exercise for the student to show by typical examples how the superposition of a carrier and sidebands may swing a constant-amplitude resultant around through an arbitrary angle (Prob. 4.10-2).

4.11 SPECTRUM OF NARROWBAND ANGLE MODULATION: ARBITRARY MODULATION

Previously we considered the spectrum, in NBFM, which is produced by sinusoidal modulation.   We found that, just as in AM, such modulation gives rise to two sidebands at frequencies $\omega_c + \omega_m$ and $\omega_c - \omega_m$.   We extend the result now to an arbitrary modulating waveform.

We may readily verify (Prob. 4.11-1) that superposition applies in narrowband angle modulation just as it does to AM.   That is, if $\beta_1 \sin \omega_1 t + \beta_2 \sin \omega_2 t$ is substituted in Eq. (4.10-1a) in place of $\beta \sin \omega_m t$, the sidebands which result are the sum of the sidebands that would be yielded by either modulation alone.   Hence even if a modulating signal of waveform $m(t)$, with a continuous distribution of spectral components, is used in either AM or narrowband angle modulation the forms of the sideband spectra will be the same in the two cases.

More formally, we have in AM, when the modulating waveform is $m(t)$, the signal is

$$v_{\text{AM}}(t) = A[1 + m(t)] \cos \omega_c t = A \cos \omega_c t + A m(t) \cos \omega_c t \tag{4.11-1}$$

Let us assume, for simplicity, that $m(t)$ is a finite energy waveform with a Fourier transform $M(j\omega)$.   We use the theorem that if the Fourier transform $\mathcal{F}[m(t)] = M(j\omega)$, then $\mathcal{F}[m(t) \cos \omega_c t]$ is as given in Eq. (3.2-4).

We then find that

$$\mathcal{F}[v_{AM}(t)] = \frac{A}{2} [\delta(\omega + \omega_c) + \delta(\omega - \omega_c)]$$

$$+ \frac{A}{2} [M(j\omega + j\omega_c) + M(j\omega - j\omega_c)] \quad (4.11\text{-}2)$$

The narrowband angle-modulation signal of Eq. (4.10-1), except of amplitude A and with phase modulation $m(t)$, may be written, for $|m(t)| \ll 1$,

$$v_{PM}(t) \cong A \cos \omega_c t - Am(t) \sin \omega_c t \quad (4.11\text{-}3)$$

so that

$$\mathcal{F}[v_{PM}(t)] = \frac{A}{2} [\delta(\omega + \omega_c) + \delta(\omega - \omega_c)]$$

$$+ \frac{A}{2} e^{-j\pi/2}[M(j\omega + j\omega_c) - M(j\omega - j\omega_c)] \quad (4.11\text{-}4)$$

Comparing Eq. (4.11-2) with Eq. (4.11-4), we observe that

$$|\mathcal{F}[v_{AM}]|^2 = |\mathcal{F}[v_{PM}]|^2 \quad (4.11\text{-}5)$$

Thus, if we were to make plots of the energy spectral densities of $v_{AM}(t)$ and of $v_{PM}(t)$, we would find them identical. Similarly, if $m(t)$ were a signal of finite power, we would find that plots of power spectral density would be the same.

4.12 SPECTRUM OF WIDEBAND FM (WBFM): ARBITRARY MODULATION[4]

In this section we engage in a heuristic discussion of the spectrum of a wideband FM signal. We shall not be able to deduce the spectrum with the precision that is possible in the NBFM case described in the previous section. As a matter of fact, we shall be able to do no more than to deduce a means of expressing approximately the power spectral density of a WBFM signal. But this result is important and useful.

Previously, to characterize an FM signal as being narrowband or wideband, we had used the parameter $\beta \equiv \Delta f/f_m$, where Δf is the frequency deviation and f_m the frequency of the sinusoidal modulating signal. The signal was then NBFM or WBFM depending on whether $\beta \ll 1$ or $\beta \gg 1$. Alternatively we distinguished one from the other on the basis of whether one or very many sidebands were produced by each spectral component of the modulating signal, and on the basis of whether or not superposition applies. We consider now still another alternative.

Let the symbol $\nu \equiv f - f_c$ represent the frequency difference between the instantaneous frequency f and the carrier frequency f_c; that is, $\nu(t) = (k/2\pi)m(t)$ [see Eq. (4.3-5)]. The period corresponding to ν is $T = 1/\nu$. As f varies, so also will ν and T. The frequency ν is the frequency with which the resultant phasor R in Fig. 4.10-1 rotates in the coordinate system in which the carrier phasor is fixed. In WBFM this resultant phasor rotates through many complete revolutions, and its speed of rotation does not change radically from revolution to revolution. Since, the resultant R is constant, then if we were to examine the plot as a function of time of the projection of R in, say, the horizontal direction, we would recognize it as a sinusoidal waveform because its frequency would be changing very slowly. No appreciable change in frequency would take place during the course of a cycle. Even a long succession of cycles would give the appearance of being of rather constant frequency. In NBFM, on the other hand, the phasor R simply oscillates about the position of the carrier phasor. Even though, in this case, we may still formally calculate a frequency ν, there is no corresponding time interval during which the phasor makes complete revolutions at approximately a constant rate ν.

Now let us consider that a carrier is wideband FM-modulated by a signal $m(t)$ such as, say, an audio signal. Let the modulation $m(t)$ be characterized by the probability density function $f(m)$. Then the fraction of the time that $m(t)$ spends in the range between m_1 and $m_1 + dm$ is the probability that $m(t)$ lies between m_1 and $m_1 + dm$, that is, $f(m_1)\,dm$. Corresponding to each value of $m(t)$, the value of the frequency deviation $\nu(t) = (k/2\pi)m(t)$. Hence, during the time $m(t)$ is in the range m_1 and $m_1 + dm$, ν is in the range ν_1 and $\nu_1 + d\nu$. As we have seen in WBFM, the frequency ν changes only relatively slowly. Thus the assignment of a frequency ν to a waveform during the interval when $m(t)$ has a value corresponding to ν has a physical as well as a purely mathematical significance. On this basis, it is reasonable to say that, of the total power in the FM waveform, the fraction of the power in the frequency range between ν_1 and $\nu_1 + d\nu$ is proportional to the time $m(t)$ spends in the range m_1 to $m_1 + dm$. With $G(\nu)$, the power spectral density of the FM waveform, we have the result that $G(\nu_1)\,d\nu$ is proportional to $f(m_1)\,dm$. Finally, since $d\nu$ is proportional to dm, we have the most important result that $G(\nu)$ is proportional to $f(m)$. Expressed in words, the *power spectral density $G(\nu)$ of a WBFM waveform is determined by, and has the same form as, the density function $f(m)$ of the modulating waveform.*

Example 4.12-1   In the WBFM signal

$$v(t) = A \cos\left[2\pi f_c t + k \int_{-\infty}^{t} m(\lambda)\,d\lambda\right] \tag{4.12-1}$$

$m(t)$ is an ergodic random process having a probability density

$$f(m) = \begin{cases} \dfrac{1}{M} & -\dfrac{M}{2} \le m \le \dfrac{M}{2} \\ 0 & \text{elsewhere} \end{cases} \tag{4.12-2}$$

Obtain an expression for $G(f)$, the power spectral density of $v(t)$.

Solution Since $G(\nu)$ (with $\nu \equiv f - f_c$) is proportional to $f(m)$, we have

$$G(\nu) = \alpha f(m) \tag{4.12-3}$$

where α is a constant of proportionality. Since $\nu(t) = km(t)/2\pi$,

$$G(\nu) = \begin{cases} \dfrac{\alpha}{M} & -\dfrac{kM}{4\pi} \le \nu \le \dfrac{kM}{4\pi} \\ 0 & \text{elsewhere} \end{cases} \tag{4.12-4}$$

Replacing ν by $f - f_c$, and expressing the power spectral density for both positive and negative frequencies (i.e., a two-sided density), we have

$$G(f) = \begin{cases} \dfrac{\alpha}{2M} & f_c - \dfrac{kM}{4\pi} \le |f| \le f_c + \dfrac{kM}{4\pi} \\ 0 & \text{elsewhere} \end{cases} \tag{4.12-5}$$

To evaluate α, we note that the power of the FM waveform is $A^2/2$. Hence,

$$\int_{-\infty}^{\infty} G(f)\, df = \frac{A^2}{2} \tag{4.12-6}$$

From Eqs. (4.12-5) and (4.12-6) we find $\alpha = \pi A^2/k$, so that

$$G(f) = \begin{cases} \dfrac{\pi A^2}{2kM} & f_c - \dfrac{kM}{4\pi} \le |f| \le f_c + \dfrac{kM}{4\pi} \\ 0 & \text{elsewhere} \end{cases} \tag{4.12-7}$$

4.13 BANDWIDTH REQUIRED FOR A GAUSSIAN MODULATED WBFM SIGNAL

Earlier, we found that when a carrier was sinusoidally modulated, Carson's rule $B = 2(\Delta f + f_m)$ given in Eq. (4.7-3) specifies the bandwidth required to transmit enough of the power (98 percent) so that the modulation may be recovered without distortion. We now make a similar calculation for the case where the modulation has a gaussian distribution.

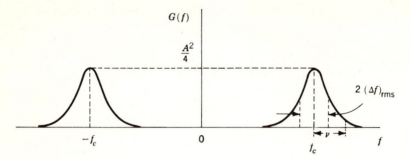

Fig. 4.13-1 The power spectral density of a carrier f_c frequency-modulated by a baseband signal with a gaussian amplitude distribution.   The variable ν measures the departure of frequency from the carrier frequency.

The result is extremely important since many physically encountered signals, while not precisely gaussian, are reasonably approximated as gaussian.

Using the result stated at the end of Sec. 4.12, we note that if $m(t)$ is gaussian, so also is $G(f)$.   Therefore the two-sided spectral density $G(f)$ has the form

$$G(f) = \frac{A^2}{4\sqrt{2\pi}\,\Delta f_{\rm rms}}\left[e^{-(f-f_c)^2/2(\Delta f_{\rm rms})^2} + e^{-(f+f_c)^2/2(\Delta f_{\rm rms})^2}\right] \qquad (4.13\text{-}1)$$

as shown in Fig. 4.13-1 where A is the amplitude of the FM waveform and $\Delta f_{\rm rms}$ is the variance of the gaussian power spectrum density.   An FM waveform of amplitude A has a power $A^2/2$, and the student may verify that, with $G(f)$ as given in Eq. (4.13-1),

$$\int_{-\infty}^{\infty} G(f)\,df = \frac{A^2}{2} \qquad (4.13\text{-}2)$$

as required.

We now ask what must be the bandwidth B of a rectangular bandpass filter centered at f_c which will pass 98 percent of the power of the FM waveform.   Recognizing that each of the terms in Eq. (4.13-1) makes equal contributions to the power and using the variable $\nu \equiv f \pm f_c$, we find that B is determined by the equation

$$\frac{1}{\sqrt{2\pi}\,\Delta f_{\rm rms}}\int_{-B/2}^{B/2} e^{-\nu^2/2(\Delta f_{\rm rms})^2}\,d\nu = 0.98 \qquad (4.13\text{-}3)$$

Letting $x \equiv \nu/(\sqrt{2}\,\Delta f_{\rm rms})$ yields

$$0.98 = \frac{2}{\sqrt{\pi}}\int_{0}^{B/(2\sqrt{2}\,\Delta f_{\rm rms})} e^{-x^2}\,dx = \operatorname{erf}\frac{B}{2\sqrt{2}\,\Delta f_{\rm rms}} \qquad (4.13\text{-}4)$$

From a table of values of the error functions we find

$$B = 2 \sqrt{2} \ (1.645)\Delta f_{rms}$$
$$= 4.6 \ \Delta f_{rms} \tag{4.13-5}$$

To recapitulate, we find that a modulating signal with a gaussian amplitude distribution gives rise to an FM waveform with a gaussian power spectral density. If the variance of the spectral density is $(\Delta f_{rms})^2$, the bandwidth required to pass 98 percent of the power of the waveform is given by Eq. (4.13-5).

There is another useful result that may be deduced for the case of a gaussian modulating signal. Suppose we have two gaussian modulating signals $m_1(t)$ and $m_2(t)$ which are related in that $\overline{m_1^2(t)} = \overline{m_2^2(t)}$ but are otherwise arbitrary. The probability density functions of these two signals are identical and therefore, by the result given at the end of Sec. 4.12, will give rise to WBFM waveforms with the same gaussian power spectral density distribution and hence bandwidth B.

4.14 ADDITIONAL COMMENTS CONCERNING BANDWIDTH IN WBFM

When an FM carrier is modulated simultaneously by a substantial number of discrete frequency spectral components, the determination of the spectral pattern that results is a very formidable task. For this reason, we find in the literature at least two *rule-of-thumb* estimates of bandwidth which are widely used. Suppose that the individual modulating signals acting alone would produce frequency deviations $(\Delta f)_1$, $(\Delta f)_2$, etc. Then one, rather pessimistic, rule says $B = 2[(\Delta f)_1 + (\Delta f)_2 + \cdots]$, while a second, rather optimistic, rule says $B = 2[\text{rms value of } (\Delta f)\text{'s}]$. As is readily verified (Prob. 4.14-1), these two rules may give widely different results.

When the modulating signal has a continuous spectral density and gives rise to an FM signal with sidebands which have a continuous spectral density, a bandwidth definition often used is defined by

$$B \equiv 2 \left[\frac{\int_{-\infty}^{\infty} \nu^2 G(\nu) \, d\nu}{\int_{-\infty}^{\infty} G(\nu) \, d\nu} \right]^{\frac{1}{2}} \tag{4.14-1}$$

This definition yields an *rms bandwidth* which is twice the *radius of gyration* of the area under the power spectral density plot. This definition often finds favor with mathematicians because of its computational ease.

A comparison of B as given by Eq. (4.14-1) with the bandwidth given by other definitions is explored in Probs. 4.14-1 and 4.14-2.

4.15 FM GENERATION: PARAMETER-VARIATION METHOD

The generator which produces the carrier of an FM waveform is, in many instances, a tuned circuit oscillator. Such oscillator circuits furnish a sinusoidal waveform whose frequency is very largely determined by, and is very nearly equal to, the resonant frequency of an inductance-capacitance combination. Thus the frequency of oscillation is $f = (2\pi \sqrt{LC})^{-1}$, in which L is the inductance and C the capacitance. Such an LC combination, a parallel combination in this case, is shown in Fig. 4.15-1. The capacitor consists here of a fixed capacitor C_0, which is shunted by a voltage-variable capacitor C_v. A voltage-variable capacitor, commonly called a *varicap*, is one whose capacitance value depends on the dc biasing voltage maintained across its electrodes. Semiconductor diodes, when operated with a reverse bias, have characteristics suitable to permit their use as voltage-variable capacitors.

In the circuit of Fig. 4.15-1 the modulating signal varies the voltage across C_v. As a consequence, the capacitance of C_v changes and causes a corresponding change in the oscillator frequency. Ordinarily the modulating frequency is very small in comparison with the oscillator frequency. Therefore the fractional change in C_v may be very small during the course of many cycles of the oscillator signal. We may consequently expect that even with this variable capacitance, the instantaneous oscillator frequency will be given by $f = (2\pi \sqrt{LC})^{-1}$. Then we have the result that the system suggested in Fig. 4.15-1 will generate an oscillator output signal whose instantaneous frequency depends on the instantaneous value of the modulating signal. Any oscillator whose frequency is controlled by the modulating-signal voltage is called a *voltage-controlled oscillator*, or VCO.

Frequency modulation may be achieved by the variation of any element or parameter on which the frequency depends. If the frequency

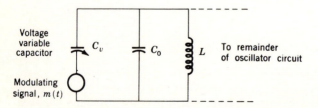

Fig. 4.15-1 A voltage-variable capacitor is used to frequency-modulate an LC oscillator.

variation is to occur in response to a modulating signal $m(t)$, then a component must be available, capacitor, resistor, or inductor, whose value can be varied with an electrical signal. We have noted that reversed-biased junctions may serve as voltage-variable capacitors. Similarly PIN diodes and FETs (field-effect transistors) have found application as variable resistors. The inductance of a magnetic-cored inductor, called a saturable reactor, can be varied by changing the dc biasing current through the winding, and thereby changing the core permeability.

There are occasions when it is appropriate to use a signal source such as a multivibrator as the carrier generator. In this case, of course, the waveform generated will not be sinusoidal. There are situations (more frequently in laboratory equipment than in a communications channel) where a sinusoidal form is not required, or, if a sinusoidal waveform is needed, it is often possible to convert the waveform to such a sinusoidal waveform by filtering, or by the use of nonlinear shaping circuits. In such cases, where the prime timing generator is a multivibrator-type circuit, there is available an additional mechanism for frequency modulation. For, in such circuits, the frequency depends not only on the values of the passive components but also on the supply voltages which are used to bias the active devices. Thus frequency modulation may be achieved by using the modulating signal to control these biasing voltages.

A principal difficulty with the parameter-variation method of frequency modulation is the difficulty it entails when we require that the carrier frequency [the signal frequency when the modulation signal $m(t) = 0$] be maintained constant to a high order of precision over extended periods of time. There is a certain measure of inconsistency in requiring that a device have *long-time* frequency stability and yet be able to respond readily to a modulating signal. We turn our attention in the next section to a system of frequency modulation in which the carrier generator is *not* required to respond to a modulating signal. The carrier generator is isolated from the remainder of the circuitry and may be designed without the need to make compromises with its frequency stability. Thus, we find that, when the frequency range is appropriate, the carrier generator is invariably a crystal-controlled oscillator.

4.16 AN INDIRECT METHOD OF FREQUENCY MODULATION (ARMSTRONG SYSTEM)

A phase-modulated waveform in which the modulating waveform is $m(t)$ is written $\cos [\omega_c t + m(t)]$. If the modulation is narrowband [$|m(t)| \ll 1$], then we may use the approximation

$$\cos [\omega_c t + m(t)] \cong \cos \omega_c t - m(t) \sin \omega_c t \qquad (4.16\text{-}1)$$

The term $m(t) \sin \omega_c t$ is a DSB-SC waveform in which $m(t)$ is the modulating waveform and $\sin \omega_c t$ the carrier. We note that the carrier of the FM waveform, that is, $\cos \omega_c t$, and the carrier of the DSB-SC waveform are in quadrature. We may note in passing that if the two carriers are in phase, the result is an AM signal since

$$\cos \omega_c t + m(t) \cos \omega_c t = [1 + m(t)] \cos \omega_c t \qquad (4.16\text{-}2)$$

A technique used in commercial FM systems to generate NBFM, which is based on our observation in connection with Eq. (4.16-1), is shown in Fig. 4.16-1. Here a balanced modulator is employed to generate the DSB-SC signal using $\sin \omega_c t$ as the carrier of the modulator. This carrier is then shifted in phase by 90° and, when added to the balanced modulator output, thereby forms an NBFM signal. However, the signal so generated will be phase-modulated rather than frequency-modulated. If we desire that the frequency rather than the phase be proportional to the modulation $m(t)$, then, as discussed in Sec. 4.3 and illustrated in Fig. 4.3-1, we need merely integrate the modulating signal before application to the modulator.

If the system of Fig. 4.16-1 is to yield an output signal whose phase deviation is directly proportional to the amplitude of the modulating signal, then the phase deviation must be kept small. That such is the case is readily to be seen in Fig. 4.10-1a. If we neglect the small second-order correction in the carrier amplitude and assume it to be of unit magnitude, we have $\tan \varphi = \Delta_1$. Since, however, $\Delta_1 \ (= \beta \sin \omega_m t)$ is proportional to the modulating signal, we actually require that $\varphi = \Delta_1$. In order that we may replace $\tan \varphi$ by φ, we require that at all times $\varphi \ll 1$. In this case $\beta \ll 1$, and then $\varphi = \beta \sin \omega_m t$.

The restriction that $\beta \ll 1$ imposes a similar constraint on the allowable frequency deviation $\Delta f \ (= \beta \omega_m / 2\pi)$ when the system of Fig. 4.16-1 is adapted for use as a frequency-modulation system by the addition of an

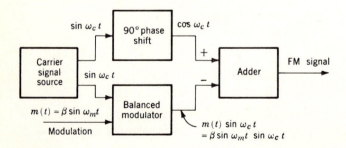

Fig. 4.16-1 Illustrating the principle of the Armstrong system of generating a PM signal.

integrator.   In the next section we discuss how the frequency deviation,
and the phase deviation as well, of a narrowband signal may be increased
by the process of frequency multiplication.

4.17   FREQUENCY MULTIPLICATION

A *frequency multiplier* is a combination of a nonlinear element and a bandpass filter.   One such possible combination is shown in Fig. 4.17-1.   We
consider the operation, qualitatively, in order to see the relevance of the
process to our present interest of increasing the frequency deviation of an
FM signal.

    Assume that the input signal to the transistor in the circuit of Fig.
4.17-1 is a periodic signal, possibly sinusoidal but not necessarily so.   The
amplitude of the input signal is large enough, and the biasing (not shown)
is such that the transistor operates nonlinearly.   Typically, the transistor
operates in the *class C* mode.   In this mode of operation the transistor
is in the cutoff region for more than half of the period of the input signal.
During intervals in the neighborhood of the peak positive excursions of
the input signal, the transistor is driven into the active region, possibly
even into saturation.   Collector current flows, not continuously, but
rather in spurts, forming pulses, one pulse for each cycle of the input
driving signal.   The collector-current waveform has the same fundamental period as has the driving signal but is rich in higher-frequency
harmonics.   The LC parallel resonant circuit is tuned to resonance at the
nth harmonic of the frequency f of the input signal.   The sharpness of
the resonance is such that the impedance presented by the resonant circuit
is very small at all harmonic frequencies except the nth.   All components
of collector current except the component at frequency nf pass through
the resonant circuit without developing appreciable voltage.   However,
in response to this nth harmonic current component, there appears across
the resonant circuit a very nearly sinusoidal voltage waveform of frequency nf.   The resonant circuit serves as a bandpass filter to selectively
single out the nth harmonic of the driving waveform.   The process of
frequency multiplication performed by the multiplier under consideration

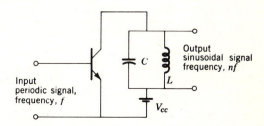

Fig. 4.17-1   A frequency-multi-
plier circuit.

is one in which a periodic signal of frequency f serves to generate a second periodic signal of frequency nf, with n an integer.

In principle, we may multiply by an arbitrary integral number n by simply tuning the resonant circuit to nf. In practice, of course, most periodic waveforms encountered in engineering applications, such as the collector-current pulse waveform of the transistor of Fig. 4.17-1, are characterized by a progressive decrease of amplitude of harmonic with increasing harmonic number. As a result, we find that as the order of multiplication increases, the output signal becomes progressively smaller. Circuits such as in Fig. 4.17-1 are commonly used for multiplication by factors from about 2 to 5. Where higher orders of multiplication are required, multipliers may be cascaded. Cascaded multipliers of order n_1, n_2, n_3, . . . yield an overall multiplication of order $n_1 n_2 n_3$ . . . .

4.18 FREQUENCY MULTIPLICATION APPLIED TO FM SIGNALS

Now let us consider the result of the application to a frequency multiplier of an FM signal. If we think of an FM signal as a "sinusoidal" signal in which the frequency changes from moment to moment, then at each instant we may expect that the output frequency will be n times the input frequency. Hence, if the input consists of a carrier of frequency f_c which ranges through a frequency deviation $\pm \Delta f$, the output will have a carrier frequency nf_c and will range through the deviation $\pm n \, \Delta f$. The multiplier multiplies both the carrier and the deviation frequency. Also, since the modulation index is proportional to the frequency deviation, for a fixed modulation frequency, the multiplier increases the modulation index by the same factor n.

By way of example, consider the case of commercial FM broadcasting in the United States. Here, the allowable frequency deviation is 75 kHz, so that for a modulation frequency $f_m = 50$ Hz, $\beta = \Delta f / f_m = 1500$. Even if we allow φ in Fig. 4.10-1a to attain a maximum value as large as $\varphi = 0.5$, then the multiplication needed is $1500/0.5 = 3000$. On the other hand, at the high-frequency end of the baseband spectrum, say, $f_m = 15$ kHz, $\beta = 75/15 = 5$. Correspondingly, with a multiplication by a factor of 3000, the phase φ in Fig. 4.10-1a need attain only a maximum value $5/3000 = 1.7 \times 10^{-3}$. Thus, it is to be seen that the required multiplication is determined by the low-frequency limit of the baseband spectrum.

4.19 AN EXAMPLE OF AN ARMSTRONG FM SYSTEM

The block diagram of Fig. 4.19-1 represents an Armstrong FM system which supplies a signal whose carrier is at 96 MHz (which is near the

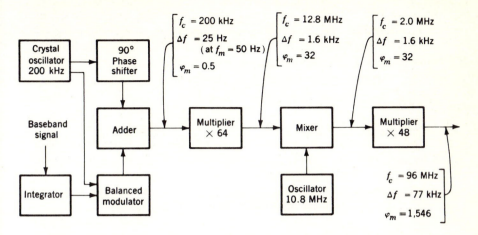

Fig. 4.19-1 Block diagram of an Armstrong system of generating an FM signal using multipliers to increase the frequency deviation.

center of the commercial FM broadcasting band). It allows direct *phase modulation* of the carrier, before multiplication, to the extent of $\varphi_m = 0.5$. Thus, at $f_m = 50$ Hz, we have $\Delta f = 25$ Hz. Note that at higher modulating frequencies, φ_m is less than 0.5 rad. The carrier frequency before multiplication has been selected at 200 kHz, a frequency at which very stable crystal oscillators and balanced modulators are readily constructed.

As already noted, if we require that $\Delta f \approx 75$ kHz, then a multiplication by a factor of 3000 is required. In Fig. 4.19-1 the multiplication is actually 3072($= 64 \times 48$). The values were selected so that the multiplication may be done by factors of 2 and 3, that is, $64 = 2^6$, $48 = 3 \times 2^4$. Direct multiplication would yield a signal of carrier frequency

$$200 \text{ kHz} \times 3072 = 614.4 \text{ MHz}$$

This signal might then be heterodyned with a signal of frequency, say, $614.4 - 96.0 = 518.4$ MHz. The difference signal output of such a mixer would be a signal of carrier frequency 96 MHz. Note particularly that a mixer, since it yields sum and difference frequencies, will translate the frequency spectrum of an FM signal but will have no effect on its frequency deviation. In the system of Fig. 4.19-1, in order to avoid the inconvenience of heterodyning at a frequency in the range of hundreds of megahertz, the frequency translation has been accomplished at a point in the chain of multipliers where the frequency is only in the neighborhood of approximately 10 MHz.

A feature, not indicated in Fig. 4.19-1 but which may be incorporated, is to derive the 10.8 MHz mixing signal not from a separate oscillator but rather through multipliers from the 0.2 MHz crystal oscillator. The multiplication required is $10.8/0.2 = 54 = 2 \times 3^3$. Such a derivation of the 10.8-MHz signal will suppress the effect of any drift in the frequency of this signal (see Prob. 4.19-3).

4.20 FM DEMODULATORS

A means of recovering the modulating signal from a frequency-modulated carrier is shown in Fig. 4.20-1. The FM signal is applied to a frequency-selective network, that is, a network whose transfer function $|H(j\omega)| = |V_o/V_i|$ is frequency-dependent. The number of such networks is, of course, limitless. A simple example of such a network is shown in Fig. 4.20-1b together with a plot of the magnitude of its transfer function (see Fig. 4.20-1c). The output signal $v_o(t)$ will be modulated not only in frequency but also in amplitude. This output signal is now applied to a diode AM demodulator as shown in the figure. The diode demodulator will yield an output which follows the amplitude modulation and is independent of the frequency modulation.

Suppose that the time constant of the network shown in Fig. 4.20-1 is selected so that at the carrier frequency the plot of $|H(j\omega)|$ has a slope $d|H(j\omega)|/df = \sigma$. Then a change Δf in the instantaneous frequency of the input will produce an amplitude change in the carrier of amount $\sigma \, \Delta f$. The diode demodulator will yield an output-voltage change of this same magnitude. If the total excursion of the instantaneous frequency encompasses a limited range, the slope of the plot of $|H(j\omega)|$ is reasonably constant. In this case a frequency-modulated signal which is given by $A \cos [\omega_c t + k\!\int\! m(\lambda) \, d\lambda]$ will yield an output $\sigma k A m(t)$.

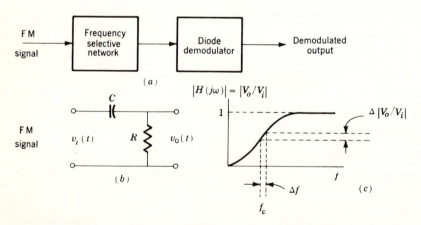

Fig. 4.20-1   Illustrating the principle of one type of FM demodulator.

Practical demodulators differ only in detail from the simple circuitry of Fig. 4.20-1 in the following respect. The simple network in Fig. 4.20-1*b* is replaced by a more elaborate circuit which provides a slope $d|H(j\omega)|/df$ which is larger in magnitude and more nearly constant over a more extended frequency range. The increased slope increases the voltage at the demodulator output, and the extended constancy of slope allows linear demodulation over an extended frequency range. In the interest of achieving linearity it is usually helpful to arrange a balanced system in which the demodulator, independent of carrier amplitude, yields zero output when the instantaneous input frequency is f_c, the carrier frequency.

Frequency demodulators of the above type are called FM *discriminators*. Other types of demodulators are the *phase-locked loop* and the *frequency demodulator using feedback*, which are discussed in Sec. 10.13.

A MODEL FOR THE DISCRIMINATOR—A DIFFERENTIATOR

Essential to the operation of frequency demodulation is the transfer of the signal through a network which has a transfer-function magnitude $|H(j\omega)|$ which varies linearly with frequency over some range. There are many networks which will provide such linear variation. One such network is a *differentiator*, i.e., a network whose output signal is proportional to the time derivative of the input signal. For if we resolve a signal into its spectral components, discrete or continuous, we may view the operation of differentiation as one in which each spectral component is multiplied by $j\omega$. Therefore an ideal differentiator has a transfer function such that $H(j\omega) = j\omega K$, where K is a constant. In such a case $|H(j\omega)| = K\omega$ as required for frequency discrimination. Hence, while a differentiator is by no means unique in its ability to serve as a discriminator, it is often a matter of great convenience to represent a discriminator, in both system diagrams and in analytical representations, as just such a differentiator. By way of example, in the network of Fig. 4.20-1*b* we have

$$H(j\omega) = \frac{j\omega RC}{1 + j\omega RC} \approx j\omega RC \qquad (4.20\text{-}1)$$

The approximation, which makes the network look like a differentiator, holds when $\omega RC \ll 1$. As shown in Fig. 4.20-1*c*, this is precisely the region in which we used the RC network to act as a discriminator.

4.21 LIMITERS

We noted, in the discussion of discriminators in the preceding section, that since the discriminator *differentiates* the incoming signal, it responds not only to a change in the instantaneous frequency of the input signal but also to its amplitude changes.   Such response to amplitude is disadvantageous.   It is, therefore, very common practice, before applying the signal to a frequency discriminator, to pass the signal through an *amplitude limiter* and thereby to remove the amplitude variation.   A commonly employed limiter consists of a transistor-amplifier stage operating into a tuned resonant collector circuit in the fashion shown in Fig. 4.17-1.   In the present instance, however, the collector circuit is tuned to the input frequency rather than to some harmonic.   At low levels of input-signal amplitude, the amplifier responds to input-amplitude changes.   With increasing input amplitude, however, a point is reached where, because of the transistor nonlinearities, the amplifier output no longer responds to changes in input amplitude.   In this range the amplifier operates as a *limiter* to generate a signal which preserves the frequency variation of the applied signal but not the amplitude variation.   Of course, in the process, because the transistor operates nonlinearly, it introduces harmonics of the input-signal frequency.   These harmonics are eliminated by the resonant collector circuit.   If the transistor was preceded by two diodes in parallel, each conducting in opposite directions, the *limiter* becomes a *hard* limiter. This means that even small amplitude variations are removed.

4.22 APPROXIMATELY COMPATIBLE SSB SYSTEMS

SSB-AM

We noted earlier in Sec. 3.12 the advantages that would accrue from the availability of compatible single-sideband systems.   While precisely compatible systems are presently impractical, approximately compatible systems are feasible, such systems are in use, and commercial equipment for such systems is available.   In a strictly compatible AM system a waveform would be generated whose envelope exactly reproduces the baseband signal.   Additionally if the highest baseband frequency component is f_M, the frequency range encompassed by the compatible signal would extend from f_c, the carrier frequency, to, say, $f_c + f_M$.   An approximately compatible system is one in which there is some relaxation of these specifications concerning envelope shape or spectral range.   On the basis of our discussion of FM-type waveforms we are now able briefly and qualitatively to discuss the principle of operation of one type of approximately compatible AM system.

We saw in Sec. 3.11 that if we suppressed one sideband component of an amplitude-modulated waveform, the envelope of the resultant waveform would still have the form of the modulating signal, provided the percentage modulation was kept small.   Consider now the phasor diagram of the AM signal shown in Fig. 4.10-1b. From this phasor diagram it is apparent that when one of the sidebands is suppressed, the resultant is a waveform which is modulated in *both amplitude and phase*. The amplitude will vary between $1 + m/2$ and $1 - m/2$. The resultant R will no longer always be parallel to the carrier phasor. Instead it will rotate clockwise by an angle φ such that $\tan \varphi = m/2$ and rotate counterclockwise by a similar angle.

Thus we find that a carrier of fixed frequency (or phase), when amplitude-modulated, gives rise to two sidebands. But an angle-modulated carrier, when amplitude-modulated, may give rise to a single sideband. Suppose then that we arrange to amplitude-modulate a carrier in such manner that its envelope faithfully reproduces the modulating signal. Is it then possible to also angle-modulate that carrier so that a single sideband results? It turns out that, to a good approximation, the answer is yes!

In the system described in Ref. 5 the modulating signal modulates not only the amplitude but the carrier phase as well. The relationship between phase and modulating signal is *nonlinear* and has been determined, at least in part experimentally, on the basis of system performance. An analysis of the system is very involved because of the two nonlinearities involved: the inherent nonlinearity of FM and the additional nonlinearity between phase and modulating signal. The system is not strictly single sideband in the sense that a modulating tone gives rise not to a single side tone but to a spectrum of side tones. However, all the side tones are on the *same side* of the carrier. The predominant side tone is separated from the carrier by the tone frequency, and the others are separated by multiples of the modulating-tone frequency.

The system is able to operate, in effect, as a single-sideband system because of the characteristics of speech or music for which it is intended, and because the side tones, other than the predominant one, fall off sharply in power content with increasing harmonic number. Most of the power in sound is in the lower-frequency ranges. High-power low-frequency spectral components in sound may give rise to numerous harmonic side tones, but because of the low frequency of the fundamental, the harmonics will still fall in the audio spectrum. On the other hand, high-frequency tones, which may give rise to harmonic side tones that may fall outside the allowed spectral range, are of very small energy. The overall result is that there may well be spectral components that fall outside the spectral range allowable in a spectrum-conserving single-sideband system. However, it has been shown experimentally that such components are small enough to cause no interference with the signal in an adjacent single-sideband channel.

SSB-FM

A compatible SSB-FM signal has frequency components extending either above or below the carrier frequency. In addition it can be demodulated using a standard limiter-discriminator. The compatible FM signal is constructed by adding amplitude modulation to the frequency-modulated signal. Since the limiter removes any and all amplitude modulation, the addition of the AM does not affect the recovery of the modulation by the discriminator. By properly adjusting the amplitude modulation, either the upper or lower sideband can be removed.

4.23 STEREOPHONIC FM BROADCASTING

In *monophonic* broadcasting of sound, a single audio baseband signal is transmitted from broadcasting studio to a home receiver. At the receiver, the audio signal is applied to a loudspeaker which then reproduces the original sound. The original source of the baseband waveform is a single microphone at the broadcasting studio. (If more than one microphone is used, as, for example, where a large orchestra is involved, the outputs of the individual microphones are combined to generate a single audio baseband signal.) In stereophonic broadcasting, two microphones (or two groups of microphones) are used. Two audio baseband signals are transmitted to the receiver where they are applied to two individual loudspeakers. At the broadcast studio, the microphones are located some distance apart from one another, and at the receiver the two loudspeakers are also physically separated. The advantage of such stereophonic broadcasting is that it yields at the receiver a more "natural" sound. The sound heard at the receiver is more nearly what the listener would hear if he were located at the broadcasting studio itself, where his two ears would receive somewhat different sounds.

The earliest commercial FM broadcasts were monophonic and conformed in transmission characteristics to standards established by the Federal Communications Commission. These standards required that the carrier frequencies of stations occupying adjacent frequency channels be separated by 200 kHz. To give reasonable assurance that there would be no interference between adjacent stations, the maximum allowable instantaneous-frequency deviation was limited by the FCC to 75 kHz. With sinusoidal modulation of the carrier frequency at the modulating frequency f_m and a frequency deviation Δf, the bandwidth requirement is $B = 2(\Delta f_m + f_m)$. If we assume a maximum audio frequency $f_m = 20$ kHz, and even if we imagine the extreme case that all the allowable baseband power is located at that frequency, then we find $B = 2(75 + 20) = 190$ kHz.

By the time commercial stereo broadcasting began to be contemplated, monophonic broadcasting was well established, and there were already many millions of monophonic FM receivers in use. Accordingly the FCC ruled that no proposed stereo scheme would be acceptable unless it were entirely *compatible* in the sense that a standard FM receiver, without modification, would be able to receive a monophonic version of a stereo transmission. Additionally the FCC required that the bandwidth occupied by a stereo transmission be no greater than the bandwidth already allocated for monophonic transmission. Many possible stereo systems were considered. We discuss now the system finally adopted in 1961 and presently in use.

TRANSMITTED SIGNAL

At the broadcast studio two microphones or microphone groups generate a left-hand audio signal $L(t)$ and a right-hand signal $R(t)$, as indicated in Fig. 4.23-1a. These signals are added and subtracted to generate $L(t) + R(t)$ and $L(t) - R(t)$. These sum and difference signals are each bandlimited to 15 kHz by filters not explicitly indicated in Fig. 4.23-1a. An oscillator makes available a sinusoidal waveform, referred to as a *pilot* carrier at a frequency $f_p = 19$ kHz. The pilot carrier is applied to a frequency doubler which generates a sinusoidal *subcarrier* at the frequency $f_{sc} = 2 \times f_p = 38$ kHz. The subcarrier and the difference signal are applied to a balanced modulator, the output of which is $[L(t) - R(t)] \cos 2\pi f_{sc} t$. By combining the modulator output, the sum signal, and the oscillator output, a composite signal $M(t)$ is formed, where

$$M(t) = [L(t) + R(t)] + [L(t) - R(t)] \cos 2\pi f_{sc} t \\ + K \cos 2\pi f_p t \quad (4.23\text{-}1)$$

Here K is a constant which determines the level of the pilot carrier in comparison with the other components of the composite signal.

The power spectral density of a typical composite signal $M(t)$ is shown in Fig. 4.23-2. The sum signal $L(t) + R(t)$ occupies the frequency range between 0 and 15 kHz. The balanced modulator output, which is the DSB-SC signal $[L(t) - R(t)] \cos 2\pi f_{sc} t$, has a lower sideband which extends from 23 (i.e., 38 − 15) to 38 kHz and an upper sideband which extends from 38 to 53 (i.e., 38 + 15). Note that there is no subcarrier at 38 kHz. The pilot carrier at 19 kHz is present, as shown in the figure. This composite-signal $M(t)$ frequency modulates a carrier, and this modulated carrier is delivered to a transmitting antenna.

OPERATION OF RECEIVER

At a stereo receiver, the composite signal $M(t)$ is recovered from the frequency-modulated carrier. One way to sort $M(t)$ into its components

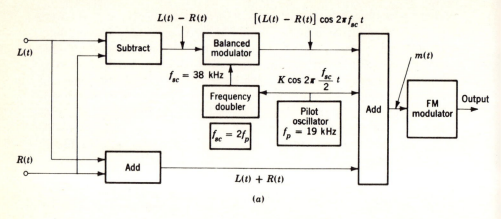

(a)

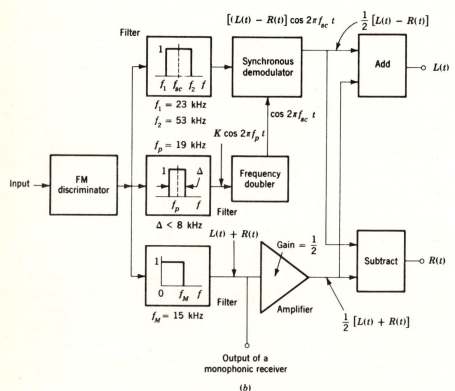

(b)

Fig. 4.23-1 Stereophonic broadcasting system. (a) Transmitter. (b) Receiver.

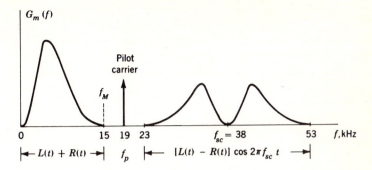

Fig. 4.23-2 Spectral density of a typical composite stereo baseband signal.

is shown in Fig. 4.23-1b. Here the individual components of the composite signal are separated by filters. The pilot carrier, applied to a frequency doubler, regenerates the subcarrier. The availability of this subcarrier now permits synchronous demodulation of the double-sideband suppressed-carrier waveform. The output of the synchronous demodulator is proportional to the difference waveform $L(t) - R(t)$, while the output of the baseband filter is proportional to $L(t) + R(t)$.

We have seen that the transmission of the pilot carrier allows us to regenerate, at the receiver, the required subcarrier waveform. We may see from Fig. 4.23-2 the reason on account of which the 38-kHz subcarrier was not itself transmitted directly. Such a subcarrier is not separated by any appreciable frequency interval from the spectral components of its accompanying sidebands. Hence, to extract such a subcarrier would require a very narrow and sharply tuned filter. On the other hand, the pilot carrier occupies an isolated place in the spectrum, there being no other spectral components present over a range of 4 kHz on either side.

Now having available the sum signal $L(t) + R(t)$, and the difference signal $L(t) - R(t)$, then, as indicated in Fig. 4.23-1b, the individual signals $L(t)$ and $R(t)$ respectively, are recovered by adding and subtracting.

The system is entirely compatible with the requirements of a monophonic receiver. In such a receiver, the sum signal $L(t) + R(t)$ passes through the baseband filter while the pilot carrier and the suppressed-carrier signal do not. Hence these latter two signals contribute nothing to the output of a monophonic receiver and neither do they interfere with the receiver's operation.

INTERLEAVING

A monophonic receiver makes use only of the $L + R$ signal in the stereo transmission. In order that the monophonic receiver output be as loud and disturbance-free as possible, it is necessary that the sum signal $V_s = L + R$, which modulates the FM carrier, be as large as possible. If the only component of the modulation waveform were the sum signal V_s, we would be at liberty to increase the peak excursion of V_s to the point where the corresponding peak instantaneous-frequency deviation of the carrier would be ± 75 kHz. However, in order to accommodate to the requirements of the stereo receiver, the composite modulating wave-form must include as well the DSB-SC signal $V_d = (L - R) \cos 2\pi f_{sc}t$. We now require that the peak excursion of $V_s + V_d$ be no larger than the peak excursion previously allowed to V_s alone, i.e., the peak must correspond to a frequency deviation no larger than ± 75 kHz. These considerations suggest that, when V_d is added to V_s, the level of V_s needs to be reduced and that, as a consequence, monophonic reception of a stereo transmission would be inferior to monophonic reception of a monophonic transmission. We shall now show that, as a matter of practice, such is not the case. We shall show that with V_s itself adjusted to produce a peak allowable frequency deviation, V_d may be added without exceeding the allowable frequency deviation. This characteristic of the stereo system under consideration is known as *interleaving*.

We recognize at the outset that, although $L(t)$ and $R(t)$ are different, they will ordinarily not be greatly different. After all, the microphones which generate the two signals are intended to represent a person's two ears. Hence, if naturalness in sound is to be achieved, presumably the placement of the microphones at the studio will take this fact into account. Thus we may expect that the levels of the signal outputs of the two microphones will be comparable. The maximum excursion L_m of $L(t)$ and the maximum excursion R_m of $R(t)$ will be about the same. We must allow for the fact that from time to time both $L(t)$ and $R(t)$ will attain a maximum at about the same time. Therefore we set the maximum of the sum signal V_{sm} at

$$V_{sm} = L_m + R_m \approx 2L_m \approx 2R_m \qquad (4.23\text{-}2)$$

Turning now to the composite signal $M(t)$ in Eq. (4.23-1), we note that $\cos 2\pi f_{sc}t$ oscillates rapidly between $+1$ and -1, and, ignoring temporarily the pilot carrier, we have that $M(t)$ oscillates rapidly between $M(t) = 2L(t)$ and $M(t) = 2R(t)$. The maximum attained by $M(t)$ is then $M_m = 2L_m$ or $M_m = 2R_m$. From Eq. (4.23-2) $M_m = V_{sm}$. Hence, in summary, we find that the addition of the difference signal V_d to the sum signal V_s does not increase the peak signal excursion.

EFFECT OF THE PILOT CARRIER

Unlike the DSB-SC signal, the pilot carrier, when added to the other components of the composite modulating signal, does produce an increase

in peak excursion.   Hence the addition of the pilot carrier calls for a
reduction in the sound signal modulation level.   A low-level pilot carrier
allows greater sound signal modulation, while a high-level pilot carrier
eases the burden of extracting the pilot carrier at the receiver.   As an
engineering compromise, the FCC standards call for a pilot carrier of such
level that the peak sound modulation amplitude has to be reduced to
about 90 percent of what would be allowed in the absence of a carrier.
This 10 percent reduction corresponds to a loss in signal level of less than
1 dB.

PROBLEMS

4.2-1. Consider the signal $\cos [\omega_c t + \varphi(t)]$ where $\varphi(t)$ is a square wave taking on
the values $\pm \pi/3$ every $2/f_c$ sec.
    (a) Sketch $\cos [\omega_c t + \varphi(t)]$.
    (b) Plot the phase as a function of time.
    (c) Plot the frequency as a function of time.

4.2-2. If the waveform $\cos (\omega_c t + k \sin \omega_m t)$ is a phase-modulated carrier, sketch
the waveform of the modulating signal.   Sketch the waveform of the modulating
signal if the carrier is frequency-modulated.

4.3-1. What are the dimensions of the constants k', k'', and k that appear in
Eqs. (4.3-1), (4.3-2), and (4.3-3)?

4.4-1. An FM signal is given by

$$v(t) = \cos \left[\omega_c t + \sum_{k=1}^{K} \beta_k \cos (k \omega_0 t + \theta_k) \right]$$

    (a) If $\theta_k = 0$ and $K = 1, 2$, find the maximum frequency deviations.
    (b) If each θ_k is an independent random variable, uniformly distributed
between $-\pi$ and π, find the rms frequency deviation.
    (c) Under the condition of (b) calculate the rms phase deviation.

4.4-2. If $v(t) = \cos \left[\omega_c t + k \int_{-\infty}^{t} m(\lambda) \, d\lambda \right]$, where $m(t)$ has a probability density

$$f(m) = \frac{1}{\sqrt{2\pi}} e^{-m^2/2}$$

calculate the rms frequency deviation.

4.4-3. A carrier which attains a peak voltage of 5 volts has a frequency of 100 MHz.
This carrier is frequency-modulated by a sinusoidal waveform of frequency 2 kHz
to such extent that the frequency deviation from the carrier frequency is 75 kHz.
The modulated waveform passes through zero and is increasing at time $t = 0$.
Write an expression for the modulated carrier waveform.

4.4-4. A carrier of frequency 10^6 Hz and amplitude 3 volts is frequency-modulated
by a sinusoidal modulating waveform of frequency 500 Hz and of peak amplitude

1 volt. As a consequence, the frequency deviation is 1 kHz. The level of the modulating waveform is changed to 5 volts peak, and the modulating frequency is changed to 2 kHz. Write the expression for the new modulated waveform.

4.5-1. A carrier is angle-modulated by two sinusoidal modulating waveforms simultaneously so that

$$v(t) = A \cos (\omega_c t + \beta_1 \sin \omega_1 t + \beta_2 \sin \omega_2 t)$$

Show that this waveform has sidebands separated from the carrier not only at multiples of ω_1 and of ω_2 but also has sidebands as well at separations of multiples of $\omega_1 + \omega_2$ and of $\omega_1 - \omega_2$.

4.6-1. Bessel functions are said to be *almost periodic* with a period of almost 2π. Demonstrate this by recording the values of β, for $J_0(\beta)$ and $J_1(\beta)$, required to make these functions equal to zero.

4.6-2. The primary difference between the Bessel functions and the sine wave is that the envelope of the Bessel function decreases.

(a) Tabulate the magnitude of all peak values of $J_0(\beta)$, positive and negative peaks, as a function of β.

(b) Plot the magnitude of the peak values obtained in part (a) versus β and draw a smooth curve through the points.

(c) Show that the magnitude decreases as $\dfrac{1}{\sqrt{\beta}}$.

4.6-3. An FM carrier is sinusoidally modulated. For what values of β does all the power lie in the sidebands (i.e., no power in the carrier)?

4.7-1. A bandwidth rule sometimes used for space communication systems is $B = (2\beta + 1)f_M$. What fraction of the signal power is passed by the filter? Consider $\beta = 1$ and 10.

4.7-2. A carrier is frequency-modulated by a sinusoidal modulating signal of frequency 2 kHz, resulting in a frequency deviation of 5 kHz. What is the bandwidth occupied by the modulated waveform? The amplitude of the modulating sinusoid is increased by a factor of 3 and its frequency lowered to 1 kHz. What is the new bandwidth?

4.7-3. Plot the spectrum of $\cos (2\pi \times 4t + 5 \sin 2\pi t)$. Note that the spectrum indicates the presence of a dc component. Plot the waveform as a function of time to indicate that the dc component is to have been expected.

4.10-1. $v(t) = \cos \omega_c t + 0.2 \cos \omega_m t \sin \omega_c t$.

(a) Show that $v(t)$ is a combination AM-FM signal.

(b) Sketch the phasor diagram at $t = 0$.

4.10-2. Consider the angle-modulated waveform $\cos (\omega_c t + 2 \sin \omega_m t)$, i.e., $\beta = 2$, so that the waveform may be approximated by a carrier and three pairs of sidebands. In a coordinate system in which the carrier phasor Δ_0 is at rest, determine the phasors Δ_1, Δ_2, and Δ_3, representing respectively the first, second, and third sideband pairs. Draw diagrams combining the four phasors for the cases $\omega_m t = 0$, $\pi/4$, $\pi/2$, $3\pi/4$, and π. For each case calculate the magnitude of the resultant phasor.

4.10-3. (a) Show with a phasor diagram that $v(t)$ given by

$$v(t) = \cos (2\pi \times 10^6 t) + 0.02 \cos [2\pi \times (10^6 + 10^3)t]$$

represents a carrier which is modulated both in amplitude and frequency.

(b) Show that, on the basis of the relative magnitudes of the two terms in $v(t)$, the amplitude and the frequency variations both vary approximately sinusoidally with time with frequency 10^3 Hz.

(c) Express $v(t)$ in the form

$$v(t) \approx (1 + m \cos 2\pi \times 10^3 t) \cos (2\pi \times 10^6 t + \beta \sin 2\pi \times 10^3 t)$$

Find m and β. Write an expression for the instantaneous frequency as a function of time.

4.10-4. Consider the angle-modulated waveform $\cos (\omega_c t + 6 \sin \omega_m t)$, i.e., $\beta = 6$, so that the waveform may be approximated by a carrier and seven pairs of sidebands. In a coordinate system in which the carrier phasor Δ_0 is at rest, determine the phasor Δ_1, Δ_2, etc., representing the first, second, etc., sideband pairs at a time when $\sin \omega_m t = 1$. For this time draw a phasor diagram showing each phasor Δ_i and the resultant phasor.

4.11-1. Verify the comment made in Sec. 4.11 that superposition applies in NBFM. To do this consider $v(t) = \cos [\omega_c t + \phi(t)]$ where $\phi(t) = \beta_1 \sin \omega_1 t + \beta_2 \sin \omega_2 t$. Let β_1 and β_2 be sufficiently small so that $|\phi(t)| \ll \pi/2$. Show that $v(t) \simeq \cos \omega_c t - (\beta_1 \sin \omega_1 t + \beta_2 \sin \omega_2 t) \sin \omega_c t$.

4.12-1. The frequency of a laboratory oscillator is varied back and forth extremely slowly and at a uniform rate between the frequencies of 99 and 101 kHz. The amplitude of the oscillator output is constant at 2 volts. Make a plot of the two-sided power spectral density of the oscillator output waveform.

4.12-2. If the probability density of the amplitude of $m(t)$ is Rayleigh:

$$f(m) = \begin{cases} me^{-m^2/2} & m \geq 0 \\ 0 & \text{elsewhere} \end{cases}$$

Find the power spectral density $G_v(f)$ of the FM signal

$$v(t) = \cos [\omega_c t + k \int_{-\infty}^{t} m(\lambda) \, d\lambda]$$

4.12-3. Repeat Prob. 4.12-2 if the probability density of $m(t)$ is

$$f(m) = \tfrac{1}{2} e^{-|m|}$$

4.12-4. The frequency of a laboratory oscillator is varied back and forth extremely slowly and in such a manner that the instantaneous frequency of the oscillator varies sinusoidally with time between the limits of 99 and 101 kHz. The amplitude of the oscillator output is constant at 2 volts.

(a) Find a function of frequency $g(f)$ such that $g(f) \, df$ is the fraction of the time that the instantaneous frequency is in the range between f and $f + df$.

(b) Make a plot of the two-sided power spectral density of the oscillator output waveform.

4.13-1. Consider that the WBFM signal having the power spectral density of Eq. (4.13-1) is filtered by a gaussian filter having the bandpass characteristic

$$|H(f)|^2 = e^{-(f-f_c)^2/2B^2} + e^{-(f+f_c)^2/2B^2}$$

Assume $f_c \gg B$:

(a) Sketch $|H(f)|^2$ as a function of f.

(b) Calculate the 3-dB bandwidth of the filter in terms of B.

(c) Find B so that 98 percent of the signal power of the WBFM signal is passed.

4.13-2. The two independent modulating signals $m_1(t)$ and $m_2(t)$ are both gaussian and both of zero mean and variance 1 volt^2. The modulating signal $m_1(t)$ is connected to a source which can be frequency-modulated in such manner that, when $m_1(t) = 1$ volt (constant), the source frequency, initially 1 MHz, increases by 3 kHz. The modulating signal $m_2(t)$ is connected in such manner that, when $m_2(t) = 1$ volt (constant), the source frequency decreases by 4 kHz. The carrier amplitude is 2 volts. The two modulating signals are applied simultaneously. Write an expression for the power spectral density of the output of the frequency-modulated source.

4.14-1. Consider the FM signal

$$v(t) = \cos\left[\omega_c t + \sum_{k=1}^{K} \beta_k \cos(\omega_k t + \theta_k)\right]$$

Let $\beta_k \omega_k = 1$ for each k.

(a) Find B if $B \equiv 2[(\Delta f)_1 + (\Delta f)_2 + \cdots + (\Delta f)_K]$.

(b) Find B if $B \equiv 2[(\Delta f)_{1\text{rms}} + (\Delta f)_{2\text{rms}} + \cdots + (\Delta f)_{K\text{rms}}]$.

4.14-2. If $G(f)$ is gaussian and is given by Eq. (4.13-1), find the rms bandwidth B. Compare your result with the value $B = 4.6\Delta f_{\text{rms}}$ given in Eq. (4.13-5).

4.15-1. In Fig. 4.15-1 the voltage-variable capacitor is a reversed-biased pn junction diode whose capacitance is related to the reverse-biasing voltage v by $C_v = (100/\sqrt{1 + 2v})$ pF. The capacitance $C_0 = 200$ pF and L is adjusted for resonance at 5 MHz when a fixed reverse voltage $v = 4$ volts is applied to the capacitor C_v. The modulating voltage is

$$m(t) = 4 + 0.045 \sin 2\pi \times 10^3 t$$

If the oscillator amplitude is 1 volt, write an expression for the angle-modulated output waveform which appears across the tank circuit.

4.17-1. (a) In the multiplier circuit of Fig. 4.17-1 assume that the transistor acts as a current source and is so biased and so driven that the collector current consists of alternate half-cycles of a sinusoidal waveform with a peak value of 50 mA. The input frequency of the driving signal is 1 MHz, and the multiplication by a factor of 3 is to be accomplished. If $C = 200$ pF and the inductor $Q = 30$, find the inductance of the inductor and calculate the amplitude of the third harmonic voltage across the tank.

(b) If multiplication by 10 is to be accomplished, calculate the amplitude of the tank voltage. Assume that the resonant impedance of the tank remains the same as in part (a).

4.18-1. (a) Consider the narrowband waveform $v(t) = \cos(\omega_c t + \beta \sin \omega_m t)$, with $\beta \ll 1$ and $\omega_m \ll \omega_c$. Show that $v(t)$, which has a frequency deviation $\Delta f = \beta f_m$, may be written approximately as

$$v(t) = \cos \omega_c t - \beta/2 \cos(\omega_c - \omega_m)t + \beta/2 \cos(\omega_c + \omega_m)t$$

and that this approximation is consistent with the general expansion for an angle-modulated waveform as given by Eq. (4.5-7). Use the approximations of Eqs. (4.6-1) and (4.6-2).

(b) Let $v(t)$ be applied as the input to a device whose output is $v^2(t)$ (i.e., the device is nonlinear and is to be used for frequency multiplication by a factor of 2). Square the approximate expression for $v(t)$ as given in part (a). Compare the spectrum of $v^2(t)$ so calculated with the exact spectrum for an angle-modulated waveform with frequency deviation $2\beta f_m$.

4.19-1. Assume that the 10.8-MHz signal in Fig. 4.19-1 is derived from the 200-kHz oscillator by multiplying by 54 and that the 200-kHz oscillator drifts by 0.1 Hz.

(a) Find the drift, in hertz, in the 10.8-MHz signal.

(b) Find the drift in the carrier of the resulting FM signal.

4.19-2. In an Armstrong modulator, as shown in Fig. 4.19-1, the crystal-oscillator frequency is 200 kHz. It is desired, in order to avoid distortion, to limit the maximum angular deviation to $\varphi_m = 0.2$. The system is to accommodate modulation frequencies down to 40 Hz. At the output of the modulator the carrier frequency is to be 108 MHz and the frequency deviation 80 kHz. Select multiplier and mixer oscillator frequencies to accomplish this end.

4.20-1. The narrowband phase modulator of Fig. 4.16-1 is converted to a frequency modulator by preceding the balanced modulator with an integrator. The input signal is a sinusoid of angular frequency ω_m.

(a) Show that, unless the frequency deviation is kept small, the modulator output, when demodulated, will yield not only the input signal but also its odd harmonics.

(b) If the modulation frequency is 50 Hz, find the allowable frequency deviation if the normalized power associated with the third harmonic is to be no more than 1 percent of the fundamental power.

4.20-2. (a) Consider the FM demodulator of Fig. 4.20-1. Let the frequency selective network be an RC integrating network. The 3-dB frequency of the network is $f_2 (= 1/2\pi RC)$. If the carrier frequency of the FM waveform is f_c, how should f_2 be selected so that the demodulator has the greatest sensitivity (i.e., greatest change in output per change in input frequency)?

(b) With f_2 selected for maximum sensitivity and with $f_c = 1$ MHz, find the change in demodulator output for a 1-Hz change in input frequency.

4.20-3. A "zero-crossing" FM discriminator operates in the following manner. The modulated waveform

$$v(t) = A \cos \left[2\pi f_c t + k \int_{-\infty}^{t} m(\lambda) \, d\lambda \right]$$

is applied to an electronic circuit which generates a narrow pulse on each occasion when $v(t)$ passes through zero. The pulses are of fixed polarity, amplitude, and duration. This pulse train is applied to a low-pass filter, say an RC low-pass network of 3-dB frequency f_2. Assume that the bandwidth of the baseband waveform $m(t)$ is f_M. Discuss the operation of this discriminator. Show that if

$$f_c \gg f_2 \gg f_M$$

the output of the low-pass network is indeed proportional to the instantaneous frequency of $v(t)$.

REFERENCES

1. Transmission Systems for Communications, Bell Telephone Laboratories, published by Western Electric Company, Tech. Pub., Winston-Salem, N.C., 1964.
2. Jahnke, E., and F. Emde: "Tables of Functions," Dover Publications Inc., New York, 1945.
3. Pipes, L. A.: "Applied Mathematics for Engineers and Physicists," McGraw-Hill Book Company, New York, 1958.
4. Blachman, N.: Calculation of the Spectrum of an FM Signal Using Woodwards Theorem, *IEEE Trans. Communication Technology*, August, 1969.
5. Kahn, L. R.: Compatible Single Sideband, *Proc. IRE*, vol. 49, pp. 1503–1527, October, 1961.

5
Pulse-modulation Systems

In Chaps. 3 and 4 we described systems with which we can transmit many signals simultaneously over a single communications channel. We found that at the transmitting end the individual signals were translated in frequency so that each occupied a separate and distinct frequency band. It was then possible at the receiving end to separate the individual signals by the use of filters.

In the present chapter we shall discuss a second method of multiplexing. This second method depends on the fact that a bandlimited signal, even if it is a continuously varying function of time, may be specified exactly by samples taken sufficiently frequently. Multiplexing of several signals is then achieved by interleaving the samples of the individual signals. This process is called time-division multiplexing. Since the sample is a pulse, the systems to be discussed are called *pulse-modulation* systems.

5.1 THE SAMPLING THEOREM. LOW-PASS SIGNALS

We consider at the outset the fundamental principle of time-division multiplexing. This principle is called the *sampling theorem:*

Let $m(t)$ be a signal which is bandlimited such that its highest frequency spectral component is f_M. Let the values of $m(t)$ be determined at regular intervals separated by times $T_s \leq 1/2f_M$, that is, the signal is periodically *sampled* every T_s seconds. Then these samples $m(nT_s)$, where n is an integer, uniquely determine the signal, and the signal may be reconstructed from these samples with no distortion.

The time T_s is called the *sampling time*. Note that the theorem requires that the *sampling rate* be rapid enough so that at least two samples are taken during the course of the period corresponding to the highest-frequency spectral component. We shall now prove the theorem by showing how the signal may be reconstructed from its samples.

The baseband signal $m(t)$ which is to be sampled is shown in Fig. 5.1-1a. A periodic train of pulses $S(t)$ of unit amplitude and of period T_s is shown in Fig. 5.1-1b. The pulses are arbitrarily narrow, having a width dt. The two signals $m(t)$ and $S(t)$ are applied to a multiplier as shown in Fig. 5.1-1c, which then yields as an output the product $S(t)m(t)$. This product is seen in Fig. 5.1-1d to be the signal $m(t)$ *sampled* at the occurrence of each pulse. That is, when a pulse occurs, the multiplier output has the same value as does $m(t)$, and at all other times the multiplier output is zero.

The signal $S(t)$ is periodic, with period T_s, and has the Fourier expansion [see Eq. (1.3-8) with $I = dt$ and $T_0 = T_s$]

$$S(t) = \frac{dt}{T_s} + \frac{2dt}{T_s}\left(\cos 2\pi \frac{t}{T_s} + \cos 2 \times 2\pi \frac{t}{T_s} + \cdots\right) \qquad (5.1\text{-}1)$$

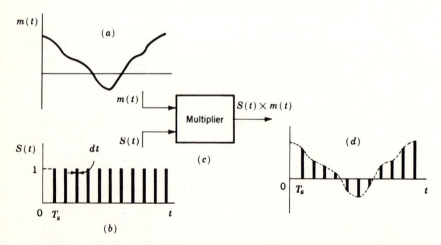

Fig. 5.1-1 (a) A signal $m(t)$ which is to be sampled. (b) The sampling function $S(t)$ consists of a train of very narrow unit amplitude pulses. (c) The sampling operation is performed in a multiplier. (d) The samples of the signal $m(t)$.

For the case $T_s = 1/2f_M$, the product $S(t)m(t)$ is

$$S(t)m(t) = \frac{dt}{T_s} m(t) + \frac{dt}{T_s} [2m(t) \cos 2\pi(2f_M)t$$
$$+ 2m(t) \cos 2\pi(4f_M)t + \cdots] \quad (5.1\text{-}2)$$

We now observe that the first term in the series is, aside from a constant factor, the signal $m(t)$ itself. Again, aside from a multiplying factor, the second term is the product of $m(t)$ and a sinusoid of frequency $2f_M$. This product then, as discussed in Sec. 3.2, gives rise to a double-sideband suppressed-carrier signal with carrier frequency $2f_M$. Similarly, succeeding terms yield DSB-SC signals with carrier frequencies $4f_M$, $6f_M$, etc.

Let the signal $m(t)$ have a spectral density $M(j\omega) = \mathfrak{F}[m(t)]$ which is as shown in Fig. 5.1-2a. Then $m(t)$ is bandlimited to the frequency range below f_M. The spectrum of the first term in Eq. (5.1-2) extends from 0 to f_M. The spectrum of the second term is symmetrical about the frequency $2f_M$ and extends from $2f_M - f_M = f_M$ to $2f_M + f_M = 3f_M$. Altogether the spectrum of the sampled signal has the appearance shown in Fig. 5.1-2b. Suppose then that the sampled signal is passed through an ideal low-pass filter with cutoff frequency at f_M. If the filter transmission were constant in the passband and if the cutoff were infinitely sharp at f_M, the filter would pass the signal $m(t)$ and nothing else.

The spectral pattern corresponding to Fig. 5.1-2b is shown in Fig. 5.1-3a for the case in which the sampling rate $f_s = 1/T_s$ is larger than $2f_M$. In this case there is a gap between the upper limit f_M of the spectrum of the baseband signal and the lower limit of the DSB-SC spectrum centered around the carrier frequency $f_s > 2f_M$. For this reason the low-

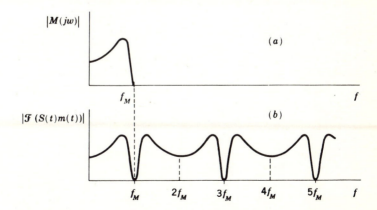

Fig. 5.1-2 (a) The magnitude plot of the spectral density of a signal bandlimited to f_M. (b) Plot of amplitude of spectrum of sampled signal.

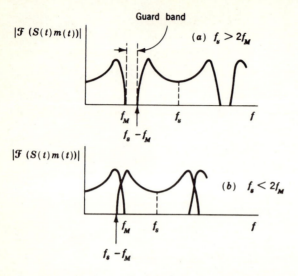

Fig. 5.1-3 (*a*) A guard band appears when $f_s > 2f_M$.
(*b*) Overlapping of spectra when $f_s < 2f_M$.

pass filter used to select the signal $m(t)$ need not have an infinitely sharp
cutoff. Instead, the filter attenuation may begin at f_M but need not
attain a high value until the frequency $f_s - f_M$. This range from f_M to
$f_s - f_M$ is called a *guard band* and is always required in practice, since a
filter with infinitely sharp cutoff is, of course, not realizable. Typically,
when sampling is used in connection with voice messages on telephone
lines, the voice signal is limited to $f_M = 3.3$ kHz, while f_s is selected at
8.0 kHz. The guard band is then $8.0 - 2 \times 3.3 = 1.4$ kHz.

The situation depicted in Fig. 5.1-3*b* corresponds to the case where
$f_s < 2f_M$. Here we find an overlap between the spectrum of $m(t)$ itself
and the spectrum of the DSB-SC signal centered around f_s. Accordingly,
no filtering operation will allow an exact recovery of $m(t)$.

We have just proved the *sampling theorem* since we have shown that,
in principle, the sampled signal can be recovered exactly when $T_s \leq
1/2f_M$. It has also been shown why the minimum allowable sampling
rate is $2f_M$. This minimum sampling rate is known as the *Nyquist rate*.
An increase in sampling rate above the Nyquist rate increases the width
of the guard band, thereby easing the problem of filtering. On the other
hand, we shall see that an increase in rate extends the bandwidth required
for transmitting the sampled signal. Accordingly an engineering compromise is called for.

An interesting special case is the sampling of a sinusoidal signal
having the frequency f_M. Here, *all* the signal power is concentrated

precisely at the cutoff frequency of the low-pass filter, and there is consequently some ambiguity about whether the signal frequency is inside or
outside the filter passband.   To remove this ambiguity, we require that
$f_s > 2f_M$ rather than that $f_s \geq 2f_M$.   To see that this condition is necessary, assume that $f_s = 2f_M$ but that an initial sample is taken at the
moment the sinusoid passes through zero.   Then all successive samples
will also be zero.   This situation is avoided by requiring $f_s > 2f_M$.

BANDPASS SIGNALS

For a signal $m(t)$ whose highest-frequency spectral component is f_M, the
sampling frequency f_s must be no less than $f_s = 2f_M$ only if the lowest-
frequency spectral component of $m(t)$ is $f_L = 0$.   In the more general
case, where $f_L \neq 0$, it may be that the sampling frequency need be no
larger than $f_s = 2(f_M - f_L)$.   For example, if the spectral range of a
signal extends from 10.0 to 10.1 MHz, the signal may be recovered from
samples taken at a frequency $f_s = 2(10.1 - 10.0) = 0.2$ MHz.

To establish the sampling theorem for such bandpass signals, let us
select a sampling frequency $f_s = 2(f_M - f_L)$ and let us initially assume
that it happens that the frequency f_L turns out to be an integral multiple
of f_s, that is, $f_L = nf_s$ with n an integer.   Such a situation is represented
in Fig. 5.1-4.   In part a is shown the two-sided spectral pattern of a signal

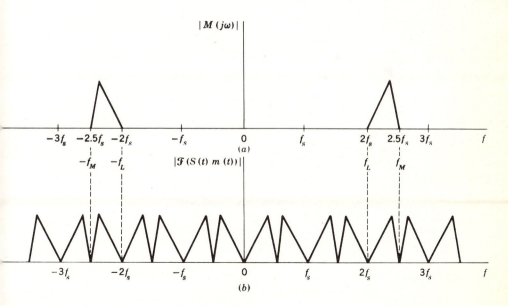

Fig. 5.1-4   (a) The spectrum of a bandpass signal.   (b) The spectrum of the sampled
bandpass signal.

$m(t)$ with Fourier transform $M(j\omega)$. Here it has been arranged that $n = 2$; that is, f_L coincides with the second harmonic of the sampling frequency, while the sampling frequency is exactly $f_s = 2(f_M - f_L)$. In part b is shown the spectral pattern of the sampled signal $S(t)m(t)$. The product of $m(t)$ and the dc term of $S(t)$ [Eq. (5.1-1)] duplicates in part b the form of the spectral pattern in part a and leaves it in the same frequency range from f_L to f_M. The product of $m(t)$ and the spectral component in $S(t)$ of frequency f_s $(= 1/T_s)$ gives rise in part b to a spectral pattern derived from part a by shifting the pattern in part a to the right and also to the left by amount f_s. Similarly, the higher harmonics of f_s in $S(t)$ give rise to corresponding shifts, right and left, of the spectral pattern in part a. We now note that if the sampled signal $S(t)m(t)$ is passed through a bandpass filter with arbitrarily sharp cutoffs and with passband from f_L to f_M, the signal $m(t)$ will be recovered exactly.

In Fig. 5.1-4 the spectrum of $m(t)$ extends over the first half of the frequency interval between harmonics of the sampling frequency, that is, from $2.0f_s$ to $2.5f_s$. As a result, there is no spectrum overlap, and signal recovery is possible. It may also be seen from the figure that if the spectral range of $m(t)$ extended over the second half of the interval from $2.5f_s$ to $3.0f_s$, there would similarly be no overlap. Suppose, however, that the spectrum of $m(t)$ were confined neither to the first half nor to the second half of the interval between sampling-frequency harmonics. In such a case, there would be overlap between the spectrum patterns, and signal recovery would not be possible. Hence the minimum sampling frequency $f_s = 2(f_M - f_L)$ is allowable provided that either f_M or f_L is a harmonic of f_s. Otherwise, a higher sampling frequency will be required.

5.2 PULSE-AMPLITUDE MODULATION

A technique by which we may take advantage of the sampling principle for the purpose of time-division multiplexing is illustrated in the idealized representation of Fig. 5.2-1. At the transmitting end on the left, a number of bandlimited signals are connected to the contact points of a rotary switch. We assume that the signals are similarly bandlimited. For example, they may all be voice signals, limited to 3.3 kHz. As the rotary arm of the switch swings around, it samples each signal sequentially. The rotary switch at the receiving end is in synchronism with the switch at the sending end. The 2 switches make contact simultaneously at similarly numbered contacts. With each revolution of the switch, one sample is taken of each input signal and presented to the correspondingly numbered contact of the receiving-end switch. The train of samples at, say, terminal 1 in the receiver, pass through low-pass filter 1, and, at the filter output, the original signal $m_1(t)$ appears reconstructed. Of course, if f_M

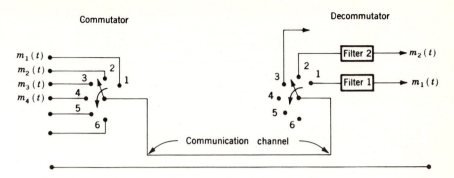

Fig. 5.2-1 Illustrating how the sampling principle may be used to transmit a number of bandlimited signals over a single communications channel.

is the highest-frequency spectral component present in any of the input signals, the switches must make at least $2f_M$ revolutions per second.

When the signals to be multiplexed vary slowly with time, so that the sampling rate is correspondingly slow, mechanical switches, indicated in Fig. 5.2-1, may be employed. When the switching speed required is outside the range of mechanical switches, electronic switching systems may be employed. In either event, the switching mechanism, corresponding to the switch at the left in Fig. 5.2-1, which samples the signals, is called the *commutator*. The switching mechanism which performs the function of the switch at the right in Fig. 5.2-1 is called the *decommutator*. The commutator samples and combines samples, while the decommutator separates samples belonging to individual signals so that these signals may be reconstructed.

The interlacing of the samples that allows multiplexing is shown in Fig. 5.2-2. Here, for simplicity, we have considered the case of the multiplexing of just two signals $m_1(t)$ and $m_2(t)$. The signal $m_1(t)$ is sampled regularly at intervals of T_s and at the times indicated in the figure. The sampling of $m_2(t)$ is similarly regular, but the samples are taken at a time different from the sampling time of $m_1(t)$. The input waveform to the filter numbered 1 in Fig. 5.2-1 is the train of samples of $m_1(t)$, and the input to the filter numbered 2 is the train of samples of $m_2(t)$. The timing in Fig. 5.2-2 has been deliberately drawn to suggest that there is room to multiplex more than two signals. We shall see shortly, in principle, how many signals may be multiplexed.

We observe that the train of pulses corresponding to the samples of each signal are *modulated in amplitude* in accordance with the signal itself. Accordingly, the scheme of sampling is called *pulse-amplitude modulation* and abbreviated PAM.

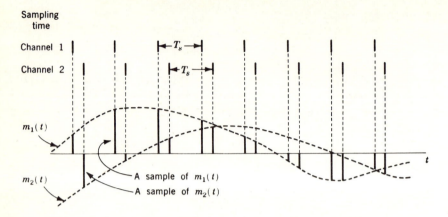

Fig. 5.2-2   The interlacing of two baseband signals.

Multiplexing of several PAM signals is possible because the various signals are kept distinct and are separately recoverable by virtue of the fact that they are sampled at different times.   Hence this system is an example of a *time-division multiplex* (TDM) system.   Such systems are the counterparts in the time domain of the systems of Chap. 3.   There, the signals were kept separable by virtue of their translation to different portions of the frequency domain, and those systems are called frequency-division multiplex (FDM) systems.

If the multiplexed signals are to be transmitted directly, say, over a pair of wires, no further signal processing need be undertaken.   Suppose, however, we require to transmit the TDM-PAM signal from one antenna to another.   It would then be necessary to amplitude-modulate or frequency-modulate a high-frequency carrier with the TDM-PAM signal; in such a case the overall system would be referred to, respectively, as PAM-AM or PAM-FM.   Note that the same terminology is used whether a single signal or many signals (TDM) are transmitted.

5.3 CHANNEL BANDWIDTH FOR A PAM SIGNAL

Suppose that we have N independent baseband signals $m_1(t)$, $m_2(t)$, etc., each of which is bandlimited to f_M.   What must be the bandwidth of the communications channel which will allow all N signals to be transmitted simultaneously using PAM time-division multiplexing?   We shall now show that, in principle at least, the channel need not have a bandwidth larger than Nf_M.

The baseband signal, say $m_1(t)$, must be sampled at intervals not longer than $T_s = 1/2f_M$.   Between successive samples of $m_1(t)$ will appear samples of the other $N - 1$ signals. Therefore the interval of separation between successive samples of different baseband signals is $1/2f_M N$. The composite signal, then, which is presented to the transmitting end of the communications channel, consists of a sequence of samples, that is, a sequence of *impulses*. If the bandwidth of the channel were arbitrarily great, the waveform at the receiving end would be the same as at the sending end and demultiplexing could be achieved in a straightforward manner.

If, however, the bandwidth of the channel is restricted, the channel response to an instantaneous sample will be a waveform which may well persist with significant amplitude long after the time of selection of the sample. In such a case, the signal at the receiving end at any particular sampling time may well have significant contributions resulting from previous samples of other signals. Consequently the signal which appears at any of the output terminals in Fig. 5.2-1 will not be a single baseband signal but will be instead a combination of many or even all the baseband signals. Such combining of baseband signals at a communication system output is called *crosstalk* and is to be avoided as far as is possible.

Let us assume that our channel has the characteristics of an ideal low-pass filter with angular cutoff frequency $\omega_c = 2\pi f_c$, unity gain, and no delay. Let a sample be taken, say, of $m_1(t)$, at $t = 0$. Then at $t = 0$ there is presented at the transmitting end of the channel an impulse of strength $I_1 = m_1(0)\,dt$. The response at the receiving end is $s_{R1}(t)$ given by (see Prob. 5.3-1)

$$s_{R1}(t) = \frac{I_1\omega_c}{\pi}\frac{\sin\omega_c t}{\omega_c t} \tag{5.3-1}$$

The normalized response $\pi s_R(t)/\omega_c$ is shown in Fig. 5.3-1 by the solid plot. At $t = 0$ the response attains a peak value proportional to the strength of the impulse $I_1 = m_1(0)\,dt$, which is in turn proportional to the value of the sample $m_1(0)$. This response persists indefinitely. Observe, however, that the response passes through zero at intervals which are multiples of $\pi/\omega_c = 1/2f_c$. Suppose, then, that a sample of $m_2(t)$ is taken and transmitted at $t = 1/2f_c$. If $I_2 = m_2(t = 1/2f_c)\,dt$,

$$s_{R2}(t) = \frac{I_2\omega_c}{\pi}\frac{\sin\omega_c(t - 1/2f_c)}{\omega_c(t - 1/2f_c)} \tag{5.3-2}$$

This response is shown by the dashed plot. Suppose, finally, that the demultiplexing is done also by instantaneous sampling at the receiving end of the channel, for $m_1(t)$ at $t = 0$ and for $m_2(t)$ at $t = 1/2f_c$. Then, in spite of the persistence of the channel response, there will be no crosstalk, and the signals $m_1(t)$ and $m_2(t)$ may be completely separated and individually recovered. Similarly, additional signals may be sampled

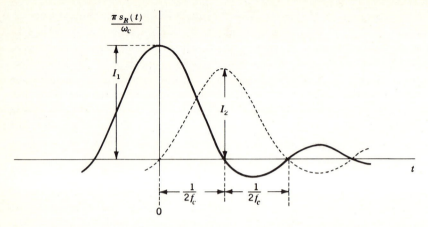

Fig. 5.3-1 The response of an ideal low-pass filter to an instantaneous sample at $t = 0$ (solid plot). The response to a sample at $t = \frac{1}{2}f_c$ (dashed plot).

and multiplexed, provided that each new sample is taken synchronously, every $1/2f_c$ sec. The sequence must, of course, be continually repeated every $1/2f_M$ sec, so that each signal is properly sampled.

We have then the result that with a channel of bandwidth f_c we need to separate samples by intervals $1/2f_c$. The sampling theorem requires that the samples of an individual baseband signal be separated by intervals not longer than $1/2f_M$. Hence the total number of signals which may be multiplexed is $N = f_c/f_M$, or $f_c = Nf_M$ as indicated earlier.

In principle then, multiplexing a number of signals by PAM time division requires no more bandwidth than would be required to multiplex these signals by frequency-division multiplexing using single-sideband transmission.

5.4 NATURAL SAMPLING

It was convenient, for the purpose of introducing some basic ideas, to begin our discussion of time multiplexing by assuming instantaneous commutation and decommutation. Such instantaneous sampling, however, is hardly feasible. Even if it were possible to construct switches which could operate in an arbitrarily short time, we would be disinclined to use them. The reason is that instantaneous samples at the transmitting end of the channel have infinitesimal energy, and when transmitted through a bandlimited channel give rise to signals having a peak value which is infinitesimally small. We recall that in Fig. 5.3-1 $I_1 = m_1(0)\, dt$. Such infinitesimal signals will inevitably be lost in background noise.

A much more reasonable manner of sampling, referred to as *natural sampling*, is shown in Fig. 5.4-1.   Here the sampling waveform $S(t)$ consists of a train of pulses having duration τ and separated by the sampling time T_s.   The baseband signal is $m(t)$, and the sampled signal $S(t)m(t)$ is shown in Fig. 5.4-1c.   Observe that the sampled signal consists of a sequence of pulses of varying amplitude whose tops are not flat but follow the waveform of the signal $m(t)$.

With natural sampling, as with instantaneous sampling, a signal sampled at the Nyquist rate may be reconstructed exactly by passing the samples through an ideal low-pass filter with cutoff at the frequency f_M, where f_M is the highest-frequency spectral component of the signal.   To

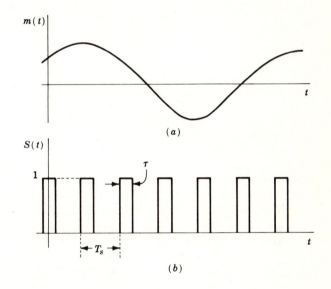

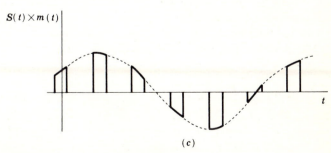

Fig. 5.4-1   (*a*) A baseband signal $m(t)$.   (*b*) A sampling signal $S(t)$ with pulses of finite duration.   (*c*) The *naturally* sampled signal $S(t)m(t)$.

prove this, we note that the sampling waveform $S(t)$ shown in Fig. 5.4-1 is given by [see Eq. (1.3-12) with $A = 1$ and $T_0 = T_s$]

$$S(t) = \frac{\tau}{T_s} + \frac{2\tau}{T_s}\left(C_1 \cos 2\pi \frac{t}{T_s} + C_2 \cos 2 \times 2\pi \frac{t}{T_s} + \cdot \cdot \right)$$

$$(5.4\text{-}1)$$

with the constant C_n given by

$$C_n = \frac{\sin (n\pi\tau/T_s)}{n\pi\tau/T_s} \tag{5.4-2}$$

This sampling waveform differs from the sampling waveform of Eq. (5.1-1) for instantaneous sampling only in that dt is replaced by τ and by the fact that the amplitudes of the various harmonics are not constant. The sampled baseband signal $S(t)m(t)$ is, for $T_s = 1/2f_M$,

$$S(t)m(t) = \frac{\tau}{T_s} m(t) + \frac{2\tau}{T_s}[m(t)C_1 \cos 2\pi(2f_M)t$$
$$+ m(t)C_2 \cos 2\pi(4f_M)t + \cdot \cdot \cdot] \quad (5.4\text{-}3)$$

Therefore, as in instantaneous sampling, a low-pass filter with cutoff at f_M will deliver an output signal $s_o(t)$ given by

$$s_o(t) = \frac{\tau}{T_s} m(t) \tag{5.4-4}$$

which is the same as is given by the first term of Eq. (5.1-2) except with dt replaced by τ.

With samples of finite duration, it is not possible to completely eliminate the crosstalk generated in a channel, sharply bandlimited to a bandwidth f_c. If N signals are to be multiplexed, then the maximum sample duration is $\tau = T_s/N$. It is advantageous, for the purpose of increasing the level of the output signal, to make τ as large as possible. For, as is seen in Eq. (5.4-4), $s_o(t)$ increases with τ. However, to help suppress crosstalk, it is ordinarily required that the samples be limited to a duration much less than T_s/N. The result is a large *guard time* between the end of one sample and the beginning of the next.

5.5 FLAT-TOP SAMPLING

Pulses of the type shown in Fig. 5.4-1c, with tops contoured to follow the waveform of the signal, are actually not frequently employed. Instead *flat-topped* pulses are customarily used, as shown in Fig. 5.5-1a. A flat-topped pulse has a constant amplitude established by the sample value of the signal at some point within the pulse interval. In Fig. 5.5-1a we have arbitrarily sampled the signal at the beginning of the pulse. In

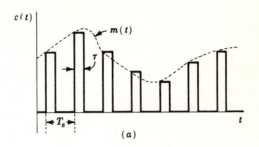

(a)

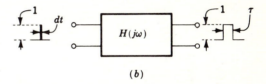

Fig. 5.5-1 (a) Flat-topped sampling. (b) A network with transform $H(j\omega)$ which converts a pulse of width dt into a rectangular pulse of like amplitude but of duration τ.

(b)

sampling of this type the baseband signal $m(t)$ cannot be recovered exactly by simply passing the samples through an ideal low-pass filter. However, the distortion need not be large. Flat-top sampling has the merit that it simplifies the design of the electronic circuitry used to perform the sampling operation.

To show the extent of the distortion, consider the signal $m(t)$ having a Fourier transform $M(j\omega)$. We have seen (see Figs. 5.1-2 and 5.1-3) how to deduce the transform of the sampled signal, when the sampling is instantaneous. The transform of the sampled signal for flat-top sampling is determined by considering that the flat-top pulse can be generated by passing the instantaneously sampled signal through a network which broadens a pulse of duration dt (an impulse) into a pulse of duration τ. The transform of a pulse of unit amplitude and width dt is

$$\mathfrak{F}[\text{impulse of strength } dt \text{ at } t = 0] = dt \qquad (5.5\text{-}1)$$

The transform of a pulse of unit amplitude and width τ is [see Eq. (1.10-20)]

$$\mathfrak{F}\left[\text{pulse, amplitude} = 1, \text{ extending from } t = -\frac{\tau}{2} \text{ to } t = \frac{\tau}{2}\right]$$

$$= \tau \frac{\sin(\omega\tau/2)}{\omega\tau/2} \qquad (5.5\text{-}2)$$

Hence, the transfer function of the network shown in Fig. 5.5-1b, is required to be

$$H(j\omega) = \frac{\tau}{dt}\frac{\sin(\omega\tau/2)}{\omega\tau/2} \qquad (5.5\text{-}3)$$

Let the signal $m(t)$, with transform $M(j\omega)$, be bandlimited to f_M and be sampled at the Nyquist rate or faster. Then in the range 0 to f_M the transform of the flat-topped sampled signal is given by the product

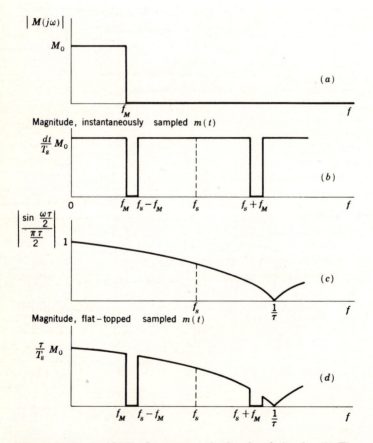

Fig. 5.5-2 (a) An idealized spectrum of a baseband signal. (b) The spectrum of the signal with instantaneous sampling. (c) The form $\lfloor(\sin x)/x$, with $x = \omega\tau/2\rfloor$ of the distortion factor (aperture effect) introduced by flat-topped sampling. (d) The spectrum of the signal with flat-topped sampling.

$H(j\omega)M(j\omega)$ or, from Eqs. (5.5-1), (5.5-2), and (5.5-3)

$$\mathfrak{F}[\text{flat-topped sampled } m(t)] = \frac{\tau}{T_s} \frac{\sin(\omega\tau/2)}{\omega\tau/2} M(j\omega)$$

$$0 \le f \le f_M \quad (5.5\text{-}4)$$

To illustrate the effect of flat-top sampling, we consider for simplicity that the signal $m(t)$ has a flat spectral density equal to M_0 over its entire range from 0 to f_M, as is shown in Fig. 5.5-2a. The form of the transform of the instantaneously sampled signal is shown in Fig. 5.5-2b. The sampling frequency $f_s = 1/T_s$ is assumed large enough to allow for a guard band between the spectrum of the baseband signal and the DSB-SC signal with carrier f_s. The spectrum of the flat-topped sampled signal is shown in Fig. 5.5-2d. We are, of course, interested only in the part of the spectrum in the range 0 to f_M. If, in this range, the spectra of the sampled signal and the original signal are identical, then the original signal may be recovered by a low-pass filter as has already been discussed. We observe, however, that such is not the case and that, as a result, distortion will result. This distortion results from the fact that the original signal was "observed" through a finite rather than an infinitesimal time "aperture" and is hence referred to as *aperture effect* distortion.

The distortion results from the fact that the spectrum is multiplied by the sampling function $Sa(x) \equiv (\sin x)/x$ (with $x = \omega\tau/2$). The magnitude of the sampling function (see Sec. 1.4) falls off slowly with increasing x in the neighborhood of $x = 0$ and does not fall off sharply until we approach $x = \pi$, at which point $Sa(x) = 0$. To minimize the distortion due to the aperture effect, it is advantageous to arrange that $x = \pi$ correspond to a frequency very large in comparison with f_M. Since $x = \pi f\tau$, the frequency f_0 corresponding to $x = \pi$ is $f_0 = 1/\tau$. If $f_0 \gg f_M$, or, correspondingly, if $\tau \ll 1/f_M$, the aperture distortion will be small. The distortion becomes progressively smaller with decreasing τ. And, of course, as $\tau \to 0$ (instantaneous sampling), the distortion similarly approaches zero.

EQUALIZATION

As in the case of natural sampling, so also in the present case of flat-top sampling, it is advantageous to make τ as large as practicable for the sake of increasing the amplitude of the output signal. If, in a particular case, it should happen that the consequent distortion is not acceptable, it may be corrected by including an *equalizer* in cascade with the output low-pass filter. An equalizer, in the present instance, is a passive network whose transfer function has a frequency dependence of the form $x/\sin x$, that is, a form inverse to the form of $H(j\omega)$ given in Eq. (5.5-3). The equalizer in combination with the aperture effect will then yield a flat overall trans-

fer characteristic between the original baseband signal and the output
at the receiving end of the system. The equalizer $x/\sin x$ cannot be
exactly synthesized, but can be approximated.

If N signals are multiplexed, $\tau \leq 1/2f_M N$, and hence for large N
$\tau \ll 1/f_M$ and $x/\sin x \simeq 1$. In this case the equalizer is not needed as
negligible distortion results.

5.6 SIGNAL RECOVERY THROUGH HOLDING

We have already noted that the maximum ratio τ/T_s, of the sample duration to the sampling interval, is $\tau/T_s = 1/N$, N being the number of
signals to be multiplexed. As N increases, τ/T_s becomes progressively
smaller, and, as is to be seen from Eq. (5.4-4), so correspondingly does
the output signal. We discuss now an alternative method of recovery
of the baseband signal which raises the level of the output signal (without
the use of amplifiers which may introduce noise). The method has the
additional advantage that rather rudimentary filtering is often quite
adequate, but has the disadvantage that some distortion must be accepted.

The method is illustrated in Fig. 5.6-1, where the baseband signal
$m(t)$ and its flat-topped samples are shown. At the receiving end, and
after demultiplexing, the sample pulses are extended; that is, the sample
value of each individual baseband signal is *held* until the occurrence of the
next sample of that same baseband signal. This operation is shown in
Fig. 5.6-1 as the dashed extension of the sample pulses. The output waveform consists then, as shown, of an up and down staircase waveform with
no blank intervals.

A method, in principle, by which this holding operation may be performed is shown in Fig. 5.6-2. The switch S operates in synchronism with

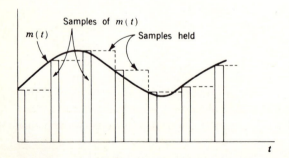

Fig. 5.6-1 Illustrating the operation of holding.

Fig. 5.6-2   Illustrating a method of performing the operation of holding.

the occurrence of input samples.   This switch, ordinarily open, closes somewhat after the occurrence of the leading edge of a sample pulse and opens somewhat before the occurrence of the trailing edge.   The amplifier, whose gain, if any, is incidental to the present discussion, has a low-output impedance.   Hence, at the closing of the switch, the capacitor C charges abruptly to a voltage proportional to the sample value, and the capacitor holds this voltage until the operation is repeated for the next sample.   In Fig. 5.6-1 we have idealized the situation somewhat by showing the output waveform maintaining a perfectly constant level throughout the sample pulse interval and its following holding interval.   We have also indicated abrupt transitions in voltage level from one sample to the next.   In practice, these voltage transitions will be somewhat rounded as the capacitor charges and discharges exponentially.   Further, if the received sample pulses are natural samples rather than flat-topped samples, there will be some departure from a constant voltage level during the sample interval itself.   As a matter of practice however, the sample interval is very small in comparison with the interval between samples, and the voltage variation of the baseband signal during the sampling interval is small enough to be neglected.

If the baseband signal is $m(t)$ with spectral density $M(j\omega) = \mathfrak{F}[m(t)]$, we may deduce the spectral density of the sampled and held waveform in the manner of Sec. 5.5 and in connection with flat-topped sampling.   We need but to consider that the flat tops have been stretched to encompass the entire interval between instantaneous samples.   Hence the spectral density is given as in Eq. (5.5-4) except with τ replaced by the time interval between samples.   We have, then,

$$\mathfrak{F}[m(t), \text{ sampled and held}] = \frac{\sin\,(\omega T_s/2)}{\omega T_s/2}\,M(j\omega)$$

$$0 \le f \le f_M \quad (5.6\text{-}1)$$

In Fig. 5.6-3 we have again assumed for simplicity that the band-limited signal $m(t)$ has a flat spectral density of magnitude M_0.   In Fig. 5.6-3a is shown the spectrum of the instantaneously sampled signal.   In Fig. 5.6-3b has been drawn the magnitude of the aperture factor $(\sin x)/x$ (with $x = \omega T_s/2$), while in Fig. 5.6-3c is shown the magnitude of the spec-

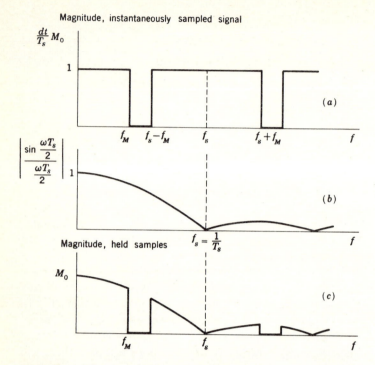

Fig. 5.6-3 (*a*) Spectrum of instantaneously sampled signal $m(t)$ with $m(t)$ having idealized spectrum shown in Fig. 5.5-2*a*. (*b*) The magnitude of the aperture effect factor. (*c*) Spectrum of sampled-and-held signal.

trum of the sampled-and-held signal. These plots differ from the plots of Fig. 5.5-2 only in the location of the nulls of the factor $(\sin x)/x$. In Fig. 5.6-3 the first null occurs at the sampling frequency f_s. We observe that, as a consequence, the aperture effect, which is responsible for the $(\sin x)/x$ term, has accomplished most of the filtering which is required to suppress the part of the spectrum of the output signal above the bandlimit f_M. Of course, the filtering is not perfect, and some additional filtering may be required. We also note that, as in the case of flat-top sampling, there will be some distortion introduced by the unequal transmission of spectral components in the range 0 to f_M. If the distortion is not acceptable, then, as before, it may be corrected by an $x/\sin x$ equalizer.

Most importantly we note in comparing Eq. (5.6-1) with Eq. (5.5-4) that, aside from the relatively small effect of the $(\sin x)/x$ terms in the two cases, the sampled-and-held signal has a magnitude larger by the factor T_s/τ than the signal of sample duration τ. This increase in amplitude is, of course, intuitively to have been anticipated.

5.7 CROSSTALK DUE TO HIGH-FREQUENCY CUTOFF OF THE CHANNEL

Suppose that a pulse, representing a sample of a baseband signal, is applied to the input of a communication channel of limited bandwidth. Then at the receiving end of the channel the waveform resulting from this input pulse will not be confined to the time interval allocated to this sample. As a result, crosstalk occurs between baseband channels. We saw in Sec. 5.3 that, in principle at least, such crosstalk may be avoided by instantaneous commutation and decommutation. We discuss now the crosstalk that may result with finite-duration sampling as a result of the upper-frequency cutoff of the communication channel. In this discussion, as well as in others, it is convenient to introduce the term *time slot*. A time slot is the entire interval allocated to an individual pulse and includes the *guard-time* intervals which separate and distinguish one pulse from its preceding and succeeding neighbors.

In order to permit an approximate calculation of crosstalk without unwarranted computational complexity, we shall assume that the communication channel may be represented by the RC circuit of Fig. 5.7-1a. This RC circuit has an upper-frequency 3-dB cutoff frequency

$$f_c = \frac{1}{2\pi RC} \tag{5.7-1}$$

In Fig. 5.7-1b we have indicated two neighboring time slots of sampling duration τ separated by a guard-time interval τ_g. We consider that

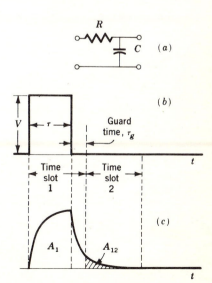

Fig. 5.7-1 (a) An RC low-pass circuit is used to represent the transmission characteristics of the channel in the neighborhood of its upper cutoff frequency. (b) Two neighboring time slots, the first of which is occupied by a pulse. (c) Showing the pulse distortion and consequent overlapping into slot 2 caused by the restricted bandwidth of the channel.

channel 1 has a signal applied to it as a result of which a sample pulse of amplitude V appears in time slot 1. The channel distorts the waveform as shown in Fig. 5.7-1c, and the signal pulse assigned to slot 1 overlaps into slot 2.

Let us assume that, after decommutation, we have chosen to recon-struct the baseband signals by the use of baseband filters. Let us further assume, at least initially, that the signal $m_1(t)$ applied to channel 1 is a dc signal. Then the input to the baseband filter of channel 2 due to channel 1 is a train of pulse waveforms of the form shown in time slot 2 in Fig. 5.7-1c. The repetition rate of this pulse train is the sampling frequency of $m_2(t)$, which is equal to or greater than $2f_M$. The pulse train can be represented by a Fourier series, consisting of a dc term and terms with the frequencies equal to or greater than $2f_M$. Thus the output of the base-band filter, with bandwidth f_M, of channel 2 due to this pulse train is the dc component of this input pulse train. This dc component is, in turn, proportional to the area under the waveform in slot 2. This area, A_{12}, shown shaded in the figure appears in time slot 2 and results from a pulse in slot 1.

If the signal applied to channel 1 is not a dc signal but varies with time, then the amplitudes of the waveforms in the slots will change some-what from sample to sample. However, the ratio of the amplitudes of output signals from channels 1 and 2 will still be proportional to the ratio of areas under the waveforms in slots 1 and 2.

It is clear from Fig. 5.7-1 that, to keep crosstalk small, we require that the channel time constant $\tau_c = RC$ must be very small in comparison with the guard interval τ_g. Since we usually find the sample duration τ comparable to or larger than τ_g, we also have that $\tau_c \ll \tau$. On this basis, it is apparent that the distortion shown in Fig. 5.7-1c is greatly exagger-ated. The pulse in slot 1 is very nearly rectangular and its area is given to a very good approximation by $A_1 = V\tau$. Assuming then that the pulse has the value V at the beginning of the guard interval, we find for A_{12} that

$$A_{12} = V\tau_c e^{-\tau_g/\tau_c}(1 - e^{-\tau/\tau_c}) \approx V\tau_c e^{-\tau_g/\tau_c} \tag{5.7-2}$$

since $\tau \gg \tau_c$.

CROSSTALK FACTOR

Because of crosstalk, a message transmitted in channel 1 is also received, at lower level, in channel 2. This crosstalk signal appears superimposed on any signal which may have been deliberately introduced into channel 2. If no signal is being transmitted in channel 2, the effect of the crosstalk is much more pronounced than if there is a strong signal in channel 2. Cross-talk is specified by a *crosstalk factor* K, which is the ratio of the signal level (area A_{12}) to the area A_2 under the sample pulse of the desired signal in channel 2. On the average, in a typical multiplexing system, we may expect that each of the multiplexed signals will be equally strong. Hence we may assume that on the average $A_1 = A_2 = V\tau$ and take K to be

$$K \equiv \frac{A_{12}}{A_2} = \frac{A_{12}}{A_1} = \frac{\tau_c}{\tau} e^{-\tau_g/\tau_c} \qquad (5.7\text{-}3)$$

As an illustration of the bandwidth required in a typical case to adequately reduce the crosstalk, let us consider the matter of multiplexing 12 speech channels in a PAM system. We noted above, in Sec. 5.1, that for telephone speech communications a sampling rate of 8 kHz was adequate. Samples are then to be taken at intervals of $\frac{1}{8000} = 125$ μsec. Since 12 channels are to be multiplexed, each sample interval together with its associated guard space may be allocated $\frac{125}{12} = 10.4$ μsec. Somewhat arbitrarily, but rather typically, we now decide that the sample interval is to be $\tau = 6.8$ μsec, leaving the guard space $\tau_g = 3.6$ μsec. The sampling interval is roughly twice the guard-space interval.

Let us now introduce a signal into the Nth channel of the multiplexing system. Suppose that we require that the crosstalk signal thereby introduced into the $(N + 1)$st channel shall be reduced in amplitude by a factor of 1000 (that is, the crosstalk signal in channel $N + 1$ is 60 dB below the desired signal in channel $N + 1$), assuming that the signal in channel $N + 1$ has the same level as the signal in N. What, then, must be the channel bandwidth? From Eq. (5.7-3) with $\tau = 6.8$ μsec, $\tau_g = 3.6$ μsec, and $K = 10^{-3}$, we find $\tau_c = 0.76$ μsec. Using Eq. (5.7-1) with $\tau_c = RC$, we find $f_c = 210$ kHz. We may note in passing that $\tau/\tau_c = 6.8/0.76 \cong 9$ and that $e^{-9} = 0.00012$, justifying the approximation indicated in Eq. (5.7-2).

When crosstalk results from the mechanism presently under discussion, the crosstalk is of significance only between adjacent channels. That such is the case is readily seen by using Eq. (5.7-3) to calculate the crosstalk between channel N and channel $N + 2$. We find that (Prob. 5.7-2), using the numerical values of the present section, the crosstalk signal in channel $N + 2$ is 180 dB below the signal in channel N, that is, an additional 120 dB below the signal in channel $N + 1$, which is already 60 dB down.

5.8 CROSSTALK DUE TO LOW-FREQUENCY CUTOFF OF THE CHANNEL

The type of pulse distortion with consequent crosstalk that results from low-frequency cutoff of the channel is shown in Fig. 5.8-1. Here we have again assumed for simplicity that the channel may be represented by a simple RC high-pass filter as shown in Fig. 5.8-1a. As shown in Fig.

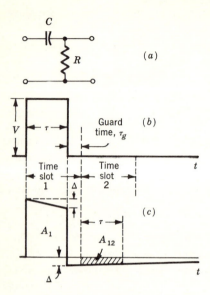

Fig. 5.8-1 (*a*) An *RC* high-pass circuit is used to represent the transmission characteristics of the channel in the neighborhood of its lower cutoff frequency. (*b*) Two neighboring time slots, the first of which is occupied by a pulse. (*c*) Showing the pulse distortion and consequent overlapping into other slots caused by the restricted bandwidth of the channel.

5.8-1*c*, the initially flat top of the pulse in Fig. 5.8-1*b* develops a tilt which accumulates to an amount $\Delta = V(1 - e^{-\tau/\tau_c})$, where again τ_c is the channel time constant $\tau_c = RC$.   The low-frequency 3-dB cutoff frequency of the channel is

$$f_1 = \frac{1}{2\pi RC} = \frac{1}{2\pi\tau_c} \tag{5.8-1}$$

As we shall verify, to reduce crosstalk, we require that $\tau \ll \tau_c$.   In this case the exponential tilt is very nearly linear, and $\Delta \cong V\tau/\tau_c$.   If $\tau_g \ll \tau_c$, as is certainly the case, then we find that the area A_{12} which measures the crosstalk signal is very nearly

$$A_{12} \approx \frac{V\tau^2}{\tau_c} \tag{5.8-2}$$

Since $\Delta \ll V$, the area $A_1 \cong V\tau$, so that the crosstalk ratio K is given by

$$K = \frac{\tau}{\tau_c} \tag{5.8-3}$$

If we then require that $K = 10^{-3}$, corresponding to a crosstalk signal 60 dB below the signal in the preceding channel, we require similarly that $\tau/\tau_c = 10^{-3}$.   In this case the approximations made above are certainly justified.

Let us use again the numerical values of the example of the previous section.   There we found that $\tau = 6.8$ μsec.   Hence for $K = 10^{-3}$,

$\tau_c = 1000$ and $\tau = 6.8 \times 10^{-3}$ sec. We then find that the low-frequency cutoff of the channel must be, from Eq. (5.8-1), $f_1 = 23$ Hz.

We noted that crosstalk which results from the upper-frequency bandwidth restriction involves only neighboring channels. The situation is different with crosstalk caused by low-frequency bandwidth restriction. In order to keep the tilt Δ small (see Fig. 5.8-1), it is necessary that $\tau_c \gg \tau$. However, the overshoot which produces the crosstalk decays very slowly, and crosstalk extends to many time slots.

5.9 PULSE-TIME MODULATION

In time-division multiplexing, a regularly recurring time slot is allocated to each baseband signal. In PAM the baseband signal modulates the amplitude of a pulse which is located within a specific time slot. As an alternative modulation scheme, we vary, rather than the *amplitude*, some feature of the *timing* of a pulse which falls within the time slot. Such *pulse-time modulation* (PTM) may be accomplished in a number of ways, two of which are indicated in Fig. 5.9-1.

A baseband modulating signal $m(t)$ is shown in Fig. 5.9-1*a*. The signal is regularly sampled, and at each sampling instant a pulse is gen-

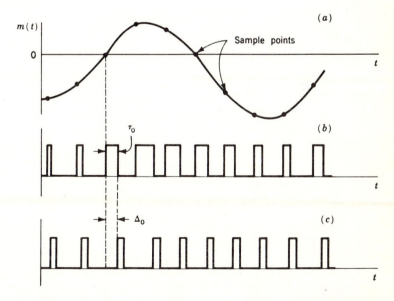

Fig. 5.9-1 (*a*) A baseband signal. (*b*) A pulse train is pulse-duration modulated by the baseband signal. (*c*) A pulse train is pulse-position modulated by the baseband signal.

erated, as shown in Fig. 5.9-1b, of fixed amplitude, but whose *duration* is modulated by $m(t)$. Corresponding to $m(t) = 0$ the unmodulated pulse width is τ_0, and the departure of pulse width from τ_0 is proportional to $m(t)$. Modulation of this type is called *pulse-width modulation* (PWM), *pulse-duration modulation* (PDM), or *pulse-length modulation* (PLM). It is, of course, clear that the duration τ_0 must be smaller than the allowable duration of the time slot allocated to a sample. In addition, the maximum duration of a pulse must fall short of the time separation between samples in order that there be a separation, i.e., a guard time, between the termination of one pulse and the beginning of the next.

Another modulating technique is shown in Fig. 5.9-1c. In this figure the modulation appears in the timing of the trailing edge of the pulse. It is also feasible to modulate, instead, the timing of the leading edge or even both edges simultaneously. Note that in Fig. 5.9-1c the baseband signal modulates the position within the time slot of a constant-duration constant-amplitude pulse. To correspond to $m(t) = 0$, we have delayed the pulse an amount Δ_0 with respect to the occurrence of the sampling point. The change in delay from Δ_0 is proportional to the modulating signal. Modulation of this type is called *pulse-position modulation* (PPM).

Pulse-duration modulation (PDM, PWM, or PLM) and pulse-position modulation (PPM) are forms of pulse-timing modulation (PTM). It is interesting to note that in a general way, PAM and PTM bear the same relationship to one another as do AM and FM. In AM, the carrier is a waveform having a regular rate of recurrence, and information is superimposed by modulating the amplitude. In FM, the carrier amplitude is constant, and information is superimposed by varying the spacings between zero crossings, i.e. varying the instantaneous frequency. Similarly in PAM the carrier, of constant repetition frequency, is varied in *amplitude*, while, as is apparent in Fig. 5.9-1, the amplitude in PTM is fixed but the *spacing* of the pulse edges varies with the modulation.

5.10 METHODS OF GENERATING
PULSE-TIME-MODULATED SIGNALS

One method of generating pulse-time-modulated signals is illustrated in Fig. 5.10-1. We begin by sampling a baseband signal and generating a *flat-topped* PAM signal. Three successive samples of such a PAM wavetrain, $S_A(t)$, are shown in Fig. 5.10-1a. We generate, synchronously with the samples, the linear ramp-type pulse waveform $R(t)$ shown in Fig. 5.10-1b. These two signals are added as shown in Fig. 5.10-1c, and the sum is applied to a comparator circuit. A comparator circuit has associated with it a "reference level" and an input terminal pair to which is applied,

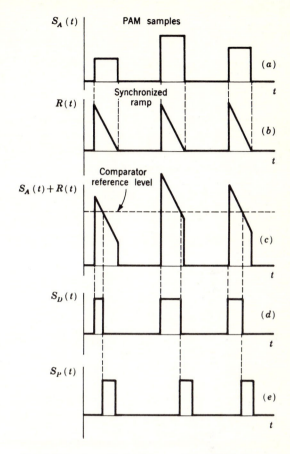

Fig. 5.10-1 A method of generating PTM signals. (a) A PAM signal $S_A(t)$. (b) An auxiliary synchronized ramp waveform $R(t)$. (c) The sum $S_A(t) + R(t)$. The comparator reference level is indicated. (d) A pulse-duration-modulated signal. (e) A pulse-position-modulated signal.

in the present case, the signal $S_A + R$. A comparator has an output terminal whose voltage may assume only two voltage levels. One voltage level is held as long as the input signal is less than the reference level, and the other level is held whenever the input exceeds the reference level. The transition at the output between these two levels occurs very abruptly.

The ramp amplitude is adjusted to be somewhat larger than the variation in amplitude of the PAM samples in Fig. 5.10-1a. The comparator reference may then be located so that it always intersects the sloping portion of the waveform $S_A + R$. The first crossing of the reference level by the waveform $S_A + R$ generates the leading edge of a pulse output of the comparator. The second crossing generates the trailing edge. These pulses, shown in Fig. 5.10-1d, constitute a pulse-duration-modulated waveform. If the ramp in Fig. 5.10-1b is linear with time, the duration modulation in Fig. 5.10-1a is also linear with time. The constant amplitude of the pulses in Fig. 5.10-1d is determined by the output-voltage levels of the comparator and is not related to the amplitude of the pulses in Fig. 5.10-1a.

One way of generating a pulse-position-modulated waveform is to use the pulse-duration waveform. We use a pulse generator which generates pulses of fixed amplitude and duration. Monostable multivibrators, delay-line controlled blocking oscillators, etc., are suitable pulse

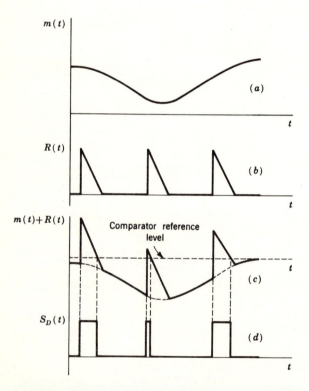

Fig. 5.10-2 A method of PTM which does not involve an initial generation of a PAM signal. (a) The baseband signal $m(t)$. (b) The ramp waveform $R(t)$. (c) The sum $m(t) + R(t)$. The comparator reference is shown. (d) The PDM signal.

generators. We then simply arrange to trigger the pulse generator with the trailing edges of the pulses in the PDM waveform, as we see in Fig. 5.10-1d. The timing of the pulses in the PPM waveform then appears as in Fig. 5.10-1e.

To generate a PTM waveform, it is not necessary first to generate a PAM waveform as is suggested above and indicated in Fig. 5.10-1. Rather, we may, as shown in Fig. 5.10-2, simply add the baseband signal directly to a regularly recurring ramp-type pulse and apply the sum to a comparator. The baseband waveform $m(t)$ is shown in Fig. 5.10-2a, and the ramp waveform in Fig. 5.10-2b. The sum is shown in Fig. 5.10-2c, and the PDM waveform output of the comparator appears in Fig. 5.10-2d. The leading edges of the pulses in Fig. 5.10-2d occur regularly, and the modulation appears on the trailing edge as before. It is of interest to note that in the present case the duration of each pulse is dependent on the magnitude of $m(t)$ at the time of occurrence of the trailing edge, rather than on the value of $m(t)$ at the regularly recurring time of the leading edge. The sampling in this case is therefore referred to as *natural sampling*, in distinction to the *uniform sampling* illustrated in the waveform of Fig. 5.10-1.

Finally we may note that if we choose to modulate the leading edge of the pulses in the PDM waveform, we may do so simply by reversing the waveform of the ramp-type pulses. Pulses with both edges modulated will result if both sides of the ramp waveform are inclined.

5.11 SPECTRA OF DURATION-MODULATION WAVEFORMS AND BASEBAND SIGNAL RECOVERY

Duration-modulated waveforms may be generated in a host of varieties. PDM waveforms may be generated with the modulation superimposed on the leading edges of the pulses, on the trailing edges, or on both edges simultaneously. These waveforms may be generated through uniform sampling or through natural sampling. The various analytic expressions for the spectra of these signals may be found in Ref. 1. We shall not reproduce these expressions here because they are quite unwieldy.

We now discuss, qualitatively, the spectrum of a naturally sampled PDM waveform with the modulation superimposed on the trailing edge of the pulse. Such a spectrum is shown qualitatively in Fig. 5.11-1 for the case of sinusoidal modulation with frequency f_m and with the sampling frequency f_s. We observe, first of all, the appearance of spectral lines at dc, at f_s, and at harmonics of f_s. These lines are, of course, to be expected because they constitute the spectrum of the unmodulated pulse train which is the "carrier" of the modulation. We find next that, with modulation, each carrier spectral line (except for the line at dc) is accom-

panied by an FM-type set of sidebands. That is, on each side of each
carrier there appears a symmetrical pattern of sideband lines, separated
from the carrier line by multiples of the modulation frequency. This
last result is not unexpected in view of the comparison suggested earlier
between PTM and FM.

Somewhat surprisingly, we find that the spectral line at dc, due to
the carrier, is not accompanied by an FM-type pattern of sideband components but rather by only a single sideband at frequency f_m. This
feature suggests that the baseband signal may be recovered by passing
the PDM signal through a low-pass filter in the manner in which the baseband signal is recovered from a PAM signal. Such is indeed the case, but
only approximately. In the PAM case, each of the lines at f_s, $2f_s$, etc.
would be accompanied by only a single sideband pair. Hence if $f_s > 2f_m$,
it is possible to adjust the cutoff frequency of a low-pass filter such that
the filter passes only the baseband spectral component. In the pattern
of Fig. 5.11-1 the sideband spectra extend indefinitely outward from each
carrier line, albeit with generally decreasing amplitude at greater distances. Therefore any low-pass filter must include some of the lower-
sideband components of the carrier at f_s and to a lesser extent the lower-
sideband components of $2f_s$, etc. It is found, however, that at least for
voice communications it is possible to keep the distortion within acceptable limits. Distortion is minimized by raising the sampling frequency
and by arranging that within any channel the variation from its unmodulated position of the modulated edge of the pulse be small in comparison
with the time interval between successive pulses. This last condition
is imposed naturally in a system in which many baseband signals are
being time-division multiplexed.

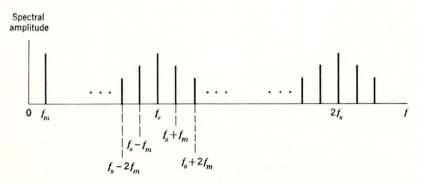

Fig. 5.11-1 The spectrum of a pulse-duration-modulated waveform generated by
natural sampling of a sinusoidal baseband signal of frequency f_m at a sampling
rate f_s.

When the sampling in PDM is done *uniformly* rather than *naturally*, it turns out that the spectrum contains, in addition to the components shown in Fig. 5.11-1, harmonics of the baseband signal itself. That is, the carrier component at *dc* develops an FM-type sideband spectrum rather than the AM-type shown in Fig. 5.11-1. In this case, the quality of baseband-signal recovery by low-pass filtering suffers somewhat.

We may see rather intuitively how the low-pass filter extracts the baseband signal from a PDM waveform. As already noted, when a train of narrow pulses of repetition frequency $2f_m$ is applied to an ideal low-pass filter having a cutoff frequency f_m, the filter output varies with the *area of the pulses*. On this basis it is a matter of indifference whether the *pulse areas* are changed by varying the pulse *amplitude* or the pulse *duration*.

5.12 RECOVERY OF BASEBAND SIGNAL BY CONVERSION OF PDM AND PPM TO PAM

We note that the low-pass filter method of baseband-signal recovery in PDM necessarily results in some distortion. While the distortion may be within acceptable limits in some cases, it is of interest to know how, in principle, the baseband signal may be precisely recovered. The method to be described is applicable both to PDM and to PPM and is illustrated in the idealized waveforms of Fig. 5.12-1. The basic idea presented in this section is that since PDM and PPM could be generated from a PAM waveform, then by applying the inverse operations we can convert the PDM and PPM waveforms back to a PAM signal and then demodulate the PAM signal.

Two successive pulses in a PDM waveform are shown in Fig. 5.12-1*a*. These pulses are used to generate the waveform shown in Fig. 5.12-1*b*. The leading edges of the pulses initiate the generation of a linear ramp whose rise is terminated at the trailing edge. The height attained by the ramp is therefore proportional to the pulse duration. The level attained by the ramp is sustained for a time (i.e., the capacitor in the ramp-forming circuit is not permitted to discharge), after which, at some rather arbitrary time, the voltage is returned to its initial level. A sequence of pulses, locally generated at the demodulator, is added to the waveform in Fig. 5.12-1*b*. These pulses are of fixed amplitude and duration and are timed so that they sit on the pedestal of the waveform in Fig. 5.12-1*b*. The waveform in Fig. 5.12-1*c* is thereby generated. This waveform is applied to a *clipping* circuit which transmits only the portion of the waveform above the reference level. The output of the clipping circuit constitutes the PAM waveform shown in Fig. 5.12-1*d*.

The baseband signal is now, of course, to be recovered by passing the PAM waveform through an appropriate low-pass filter. In order

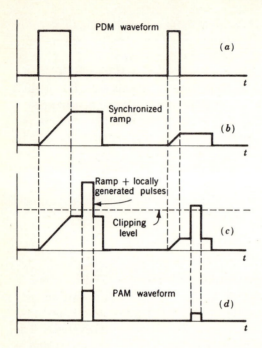

Fig. 5.12-1 Baseband signal recovery from a PDM waveform through conversion to PAM. (*a*) The PDM signal. (*b*) A synchronized ramp-pedestal waveform. (*c*) Locally generated pulses of fixed amplitude are added. The clipping level is indicated. (*d*) The PAM waveform.

that distortion be avoided, it is necessary that the PAM samples be able to be approximated by impulses. If many baseband signals are being multiplexed, the time slot available for a particular sample pulse is usually very small in comparison with the time between successive samples of the same baseband signal, that is, $1/2f_M$. In this case, the approximation is valid, and the modulation can be perfectly reconstructed.

The same technique may be used to convert a PPM waveform into a PAM waveform prior to signal recovery through the use of a low-pass filter. Considering again a single-baseband signal, we arrange to initiate a ramp at the beginning of the time slot. The rise of the ramp is terminated by the occurrence of the leading edge of the PPM sample pulse, and the remainder of the procedure is the same as before.

5.13 CROSSTALK IN PTM SYSTEMS

In PTM, just as in PAM, if the channel used to transmit the multiplexed signal is not of infinite bandwidth, crosstalk will occur between baseband channels. The origin of the crosstalk is much the same in PTM as in PAM. However, we shall discuss crosstalk in PTM qualitatively in order to point out a distinction of interest.

As noted in the previous section, the only feature of a sample pulse in PTM which is of consequence is its timing. In addition, the only information we need extract from the pulse is the time of occurrence of its time-modulated edge. (In PPM, where both edges are modulated, either edge will serve.) The pulse amplitude and duration are of no consequence, provided that they are large enough to permit the pulse to stand out distinctly against the background noise. We note in passing that for this reason PDM suffers a disadvantage with respect to PPM. In PPM the pulse of fixed width can always be kept as narrow as is consistent with the background noise. In PDM the pulse width varies and is narrowest at only one extreme of the modulation cycle.

A standard procedure for noting the time of a pulse edge is to apply the pulse to a comparator-type circuit whose reference voltage level is set to be nominally midway between the top and bottom of the pulse. At the moment the pulse voltage crosses the reference level, the comparator output makes a large and abrupt change.

Now let us consider a PPM multiplexed signal being transmitted over a channel which is limited in bandwidth at its high-frequency end. Then the initially rectangular pulses will arrive at the demodulator with finite rise and fall times. Two such successive pulses in the pulse train are shown in Fig. 5.13-1. The first is a sample pulse of baseband channel N, the second of baseband channel $N + 1$. The solid waveform corresponds to the case where neither baseband channel is modulated. The times t_N and t_{N+1} are the times of occurrence of the leading edges of the pulses in the unmodulated case. The comparator responds at t_{rN} and $t_{r(N+1)}$ which occur later than t_N and t_{N+1} by a *fixed* interval, and therefore no difficulty is introduced. Suppose, now, that channel N is modulated, as a consequence of which the pulse of channel N is delayed by a time ΔT. This delayed pulse is shown by the dashed waveforms in Fig. 5.13-1. It then appears that even though channel $N + 1$ was not modulated, the time at which the pulse of channel $N + 1$ crosses the reference level has been advanced by the time Δt. A measure of the crosstalk between channels is the ratio $\Delta t/\Delta T$.

The crosstalk time Δt in Fig. 5.13-1 depends, of course, on the amount by which the delayed *tail* of the preceding pulse has *lifted* the initiation point I of the succeeding pulse. However, Δt also depends on the rate of rise of this succeeding pulse as it crosses the reference level. Thus, suppose it were possible to have a type of pulse distortion which resulted in an extended exponential tail as in Fig. 5.13-1 but which left the leading edge of the pulse entirely vertical, that is, with zero rise time. In such a case the reference-level crossing would always take place exactly at the time of initiation of the pulse, Δt would always be zero, and there would be no crosstalk. Note that this is not the case in PAM, where signal information is extracted from the area under a pulse rather than its timing.

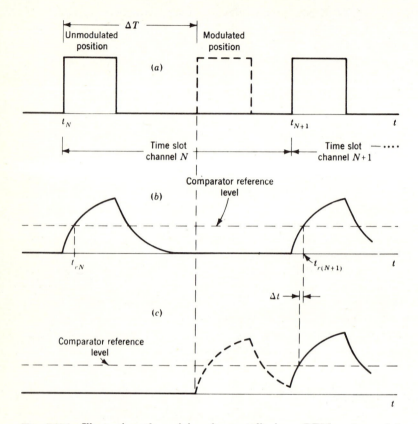

Fig. 5.13-1 Illustrating the origin of crosstalk in a PTM system. (*a*) Unmodulated positions of pulses in channel *N* and *N* + 1 and modulated position (dashed) of channel *N* pulse. (*b*) Showing distortion in pulse waveforms due to limited bandwidth. (*c*) Showing timing error introduced in channel *N* + 1 due to modulation of channel *N*.

We observe that when the bandwidth of the multiplex channel is extended, crosstalk of the type contemplated in Fig. 5.13-1 decreases, both because the tail of the preceding pulse decays more rapidly and also because the leading edge of the succeeding pulse rises more rapidly. As a consequence, the crosstalk decreases with increasing bandwidth more rapidly in PPM than in PAM.

When the bandwidth of the channel does not extend to adequately low frequencies, the pulse waveform will develop *tilts* on its nominally flat portions as is illustrated for a PAM signal in Fig. 5.8-1. When the sample pulse of one multiplex channel is modulated, the succeeding pulses will ride up and down on the slowly decaying backswing. In the case of PAM, where pulse-area changes represent signal, crosstalk will result and will not depend particularly on the rise time of the pulses. Note that in Fig. 5.8-1 we assumed zero rise time. In the case of PPM, however, if the rise time is zero, i.e., if the upper-frequency cutoff of the multiplex channel is arbitrarily large, there will be no crosstalk even in the presence of pulse distortion due to inadequate low-frequency response.

5.14 SYNCHRONIZATION OF TRANSMITTER AND RECEIVER

In any type of time-division multiplexing system it is required that the commutation at the transmitting end and decommutation at the receiving end be in step, that is, in synchronism with each other. Thus, for example, in the PAM system of Fig. 5.2-1, when the commutator switch is at position 3, the decommutator switch must also be in position 3. At the transmitter, there is a source which generates a regularly recurring waveform to switch the commutator from one time slot to the next. This source of time is called a "clock" and may be a sinusoidal oscillator from which, using various waveshaping circuits, the required gating voltages are derived. The repetition frequency of this clock may well be equal to the product of the sampling frequency and the number of channels being multiplexed, thereby assigning one clock cycle per time slot.

At the demodulator end, a clock signal is required to keep the decommutator running at the same rate as the commutator. In addition, there must be agreement in the baseband channel selected at the two ends. In order to keep the demodulator clock running at the proper rate and in order to keep the channel selection the same at the two ends, it is necessary to periodically transmit, along with the multiplexed signal, information to tell the demodulator where, in its cycle of operations, the commutator is to be found.

The time interval, from the beginning of the time slot allocated to a particular channel until the next time of recurrence of that time slot, is called a *frame*. A very common arrangement is the allowance of one time slot per frame for the purpose of transmitting synchronization information. If, say, N baseband channels are to be multiplexed, $N + 1$ time slots are provided, N for the signal samples and one for synchronizing information. If many baseband signals are to be multiplexed (say several thousand, as in a telephone multiplex system), the time duration of a frame may be so long that several time slots per frame are allocated to synchronization.

The synchronization time slot is distinguished from the signal sample slot by the distinctive nature of its pulse. In PAM the frame synchronization pulse is distinguished by its amplitude, which is made larger than the amplitude of any signal pulse. Since the synchronization pulse is

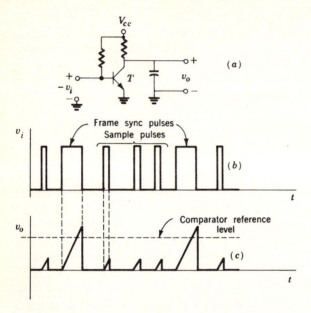

Fig. 5.14-1 (*a*) A circuit used to distinguish a frame synchronizing pulse in a PPM signal. (*b*) A PPM signal. Three channels are being multiplexed. (*c*) The circuit output voltage v_o with comparator reference indicated.

used only for timing, its time of occurrence rather than area is of significance. This time may be the time at which the synchronization-pulse voltage passes through the reference level of a comparator, whose reference level is set somewhat above the maximum attained by the signal pulses.

SYNCHRONIZATION IN PTM

A representative scheme which allows identification of the time of occurrence of the synchronization slot in PTM is shown in Fig. 5.14-1. A three-channel PPM pulse train is shown in Fig. 5.14-1*b*. The frame synchronization pulse is wider than the sample pulses. The pulse train, inverted, is applied to the base of transistor T, which is used as a switch. During the pulse, the transistor is cut off, and the capacitor charges. At the end of a pulse the transistor turns on, and the capacitor discharges rapidly. The waveform v_o is shown in Fig. 5.14-1*c*. Since the synchronization pulses are wider than the sample pulses, only at the occurrence of a synchronization pulse does the waveform v_o rise to a level adequate to cause a response by the comparator.

PROBLEMS

5.1-1. The signal $v(t) = \cos 5\pi t + 0.5 \cos 10\pi t$ is instantaneously sampled. The interval between samples is T_s.

(a) Find the maximum allowable value for T_s.

(b) If the sampling signal is $S(t) = 5 \displaystyle\sum_{k=-\infty}^{\infty} \delta(t - 0.1k)$, the sampled signal $v_s(t) = v(t)S(t)$ consists of a train of impulses, each with a different strength

$$v_s(t) = \sum_{k=-\infty}^{\infty} I_k \, \delta(t - 0.1k)$$

Find I_0, I_1, and I_2, and show that $I_k = I_{4+k}$.

(c) To reconstruct the signal $v_s(t)$ is passed through a rectangular low-pass filter. Find the minimum filter bandwidth to reconstruct the signal without distortion.

5.1-2. We have the signal $v(t) = \cos 2\pi f_0 t + \cos 2 \times 2\pi f_0 t + \cos 3 \times 2\pi f_0 t$. Our interest extends, however, only to spectral components up to and including $2f_0$. We therefore sample at the rate $5f_0$ which is adequate for the $2f_0$ component of the signal.

(a) If sampling is accomplished by multiplying $v(t)$ by an impulse train in which the impulses are of unit strength, write an expression for the sampled signal.

(b) To recover the part of the signal of interest, the sampled signal is passed through a rectangular low-pass filter with passband extending from 0 to slightly beyond $2f_0$. Write an expression for the filter output. Is the part of the signal of interest recovered exactly? If we want to reproduce the first two terms of $v(t)$ without distortion, what operation must be performed at the very outset?

5.1-3. The bandpass signal $v(t) = \cos 10\omega_0 t + \cos 11\omega_0 t + \cos 12\omega_0 t$ is sampled by an impulse train $S(t) = I \displaystyle\sum_{k=-\infty}^{\infty} \delta(t - kT_s)$.

(a) Find the maximum time between samples, T_s, to ensure reproduction without error.

(b) Using the result obtained in (a), obtain an expression for $v_s(t) = S(t)v(t)$.

(c) The sampled signal $v_s(t)$ is filtered by a rectangular low-pass filter with a bandwidth $B = 2f_0$. Obtain an expression for the filter output.

(d) The sampled signal $v_s(t)$ is filtered by a rectangular bandpass filter extending from $2f_0$ to $4f_0$. Obtain an expression for the filter output.

5.1-4. The bandpass signal $v(t) = \cos 10\omega_0 t + \cos 11\omega_0 t + \cos 12\omega_0 t$ is sampled by an impulse train, $S(t) = I \displaystyle\sum_{k=-\infty}^{\infty} \delta(t - k/8f_0)$. The sampled signal $v_s(t) = S(t)v(t)$ is then filtered by a rectangular low-pass filter having a bandwidth $B = 2f_0$. Obtain an expression for the filter output.

5.1-5. Let us view the waveform $v(t) = \cos \omega_0 t$ as a bandpass signal occupying an arbitrarily narrow frequency band. On this basis we find that the required sampling rate is $f_s = 0$. Discuss.

5.2-1. The TDM system shown in Fig. 5.2-1 is used to multiplex the four signals $m_1(t) = \cos \omega_0 t$, $m_2(t) = 0.5 \cos \omega_0 t$, $m_3(t) = 2 \cos 2\omega_0 t$, and $m_4(t) = \cos 4\omega_0 t$.

(a) If each signal is sampled at the same sampling rate, calculate the minimum sampling rate f_s.

(b) What is the commutator speed in revolutions per second.

(c) Design a commutator which will allow each of the four signals to be sampled at a rate no faster than is required to satisfy the Nyquist criterion for the individual signal.

5.2-2. Three signals m_1, m_2, and m_3 are to be multiplexed. m_1 and m_2 have a 5-kHz bandwidth, and m_3 has a 10-kHz bandwidth. Design a commutator switching system so that each signal is sampled at its Nyquist rate.

5.3-1. Show that the response of a rectangular low-pass filter, with a bandwidth f_c, to the impulse function $I\,\delta(t - k/2f_c)$ is

$$S_R(t) = \frac{I\omega_c}{\pi} \frac{\sin \omega_c(t - k/2f_c)}{\omega_c(t - k/2f_c)}$$

Assume that in its passband the filter has $H(f) = 1$.

5.3-2. Four signals, $m_1(t) = 1 \cos \omega_0 t$, $m_2(t) = 1 \sin \omega_0 t$, $m_3(t) = -1 \sin \omega_0 t$, and $m_4(t) = -1 \cos \omega_0 t$ are sampled every $1/2f_0$ sec by the sampling function

$$S(t) = 1 \sum_{k=-\infty}^{\infty} \delta\left(t - \frac{k}{2f_0}\right)$$

The signals are then time-division multiplexed. The TDM signal is filtered by a rectangular low-pass filter having a bandwidth $f_c = 4f_0$ and then decommutated.

(a) Sketch the four outputs of the decommutator.

(b) Each of the four output signals is filtered by a rectangular low-pass filter having a bandwidth f_0. Show that the four signals are reconstructed without error.

5.3-3. The four signals of Prob. 5.3-2 are sampled, as indicated in that problem, and time-division multiplexed. The TDM signal is filtered by a rectangular low-pass filter having a bandwidth $f_c = 2f_0$ and then decommutated. Sketch the output at the decommutator switch segment where the samples of $m_1(t)$ should appear and show that $m_1(t)$ cannot be recovered.

5.4-1. The signal $v(t) = \cos \omega_0 t + \cos 8\omega_0 t$ is sampled by using *natural sampling*.

(a) Determine the minimum sampling rate f_s.

(b) Sketch $v_s(t) = S(t)v(t)$ if $S(t)$ is a train of pulses having unit height, occurring at the rate f_s; and $S(t) = 1$ for $nT - \tau/2 \leq t \leq nT + \tau/2$. The pulse duration is $\tau = 1/32f_0$.

(c) Repeat (b) if $\tau = 1/320f_0$.

5.5-1. Show that an impulse function $I\delta(t)$ can be stretched to have a width τ by passing the impulse function through a filter $(1 - e^{-j\omega\tau})/j\omega$. Show that this operation is identical with integrating the impulse for τ sec, that is, that the output $v_o(t)$ is given by

$$v_o(t) = \int_0^t I\,\delta(t)\,dt \qquad 0 \leq t \leq \tau$$
$$= 0 \qquad\qquad\qquad \text{otherwise}$$

5.5-2. The signal $v(t) = \cos 5\pi t + 0.5 \cos 10\pi t$ is flat-topped sampled at the Nyquist rate. Let $p(t)$ represent a pulse of unit amplitude extending from $t = 0$ to $t = \tau$. The sampling signal $S(t) = \sum\limits_{k=-\infty}^{\infty} p(t - kT)$ where T is the period of the pulses. Find T. Write an expression for the flat-topped sampled signal $v_s(t)$. Show that $v_s(t)$ is periodic.

5.5-3. A signal bandlimited to f_M is flat-topped sampled. In the receiver it is passed through an equalizing filter before being low-pass filtered. If the pulse width $\tau = 1/4f_M$, sketch the magnitude of the transfer function of the equalizing filter.

5.5-4. In Prob. 5.5-2 let $\tau = \frac{1}{20}$ sec.

(a) Expand $v_s(t)$ in a Fourier series.

(b) $v_s(t)$ is passed through a rectangular low-pass filter having a bandwidth $B = 5$ Hz. Find an expression for the output of the low-pass filter and compare with the original signal.

5.6-1. One thousand signals each having a 5-kHz bandwidth are sampled at the Nyquist rate and then time-division multiplexed. The sampling waveform is an impulse train $S(t) = I \sum\limits_{k=-\infty}^{\infty} \delta(t - kT_s)$.

(a) Find T_s.

(b) The sample impulses $v_s(t)$ are stretched to form pulses of duration τ. Find the maximum value of τ.

5.7-1. The signals $m_1(t) = \sin \omega_0 t$ and $m_2(t) = \sin \omega_0(t - 1/8f_0)$ are each sampled at the rate $4f_0$ (twice the Nyquist rate). $m_1(t)$ is sampled at times $t = k/4f_0$ $(k = 0, 1, 2, \ldots)$ and $m_2(t)$ is sampled at times $t = 1/8f_0 + k/4f_0$. The sampling waveforms are impulse trains, $S_1(t)$ sampling $m_1(t)$, and $S_2(t)$ sampling $m_2(t)$, where

$$S_1(t) = 1 \sum_{k=-\infty}^{\infty} \delta\left(t - \frac{k}{4f_0}\right)$$

and

$$S_2(t) = 1 \sum_{k=-\infty}^{\infty} \delta\left(t - \frac{1}{8f_0} - \frac{k}{4f_0}\right)$$

The signal samples are interleaved and transmitted over a channel which can be represented by a cascade of an RC filter having a 3-dB bandwidth $f_{c1} = 2f_0$, and a rectangular low-pass filter having a cutoff frequency $f_{c2} = 1.1f_0$.

(a) Sketch the waveform present at the output of the channel.

(b) The channel output is decommutated by sampling the waveform every $1/8f_0$ sec. The decommutator switch arm dwells for an arbitrarily short time on each decommutator segment. Will the signals be reconstructed without error in amplitude and phase? Discuss your result.

5.7-2. For the numerical example presented in Sec. 5.7 show that the crosstalk in channel $N + 2$ due to channel N is 180 dB below the signal in channel N.

5.8-1. Two signals $m_1(t)$ and $m_2(t)$ are time-division multiplexed after flat-top sampling. The pulse duration τ of each sample completely fills the time slot leaving no guard band. During a particular time interval lasting for 100 samples, $m_1(t) = 10\ V$ and $m_2(t) = 0$. The TDM signal is passed through a channel which can be represented by an RC high-pass filter with a 3-dB frequency $f_c = 100$ Hz. Assume that the Nyquist rate for each signal is 10 kHz.

(a) Find τ.

(b) Calculate Δ.

(c) Find the area A_{12}.

(d) Find the crosstalk ratio K.

5.10-1. Devise a scheme whereby both edges are simultaneously modulated and thereby form a PTM signal. How would you recover the signal?

5.10-2. Devise a scheme to form a PTM signal by modulating the *leading* edge of a pulse.

5.10-3. Show block diagrams of the circuits required to form a pulse-duration-modulated signal.

5.10-4. Show block diagrams of the circuits required to form a pulse-position-modulated signal.

5.10-5. Show block diagrams of the circuits required to form a pulse-duration-modulated signal using the technique shown in Fig. 5.10-2.

5.11-1. A sinusoidal signal 1 cos $\omega_m t$ is pulse-duration modulated using the technique shown in Fig. 5.10-1. Assume that sampling occurs every $T_s = 1/4f_m$ and that one sample occurs at $t = 0$.

(a) Select the height of the sawtooth waveform $R(t)$ and the comparator level so that the duration-modulated pulse train $S_D(t)$ can have pulses that vary between $t = 0$ and $t = T_s/2$.

(b) Sketch $S_D(t)$.

(c) Observe that $S_D(t)$ is periodic. Expand $S_D(t)$ into a Fourier series and sketch the spectrum for $0 \le f \le 10f_m$.

5.12-1. Illustrate in detail how to convert a PPM waveform back to a PAM waveform prior to signal recovery.

5.13-1. Two signals $m_1(t)$ and $m_2(t)$ are pulse-position modulated and then time-division multiplexed. Both $m_1(t)$ and $m_2(t)$ have the same sampling rate and are, therefore, interleaved. During a certain interval of time, $m_1(t) = 0$ so that each sample of $m_1(t)$ is a pulse occurring $t = T/2$ sec after the start of its sampling interval. The sampling interval lasts T sec. During that same time interval

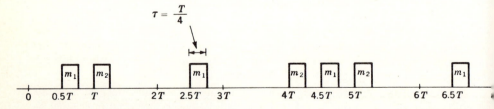

Fig. P5.13-1 A time-division multiplexed PPM signal.

$m_2(t)$ is periodic and results in a PPM signal which alternates between $t = 0$ and $t = T$ sec after the start of its sampling interval. The resulting TDM signal is shown in Fig. P5.13-1. We note from Fig. P5.13-1 that our choice of $m_1(t)$ and $m_2(t)$ results in the TDM signal being periodic.

Let us select $\tau = T/4$ and assume that the channel can be represented by an RC low-pass circuit having a 3-dB frequency $f_h = 2/\pi T$ ($RC = T/4$). In the receiver a threshold is set at 0.5 volt. When the input TDM signal exceeds the threshold, a pulse results.

(a) Sketch the TDM waveform at the channel output.

(b) Indicate the times at which the threshold is crossed.

(c) Where is the timing error worse?

(d) The receiver produces a voltage which is proportional to the time between the start of the interval and the time that the threshold is crossed. If the channel has an infinite high-frequency response, sketch $m_1(t)$ and $m_2(t)$ at the output of this voltage-timing circuit. Neglect subsequent low-pass filtering.

5.14-1. How does the providing of a synchronization pulse every N pulses tell us when each of the N pulses occurs? Explain by sketching a block diagram of your system.

REFERENCE

1. Black, H. S.: "Modulation Theory," D. Van Nostrand Co. Inc., New York, 1953.

6

Pulse-code Modulation

All of the modulation systems considered, up to this point, are *analog* systems. The carrier is a sinusoidal or pulse waveform and the modulation *continuously* varies the amplitude or timing of the carrier. We consider in this chapter a type of modulation which is *digital* rather than *analog* and in which the digital message is *encoded* before transmission.

6.1 NOISY COMMUNICATIONS CHANNELS

We consider a basic problem associated with the transmission of a signal over a noisy communication channel. For the sake of being specific, suppose we require that a telephone conversation be transmitted from New York to Los Angeles. If the signal is transmitted by radio, then, when the signal arrives at its destination, it will be greatly attenuated and also combined with noise due to all manner of random electrical disturbances which are added to the radio signal during its propagation across country. (We neglect as irrelevant, for the present discussion, whether such direct radio communication is reliable over such long channel distances.)   As a result, the received signal may not be distinguishable against its background of noise.   The situation is not fundamentally different if the signal is transmitted over wires.   Any physical wire transmission path will both attenuate and distort a signal by an amount which increases with path length.   Unless the wire path is completely and perfectly shielded, as in the case of a perfect coaxial cable, electrical noise and crosstalk disturbances from neighboring wire paths will also be picked up in amounts increasing with the path length.   In this connection it is of interest to note that even coaxial cable does not provide complete freedom from crosstalk.   External low-frequency magnetic fields will penetrate the outer conductor of the coaxial cable and thereby induce signals on the cable. In telephone cable, where coaxial cables are combined with parallel wire signal paths, it is common practice to wrap the coax in Permalloy for the sake of magnetic shielding.

One way of resolving this problem is simply to raise the signal level at the transmitting end to so high a level that, in spite of the attenuation, the received signal substantially overrides the noise.   (Signal distortion may be corrected separately by equalization.)   Such a solution is hardly feasible on the grounds that the signal power and consequent voltage levels at the transmitter would be simply astronomical and beyond the range of amplifiers to generate, and cables to handle.   For example, at 1 kHz, a telephone cable may be expected to produce an attenuation of the order of 1 dB per mile.   For a 3000-mile run, even if we were satisfied with a received signal of 1 mV, the voltage at the transmitting end would have to be 10^{147} volts.

An amplifier at the receiver will not help the above situation, since at this point both signal and noise levels will be increased together.   But suppose that a *repeater* (repeater is the term used for an amplifier in a communications channel) is located at the midpoint of the long communications path.   This repeater will raise the signal level; in addition, it will raise the level of only the noise introduced in the first half of the communications path.   Hence, such a midway repeater, as contrasted with an amplifier at the receiver, has the advantage of improving the received signal-to-noise ratio.   This midway repeater will relieve the burden imposed on transmitter and cable due to higher power requirements when the repeater is not used.

The next step is, of course, to use additional repeaters, say initially at the one-quarter and three-quarter points, and thereafter at points in between.   Each added repeater serves to lower the maximum power level encountered on the communications link, and each repeater improves the signal-to-noise ratio over what would result if the corresponding gain were introduced at the receiver.

In the limit we might, conceptually at least, use an infinite number of repeaters. We could even adjust the gain of each repeater to be infinitesimally greater than unity by just the amount to overcome the attenuation in the infinitesimal section between repeaters. In the end we would thereby have constructed a cable which had no attenuation. The signal at the receiving terminal of the channel would then be the unattenuated transmitted signal. We would then, in addition, have at the receiving end all the noise introduced at all points of the channel. This noise is also received without attenuation, no matter how far away from the receiving end the noise was introduced. If now, with this finite array of repeaters, the signal-to-noise ratio is not adequate, there is nothing to be done but to raise the signal level or to make the channel quieter.

The situation is actually somewhat more dismal than has just been intimated, since each repeater (transistor amplifier) introduces some noise on its own accord. Hence, as more repeaters are cascaded, each repeater must be designed to more exacting standards with respect to *noise figure* (see Sec. 14.10).

6.2 QUANTIZATION OF SIGNALS

The limitation of the system we have been describing for communicating over long channels is that once noise has been introduced any place along the channel, we are "stuck" with it. We now describe how the situation is modified by subjecting a signal to the operation of *quantization*. In quantizing a signal $m(t)$, we create a new signal $m_q(t)$ which is an approximation to $m(t)$. However, the quantized signal $m_q(t)$ has the great merit that it is, in large measure, separable from additive noise.

The operation of quantization is illustrated in Fig. 6.2-1. A baseband signal $m(t)$ is shown in Fig. 6.2-1a. This signal, which is called v_i, is applied to the quantizer input. The output of the quantizer is called v_o. The quantizer has the essential feature that its input-output characteristic has the *staircase* form shown in Fig. 6.2-1b. As a consequence, the output v_o, shown in Fig. 6.2-1c, is the quantized waveform $m_q(t)$. It is observed that while the input $v_i = m(t)$ varies smoothly over its range, the quantized signal $v_o = m_q(t)$ holds at one or another of a number of fixed levels . . . m_{-2}, m_{-1}, m_0, m_1, m_2, . . . , etc. Thus, the signal $m_q(t)$ either does not change or it changes abruptly by a quantum jump S called the *step size*.

The waveform $m'(t)$ shown dotted in Fig. 6.2-1c represents the output waveform, assuming that the quantizer is linearly related to the input. If the factor of proportionality is unity, $v_o = v_i$, and $m'(t) = m(t)$. We see then that the level held by the waveform $m_q(t)$ is the level to which $m'(t)$ is *closest*. The transition between one level and the next occurs at

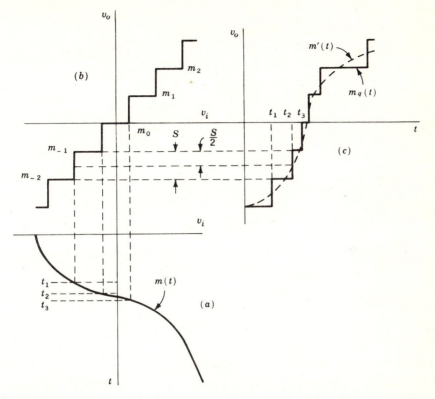

Fig. 6.2-1 Illustrating the operation of quantization. The step size is S. (a) The baseband signal $m(t)$. (b) The input-output characteristic of the quantizer. (c) The quantizer output (solid line) response to $m(t)$. The dashed waveform $m'(t)$ shows the waveform of the output signal for a linear characteristic.

the instant when $m'(t)$ crosses a point midway between two adjacent levels.

We see, therefore, that the quantized signal is an approximation to the original signal. The quality of the approximation may be improved by reducing the size of the steps, thereby increasing the number of allowable levels. Eventually, with small enough steps, the human ear or the eye will not be able to distinguish the original from the quantized signal. To give the reader an idea of the number of quantization levels required in a practical system, we note that 512 levels can be used to obtain the quality of commercial color TV, while 64 levels gives only fairly good color TV performance.

Now let us consider that our quantized signal has arrived at a repeater somewhat attenuated and corrupted by noise. This time our repeater consists of a quantizer and an amplifier. There is noise super-imposed on the quantized levels of $m_q(t)$. But suppose that we have placed the repeater at a point on the communications channel where the instantaneous noise voltage is almost always less than half the separation between quantized levels. Then the output of the quantizer will consist of a succession of levels duplicating the original quantized signal and *with the noise removed*. In rare instances the noise results in an error in quanti-zation level. A noisy quantized signal is shown in Fig. 6.2-2a. The allowable quantizer output levels are indicated by the dashed lines sepa-rated by amount S. The output of the quantizer is shown in Fig. 6.2-2b. The quantizer output is the level to which the input is closest. There-fore, as long as the noise has an instantaneous amplitude less than $S/2$, the noise will not appear at the output. One instance in which the noise does exceed $S/2$ is indicated in the figure, and, correspondingly, an error in level does occur. The statistical nature of noise is such that even if the average noise magnitude is much less than $S/2$, there is always a finite probability that, from time to time, the noise magnitude will exceed $S/2$. Note that it is never possible to suppress completely level errors such as the one indicated in Fig. 6.2-2.

We have shown that through the method of signal quantization, the effect of additive noise can be significantly reduced. By decreasing the spacing of the repeaters, we decrease the attenuation suffered by $m_q(t)$. This effectively decreases the relative noise power and hence decreases the probability P_q of an error in level. P_q can also be reduced by increasing the step size S. However, increasing S results in an increased discrepancy between the true signal $m'(t)$ and the quantized signal $m_q(t)$. This difference $m'(t) - m_q(t)$ can be regarded as noise and

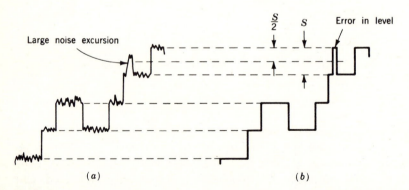

Fig. 6.2-2 (a) A quantized signal with added noise. (b) The signal after requantization. One instance is recorded in which the noise level is so large that an error results.

is called *quantization noise*. Hence, the received signal is not a perfect replica of the transmitted signal $m(t)$. The difference between them is due to errors caused by additive noise and quantization noise. These noises are discussed further in Chap. 12.

6.3   QUANTIZATION ERROR

It has been pointed out that the quantized signal and the original signal from which it was derived differ from one another in a random manner. This difference or error may be viewed as a noise due to the quantization process and is called *quantization error*. We now calculate the mean-square quantization error $\overline{e^2}$, where e is the difference between the original and quantized signal voltages.

Let us divide the total peak-to-peak range of the message signal $m(t)$ into M equal voltage intervals, each of magnitude S volts. At the center of each voltage interval we locate a quantization level m_1, m_2, . . . , m_M as shown in Fig. 6.3-1a. The dashed level represents the instantaneous value of the message signal $m(t)$ at a time t. Since, in this figure, $m(t)$ happens to be closest to the level m_k, the quantizer output will be m_k, the voltage corresponding to that level. The error is $e = m(t) - m_k$.

Let $f(m)\, dm$ be the probability that $m(t)$ lies in the voltage range $m - dm/2$ to $m + dm/2$. Then the mean-square *quantization error* is

$$\overline{e^2} = \int_{m_1-S/2}^{m_1+S/2} f(m)(m - m_1)^2\, dm$$
$$+ \int_{m_2-S/2}^{m_2+S/2} f(m)(m - m_2)^2\, dm + \cdots \quad (6.3\text{-}1)$$

Now, ordinarily the probability density function $f(m)$ of the message signal $m(t)$ will certainly not be constant. However, suppose that the number M of quantization levels is large, so that the step size S is small in comparison with the peak-to-peak range of the message signal. In this case, it is certainly reasonable to make the approximation that $f(m)$ is constant within each quantization range. Then in the first term of Eq. (6.3-1) we set $f(m) = f^{(1)}$, a constant. In the second term $f(m) = f^{(2)}$, etc. We may now remove $f^{(1)}$, $f^{(2)}$, etc., from inside the integral sign. If we make the substitution $x \equiv m - m_k$, Eq. (6.3-1) becomes

$$\overline{e^2} = (f^{(1)} + f^{(2)} + \cdots) \int_{-S/2}^{S/2} x^2\, dx = (f^{(1)} + f^{(2)} + \cdots)\frac{S^3}{12}$$
$$(6.3\text{-}2a)$$

$$= (f^{(1)}S + f^{(2)}S + \cdots)\frac{S^2}{12} \qquad (6.3\text{-}2b)$$

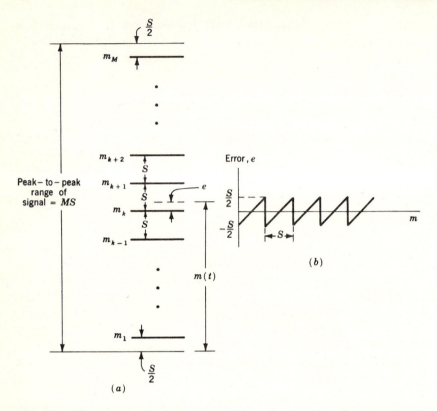

Fig. 6.3-1 (*a*) A range of voltage over which a signal $m(t)$ makes excursions is divided into M quantization ranges each of size S. The quantization levels are located at the center of the range. (*b*) The error voltage $e(t)$ as a function of the instantaneous value of the signal $m(t)$.

Now $f^{(1)}S$ is the probability that the signal voltage $m(t)$ will be in the first quantization range, $f^{(2)}S$ is the probability that m is in the second quantization range, etc. Hence the sum of terms in the parentheses in Eq. (6.3-2*b*) has a total value of unity. Therefore, the mean-square quantization error is

$$\overline{e^2} = \frac{S^2}{12} \tag{6.3-3}$$

6.4 COMPANDING[1]

We see from Eq. (6.3-3) that quantization noise (error) depends on the step size S. Hence if the steps are uniform in size, small-amplitude signals will have a poorer signal-to-quantization-noise ratio than large-amplitude signals. To correct this situation within the constraint of a fixed number of levels, it is advantageous to taper the step size so that the steps are close together at low signal amplitudes and further apart at large amplitudes. Such variation of step size yields a signal-to-noise ratio improvement for small signals, although strong signals will be impaired. Tapering of steps is very useful in connection with speech signals. It has been determined experimentally that, rather typically, instantaneous speech signal amplitudes are less than one-fourth of the rms signal value 50 percent of the time.

While it is possible to build a quantizer with tapered steps, it is more feasible to achieve an equivalent effect by distorting the signal before application to the transmitting quantizer. An inverse distortion is introduced at the receiving end so that the overall transmission is distortionless. Thus, before application to the quantizer, the signal is passed through a nonlinear network which has an input-output characteristic as shown in Fig. 6.4-1. At low amplitudes its slope is larger than at high amplitudes. Consequently a given signal change at low amplitude will carry the quantizer through more steps than will be the case at large amplitudes. A signal transmitted through a network with the characteristic shown in Fig. 6.4-1 will have the extremities of its waveform compressed, the compression being more pronounced with increasing amplitude. Hence such a network is called a *compressor*. The inverse

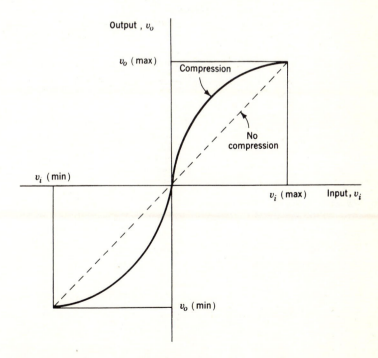

Fig. 6.4-1 An input-output characteristic which provides compression.

operation is performed by an *expander*. The combination of a compressor and an expander is called a *compander*, which then performs the operation of *companding*.

6.5 PULSE-CODE MODULATION (PCM)

A signal which is to be quantized prior to transmission is usually sampled as well. The quantization is used to reduce the effects of noise, and the sampling allows us to time-division multiplex a number of messages if we choose to do so. The combined operations of sampling and quantizing generate a quantized PAM waveform, that is, a train of pulses whose amplitudes are restricted to a number of discrete magnitudes.

We may, if we choose, transmit these quantized sample values directly. Alternatively we may represent each quantized level by a code number and transmit the code number rather than the sample value itself. The merit of so doing will be developed in the subsequent discussion. Most frequently the code number is converted, before transmission, into its representation in binary arithmetic, i.e., base-2 arithmetic. The digits of the binary representation of the code number are transmitted as pulses. Hence the system of transmission is called (binary) *pulse-code modulation* (PCM).

We review briefly some elementary points about binary arithmetic. The binary system uses only two digits, 0 and 1. An arbitrary number N is represented by the sequence $\cdots k_2 k_1 k_0$, in which the k's are determined from the equation

$$N = \cdots + k_2 2^2 + k_1 2^1 + k_0 2^0 \tag{6.5-1}$$

with the added constraint that each k has the value 0 or 1. The binary representations of the decimal numbers 0 to 15 are given in Table 6.5-1. Observe that to represent the four (decimal) numbers 0 to 3, we need only two binary digits k_1 and k_0. For the eight (decimal) numbers from 0 to 7 we require only three binary places, and so on. In general, if M numbers $0, 1, \ldots, M-1$ are to be represented, then an N binary digit sequence $k_{N-1} \cdots k_0$ is required, where $M = 2^N$.

The essential features of binary PCM are shown in Fig. 6.5-1. We assume that the analog message signal $m(t)$ is limited in its excursions to the range from -4 to $+4$ volts. We have set the step size between quantization levels at 1 volt. Eight quantization levels are employed, and these are located at $-3.5, -2.5, \ldots, +3.5$ volts. We assign the code number 0 to the level at -3.5 volts, the code number 1 to the level at -2.5 volts, etc., until the level at $+3.5$ volts, which is assigned the code number 7. Each code number has its representation in binary arithmetic ranging from 000 for code number 0 to 111 for code number 7.

In Fig. 6.5-1, in correspondence with each sample, we specify the sample value, the nearest quantization level, and the code number and

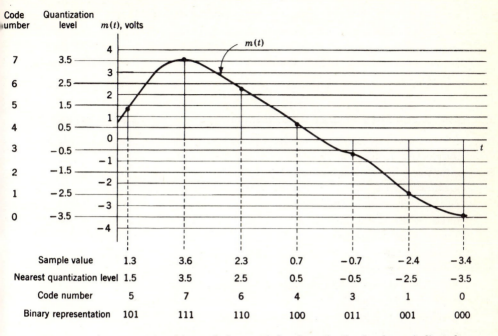

Code number	Quantization level $m(t)$, volts						
Sample value	1.3	3.6	2.3	0.7	−0.7	−2.4	−3.4
Nearest quantization level	1.5	3.5	2.5	0.5	−0.5	−2.5	−3.5
Code number	5	7	6	4	3	1	0
Binary representation	101	111	110	100	011	001	000

Fig. 6.5-1 A message signal is regularly sampled. Quantization levels are indicated. For each sample the quantized value is given and its binary representation is indicated.

Table 6.5-1 Equivalent numbers in decimal and binary representation

	Binary			Decimal
k_3	k_2	k_1	k_0	
0	0	0	0	0
0	0	0	1	1
0	0	1	0	2
0	0	1	1	3
0	1	0	0	4
0	1	0	1	5
0	1	1	0	6
0	1	1	1	7
1	0	0	0	8
1	0	0	1	9
1	0	1	0	10
1	0	1	1	11
1	1	0	0	12
1	1	0	1	13
1	1	1	0	14
1	1	1	1	15

its binary representation. If we were transmitting the analog signal, we would transmit the sample values 1.3, 3.6, 2.3, etc. If we were transmitting the quantized signal, we would transmit the quantized sample values 1.5, 3.5, 2.5, etc. In binary PCM we transmit the binary representations 101, 111, 110, etc.

6.6 ELECTRICAL REPRESENTATIONS OF BINARY DIGITS

As intimated in the previous section, we may represent the binary digits by electrical pulses in order to transmit the code representations of each quantized level over a communication channel. Such a representation is shown in Fig. 6.6-1. Pulse time slots are indicated at the top of the figure, and, as shown in Fig. 6.6-1a, the binary digit 1 is represented by a pulse, while the binary digit 0 is represented by the absence of a pulse. The row of three-digit binary numbers given in Fig. 6.6-1 is the binary representation of the sequence of quantized samples in Fig. 6.5-1. Hence the pulse pattern in Fig. 6.6-1a is the (binary) PCM waveform that would be transmitted to convey to the receiver the sequence of quantized samples of the message signal $m(t)$ in Fig. 6.5-1. Each three-digit binary number that specifies a quantized sample value is called a *word*. The spaces between words allow for the multiplexing of other messages.

At the receiver, in order to reconstruct the quantized signal, all that is required is that a determination be made, within each pulse time slot, about whether a pulse is present or absent. The exact amplitude of the pulse is not important. There is an advantage in making the pulse width as wide as possible since the pulse energy is thereby increased and it becomes easier to recognize a pulse against the background noise.

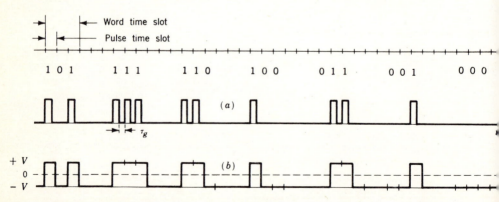

Fig. 6.6-1 (*a*) Pulse representation of the binary numbers used to code the samples in Fig. 6.5-1. (*b*) Representation by voltage levels rather than pulses.

Suppose then that we eliminate the guard time τ_g between pulses. We would then have the waveform shown in Fig. 6.6-1b. We would be rather hard put to describe this waveform as either a sequence of positive pulses or of negative pulses. The waveform consists now of a sequence of transitions between two levels. When the waveform occupies the lower level in a particular time slot, a binary 0 is represented, while the upper voltage level represents a binary 1.

Suppose that the voltage difference of $2V$ volts between the levels of the waveform of Fig. 6.6-1b is adequate to allow reliable determination at the receiver of which digit is being transmitted. We might then arrange, say, that the waveform make excursions between 0 and $2V$ volts or between $-V$ volts and $+V$ volts. The former waveform will have a dc component, the latter waveform will not. Since the dc component wastes power and contributes nothing to the reliability of transmission, the latter alternative is preferred and is indicated in Fig. 6.6-1b.

In Sec. 5.7, where we discussed unquantized PAM, we found it necessary to leave a guard time between pulses to minimize crosstalk between channels. In the present discussion we suggest that the guard time is profitably eliminated. The reason for the distinction has to do, of course, precisely with the quantization and is discussed in Sec. 6.8.

6.7 THE PCM SYSTEM

THE ENCODER

A PCM communication system is represented in Fig. 6.7-1. The analog signal $m(t)$ is sampled, and these samples are subjected to the operation of quantization. The quantized samples are applied to an *encoder*. The encoder responds to each such sample by the generation of a unique and identifiable binary pulse (or binary level) pattern. In the example of Figs. 6.5-1 and 6.6-1 the pulse pattern happens to have a numerical significance which is the same as the order assigned to the quantized levels. However, this feature is not essential. We could have assigned any pulse pattern to any level. At the receiver, however, we must be able to identify the level from the pulse pattern. Hence it is clear that not only does the encoder number the level, it also assigns to it an identification code.

The combination of the quantizer and encoder in the dashed box of Fig. 6.7-1 is called an *analog-to-digital converter*, usually abbreviated A-to-D converter. In commercially available A-to-D converters there is normally no sharp distinction between that portion of the electronic circuitry used to do the quantizing and that portion used to accomplish the encoding. In summary, then, the A-to-D converter accepts an analog signal and replaces it with a succession of *code* symbols, each symbol con-

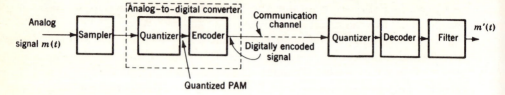

Fig. 6.7-1 A PCM communication system.

sisting of a train of pulses in which each pulse may be interpreted as the representation of a *digit* in an arithmetic system.   Thus the signal transmitted over the communications channel in a PCM system is referred to as a digitally encoded signal.

THE DECODER

When the digitally encoded signal arrives at the receiver (or repeater), the first operation to be performed is the separation of the signal from the noise which has been added during the transmission along the channel. As noted previously, separation of the signal from the noise is possible because of the quantization of the signal.   Such an operation is again an operation of *requantization;* hence the first block in the receiver in Fig. 6.7-1 is termed a quantizer.   A feature which eases the burden on this quantizer is that for each pulse interval it has only to make the relatively simple decision of whether a pulse has or has not been received or which of two voltage levels has occurred.   Suppose the quantized sample pulses had been transmitted instead, rather than the binary-encoded codes for such samples.   Then this quantizer would have had to have yielded, in each pulse interval, not a simple yes or no decision, but rather a more complicated determination about which of the many possible levels had been received.   In the example of Fig. 6.6-1, if a quantized PAM signal had been transmitted, the receiver quantizer would have to decide which of the levels 0 to 7 was transmitted, while with a binary PCM signal the quantizer need only distinguish between two possible levels.   The relative reliability of the yes or no decision in PCM over the multivalued decision required for quantized PAM constitutes an important advantage for PCM.

The receiver quantizer then, in each pulse slot, makes an educated and sophisticated guess about whether a positive pulse or a negative pulse was received and transmits its decisions, in the form of a reconstituted or regenerated pulse train, to the decoder.   (If repeater operation is intended, the regenerated pulse train is simply raised in level and sent along the next section of the transmission channel.)   The decoder, also called a *digital-*

to-analog (*D*-to-*A*) converter, performs the inverse operation of the encoder. The decoder output is the sequence of quantized multilevel sample pulses. The quantized PAM signal is now reconstituted. It is then filtered to reject any frequency components lying outside of the baseband. The final output signal $m'(t)$ is identical with the input $m(t)$ except for quantization noise and the occasional error in yes-no decision making at the receiver.

In Fig. 6.7-1 we have omitted any provision for companding. If companding is to be used, the compressor precedes the sampler, and the expander follows the filter. Also omitted from the figure is provision for bit (pulse) synchronization. The receiver must, of course, be given timing information identifying the beginning and end of a pulse time slot. Furthermore, if a number of baseband signals are being multiplexed, frame synchronization information must also be transmitted.

6.8   INTERSYMBOL INTERFERENCE

When the bandwidth of the PCM communication channel is restricted, the waveform of Fig. 6.6-1*b* will be distorted. Errors may then be made at the receiver in determining the voltage level transmitted within the time slot associated with each bit. (The term *bit* is a generally used abbreviation for *binary digit*.) The situation is entirely analogous to crosstalk encountered in unquantized PAM and discussed in Sec. 5.7. In unquantized PAM, adjacent time slots are often associated with different message channels, and the term crosstalk is appropriate. In PCM adjacent bits are more generally symbols in the code representation of a single quantized sample. Hence the term *intersymbol interference*. We discuss now the intersymbol interference in a binary PCM system caused by the high-frequency cutoff of the communications channel.

Let us represent the transmission characteristics of the channel by a low-pass resistance-capacitance filter of angular cutoff frequency ω_c. The worst case for intersymbol interference occurs when the two-level waveform of Fig. 6.6-1*b* has persisted at one level, say $,-V$, for a time long enough for the capacitor of the filter to have charged completely to the voltage $-V$. At this point let there be a change in voltage level to $+V$ as indicated in Fig. 6.8-1. The response of the filter is shown by the dashed plot $v(t)$.

Now, the manner in which we shall determine whether the transmitted level was $+V$ or $-V$ is to measure the area under the filter output within each time-slot interval τ. We judge that the transmitted level was $+V$ or $-V$ depending on whether the area is positive or negative. The merit of such a method of determination is that the process of integrating to find the area may be expected to average out to near zero the random

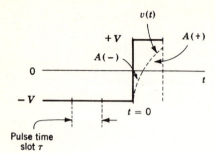

Pulse time
slot τ

Fig. 6.8-1 Illustrating intersymbol interference.

noise which accompanies the signal.   We shall discuss this method of level identification in greater detail in Sec. 11.3, where the matter of the *matched filter* is studied.

We observe in Fig. 6.8-1 that the response waveform $v(t)$ generates both a negative area $A(-)$ and a positive area $A(+)$.   The waveform is given by

$$v(t) = V(1 - 2e^{-\omega_c t}) \qquad 0 \leq t \leq \tau \tag{6.8-1}$$

It is easily verified that the area $A = A(-) + A(+)$ will be positive, provided $\omega_c \tau > 1.59$.   Thus if $\tau = 100$ nsec, $f_c \equiv \omega_c/2\pi \geq 2.53$ MHz, to obtain a correct decision.

It is of interest to note how the channel bandwidth affects the number of quantized levels of the message signal.   Suppose we have one system in which M_1 quantized levels are used.   Then in binary PCM each word would have N_1 time slots of duration τ_1 with $M_1 = 2^{N_1}$.   (We assume M_1 is selected such that N_1 is an integer.)   Let a second system *with the same word length* use $M_2 = 2^{N_2}$ quantization levels.   This system has a time-slot duration τ_2.   Since the word length is fixed

$$N_1\tau_1 = N_2\tau_2 \tag{6.8-2}$$

The first system requires a bandwidth f_{c1}, and the second a bandwidth f_{c2}. Since $\tau_1 f_{c1} = \tau_2 f_{c2}$, to achieve the same degree of error immunity (see Eq. 6.8-1), we have from Eq. (6.8-2) that

$$\frac{f_{c2}}{f_{c1}} = \frac{N_2}{N_1} = \frac{\log_2 M_2}{\log_2 M_1} \tag{6.8-3}$$

or

$$M_2 = (M_1)^{f_{c2}/f_{c1}} \tag{6.8-4}$$

Thus, suppose our first system had a bandwidth which allowed, say, 8 levels, $M_1 = 8$.   Then a doubling in bandwidth, $f_{c2}/f_{c1} = 2$, would allow an increase in the number of quantization levels to 64.   This, in turn, results in a decrease in quantization error.

To see how the quantization error is reduced, refer to Fig. 6.5-1. Eight quantization levels are employed, and the step size $S = 1$ volt.   If 64 quantization levels are used, $S = 125$ mV.   The variance of the quantization error $\overline{e^2} = S^2/12$ is therefore reduced by 64.

6.9 EYE PATTERNS

We have noted that intersymbol interference may cause errors in reading the digits of the bit stream being transmitted in a PCM system.   Errors of this type may be avoided by extending the bandwidth.   Errors will still occur, however, because of the additive noise.   More generally, even when the intersymbol interference or the additive noise, acting individually, is too small to cause an error, the two together may combine to cause an error.

One way to obtain a good qualitative indication of the performance of a PCM system is to examine the bit stream on a cathode ray oscillo-

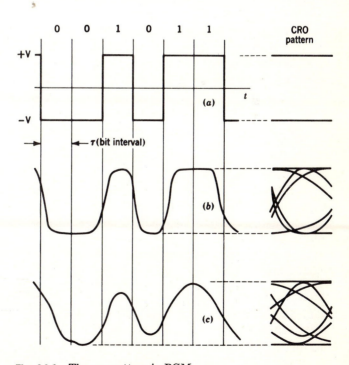

Fig. 6.9-1   The eye pattern in PCM.

scope. A manner of making such an examination in a way to yield a great deal of information at a glance is to set the time base of the scope so that it triggers at the bit rate and yields a sweep lasting 1 time-slot duration. In the ideal case of no noise and no bandwidth restriction the bit-train waveform would appear as at the left in Fig. 6.9-1a and the scope pattern as at the right. The scope pattern would consist of two horizontal lines. In Fig. 6.9-1b the bit-train waveform illustrates the effects of limited bandwidth and the scope pattern suggests an *eye* and is called an *eye pattern*.[2] A longer bit train would add more traces to the scope pattern, generally filling in the periphery and leaving an opening, an eye, in the center of the figure. In Fig. 6.9-1c further bandwidth limiting is indicated, and the open space at the center of the figure, the eye, has closed somewhat. The addition of noise to the bit stream would close the eye still further.

6.10 EQUALIZATION

An *equalizer*[3] is a frequency-selective network introduced into a system to undo distortion produced elsewhere in the system. We encountered an equalizer in Sec. 5.5, where equalization was used to correct distortion produced by flat-top sampling.

The same principle can be used in PCM. Here the *equalizer* filter $H_e(\omega)$ is designed as the inverse of the channel filter $H_c(\omega)$. Thus, if the transfer function of the channel $H_c(\omega)$ is known, it is always possible to construct its inverse and thereby eliminate the intersymbol interference. A problem arises when, as in ordinary telephone systems, the telephone lines connecting two parties vary from call to call. Thus, we do not know a priori which channel will be connected between the transmitter and receiver. In this case the equalizer filter is designed to be *adaptive*. This means that the filter adjusts itself by using feedback techniques to reduce the distortion.

To make such adaptive equalization possible, it is often necessary to precede each message by the transmission of a *test bit stream*. If we were adjusting the filter manually, we might observe the eye pattern and adjust for maximum eye opening. Systems are available in which the adjustment of the equalization is performed automatically, without human intervention.

6.11 SYNCHRONOUS TIME-DIVISION MULTIPLEXING WITH PCM

To illustrate multiplexing with PCM, we shall consider an example in which 1000 voice message signals and a single television signal are multiplexed.   We assume, as is customary for good fidelity of signal reproduction, that 8-bit PCM is used.   Eight bits correspond to $2^8(=256)$ quantization levels.   A bandwidth of 4 kHz is used for the voice messages, while the television signal requires a bandwidth of 4 MHz.   The multiplexing system, shown in Fig. 6.11-1a, uses two commutator switches.

Let us assume that the commutator dwells on a particular contact point only for the duration of a *single* bit.   Further, we arrange that the time between breaking contact at one switch point and making contact at the next is similarly equal to the time of one bit.   The output waveform of commutator 1 is as shown in Fig. 6.11-1.   The pulse marked

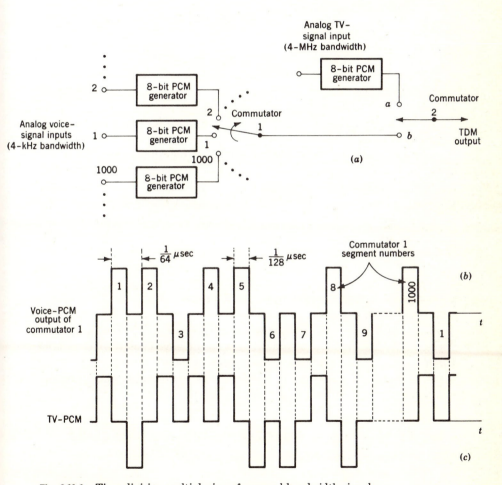

Fig. 6.11-1   Time-division multiplexing of unequal bandwidth signals.

1 is, say, the first bit of 8 which will specify a sample value in channel 1. Pulse 2 is the first bit of 8 for a sample value in channel 2, etc.

Since the voice signals have a 4-kHz bandwidth, the sampling rate must be 8 kHz or the sampling time $\frac{1}{8} \times 10^{-3}$ sec. There are 8 bits per sample, so the commutator must make 8 revolutions in $\frac{1}{8} \times 10^{-3}$ sec, or 1 revolution in $\frac{1}{64} \times 10^{-3}$ sec. Since there are 1000 voice signals, the time between bits in adjacent channels is $\frac{1}{64} \times 10^{-6}$ sec. The bit duration shown in Fig. 6.11-1 is $\frac{1}{128}$ μsec ($=7.8$ nsec), leaving an equal interval between bits.

The television signal with its 4-MHz bandwidth and 8 bits per sample must generate a bit stream with bits separated by

$$10^{-6}/8(2 \times 4) \text{ sec} = \frac{1}{64} \mu\text{sec}$$

Such a television bit stream is shown in Fig. 6.11-1c, and we observe that the bit pulses in Fig. 6.11-1c occur just at times when there is no bit pulse in the waveform in Fig. 6.11-1b. Thus the two pulse trains may be combined to form a single composite pulse train carrying the television signal as well as all the voice messages.

It is, of course, necessary that all the various operations be affected in synchronism. Thus commutator 2 must make contact with point b when commutator 1 is on a contact point. And when commutator 1 is between contacts, commutator 2 must make contact with point a. Similarly each pulse stream from each of the 1001 PCM generators must be timed so that individual bits fall in proper time slots without overlapping. Finally we may note that the success of the system depends on the fortuitous circumstances that the product 4 kHz/channel $\times$ 1000 channels yields 4 MHz, which is precisely the bandwidth of the television signal. Alternatively, we may consider that the need for such a relationship, in order to allow easy interleaving of voice and television signals, is precisely the consideration which suggests 1000 voice channels rather than some other number.

It is to be noted that this multiplexing arrangement could have been designed in an alternative manner. We might have chosen to allow both commutators to have dwelled on a contact point for an entire word interval rather than for a bit interval. As a matter of fact, such an arrangement would result in a decreased commutator speed.

6.12 ASYNCHRONOUS TIME-DIVISION MULTIPLEXING

When the signals to be time-division multiplexed are sampled at arbitrary sampling frequencies they cannot, in general, be combined using the technique described in Sec. 6.11. A technique used to combine a group of *asynchronously sampled* time-division multiplexed signals is called *pulse stuffing.*

Essential to the multiplexing of asynchronous signals is the use of a device which can store a digital bit stream and which may be operated in such a manner that the bit stream may be read out of storage at a rate different from the rate at which the bit stream was written in. Such a storage device is called an *elastic store.* One example, among many, of such a storage device is a tape recorder. After a bit stream has been recorded on the tape, the rate at which the bit stream is read out of the tape may be adjusted to suit our convenience simply by adjusting the mechanical speed with which the tape moves during playback.

Consider now that different experiments are being performed on board a satellite. Each experiment provides a signal which will eventually be transmitted to earth. Since each signal has a different bandwidth, the sampling frequencies are different. Each of these signals can be sampled, and the samples stored in electronic digital storage devices. Note that the recording rate of each storage device is different, since the sampling frequency of each signal is different. Just prior to transmission from the satellite to earth, the experiments are temporarily halted and the recording ends (the experiments begin again after transmission has ceased). Each storage device is now *played back* at the same rate. Thus, these signals can be synchronously time-division multiplexed, and then the single TDM signal transmitted to earth. The earth-stationed receiver reverses this process to recover each signal.

There is one major problem associated with this procedure. Let us illustrate the problem by means of an example employing only two signals, the first having a bandwidth of 4 kHz and the second having a bandwidth of 5 kHz. Thus the *word* duration of the first signal is 125 μsec, while the *word* duration of the second signal is 100 μsec. (If each word contains N bits, i.e., if there are $M = 2^N$ quantization levels, the duration of each bit is $125/N$ μsec and $100/N$ μsec, respectively.)

Consider now that the signals are recorded onto the digital storage device for 1 sec. Then 8000 words of the first signal are recorded on its storage unit, while 10,000 words of the second signal are recorded on the second storage unit. In order to time-division multiplex these two signals, each storage unit is played back at the same rate (note that the rate does not matter). The first 8000 words of each signal are now multiplexed without any trouble. However, during the multiplexing of the final 2000 words of the second signal, there is *no* contribution from the first signal. Because of noise present during the recording and the playback for multiplexing, and as a result of the presence of thermal noise in the receiver, the receiver will read "words" even when none are being transmitted.

One procedure that may be used to avoid such erroneous interpretation of noise as signal is to fill in the time slots corresponding to the first message, which has already terminated, with a highly improbable sequence of digits. For example, the time slots may be filled with a sequence of bits all of which are 1s. Such a sequence corresponds to the transmission of direct current, which is normally not transmitted. It is therefore possible so to program the receiver that the receiver recognizes the transmission of direct current as an indication that actually no message is being transmitted.

Since the above technique requires that pulses be *stuffed* into the spaces provided for the missing bits, the technique is called *pulse stuffing*[4].

6.13 DELTA MODULATION

Delta modulation[5] (DM) is a technique by which an analog signal can be *encoded* into binary digits (bits). Hence DM is a PCM system. We shall, however, restrict the name PCM to the encoding technique described in Sec. 6.5. DM has the merit that the electronic circuitry required for modulation at the transmitter and particularly for demodulation at the receiver is substantially simpler than the corresponding hardware needed for PCM.

A delta-modulation system is shown in block diagram form in Fig. 6.13-1. The pulse generator furnishes a regularly recurring train of pulses $p_i(t)$ of fixed amplitude and polarity. To simplify the explanation, we shall assume that these pulses are arbitrarily narrow but still of finite area, i.e., impulses. The modulator receives these input pulses $p_i(t)$ as well as a signal $\Delta(t)$. The modulator output $p_o(t)$ is the input pulse train

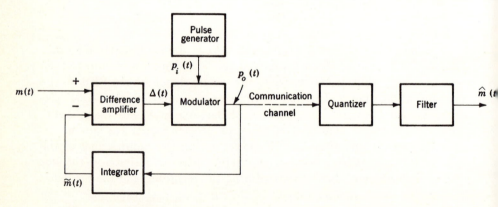

Fig. 6.13-1 A delta-modulation system of communication.

$p_i(t)$ multiplied by $+1$ or by -1 depending on the polarity only (not the magnitude) of $\Delta(t)$. If $\Delta(t)$ is positive when $p_i(t)$ occurs, the multiplication is by $+1$, and if $\Delta(t)$ is negative, by -1. [We disregard as being too unlikely the possibility that $\Delta(t)$ is so close to zero that the modulator cannot decide. The probability of $\Delta(t)$ being exactly zero when $p_i(t)$ occurs is, of course, zero.]

The waveform $p_o(t)$ is applied to an integrator the output of which is designated $\tilde{m}(t)$. As we shall see, $\tilde{m}(t)$ is an approximation to the input signal $m(t)$. The signals $m(t)$ and $\tilde{m}(t)$ are compared in a difference amplifier. The amplifier output $\Delta(t)$ is given by $\Delta(t) = m(t) - \tilde{m}(t)$.

The operation of this modulator may now be seen using the waveforms shown in Fig. 6.13-2. In this figure, $t = 0$ has been selected to occur midway between pulse occurrences. The initial values of $m(t)$ and $\tilde{m}(t)$ have been selected arbitrarily. At the time t_1 of the first pulse, it happens that $m(t)$ is larger than $\tilde{m}(t)$. Therefore the modulator-output pulse is positive. The integrator response to this pulse (impulse) is an abrupt positive step as shown. At time t_2, $\Delta(t) = m(t) - \tilde{m}(t)$ is still positive with the result that $\tilde{m}(t)$ steps positively again. The waveform $\tilde{m}(t)$ continues its stepwise approach to $m(t)$ through the fourth pulse, at which time $\tilde{m}(t)$ overshoots its mark. Hence immediately after the fourth pulse $\Delta(t)$ is negative, and the next pulse in the modulator output is of negative polarity. The first part of $m(t)$ has been indicated without time variation in order to display the initial transient approach of $\tilde{m}(t)$ to $m(t)$ and to show, as well, the hunting of $\tilde{m}(t)$ when $m(t)$ does not vary.

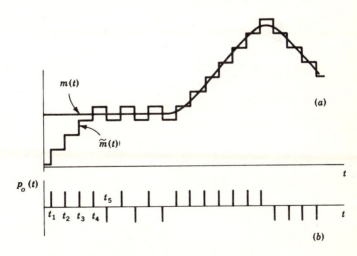

Fig. 6.13-2 Waveforms of the delta-modulation system of Fig. 6.13-1. (*a*) The signal $m(t)$ and its approximation $\tilde{m}(t)$. (*b*) The transmitted pulse train.

Also to be seen in Fig. 6.13-2 is the stepwise approximation of $m(t)$ by $\tilde{m}(t)$ when there is signal variation.

The signal which is transmitted over the communications channel is the pulse waveform $p_o(t)$. In practice each pulse is widened to increase the energy in the transmitted bit. We observe that in transmitting $p_o(t)$, we transmit, in *coded* form, not the signal level itself but rather information about the *difference* between the waveform $m(t)$ and its approximation $\tilde{m}(t)$. Hence the name delta modulation.

The quantizer in Fig. 6.13-1 serves to perform, in principle, the same function performed by the receiver quantizer in a PCM system. It is hoped that, if the noise added to the signal in the communications channel is not overwhelming, the quantizer will almost always be able to distinguish a positive pulse from a negative pulse. The pulse train of Fig. 6.13-2b is then available at the receiver. We may therefore reconstruct the waveform $\tilde{m}(t)$ by using an integrator as at the transmitter. We follow the integrator by a low-pass filter in order to suppress the jumps in $\tilde{m}(t)$ and thereby smooth the signal so that it will follow, more smoothly, the original $m(t)$. However, we recognize that a low-pass filter will, by itself, provide an approximate measure of integration. Hence we can dispense with the receiver integrator and depend on the filter alone. The output of the filter is $\hat{m}(t)$, which differs from the transmitter input $m(t)$ only because of the effects of the stepwise approximation of delta modulation and because of errors made by the receiver quantizer due to noise.

Having established the principle of operation of delta modulation, we may now make some concessions to practical reality. In the first place, it is hardly necessary that the stepped waveform $\tilde{m}(t)$ display abrupt rising and falling edges and exact constancy between jumps. After all, as we have noted, even if we were to have reproduced this waveform at the receiver, we would thereafter filter it to smooth it out. Hence the transmitter integrator need not be a precise integrator. Instead a simple RC low-pass filter, which constitutes a quasi-integrator, will serve adequately. By the same token, the pulse generator need not furnish impulses but only pulses which are moderately short in comparison with the interval between pulses.

The essential feature of a DM system is that it transmits information about the *difference signal* $\Delta(t)$. Note that this signal $\Delta(t)$ is available at the output of the difference amplifier in Fig. 6.13-1. In the system of Fig. 6.13-1 we do not transmit $\Delta(t)$ but transmit only its polarity as determined at each sampling time. We may, however, if we choose to do so, transmit $\Delta(t)$ itself. Thus suppose that $\Delta(t)$ were applied as the signal input to a PCM system. Specifically, suppose that $\Delta(t)$ were applied as the input waveform $m(t)$ in the PCM system of Fig. 6.7-1. Such a system is entirely feasible and is referred to as a delta PCM or simply DPCM.

6.14   LIMITATIONS OF DELTA MODULATION DUE TO FIXED STEP SIZE

The fixed step in $\tilde{m}(t)$ imposed on DM a limitation not encountered in other pulse-modulation schemes and results in *overloading* when the signal changes too rapidly.

All modems (systems consisting of modulation and demodulation, as noted earlier), including delta modulation, overload when the amplitude of the modulating baseband signal exceeds the range of the active devices used to process the signal.   But delta modulation exhibits an additional type of overload not encountered in other modems.   This overload appears when the modulating signal changes, between samplings, by an amount greater than the size of a step.   Hence this type of overload is not determined by the amplitude of the modulating signal but rather by its *slope*.   In Fig. 6.14-1a a signal $m(t)$ is shown which is changing about as rapidly as the modulator can follow.   In Fig. 6.14-1b we see a signal of the same peak-to-peak amplitude but of increased rate of rise (increased slope). We observe in this case that the slope of the modulator output increases no more rapidly than it did in the previous case.   Thus $\tilde{m}'(t)$ does not follow $m'(t)$ in this region of increased slope.   The DM system is therefore said to be *slope-overloaded*.

By way of example, let the modulating signal $m(t)$ be a sinusoid of amplitude A and frequency f.   Then the maximum rate of rise, which occurs as the waveform passes through zero phase, is $2\pi fA$.   If the step size is S and the sampling frequency (the pulse rate in Fig. 6.13-2) is $f_s^{(\Delta)}$, then the average *rate of rise* (slope) of the feedback signal $\tilde{m}(t)$, in the interval between pulses, is $Sf_s^{(\Delta)}$.   The point of slope-overload will occur when the rates of rise (slopes) are equal, i.e., when $2\pi fA = Sf_s^{(\Delta)}$, or at a

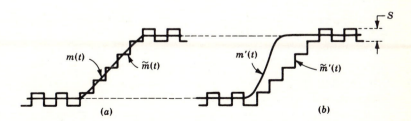

Fig. 6.14-1   Illustrating rate-of-rise overloading in delta modulation.   The signals $m(t)$ and $m'(t)$ in (a) and (b) have the same amplitude.   However, because of the greater rate of rise in (b) the approximate $\tilde{m}'(t)$ cannot follow $m'(t)$.

peak signal amplitude

$$A = \frac{S f_s{}^{(\Delta)}}{2\pi f} \tag{6.14-1}$$

Equation (6.14-1) has an interesting bearing on the application of delta modulation for the transmission of speech. It has been determined experimentally that delta modulation will transmit speech without noticeable slope overload provided that the amplitude of the speech signal does not exceed the maximum sinusoidal amplitude, as given in Eq. (6.14-1), corresponding to a frequency $f = 800$ Hz.

6.15 ADAPTIVE DELTA MODULATION

During any period of time when the changes in the signal $m(t)$ are less than the step size, the modulator no longer follows the signal, and the modulator produces a train of alternating positive and negative pulses. Similarly the modulator overloads when the slope of the signal is too high. Both of these limitations may be relieved by adjusting the step size in accordance with the signal being encountered. When the signal changes are small, we would like the step size to be reduced, while to avoid slope overload, we would like the step size to increase.

A DM system which adjusts its step size and is hence described as being *adaptive* is shown in Fig. 6.15-1. The amplifier has a variable gain. That is, its gain is a function of the voltage applied at its gain-control terminal. We assume that the characteristics of the amplifier are such that when the gain-control voltage is zero, its gain is low, and that the gain increases with increasing positive gain-control voltage. The resistor-capacitor combination serves as an integrator, the voltage across C being proportional to the integral of the pulse signal $p_o(t)$. The voltage across C is used to control the gain of the amplifier. The square-law device ensures that whatever the polarity of the voltage across C, a positive voltage will be applied to the gain-control terminal of the amplifier.

Assume now that $m(t)$ is making only small excursions so that the modulator does not follow. The output $p_o(t)$ consists then of alternate polarity pulses. These pulses, when integrated, yield an average output of almost zero. The gain-control input is, hence, almost zero, the gain is low, and consequently the step size is reduced. Next consider the case of slope overload. If $m(t)$ increases positively or negatively at too rapid a rate, $\tilde{m}(t)$ cannot follow. The output $p_o(t)$ is then a train of all positive or all negative pulses. The integrator averages and provides a large voltage to increase the gain of the amplifier. Because of the squaring circuit the amplifier gain will increase no matter what the polarity of the integrator voltage. The end result is an increase in step size and a reduction in slope

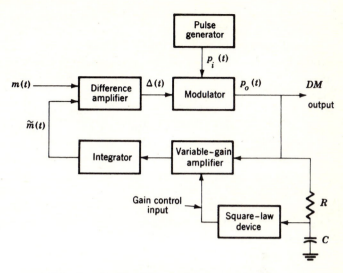

Fig. 6.15-1   An adaptive delta modulator.

overload.   It is of course, necessary that there be an adaptive adjustment of the step size at the receiver, as well.   For this purpose a variable
gain amplifier may be interposed, in Fig. 6.13-1, before the final output
filter.   The gain-control input of this variable amplifier is determined, as
at the receiver, by an integrator and squaring circuit, the integrator input
being the received signal.

6.16   INTELLIGIBILITY COMPARISON:
PCM AND DELTA MODULATION

In PCM the signal is sampled at a rate f_s (Nyquist or higher), and each
sample is encoded into an N bit code.   The binary pulse train (pulse time
slots) generated and transmitted over the communications channel has a
maximum frequency Nf_s.   In delta modulation, the transmitted signal
may also take the form of a binary pulse train, and its frequency will be
$f_s^{(\Delta)}$.   An increase in $f_s^{(\Delta)}$ will allow the step-wise approximation $\tilde{m}(t)$ to
follow the signal $m(t)$ more closely, especially if such an increase is accom-
panied by a decrease in step size, thereby maintaining a constant overload
slope.   Further, as $f_s^{(\Delta)}$ increases, it becomes easier for the receiver filter to
smooth out the received quantized signal.

The transmission-path bandwidths in the two cases are proportional
to Nf_s and $f_s^{(\Delta)}$.   It is of interest to compare the bandwidths required for
comparable quality of transmission.   It has been found experimentally,

for speech transmission, that when the quality of transmission is good, PCM requires less bandwidth than does delta modulation. With the use of PCM, speech transmission is found to be of high quality when samples, taken at a rate of 8 kHz, are encoded into an eight-bit code. The corresponding transmitted pulse train has a frequency equal to $8 \times 8\,\text{kHz} = 64$ kHz. To obtain comparable quality using delta modulation requires a sampling rate of about 100 kHz. On the other hand, when in the interest of bandwidth conservation, compromises are allowable with quality and intelligibility, it appears that delta modulation performs better.

As indicated, when high-quality transmission is required, there is a bandwidth advantage of PCM over DM. On the other hand, DM has the advantage that the hardware required for its physical implementation is very much simpler than that required for PCM. We may therefore note that delta PCM allows a compromise between the two systems. The greater the number of levels into which the difference signal $\Delta(t)$ is quantized, the greater will be the quality of system performance, but the greater also will by the physical complexity of the system.

6.17 PHASE-SHIFT KEYING (PSK)

When it becomes necessary to superimpose a binary PCM waveform on a carrier, then amplitude modulation, phase modulation, or frequency modulation may be used. As a matter of practice, straightforward AM is rarely used. Phase and frequency modulation are commonly employed. Because of the special (two-level) nature of the carrier modulating signal, phase modulation is referred to as *phase-shift keying* (PSK), and frequency modulation is called *frequency-shift keying* (FSK). In this section we discuss PSK.

Consider that a binary signal $v(t)$, which takes on the values $v(t) = +V$ or $v(t) = -V$, is to be the modulating waveform in a PSK system. The PSK waveform is

$$v_{\text{PSK}}(t) = A \cos [\omega_0 t + \varphi(t)] \tag{6.17-1}$$

in which A is a fixed amplitude and $\varphi = 0$ for, say, $v(t) = +V$ and $\varphi = \pi$ for $v(t) = -V$. Equation (6.17-1) may be written in the alternative form

$$v_{\text{PSK}}(t) = \frac{v(t)}{V} A \cos \omega_0 t \tag{6.17-2}$$

so that $v_{\text{PSK}}(t)$ is $A \cos \omega_0 t$ or $-A \cos \omega_0 t$ for $v(t) = +V$ or $v(t) = -V$. The waveform of Eq. (6.17-2) may be generated, as in Fig. 6.17-1, by applying the waveform $v(t)$ and the carrier $\cos \omega_0 t$ to a balanced modulator.   A balanced modulator, it will be recalled, yields an output waveform, which, aside from a constant factor, is the product of its input waveforms.

The received signal has the form

$$v_{\text{PSK}}(t) = \frac{v(t)}{V} A \cos (\omega_0 t + \theta) \qquad (6.17\text{-}3)$$

in which θ is a phase angle which depends on the effective length of path between transmitter and receiver.   Demodulation must be performed synchronously; hence we require the waveform $\cos (\omega_0 t + \theta)$ at the receiver.   A synchronizing circuit which can extract the waveform $\cos (\omega_0 t + \theta)$ from the received signal itself is shown in Fig. 6.17-1.   The output of the square-law device is $A^2 \cos^2 (\omega_0 t + \theta)$ since $v^2(t)/V^2$ is always $+1$.   Now $A^2 \cos^2 (\omega_0 t + \theta) = A^2/2 + (A^2/2) \cos 2(\omega_0 t + \theta)$.   Hence a bandpass filter may be used as indicated to separate the waveform $\cos 2(\omega_0 t + \theta)$.   The frequency divider divides the frequency by 2, yielding $\cos (\omega_0 t + \theta)$ as required.   (This system of extracting a synchronous carrier was earlier encountered in Sec. 3.3.)

In the synchronous demodulator the signal of Eq. (6.17-3) is multiplied by the locally recovered carrier $\cos (\omega_0 t + \theta)$.   This product is

$$\cos (\omega_0 t + \theta) v_{\text{PSK}}(t) = \frac{1}{2} \frac{v(t)A}{V} + \frac{1}{2} \frac{v(t)A}{V} \cos 2(\omega_0 t + \theta) \qquad (6.17\text{-}4)$$

Our interest is in $v(t)$.   If $v(t)$ were a bandlimited signal, then we might recover $v(t)$ precisely through the use of a low-pass filter.   However, in principle at least, the waveform $v(t)$ is not bandlimited because of the

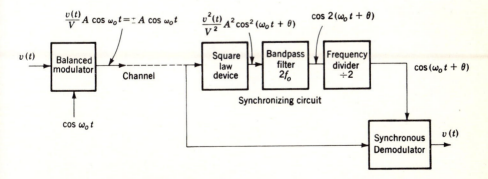

Fig. 6.17-1   A binary PSK system.

abrupt transitions in its waveform. Hence a low-pass filter will introduce some distortion in $v(t)$ and will also pass some of the sidebands of the double-frequency carrier in Eq. (6.17-4). It is to be noted, however, that we are not really interested in recovering $v(t)$ but only in knowing whether $v(t) = +V$ or $-V$ in each bit interval. If a bit interval extends over many cycles of the carrier $\cos \omega_0 t$, then it will be easy to find a low-pass filter which will effect an adequate separation of the terms in Eq. (6.17-4) to allow such a determination to be made. We shall return again to this matter in Sec. 11.7, where we shall consider how best to design a filter to allow such determination to be made with best reliability in the presence of noise.

6.18 DIFFERENTIAL PHASE-SHIFT KEYING (DPSK)

Differential phase-shift keying (DPSK) is a modification of PSK which avoids the necessity of providing the synchronous carrier required at the receiver for demodulating a PSK signal. The generation of DPSK signals is illustrated in Fig. 6.18-1. The binary encoded message to be transmitted is represented by the binary sequence $b'(t)$ indicated in Fig. 6.18-1a. An auxiliary binary sequence $b(t)$, as shown, is generated. Note that the sequence $b(t)$ has one more digit than does the sequence

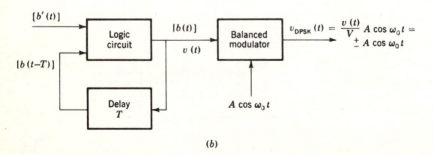

Fig. 6.18-1 (a) The bit stream $b'(t)$ represents the message to be transmitted. $b(t)$ is an auxiliary bit stream generated for DPSK transmission. The relative phases of the carrier in each bit interval are also indicated. (b) A method of generating the DPSK signal.

previously in connection with PSK, and so here, a low-pass filter may separate this signal waveform from the double-frequency carrier term in Eq. (6.18-2) adequately to allow a determination, in each bit interval, of whether a 1 or a 0 was transmitted.

As appears in Eq. (6.18-2), in order that the signal output be as large as possible, we should select T so that $\cos \omega_0 T = \pm 1$. Thus, the carrier frequency $\omega_0/2\pi$ should be selected so that the bit duration is an integral number of half cycles in duration.

The *differentially coherent* system, DPSK, has a clear advantage over the coherent system, PSK, in that the former avoids the need for complicated circuitry used to generate a local carrier at the receiver. To see the relative disadvantage of DPSK in comparison with PSK, consider that during some bit interval the received signal is so contaminated by noise that a PSK system makes an error in the determination of whether the transmitted bit was a 1 or a 0. In DPSK a bit determination is made on the basis of the signal received in two successive bit intervals. Hence noise in one bit interval may cause errors to two bit determinations. The error rate in DPSK is therefore greater than in PSK, and, as a matter of fact, there is a tendency for bit errors to occur in pairs.

6.19   FREQUENCY-SHIFT KEYING (FSK)

In frequency-shift keying (FSK) the binary signal $b(t)$ is used to generate a waveform

$$v_{\text{FSK}}(t) = A \cos (\omega_c \pm \Omega)t \tag{6.19-1}$$

in which the plus sign or the minus sign applies, depending on whether the bit is a 1 or a 0. The transmitted signal, then, is of amplitude A and has an angular frequency $\omega_c + \Omega$ or $\omega_c - \Omega$, with Ω a constant angular frequency offset from the carrier frequency ω_c.

An entirely straightforward way in which such an FSK signal might be demodulated would consist in applying the signal simultaneously to two sharply tuned filters, one tuned to angular frequency $\omega_c + \Omega$, the other tuned to $\omega_0 - \Omega$. We would then determine that a 1 or a 0 had been transmitted in any bit interval depending on which filter yielded the larger output signal.

A synchronous method of demodulation is shown in Fig. 6.19-1. This scheme has the advantage that, as is discussed in Chap. 11, it may readily be adapted to yield optimum performance in the presence of noise. Observe that two synchronous local carriers are required here, at angular frequencies $\omega_c \pm \Omega$. If the received signal is $A \cos (\omega_c + \Omega)t$, then the output of the difference amplifier will be

$$v_D = \tfrac{1}{2}A - \tfrac{1}{2}A[\cos 2\Omega t + \cos 2\omega_c t - \cos 2(\omega_c + \Omega)t] \tag{6.19-2}$$

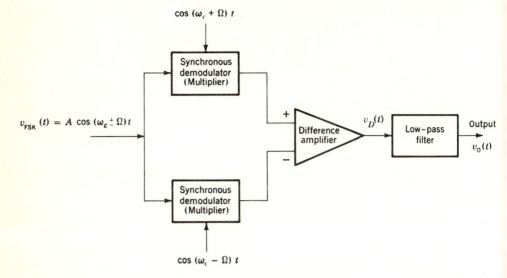

Fig. 6.19-1 Synchronous demodulation of an FSK signal.

If the received signal is $A \cos (\omega_c - \Omega)t$, the output of the difference amplifier will be

$$v_D = -\tfrac{1}{2}A + \tfrac{1}{2}A[\cos 2\Omega t + \cos 2\omega_c t - \cos 2(\omega_c - \Omega)t] \quad (6.19\text{-}3)$$

As in the case of PSK, a low-pass filter may be used to effect an adequate separation of the dc terms in Eq. (6.19-2) or (6.19-3) to allow a determination of whether a 1 or a 0 has been transmitted. Ordinarily $\omega_c \gg \Omega$; hence the lowest frequency, by far, in the brackets in Eqs. (6.19-2) and (6.19-3) is the term $\cos 2\Omega t$. To allow an easy separation by filter of the dc terms, it is therefore necessary that the bit interval extend over a time T such that, in this interval, $\cos 2\Omega t$ includes many cycles. Hence we require that $2\Omega T \gg 2\pi$.

PROBLEMS

6.1-1. (a). If the attenuation in a telephone cable is 1 dB per mile and the transmitted voltage is 1 volt rms, calculate the voltage received 1000 miles away.

(b) If repeaters are employed which yield a maximum rms output of 1 volt and have a voltage gain of 100, calculate the spacing between repeaters and the number of repeaters needed if a 1-volt rms signal is to be received 1000 miles away.

6.2-1. For the quantizer characteristic shown in Fig. 6.2-1b.

(a) Plot the error characteristic $e = v_o - v_i$ versus v_i. Assume that $S = 1$, that is, $m_{k+1} - m_k = 1$ volt.

(b) Since the error $e = v_o - v_i$ is periodic, it can be expanded in a Fourier series. Write the Fourier series for the error $e = e(v_i)$.

(c) If $v_i = S \sin \omega_o t$, find the component of the error e at the angular frequency ω_o.

6.3-1. Show that if the signal is uniformly distributed, Eq. (6.3-3) results even if M is not large.

6.3-2. Consider a signal having a probability density

$$f(v) = \begin{cases} Ke^{-|v|} & -4 < v < 4 \\ 0 & \text{elsewhere} \end{cases}$$

(a) Find K.

(b) Determine the step size S if there are four quantization levels.

(c) Calculate the variance of the quantization error when there are four quantization levels. Do not assume that $f(v)$ is constant over each level. Compare your result with Eq. (6.3-3).

6.3-3. Consider a signal having a probability density

$$f(v) = K(1 - |v|) \qquad -1 \leq v \leq 1$$

Calculate (a) to (c) of Prob. 6.3-2.

6.4-1. The compressor shown in Fig. 6.4-1 has the characteristic $v_o = C(v)$ where v is the input signal and v_o is the compressed signal. Thus, if no compression is employed, $v_o = C(v) = v$.

(a) Redraw Fig. 6.4-1 and show that as a result of compression a uniform step size of S volts in the output voltage v_o results in nonuniform quantization, i.e., varying step size, of the input voltage v. Do this by dividing v_o into 8 equal quantization steps. Label the input steps $S_1, S_2, \ldots, S_8$.

(b) Show that the quantization error is now

$$\overline{e_c^2} = \int_{v_1 - S_1/2}^{v_1 + S_1/2} (v - v_1)^2 f(v) \, dv + \int_{v_2 - S_2/2}^{v_2 + S_2/2} (v - v_2)^2 f(v) \, dv$$

$$+ \cdots + \int_{v_8 - S_1/2}^{v_8 + S_1/2} (v - v_8)^2 f(v) \, dv$$

where

$$v_1 - S_1/2 = v_{\min}$$
$$v_i + S_i/2 = v_{i+1} - (S_{i+1})/2$$

and

$$v_8 + S_8/2 = v_{\max}$$

(c) If there are a large number of quantization levels, show that

$$S_i = \frac{S}{C'(v_i)}$$

Hint: Note that $\Delta v_o / \Delta v \equiv C'(v)$.

(d) If $f(v)$ is approximately constant throughout each step, show that $\overline{e_c^2}$ becomes

$$\overline{e_c^2} \simeq \frac{1}{12}[S_1^3 f(v_1) + S_2^3 f(v_2) + \cdots + S_8^3 f(v_8)]$$

(e) Using the result of (c), show that if there are many quantization levels, that is, if S is small,

$$\overline{e_c^2} = \frac{S^2}{12} \left\{ \sum_i \frac{S_i f(v_i)}{[C'(v_i)]^2} \right\} \simeq \frac{S^2}{12} \int_{v_{min}}^{v_{max}} \frac{f(v)dv}{[C'(v)]^2}$$

(f) The companding improvement factor I_c is defined as the ratio of the quantization error with no companding to the quantization error with companding. Obtain an expression for I_c.

6.4-2. Logarithmic companding is often used for speech communication. When employed,

$$v_0 = \begin{cases} v_{max} \dfrac{\ln\left(1 + \mu v/v_{max}\right)}{\ln\left(1 + \mu\right)} & 0 \le v \le v_{max} \\[3mm] -v_{max} \dfrac{\ln\left(1 - \mu v/v_{max}\right)}{\ln\left(1 + \mu\right)} & -v_{max} \le v \le 0 \end{cases}$$

(a) Sketch the resulting compressor characteristic; choose $\mu = 0,\ 3,\ 100$.

(b) Plot the corresponding expander characteristic.

(c) If there are 32 quantization levels, discuss the variation of step size vs. input voltage v.

(d) If the signal v is uniformly distributed between $-v_{max}$ and $+v_{max}$, calculate the companding improvement factor I_c. (See Prob. 6.4-1f.)

6.5-1. Show that the numbers 0 to 7 can be written using 3 binary digits (bits). How many bits are required to write the numbers 0 to 5?

6.6-1. Consider that the signal $\cos 2\pi t$ is quantized into 16 levels. The sampling rate is 4 Hz. Assume that the sampling signal consists of pulses each having a unit height and duration dt. The pulses occur every $t = k/4$ sec, $-\infty < k < \infty$.

(a) Sketch the binary signal representing each sample voltage.

(b) How many bits are required per sample?

6.7-1. A D-to-A converter is shown in Fig. P6.7-1. Using the set-reset flip-flops shown explain the operation of the device.

6.7-2. An A-to-D converter is shown in Fig. P6.7-2. Using *trigger* flip-flops as indicated explain the operation of the device.

6.8-1. Show that if $v(t)$ is given by Eq. (6.8-1), the area $A = A(-) + A(+)$ is positive if $\omega_c \tau > 1.59$.

6.8-2. Instead of using 64 quantization levels, we want to use 512 quantization levels. How much must the channel bandwidth be increased if both systems are to enjoy the same noise immunity?

6.9-1. A PCM signal is passed through a channel which can be represented by an RC low-pass filter with a 3-dB frequency f_c. The bit duration of the PCM signal is $T = 1/f_c$. Sketch the eye patterns for the following signals:

(a) An alternating 1010 sequence.

(b) An alternating 11001100 sequence.

(c) An alternating 111110111110 sequence.

Can you tell from the eye patterns where the probability of error is greatest?

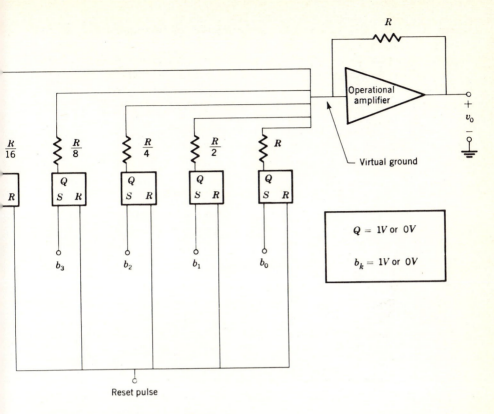

Fig. P6.7-1 A 5-bit D to A converter.

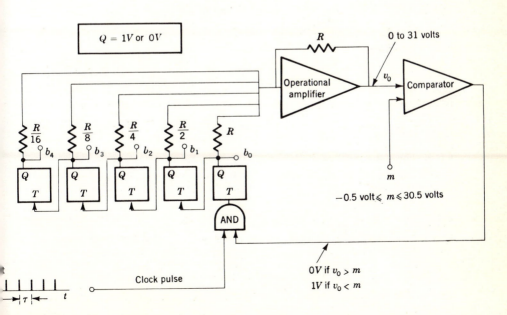

Fig. P6.7-2 A 32 quantization level A to D converter.

6.11-1. Using block diagrams design a PCM multiplexing system to permit the combining of 3 signals m_1, m_2, m_3. The signals are bandlimited to 5 kHz, 10 kHz, and 5 kHz, respectively. Each signal is to be sampled at its Nyquist rate and quantized into 256 levels ($N = 8$). The commutator dwells on each signal so that the 8 bits are transmitted simultaneously.

    (*a*) Sketch the system. What is the maximum bit duration that may be used?

    (*b*) What is the channel bandwidth required to pass the PCM signal?

    (*c*) What is the commutator speed in revolutions per second?

6.11-2. Repeat Prob. 6.11-1. Now, however, the commutator dwells on each signal only long enough to record just one bit of each sample.

6.12-1. Three signals m_1, m_2, m_3 having the bandwidths 5 kHz, 10 kHz, and 20 kHz are each quantized into 256 levels. Each signal lasts for 10 min. If each signal is sampled at its Nyquist rate,

    (*a*) How many samples are taken from each signal?

    (*b*) How many bits are produced from each signal, if each signal is PCM-encoded using 8 bits per sample?

6.13-1. (*a*) Write the nonlinear difference equation of the delta modulator shown in Fig. 6.13-1. Let $m(t) = 18 \times 10^{-3} \sin 2\pi t$ and let the samples occur every 0.05 sec starting at $t = 0.01$ sec. The step size is 5 mV.

    (*b*) Write and run a computer program to solve for $\Delta(t)$.

    (*c*) Repeat (*b*) if the samples occur every 0.1 sec.

    (*d*) Compare the results of (*b*) and (*c*).

6.13-2. The input to a DM is $m(t) = 0.01t$. The DM operates at a sampling frequency of 20 Hz and has a step size of 2 mV. Sketch the delta modulator output, $\Delta(t)$ and $\tilde{m}(t)$.

6.13-3. Consider the delta PCM system shown in Fig. P6.13-3.

    (*a*) Explain its operation.

    (*b*) Sketch the receiver.

    (*c*) If $m(t) = 0.05 \sin 2\pi t$, find $\tilde{m}(t)$ and $\Delta(t)$ graphically. $\tau = \frac{1}{8}$ sec.

6.14-1. The input to a DM is $m(t) = kt$. Prove, by graphically determining $\tilde{m}(t)$, that slope overload occurs when k exceeds a specified value. What is this value in terms of the step size S and the sampling frequency f_s?

6.14-2. If the step size is S, the sampling frequency is $f_s^{(\Delta)}$, and $m(t) = M \sin \omega t$, explain what happens to $\tilde{m}(t)$ if $2M < S$. This is called *step-size limiting*.

6.14-3. The signal $m(t) = M \sin \omega_0 t$ is to be encoded by using a delta modulator. If the step size S and sampling frequency $f_s^{(\Delta)}$ are selected so as to ensure that neither slope overloads [Eq. (6.14-1)] nor step-size limiting (Prob. 6.14-2) occurs, show that $f_s^{(\Delta)} > 3f_0$.

6.15-1. An adaptive delta modulator is shown in Fig. P6.15-1. The gain K is variable and is adjusted using the following logic. If $p_0(t)$ alternates between $+1$ and -1, $K = 1$; if a sequence of N positive or N negative pulses occurs, K increases by N; if after a sequence of N pulses the polarity changes, $|K|$ decreases by 2. Consider the sequence $1, 1, 1, -1, 1$. Then $K = 1, 2, 3, 1, 1$. $K_{\min} = 1$. Find $\tilde{m}(t)$ if $m(t) = \sin 2\pi t$. Perform the analysis graphically. Consider a

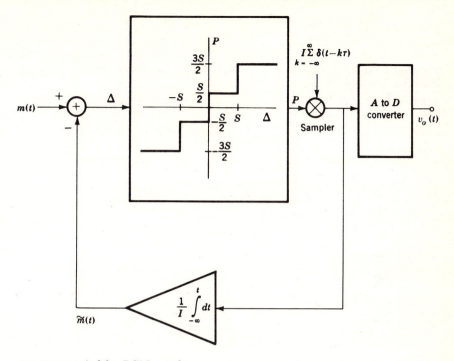

Fig. P6.13-3   A delta PCM encoder.

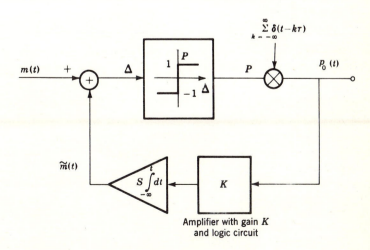

Fig. P6.15-1   An adaptive delta modulator.

sampling time of 0.05 sec, and at $t = 0.01$ sec a sample occurs. At this time the step size $S = 1$ volt when $K = 1$.

6.17-1. The binary signal $v(t) = +V, -V, +V, -V, -V, -V$ is transmitted by PSK using the format given in Eq. (6.17-2). If each binary digit in $v(t)$ has a duration $T = 1/f_c$, sketch $v_{\text{PSK}}(t)$.

6.18-1. The bit stream $b'(t) = 101000110111011$ is transmitted by DPSK. Find the encoded sequence $b(t)$.

6.18-2. Using the DPSK receiver to receive the DPSK signal obtained in Prob. 6.18-1, show that $b'(t)$ is recovered.

6.19-1. The bit stream $b(t) = 10100011$ is transmitted by FSK. The frequency $f_0 + \Omega/2\pi$ is transmitted when $b(t) = 1$, and $f_0 - \Omega/2\pi$ is transmitted when $b(t) = 0$. If the duration of each binary digit is 10^{-6} sec, and $f_0 = 2$ MHz and $\Omega/2\pi = 1$ MHz, sketch v_{FSK} [use the format of Eq. (6.19-1)].

REFERENCES

1. Transmission Systems for Communications, Bell Telephone Laboratories, published by Western Electric Company, *Tech. Pub.*, Winston-Salem, N.C., 1964.
2. Lucky, R. W., J. Salz, and E. J. Weldon, Jr.: "Principles of Data Communication," pp. 61–87 McGraw-Hill Book Company, New York, 1968.
3. Lucky, R. W., J. Salz, and E. J. Weldon, Jr.: "Principles of Data Communication," pp. 128–165, McGraw-Hill Book Company, New York, 1968.
4. Mayo, J. S., et al.: "An Experimental 224 mb/s PCM Transmission System," BSTJ Monograph 5097.
5. de Jager, F.: Deltamodulation: A Method of PCM Transmission Using a 1-unit Code, *Philips Res. Rept.* 7, pp. 442–446, 1952.

7
Mathematical Representation of Noise

All the communication systems discussed in the preceding chapters accomplish the same end. They allow us to reproduce the signal, impressed on the communication channel at the transmitter, at the demodulator output. Our only basis for comparison between systems, up to this point, has been bandwidth occupancy, convenience of multiplexing, and ease of implementation of the physical hardware. We have neglected, however, in our preceding discussions, the very important and fundamental fact that, in any real physical system, when the signal voltage arrives at the demodulator, it will be accompanied by a voltage waveform which varies with time in an entirely *unpredictable* manner. This unpredictable voltage waveform is a random process called *noise*. A signal accompanied by such a waveform is described as being *contaminated* or *corrupted* by noise. We now find that we have a new basis for system comparison, that is, the extent to which a communication system is able to distinguish the signal from the noise and thereby yield a *distortion-free* and *error-free* reproduction of the original signal.

In the present chapter we shall make only a brief reference to the sources of noise which are discussed more extensively in Chap. 14. Here we shall be concerned principally with a discussion of the mathematical representation and statistical characterizations of noise.

7.1 SOME SOURCES OF NOISE

One source of noise is the constant agitation which prevails throughout the universe at the molecular level. Thus, a piece of solid metal may appear to our gross view to be completely at rest. We know, however, that the individual molecules are vibrating about their positions of equilibrium in a crystal lattice, and that the conduction electrons of the metal are wandering randomly throughout the volume of the metal. Similarly the molecules of an enclosed gas are in constant motion, colliding with one another and colliding also with the walls of the container. These agitations of molecules are called *thermal* agitations because they increase with temperature.

Let us consider a simple resistor. It is a resistor, or rather a conductor, because there are within it conduction electrons which are free to wander randomly through the entire volume of the resistor. On the average these electrons will be uniformly distributed through the volume, as will positive ions, and the entire structure will be electrically neutral. However, because of the random and erratic wanderings of the electrons, there will be *statistical fluctuations* away from neutrality. Thus at one time or another the distribution of charge may not be uniform, and a voltage difference will appear between the resistor terminals. The random, erratic, unpredictable voltage which so appears is referred to as *thermal resistor noise*. As is to be expected, thermal resistor noise increases with temperature. Resistor noise also increases with the resistance value of the resistor, being zero in a perfect conductor.

A second type of noise results from a phenomenon associated with the flow of current across semiconductor junctions. The charge carriers, electrons or holes, enter the junction region from one side, drift or are accelerated across the junction, and are collected on the other side. The average junction current determines the average interval that elapses between the times when two successive carriers enter the junction. However, the exact interval that elapses is subject to random statistical fluctuations. This randomness gives rise to a type of noise which is referred to as *shot noise*. Shot noise is also encountered as a result of the randomness of emission of electrons from a heated surface and is consequently also associated with thermionic devices.

When a signal reaches a receiver it may well arrive very greatly attenuated. It is therefore necessary to provide amplification. This amplification is accomplished in circuits using active devices (transistors, etc.) and resistors.   Hence the signal becomes corrupted by *thermal* and *shot* noise.   Even more, the signal may have been contaminated by noise as a result of many types of random disturbances superimposed on the signal during the course of its transfer over the communication channel. The contamination of the signal may take several forms.   The noise may be added to the signal, in which case it is called *additive* noise, or the noise may multiply the signal, in which case the effect is called *fading*. In this text we consider additive noise only.

We shall confine our interest, for the most part, albeit not exclusively, to noise which may be described as an ergodic random process. The characteristic of ergodicity of interest here is that an ergodic process is also stationary, that is, statistical averages taken over an ensemble representing the processes yield a result that is independent of the time at which the averages are evaluated.   We shall further assume, except where specifically noted, that the probability density of the noise is gaussian.   In very many communication systems and in a wide variety of circumstances the assumption of a gaussian density is justifiable.   On the other hand, it needs to be noted that such an assumption is hardly universally valid.   For example, if gaussian noise is applied to the input of a rectifier circuit, the output is not gaussian.   Similarly, it may well be that the noise encountered on a telephone line or on other channels consists of short, pulse-type disturbances whose amplitude distribution is decidedly not gaussian.

7.2   A FREQUENCY-DOMAIN REPRESENTATION OF NOISE

In a communication system noise is often passed through filters.   These filters are usually described in terms of their characteristics in the frequency domain.   Hence, to determine the influence of these filters on the noise, it is convenient to have a *frequency-domain characterization* of the noise.[1]   We shall now establish such a frequency-domain characterization.   On the basis of this representation we shall be able to define a *power spectral density* for a noise waveform which has characteristics similar to those of the power spectral density of a deterministic waveform.   Our discussion will be somewhat heuristic.

Let us select a particular sample function of the noise, and select from that sample function an interval of duration T extending, say, from $t = -T/2$ to $t = T/2$.   Such a noise sample function $n^{(s)}(t)$ is shown in Fig. 7.2-1a.   Let us generate, as in Fig. 7.2-1b, a periodic waveform in which the waveform in the selected interval is repeated every T sec. This periodic waveform $n_T^{(s)}(t)$ can be expanded in a Fourier series, and such a series will properly represent $n^{(s)}(t)$ in the interval $-T/2$ to $T/2$.

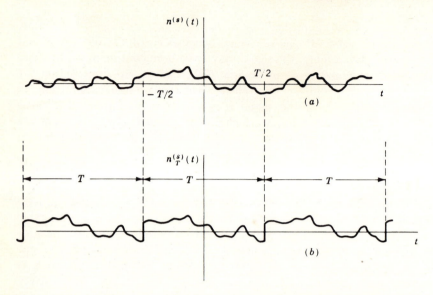

Fig. 7.2-1 (*a*) A sample noise waveform. (*b*) A periodic waveform is generated by repeating the interval in (*a*) from $-T/2$ to $T/2$.

The fundamental frequency of the expansion is $\Delta f = 1/T$, and, assuming no dc component, we have

$$n_T^{(s)}(t) = \sum_{k=1}^{\infty} (a_k \cos 2\pi k \, \Delta ft + b_k \sin 2\pi k \, \Delta ft) \qquad (7.2\text{-}1)$$

or alternately

$$n_T^{(s)}(t) = \sum_{k=1}^{\infty} c_k \cos (2\pi k \, \Delta ft + \theta_k) \qquad (7.2\text{-}2)$$

in which a_k, b_k, and c_k are the constant coefficients of the spectral terms and θ_k is a phase angle. Of course,

$$c_k^2 = a_k^2 + b_k^2 \qquad (7.2\text{-}3)$$

and

$$\theta_k = -\tan^{-1} \frac{b_k}{a_k} \qquad (7.2\text{-}4)$$

The power spectrum of the expansion is shown in Fig. 7.2-2. The power associated with each spectral term is $c_k^2/2 = (a_k^2 + b_k^2)/2$. Since a two-sided spectrum is shown, each power spectral line is of height $c_k^2/4$. The frequency axis has been marked off in intervals Δf, and a power

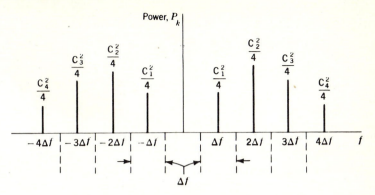

Fig. 7.2-2 The power spectrum of the waveform $n_T^{(s)}$.

spectral line is located at the center of each interval.   We now define the power spectral density at the frequency $k \, \Delta f$ as the quantity

$$G_n(k \, \Delta f) \equiv G_n(-k \, \Delta f) \equiv \frac{c_k^2}{4\Delta f} = \frac{a_k^2 + b_k^2}{4\Delta f} \tag{7.2-5}$$

The total power P_k associated with the frequency interval Δf at the frequency $k \, \Delta f$ is

$$P_k = 2G_n(k \, \Delta f) \, \Delta f \tag{7.2-6}$$

One-half of the power, $P_k/2 = G_n(k \, \Delta f) \, \Delta f$, is associated with a spectral line at frequency $k \, \Delta f$, the other half with a line at frequency $- \, k \, \Delta f$. Thus $G_n(k \, \Delta f) \equiv G_n(-k \, \Delta f)$ is equal to the power in the positive or negative interval divided by the size of the interval.   Hence $G_n(k \, \Delta f)$ is the (two-sided) *mean power spectral density* within each interval.

A particular selection of coefficients a_k and b_k in Eq. (7.2-1) [or coefficients c_k and angles θ_k in Eq. (7.2-2)] will represent a particular sample function of periodic noise.   A different selection will represent a different sample function.   If we propose that the representations of Eq. (7.2-1) or (7.2-2) are to represent generally the random process under discussion, i.e., the periodic noise, we need but consider that the a_k's, b_k's, c_k's, and θ_k's are not fixed numbers but are instead *random variables.*† Finally, let us allow that $T \to \infty$ ($\Delta f \to 0$), so that the periodic sample functions of the noise revert to the actual noise sample functions.   We

---

† In Chap. 2, following accepted practice, we used capital symbols to represent random variables (e.g., X, Y, etc.) and lowercase symbols (x, y, etc.) to represent possible particular values attainable by the random variables.   Since it is inconvenient to do so, we shall no longer persist in this precision of notation.   No confusion will result thereby since the text will always make clear the significance of a symbol.

then have that the noise $n(t)$ is to be represented as

$$n(t) = \lim_{\Delta f \to 0} \sum_{k=1}^{\infty} (a_k \cos 2\pi k \, \Delta f t + b_k \sin 2\pi k \, \Delta f t) \tag{7.2-7}$$

or as

$$n(t) = \lim_{\Delta f \to 0} \sum_{k=1}^{\infty} c_k \cos (2\pi k \, \Delta f t + \theta_k) \tag{7.2-8}$$

We continue to accept Eq. (7.2-5) as the definition of the power spectral density of the noise $n(t)$, except that we replace c_k^2 by $\overline{c_k^2}$, that is, by the expected or ensemble average value of the square of the random variable c_k. Further, as $\Delta f \to 0$, the discrete spectral lines get closer and closer, finally forming a continuous spectrum. In Eq. (7.2-5) we therefore replace $k \, \Delta f$ by the continuous frequency variable f. From Eq. (7.2-3) we also have that

$$\overline{c_k^2} = \overline{a_k^2} + \overline{b_k^2} \tag{7.2-9}$$

so that finally Eq. (7.2-5) becomes

$$G_n(f) = \lim_{\Delta f \to 0} \frac{\overline{c_k^2}}{4\Delta f} = \lim_{\Delta f \to 0} \frac{\overline{a_k^2} + \overline{b_k^2}}{4\Delta f} \tag{7.2-10}$$

Note that the power in the frequency range from f_1 to f_2 is

$$P(f_1 \to f_2) = \int_{-f_2}^{-f_1} G_n(f) \, df + \int_{f_1}^{f_2} G_n(f) \, df = 2 \int_{f_1}^{f_2} G_n(f) \, df \tag{7.2-11}$$

while the total power P_T is

$$P_T = \int_{-\infty}^{\infty} G_n(f) \, df = 2 \int_{0}^{\infty} G_n(f) \, df \tag{7.2-12}$$

7.3 THE EFFECT OF FILTERING ON THE PROBABILITY DENSITY OF GAUSSIAN NOISE

We shall now show that if, as in Fig. 7.3-1a, gaussian noise $n_i(t)$ is applied to the input of a filter, the output noise $n_o(t)$ is also gaussian. Let the impulse response of the filter be $h(t)$. Then, applying Eq. (1.11-8) to the present situation, we have that

$$n_o(t) = \int_{-\infty}^{\infty} n_i(\tau) h(t - \tau) \, d\tau = \int_{-\infty}^{t} n_i(\tau) h(t - \tau) \, d\tau \tag{7.3-1}$$

In Eq. (7.3-1) we have taken cognizance of the fact that, in general, the upper limit may actually be set at $\tau = t$, since the variable of integration is τ and $h(t - \tau) = 0$ for $\tau > t$. Equation (7.3-1) expresses the output $n_o(t)$ as the superposition of a succession of impulses of strength $n_i(\tau) \, d\tau$

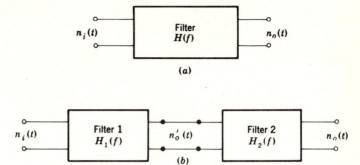

Fig. 7.3-1 (*a*) Gaussian noise $n_i(t)$ is applied to a linear filter whose output is $n_o(t)$. (*b*) The filter in (*a*) is split into two parts.

applied at the filter input. This point is emphasized by rewriting Eq. (7.3-1) in the form

$$n_o(t) = \lim_{\Delta\tau\to0} \sum_{k=-\infty}^{k=t/\Delta\tau} n_i(k\,\Delta\tau)h(t - k\,\Delta\tau)\,\Delta\tau \qquad (7.3\text{-}2)$$

in which k ranges over integral values. In the limit $k\,\Delta\tau$ reverts to the continuous variable τ, and the summation reverts to the integral. In Eq. (7.3-2), $n_i(k\,\Delta\tau)$ is a random variable. That is, the noise is represented by an ensemble of sample functions, and the value of $n_i(k\,\Delta\tau)$ depends on the sample function we consider. On the other hand, the quantities $h(t - k\,\Delta\tau)$ are fixed, determininistic numbers, regulated by the nature of the filter.

Let us assume that the input noise is white and gaussian. Then, as discussed in Sec. 2.20, the past history of the noise waveform provides no information about its future behavior. That is to say, on any one sample noise waveform of the ensemble, successive noise voltage values are independent of one another no matter how small the interval $\Delta\tau$ between voltage determinations. Hence, the random variables $n_i(k\,\Delta\tau)$ and $n_i(l\,\Delta\tau)$ are independent except for the case where $k = l$. Thus, at any time, with $t = t_0$, $n_o(t) = n_o(t_0)$, as given by Eq. (7.3-2), is a linear superposition of independent gaussian random variables. Hence, as discussed in Sec. 2.16, $n_o(t_0)$ is a gaussian random variable, and $n_o(t)$ is a gaussian random process.

Now suppose that the filter of transfer function $H(f)$ is split into two parts, as shown in Fig. 7.3-1*b*, with $H(f) = H_1(f)H_2(f)$. Then if $n_i(t)$ is white and gaussian, both $n_o'(t)$ and $n_o(t)$ are, on the basis of the above discussion, gaussian, albeit, in general, not white. Further, since both

$n'_o(t)$ and $n_o(t)$ are gaussian, we have the result that nonwhite gaussian noise, when applied to a linear filter, yields an output noise which is again gaussian. Strictly, we have established this last result only for the case where the nonwhite gaussian noise input is itself derived by filtering white noise. But this result is quite general enough for our purposes.

Equation (7.3-2) may be applied directly to filter 2 in Fig. 7.3-1b by simply replacing the impulses $n_i(k \, \Delta\tau)$ by the impulses $n'_o(k \, \Delta\tau)$. However, since $n'_o(t)$ is not white noise, successive noise voltage values are *not* independent of one another; that is, the random variable $n'_o(k \, \Delta\tau)$ and $n'_o(l \, \Delta\tau)$ are gaussian but not independent. Nonetheless, we found that $n_o(t)$ is gaussian. Hence we deduce that a linear superposition of *dependent* gaussian random variables is still gaussian.

7.4   SPECTRAL COMPONENTS OF NOISE

We have represented noise $n(t)$ as a superposition of noise spectral components. The spectral component associated with the kth frequency interval is given, in the limit as $\Delta f \to 0$, by $n_k(t)$ written as

$$n_k(t) = a_k \cos 2\pi k \, \Delta f t + b_k \sin 2\pi k \, \Delta f t \qquad (7.4\text{-}1)$$

or as

$$n_k(t) = c_k \cos (2\pi k \, \Delta f t + \theta_k) \qquad (7.4\text{-}2)$$

The spectral components which compose a deterministic waveform are themselves deterministic. The spectral components, as in the present case, which compose noise, are themselves random processes. Thus, in Eqs. (7.4-1) and (7.4-2), a_k, b_k, c_k, and θ_k are random variables, and $n_k(t)$ represents an ensemble of sample functions, one sample function for each possible set of values of a_k and b_k or of c_k and θ_k. The sample functions are each deterministic waveforms. They are, as a matter of fact, pure sinusoids differing from one another in phase and amplitude, depending on the value of θ_k and c_k. The random process $n_k(t)$ is stationary; that is, its statistical properties do not change with time. However, it is not ergodic; that is, the time averages of the individual sample function of the ensemble are different from one another.

We look now at some of the properties of the random variables a_k and b_k. The normalized power P_k (variance) of $n_k(t)$ is determined by taking the average over the *ensemble* of $[n_k(t)]^2$. We find, from Eq. (7.4-1), that

$$P_k = \overline{[n_k(t)]^2} = \overline{a_k^2} \cos^2 2\pi k \, \Delta f t + \overline{b_k^2} \sin^2 2\pi k \, \Delta f t$$
$$+ \overline{2a_k b_k} \sin 2\pi k \, \Delta f t \cos 2\pi k \, \Delta f t \quad (7.4\text{-}3)$$

As noted, $n_k(t)$ is a stationary process, so that $\overline{[n_k(t)]^2}$ does not depend on the time selected for its evaluation. We are therefore at liberty to evaluate P_k by substituting in Eq. (7.4-3) a value $t = t_1$ for which $\cos 2\pi k\, \Delta f t_1 = 1$, in which case $\sin 2\pi k\, \Delta f t_1 = 0$. We then have

$$P_k = \overline{a_k^2} \tag{7.4-4}$$

Similarly we may show that $P_k = \overline{b_k^2}$ so that

$$\overline{a_k^2} = \overline{b_k^2} \tag{7.4-5}$$

From Eqs. (7.2-6), (7.2-9), (7.4-4), and (7.4-5) we have

$$P_k = 2G_n(k\, \Delta f)\, \Delta f = 2G_n(-k\, \Delta f)\, \Delta f = \overline{a_k^2} = \overline{b_k^2} = \frac{\overline{a_k^2}}{2} + \frac{\overline{b_k^2}}{2} = \frac{\overline{c_k^2}}{2} \tag{7.4-6}$$

Since $\overline{a_k^2} = \overline{b_k^2}$, Eq. (7.4-3) may be written

$$P_k = \overline{a_k^2}(\cos^2 2\pi k\, \Delta f t + \sin^2 2\pi k\, \Delta f t)$$
$$+ \overline{2a_k b_k} \sin 2\pi k\, \Delta f t \cos 2\pi k\, \Delta f t \tag{7.4-7a}$$

$$P_k = \overline{a_k^2} + \overline{2a_k b_k} \sin 2\pi k\, \Delta f t \cos 2\pi k\, \Delta f t \tag{7.4-7b}$$

We note, however, from Eq. (7.4-4) that $P_k = \overline{a_k^2}$ independently of time. In order that Eq. (7.4-7b) be consistent with this result, we require that

$$\overline{a_k b_k} = 0 \tag{7.4-8}$$

Thus, the coefficients a_k and b_k are *uncorrelated*.

We may also establish that the coefficients a_k and b_k are gaussian. For this purpose we use Eq. (7.4-1) and substitute for t a value $t = t_1$, for which $\cos 2\pi k\, \Delta f t_1 = 1$ and $\sin 2\pi k\, \Delta f t_1 = 0$. Then

$$n_k(t_1) = a_k \tag{7.4-9}$$

Now $n_k(t_1)$ is a gaussian random variable. We have this result from the consideration that $n_k(t)$ may be viewed as the output of a very narrowband filter whose input is gaussian noise. As discussed in Sec. 7.3 such a noise output is gaussian, and the noise voltage at any time, as at $t = t_1$ above, has a gaussian probability density. Hence, Eq. (7.4-9) says that a_k is also a gaussian random variable. It is, of course, similarly established that b_k is gaussian.

We note also that since $n_k(t)$ is the output of a narrowband filter at frequency $k\, \Delta f$, it has no dc component. Hence from Eq. (7.4-9) a_k has no dc component; that is, $\overline{a_k} = 0$, and, of course, similarly $\overline{b_k} = 0$.

Finally, let us consider two spectral components of noise, one at frequency $k \, \Delta f$ and the other at frequency $l \, \Delta f$.   Then

$$n_k(t) = a_k \cos 2\pi k \, \Delta ft + b_k \sin 2\pi k \, \Delta ft \qquad (7.4\text{-}10)$$

and

$$n_l(t) = a_l \cos 2\pi l \, \Delta ft + b_l \sin 2\pi l \, \Delta ft \qquad (7.4\text{-}11)$$

If we form the product $n_k(t)n_l(t)$ from Eqs. (7.4-10) and (7.4-11), we find that the product has four terms (products of sinusoids), all of which are time-dependent, provided that $k \neq l$.   The coefficients of these terms are $a_k a_l$, $a_k b_l$, $b_k a_l$, and $b_k b_l$.   Now let us take the ensemble average of the product, that is, $\overline{n_k(t)n_l(t)}$.   Then, again, because of the stationary character of the random processes involved, this ensemble average must be time-independent.   Hence we have that

$$\overline{a_k a_l} = \overline{a_k b_l} = \overline{b_k a_l} = \overline{b_k b_l} = 0 \qquad (7.4\text{-}12)$$

That is, each of the coefficients a_k and b_k is uncorrelated with each of the coefficients a_l and b_l.

In summary, we find that we have described noise in the following manner.   Noise $n(t)$ is a gaussian, ergodic, random process.   It may be represented as a linear superposition of spectral components of the form given in Eq. (7.4-1) with the understanding, of course, that the description becomes more precise as $\Delta f \rightarrow 0$.   The coefficients a_k and b_k are gaussian random variables of average value zero and equal variance (normalized power) related to the two-sided power spectral density as in Eq. (7.4-6).   The coefficients a_k and b_k are uncorrelated with one another and are uncorrelated also with the coefficients of a spectral component at a different frequency.

We may now deduce some statistical characteristics of interest concerning the c_k's and θ_k's in the noise spectral component representation of Eq. (7.4-2).   We note from Eqs. (7.2-3) and (7.2-4) that the c_k's and θ_k's are related to the a_k's and b_k's in precisely the manner in which the random variables R and Θ of Sec. 2.14 are related to the gaussian variables X and Y.   From the results of Sec. 2.14 we then find (Prob. 7.4-2) that the c_k's have a Rayleigh probability density

$$f(c_k) = \frac{c_k}{P_k} \, e^{-c_k{}^2/2P_k} \qquad c_k \geq 0 \qquad (7.4\text{-}13)$$

where P_k[Eq. (7.4-6)] is the normalized power in the spectral range Δf at the frequency $k \, \Delta f$.   Similarly, the angle θ_k has a uniform probability density

$$f(\theta_k) = \frac{1}{2\pi} \qquad -\pi \leq \theta_k \leq \pi \qquad (7.4\text{-}14)$$

Furthermore, the amplitude c_k and phase θ_k are independent of one another and are independent as well of the amplitude and phase of a spectral component at a different frequency.

7.5 RESPONSE OF A NARROWBAND FILTER TO NOISE

If the representation of noise as a superposition of spectral components is a reasonable one, we should expect that when noise is passed through a narrowband filter the output of the filter should look rather like a sinusoid. We find that such is indeed the case, for we find that the output of a narrowband filter with noise input has the appearance shown in Fig. 7.5-1. The output waveform looks like a sinusoid except that, as expected, the amplitude varies randomly. The spectral range of the *envelope* of the filter output encompasses the spectral range from $-B/2$ to $B/2$, where B is the filter bandwidth. The average frequency of the waveform is the center frequency f_c of the filter. If $B \ll f_c$, the envelope changes very "slowly" and makes an appreciable change only over very many cycles. Thus, while the spacings of the *zero crossings* of the waveform are not precisely constant, the change from cycle to cycle is small, and when averaged over many cycles is quite constant at the value $1/2f_c$. Finally, we may note that as B becomes progressively smaller, so also does the average amplitude, and the waveform becomes more and more sinusoidal.

7.6 EFFECT OF A FILTER ON THE POWER SPECTRAL DENSITY OF NOISE

Let a spectral component of noise $n_{k_i}(t)$ given by Eq. (7.4-1) be the input to a filter whose transfer function at the frequency $k\,\Delta f$ is

$$H(k\,\Delta f) = |H(k\,\Delta f)|e^{j\varphi_k} = |H(k\,\Delta f)|\underline{/\varphi_k} \qquad (7.6\text{-}1)$$

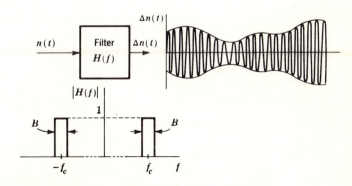

Fig. 7.5-1 Response of a narrowband filter to noise.

The corresponding output spectral component of noise will be $n_{k_o}(t)$

$$n_{k_o}(t) = |H(k \, \Delta f)| a_k \cos (2\pi k \, \Delta f t + \varphi_k)$$
$$+ |H(k \, \Delta f)| b_k \sin (2\pi k \, \Delta f t + \varphi_k) \quad (7.6\text{-}2)$$

The power P_{k_i} associated with $n_{k_i}(t)$ is, from Eq. (7.4-6),

$$P_{k_i} = \frac{\overline{a_k^2} + \overline{b_k^2}}{2} \tag{7.6-3}$$

Since $|H(k \, \Delta f)|$ is a deterministic function, $\overline{[|H(k \, \Delta f)| a_k]^2} = |H(k \, \Delta f)|^2 \overline{a_k^2}$ and $\overline{[|H(k \, \Delta f)| b_k]^2} = |H(k \, \Delta f)|^2 \overline{b_k^2}$. Hence, comparing Eq. (7.6-2) with Eq. (7.4-1), we find that the power P_{k_o} associated with $n_{k_o}(t)$ is

$$P_{k_o} = |H(k \, \Delta f)|^2 \frac{\overline{a_k^2} + \overline{b_k^2}}{2} \tag{7.6-4}$$

Finally, then, from Eqs. (7.6-3) and (7.6-4), using also Eq. (7.4-6), we have that the power spectral densities at input and output, $G_{n_i}(k \, \Delta f)$ and $G_{n_o}(k \, \Delta f)$, are related by

$$G_{n_o}(k \, \Delta f) = |H(k \, \Delta f)|^2 G_{n_i}(k \, \Delta f) \tag{7.6-5}$$

In the limit as $\Delta f \to 0$ and $k \, \Delta f$ is replaced by a continuous variable f, Eq. (7.6-5) becomes

$$G_{n_o}(f) = |H(f)|^2 G_{n_i}(f) \tag{7.6-6}$$

Note the similarity between this result and Eq. (1.9-7), which applies to a deterministic waveform.

7.7 SUPERPOSITION OF NOISES

The concept of a power spectrum is useful because it allows us to resolve a deterministic waveform, or a random process, $f(t)$ into a sum

$$f(t) = f_1(t) + f_2(t) + \cdots \tag{7.7-1}$$

in such manner that *superposition of power* applies, i.e., the power of $f(t)$ is the sum of the powers of $f_1(t)$, $f_2(t)$, ... When a deterministic waveform is resolved into a series of spectral components, superposition of power applies because of the orthogonality of spectral components of different frequencies. This point was discussed in Sec. 1.7.

We have also represented a noise waveform as a superposition of spectral components, all of which are harmonics of some fundamental frequency Δf which, in the limit, approaches zero. Hence, up to the present point, this feature alone would have been enough to justify super-position of noise power as expressed in Eqs. (7.2-11) and (7.2-12). But suppose that we have two noise processes $n_1(t)$ and $n_2(t)$, whose spectral ranges overlap in part or in their entirety. Then the power P_{12} of the sum $n_1(t) + n_2(t)$ would be

$$P_{12} = E\{[n_1(t) + n_2(t)]^2\} = E[n_1^2(t)] + E[n_2^2(t)] + 2E[n_1(t)n_2(t)]$$
$$(7.7\text{-}2a)$$

$$= P_1 + P_2 + 2E[n_1(t)n_2(t)] \qquad (7.7\text{-}2b)$$

where P_1 and P_2 are the powers, respectively, of the noise processes $n_1(t)$ and $n_2(t)$, and $E[n_1(t)n_2(t)]$, which is the expected value of the product, is the *cross correlation* of the processes. Thus superposition of power, $P_{12} = P_1 + P_2$, continues to apply, provided the processes are *uncorrelated*. Such would be the case, for example, if $n_1(t)$ and $n_2(t)$ were the thermal noises of two different resistors.

7.8   MIXING INVOLVING NOISE

NOISE WITH SINUSOID

A situation often encountered in communication systems is one in which noise is mixed with (i.e., multiplied by) a deterministic sinusoidal waveform. Let the sinusoidal waveform be $\cos 2\pi f_0 t$. Then the product of this waveform with a spectral noise component, as given by Eq. (7.4-1), yields

$$n_k(t) \cos 2\pi f_0 t = \frac{a_k}{2} \cos 2\pi (k\,\Delta f + f_0)t + \frac{b_k}{2} \sin 2\pi (k\,\Delta f + f_0)t$$

$$+ \frac{a_k}{2} \cos 2\pi (k\,\Delta f - f_0)t + \frac{b_k}{2} \sin 2\pi (k\,\Delta f - f_0)t$$
$$(7.8\text{-}1)$$

Thus the mixing gives rise to two noise spectral components, one at the sum frequency $f_0 + k\,\Delta f$ and one at the difference frequency $f_0 - k\,\Delta f$. In addition, the amplitudes of each of the two noise spectral components generated by mixing has been reduced by a factor of 2 with respect to the original noise spectral component. Hence the variances (normalized power) of the two new noise components are smaller by a factor of 4. Accordingly, if the power spectral density of the original noise component at frequency $k\,\Delta f$ is $G_n(k\,\Delta f)$, then, from Eq. (7.8-1), the new components have spectral densities

$$G_n(k\,\Delta f + f_0) = G_n(k\,\Delta f - f_0) = \frac{G_n(k\,\Delta f)}{4} \qquad (7.8\text{-}2)$$

In the limit as $\Delta f \to 0$, we replace $k\,\Delta f$ by the continuous variable f, and Eq. (7.8-2) becomes

$$G_n(f + f_0) = G_n(f - f_0) = \frac{G_n(f)}{4} \qquad (7.8\text{-}3)$$

In words: given the power spectral density plot $G_n(f)$ of a noise waveform $n(t)$, the power spectral density of $n(t) \cos 2\pi f_0 t$ is arrived at as follows: divide $G_n(f)$ by 4, shift the divided plot to the left by amount f_0, to the right by amount f_0, and add the two shifted plots.

Now consider the following situation. We have noise $n(t)$ from which we single out two spectral components, one at frequency $k \, \Delta f$ and one at frequency $l \, \Delta f$. We mix with a sinusoid at frequency f_0, with f_0 selected to be midway between $k \, \Delta f$ and $l \, \Delta f$; that is, $f_0 = \frac{1}{2}(k + l) \, \Delta f$. Then the mixing will give rise to *four* spectral components, two difference-frequency components and two sum-frequency components. The two difference-frequency components will be at the same frequency $p \, \Delta f = f_0 - k \, \Delta f = l \, \Delta f - f_0$. However, we now show that these difference-frequency components are uncorrelated. Representing the spectral components at $k \, \Delta f$ and at $l \, \Delta f$ as in Eqs. (7.4-10) and (7.4-11), we find that the difference-frequency components are

$$n_{p_1}(t) = \frac{a_k}{2} \cos 2\pi p \, \Delta f t - \frac{b_k}{2} \sin 2\pi p \, \Delta f t \tag{7.8-4}$$

and

$$n_{p_2}(t) = \frac{a_l}{2} \cos 2\pi p \, \Delta f t + \frac{b_l}{2} \sin 2\pi p \, \Delta f t \tag{7.8-5}$$

where $n_{p_1}(t)$ is the difference component due to the mixing of frequencies f_0 and $k \, \Delta f$, while $n_{p_2}(t)$ is the difference component due to the mixing of frequencies f_0 and $l \, \Delta f$. If we now take into account, as established in Sec. 7.4, that $\overline{a_k a_l} = \overline{a_k b_l} = \overline{b_k a_l} = \overline{b_k b_l} = 0$, then we find from Eqs. (7.8-4) and (7.8-5) that

$$E[n_{p_1}(t) n_{p_2}(t)] = 0 \tag{7.8-6}$$

Thus, as discussed in connection with Eq. (7.7-2), superposition of power applies, and the power at the difference frequency due to the superposition of $n_{p_1}(t)$ and $n_{p_2}(t)$ is

$$E\{[n_{p_1}(t) + n_{p_2}(t)]^2\} = E\{[n_{p_1}(t)]^2\} + E\{[n_{p_2}(t)]^2\} \tag{7.8-7}$$

Thus, mixing noise with a sinusoidal signal results in a frequency shifting of the original noise by f_0. The variance of the shifted noise is found by adding the variance of each new noise component. Thus we see that the principle stated immediately after Eq. (7.8-3) applies even when there is overlap in the two shifted power spectral density plots.

NOISE-NOISE MIXING

We consider now the result of multiplying two spectral components of noise. Representing the spectral components as in Eq. (7.4-2), we find that the product of two components, one at the frequency $k\,\Delta f$, the other at the frequency $l\,\Delta f$ is

$$n_k(t)n_l(t) = \tfrac{1}{2}c_kc_l \cos[2\pi(k+l)\,\Delta ft + \theta_k + \theta_l]$$
$$+ \tfrac{1}{2}c_kc_l \cos[2\pi(k-l)\,\Delta ft + \theta_k - \theta_l] \quad (7.8\text{-}8)$$

The multiplication thus gives rise to two new spectral components of noise, one at the sum frequency $(k+l)\,\Delta f$ and one at the difference frequency $(k-l)\,\Delta f$. The terms in Eq. (7.8-8) are of the form of Eq. (7.4-2), except that $\tfrac{1}{2}c_kc_l$ replaces c_k, and θ_k in Eq. (7.4-2) is replaced by $\theta_k + \theta_l$ in one case and $\theta_k - \theta_l$ in the other. Since θ_k and θ_l have uniform probability densities, it is intuitively apparent from the principle of *minimum astonishment* that $\theta_k + \theta_l$ and $\theta_k - \theta_l$ also have uniform densities. (A more formal proof is possible; see Prob. 7.8-3.) Hence the normalized power associated with each of the terms in Eq. (7.8-8) may be deduced as for the spectral component of Eq. (7.4-2). We then have, using Eq. (7.4-6),

$$P_{k+l} = P_{k-l} = \tfrac{1}{2}\overline{(\tfrac{1}{2}c_kc_l)^2} \qquad (7.8\text{-}9)$$

Since c_k and c_l are independent random variables,

$$P_{k+l} = P_{k-l} = \tfrac{1}{8}\overline{c_k^2}\,\overline{c_l^2} = \tfrac{1}{2}P_kP_l \qquad (7.8\text{-}10)$$

7.9 LINEAR FILTERING

Thermal noise has a power spectral density which is quite uniform up to frequencies of the order of 10^{13} Hz. Shot noise has a power spectral density which is reasonably constant up to frequencies which are of the order of the reciprocal of the transit time of charge carriers across the junction. Other noise sources similarly have very wide spectral ranges. We shall assume, in discussing the effect of noise on communication systems, that we have to contend with *white* noise. White noise is noise whose power spectral density is uniform over the entire frequency range of interest. The term *white* is used in analogy with white light, which is a superposition of all visible spectral components. Thus we assume, as shown in Fig. 7.9-1, that over the entire spectrum, including positive and

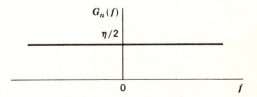

Fig. 7.9-1 Power spectral density of white noise.

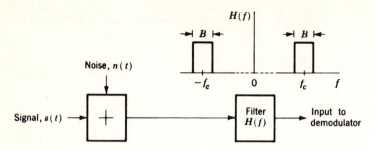

Fig. 7.9-2 A filter is placed before a demodulator to limit the noise power input to the demodulator.

negative frequencies,

$$G_n(f) = \frac{\eta}{2} \tag{7.9-1}$$

in which η is a constant.

In order to minimize the noise power that is presented to the demodulator of a receiving system, we introduce a filter before the demodulator as indicated in Fig. 7.9-2. The bandwidth B of the filter is made as narrow as possible so as to avoid transmitting any unnecessary noise to the demodulator. For example, in an AM system in which the baseband extends to a frequency of f_M, the bandwidth $B = 2f_M$. In a wideband FM system the bandwidth is proportional to twice the frequency deviation.

It is useful to consider the effect of certain types of filters on the noise. One of the filters most often used is the simple RC low-pass filter.

THE RC LOW-PASS FILTER

An RC low-pass filter with a 3-dB frequency f_c has the transfer function

$$H(f) = \frac{1}{1 + jf/f_c} \tag{7.9-2}$$

If the input noise to this filter has a power spectral density $G_{n_i}(f)$ and the power spectral density of the output noise is $G_{n_o}(f)$, then, using Eq. (7.6-6), we have

$$G_{n_o}(f) = G_{n_i}(f)|H(f)|^2 \tag{7.9-3}$$

If the noise is white, $G_{n_i}(f) = \eta/2$ for all frequencies, Eq. (7.9-3) becomes

$$G_{n_o}(f) = \frac{\eta}{2} \frac{1}{1 + (f/f_c)^2} \tag{7.9-4}$$

The noise power at the filter output N_o is

$$N_o = \int_{-\infty}^{\infty} G_{n_o}(f)\, df = \frac{\eta}{2} \int_{-\infty}^{\infty} \frac{df}{1 + (f/f_c)^2} \tag{7.9-5}$$

Changing variables to $x \equiv f/f_c$, and noting that $\int_{-\infty}^{\infty} dx/(1 + x^2) = \pi$, we have

$$N_o = \frac{\pi}{2} \eta f_c \tag{7.9-6}$$

THE RECTANGULAR (IDEAL) LOW-PASS FILTER

A *rectangular* low-pass filter has the transfer function

$$H(f) = \begin{cases} 1 & |f| \leq B \\ 0 & \text{elsewhere} \end{cases} \tag{7.9-7}$$

Assuming that the noise input to the filter is *white*, the output-power spectral density is

$$G_{n_o}(f) = \begin{cases} \dfrac{\eta}{2} & -B \leq f \leq B \\ 0 & \text{elsewhere} \end{cases} \tag{7.9-8}$$

The output noise power is

$$N_o = \eta B \tag{7.9-9}$$

A RECTANGULAR BANDPASS FILTER

A rectangular bandpass filter is shown in Fig. 7.9-3. The filter bandwidth is $f_2 - f_1$. Then, with a *white* noise input, the output-noise power is

$$N_o = 2\frac{\eta}{2}(f_2 - f_1) = \eta(f_2 - f_1) \tag{7.9-10}$$

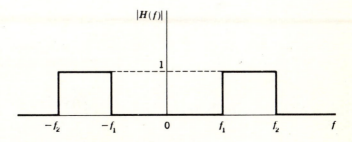

Fig. 7.9-3   A rectangular bandpass filter.

A DIFFERENTIATING FILTER

A differentiating filter is a network which yields at its output a waveform which is proportional to the time derivative of the input waveform.   As discussed in Sec. 1.9, such a network has a transfer function $H(f)$ which is proportional to the frequency; that is,

$$H(f) = j2\pi\tau f \tag{7.9-11}$$

where τ is a constant factor of proportionality.   If white noise with $G_{n_i}(f) = \eta/2$ is passed through such a filter, then the output-noise-power spectral density is

$$G_{n_o}(f) = |H(f)|^2 G_{n_i}(f) = 4\pi^2\tau^2 f^2 \frac{\eta}{2} \tag{7.9-12}$$

If the differentiator is followed by a rectangular low-pass filter having a bandwidth B, as described by Eq. (7.9-7), the noise power at the output of the low-pass filter is

$$N_o = \int_{-B}^{B} 4\pi^2\tau^2 f^2 \frac{\eta}{2}\, df = \frac{4\pi^2}{3}\, \eta\tau^2 B^3 \tag{7.9-13}$$

AN INTEGRATOR

Let noise $n(t)$ be applied to the input of an integrator at time $t = 0$.   We calculate the noise power at the integrator output at a time $t = T$.   The result will be of interest in connection with the discussion of the *matched filter* in Chap. 11.

A network which performs the operation of integration has a transfer function $1/j\omega\tau$.   A delay by an interval T is represented by a factor $e^{-j\omega T}$.   Hence a network which performs an integration over an interval T may be represented by a network whose transfer function is

$$H(f) = \frac{1}{j\omega\tau} - \frac{e^{-j\omega T}}{j\omega\tau} = \frac{1 - e^{-j\omega T}}{j\omega\tau} \tag{7.9-14}$$

where τ is a constant.   We find, with $\omega = 2\pi f$, that

$$|H(f)|^2 = \left(\frac{T}{\tau}\right)^2 \left(\frac{\sin\pi T f}{\pi T f}\right)^2 \tag{7.9-15}$$

The noise power output of such a filter with white input noise of power spectral density $\eta/2$ is (using $x \equiv \pi\tau f$)

$$N_o = \int_{-\infty}^{\infty} \frac{\eta}{2} |H(f)|^2\, df = \frac{\eta}{2}\left(\frac{T}{\tau}\right)^2 \int_{-\infty}^{\infty} \left(\frac{\sin\pi T f}{\pi T f}\right)^2 df \tag{7.9-16a}$$

$$= \frac{\eta T}{2\pi\tau^2} \int_{-\infty}^{\infty} \left(\frac{\sin x}{x}\right)^2 dx \tag{7.9-16b}$$

The definite integral in Eq. (7.9-16b) has the value π, so that finally

$$N_o = \frac{\eta T}{2\tau^2} \qquad (7.9\text{-}17)$$

7.10 NOISE BANDWIDTH

Consider that white noise is present at the input to a receiver. Suppose also that a filter with transfer function $H(f)$ centered at f_0, such as is indicated by the solid plot of Fig. 7.10-1, is being used to restrict the noise power actually passed on to the receiver. Now contemplate a rectangular filter as shown by the dotted plot in Fig. 7.10-1. This filter is also centered at f_0. Let the rectangular filter bandwidth B_N be adjusted so that the real filter and the rectangular filter transmit the same noise power. Then the bandwidth B_N is called the *noise bandwidth* of the real filter. The noise bandwidth, then, is the bandwidth of an idealized (rectangular) filter which passes the same noise power as does the real filter.

We illustrate the concept of noise bandwidth by considering the case of the low-pass RC filter with transfer function as given by Eq. (7.9-2). For this filter $H(f)$ attains its maximum value $H(f) = 1$ at $f = 0$. As given by Eq. (7.9-6), with white-noise input of power spectral density $\eta/2$, the noise output of the filter is

$$N_o(RC) = \frac{\pi}{2}\, \eta f_c \qquad (7.10\text{-}1)$$

In the presence of such noise, a rectangular low-pass filter with $H(f) = 1$ over its bandpass B_N would yield an output-noise power

$$N_o(\text{rectangular}) = \frac{\eta}{2}\, 2B_N = \eta B_N \qquad (7.10\text{-}2)$$

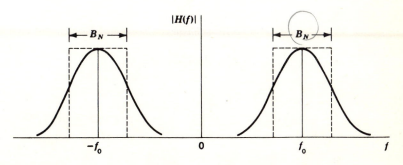

Fig. 7.10-1 Illustration of the noise bandwidth of a filter.

Setting $N_o(RC) = N_o$ (rectangular), we find the noise bandwidth to be

$$B_N = \frac{\pi}{2} f_c \qquad (7.10\text{-}3)$$

Thus, the noise bandwidth of the RC filter is $\pi/2$ ($= 1.57$) times its 3-dB bandwidth f_c.

7.11   QUADRATURE COMPONENTS OF NOISE

We have represented noise $n(t)$, as in Eq. (7.2-7), as the superposition of spectral components of noise in the expression

$$n(t) = \lim_{[\Delta f \to 0} \sum_{k=1}^{\infty} (a_k \cos 2\pi k \, \Delta ft + b_k \sin 2\pi k \, \Delta ft) \qquad (7.11\text{-}1)$$

It is sometimes more advantageous to represent the noise through an alternative representation given by

$$n(t) = n_c(t) \cos 2\pi f_0 t - n_s(t) \sin 2\pi f_0 t \qquad (7.11\text{-}2)$$

in which f_0 is an arbitrary frequency.   The representation of Eq. (7.11-2) is frequently used with great convenience in dealing with noise confined to a relatively narrow frequency band in the neighborhood of f_0.   For this reason Eq. (7.11-2) is often referred to as the *narrowband* representation. The term *quadrature component* representation is also often used because of the appearance in the equation of sinusoids in quadrature.

We may readily transform Eq. (7.11-1) into Eq. (7.11-2) and, in so doing, arrive at explicit expressions for $n_c(t)$ and $n_s(t)$.   Let us select f_0 to correspond to $k = K$; that is, we set

$$f_0 = K \, \Delta f \qquad (7.11\text{-}3)$$

Adding $2\pi f_0 t - 2\pi K \, \Delta ft = 0$ to the arguments in Eq. (7.11-1), we have

$$n(t) = \lim_{\Delta f \to 0} \sum_{k=1}^{\infty} \{a_k \cos 2\pi[f_0 + (k - K) \, \Delta f]t$$
$$+ b_k \sin 2\pi[f_0 + (k - K) \, \Delta f]t\} \qquad (7.11\text{-}4)$$

Using the trigonometric identities for the cosine of the sum of two angles and for the sine of the sum of two angles, it is readily verified that $n(t)$ is indeed given by Eq. (7.11-2), provided that $n_c(t)$ and $n_s(t)$ are taken to be

$$n_c(t) = \lim_{\Delta f \to 0} \sum_{k=1}^{\infty} [a_k \cos 2\pi(k - K) \, \Delta ft$$
$$+ b_k \sin 2\pi(k - K) \, \Delta ft] \qquad (7.11\text{-}5)$$

and

$$n_s(t) = \lim_{\Delta f \to 0} \sum_{k=1}^{\infty} [a_k \sin 2\pi(k - K) \, \Delta ft$$
$$- b_k \cos 2\pi(k - K) \, \Delta ft] \qquad (7.11\text{-}6)$$

Like $n(t)$, so also $n_c(t)$ and $n_s(t)$ are stationary random processes which are represented as linear superpositions of spectral components. We recall, from Sec. 7.4, that the a_k's and b_k's are gaussian random variables of zero mean and equal variance and that, further, the a_k's and b_k's are uncorrelated. We may then readily establish (Probs. 7.11-1 to 7.11-4) that $n_c(t)$ and $n_s(t)$ are gaussian random processes of zero mean value and of equal variance and that, further, $n_c(t)$ and $n_s(t)$ are uncorrelated.

To see the significance of the quadrature representation of noise, let us use it in connection with narrowband noise. We observe in Eqs. (7.11-5) and (7.11-6) that a noise spectral component in $n(t)$ of frequency $f = k\,\Delta f$ gives rise in $n_c(t)$ and in $n_s(t)$ to a spectral component of frequency $(k - K)\,\Delta f = f - f_0$. Suppose then that the noise $n(t)$ is narrowband, extending over a bandwidth B. And suppose that f_0 is selected midway in the frequency range of the noise. Then the spectrum of the noise $n(t)$ extends over the range $f_0 - B/2$ to $f_0 + B/2$. On the other hand, the spectrum of $n_c(t)$ and of $n_s(t)$ extends over only the range from $-B/2$ to $B/2$. By way of example, if the noise $n(t)$ is confined to a frequency band of only 10 kHz centered around $f_0 = 10$ MHz, then while $n(t)$ is a superposition of spectral components around the 10-MHz frequency, $n_c(t)$ and $n_s(t)$ change only insignificantly during the time the sinusoid of frequency f_0 executes a full cycle.

In view of the slow variations of $n_c(t)$ and $n_s(t)$ relative to the sinusoid of frequency f_0, it is reasonable and useful to give the quadrature representation of noise an interpretation in terms of phasors and a phasor diagram. Thus, in Eq. (7.11-2) the term $n_c(t)\cos 2\pi f_0 t$ is of frequency f_0 and of relatively slowly varying amplitude $n_c(t)$. Similarly, the term $-n_s(t)\sin 2\pi f_0 t$ is in quadrature with the first term and has a relatively slowly varying amplitude $n_s(t)$. In a coordinate system rotating counterclockwise with angular velocity $2\pi f_0$, these phasors are as represented in Fig. 7.11-1. These two phasors of varying amplitude give rise to a resultant phasor of amplitude $r(t) = [n_c^2(t) + n_s^2(t)]^{1/2}$ which makes an angle

$$\theta(t) = \tan^{-1}[n_s(t)/n_c(t)] \qquad (7.11\text{-}7)$$

with the horizontal. With the passage of time, the end point of this resultant phasor wanders about randomly over the phasor diagram.

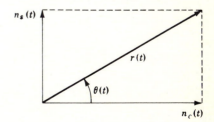

Fig. 7.11-1 A phasor diagram of the quadrature representation of noise.

We shall find the quadrature representation useful generally in the analysis of noise and shall find the phasor interpretation especially useful in discussing angle modulation communications systems.

7.12 POWER SPECTRAL DENSITY OF $n_c(t)$ AND $n_s(t)$

To determine the power spectral density of $n_c(t)$, let us select from the original noise $n(t)$ those spectral components corresponding to $k = K + \lambda$ and $k = K - \lambda$, where λ, like k and K, is an integer. Since $k = K$ corresponds to the frequency f_0, the selected components correspond to frequencies $f_0 + \lambda \, \Delta f$ and $f_0 - \lambda \, \Delta f$. These two frequencies give rise to four power spectral lines in a two-sided power spectral density plot as shown in Fig. 7.12-1. In this figure we have assumed bandlimited noise. However, for the sake of generality we have not assumed that the power spectral density is uniform in the band, nor have we assumed that $f_0 = K \, \Delta f$ is located at the center of the band.

We now select from $n_c(t)$, as given by Eq. (7.11-5), that part, $\Delta n_c(t)$, corresponding to our selection of frequencies, $f_0 \pm \lambda \, \Delta f$, from $n(t)$. We find

$$\Delta n_c(t) = a_{K-\lambda} \cos 2\pi\lambda \, \Delta ft - b_{K-\lambda} \sin 2\pi\lambda \, \Delta ft$$
$$+ a_{K+\lambda} \cos 2\pi\lambda \, \Delta ft + b_{K+\lambda} \sin 2\pi\lambda \, \Delta ft \quad (7.12\text{-}1)$$

Note that all four terms in Eq. (7.12-1) are of the same frequency $\lambda \, \Delta f$. These four terms represent four uncorrelated random processes, since the a's and b's are uncorrelated random variables. Hence we may find the power P_λ of $\Delta n_c(t)$ by determining the ensemble average of $[\Delta n_c(t)]^2$. Since $\Delta n_c(t)$ is a stationary random process, the ensemble average can be calculated at any time $t = t_1$. Following the procedure employed in Sec. 7.4, we choose t_1 so that $\lambda \, \Delta ft_1$ is an integer. Then

$$\Delta n_c(t_1) = a_{K-\lambda} + a_{K+\lambda} \quad (7.12\text{-}2)$$

and

$$P_\lambda = E\{[\Delta n_c(t_1)]^2\} = E[(a_{K-\lambda} + a_{K+\lambda})^2] \quad (7.12\text{-}3)$$

Using Eq. (7.4-12), which says that $E(a_{K-\lambda}a_{K+\lambda}) = 0$, we have, from Eq. (7.12-3), that

$$P_\lambda = \overline{a_{K-\lambda}^2} + \overline{a_{K+\lambda}^2} \quad (7.12\text{-}4)$$

From Eqs. (7.4-6) and (7.12-4) we then find

$$P_\lambda = 2G_{n_c}(\lambda \, \Delta f) \, \Delta f = 2G_n[(K - \lambda) \, \Delta f] \, \Delta f$$
$$+ 2G_n[(K + \lambda) \, \Delta f] \, \Delta f \quad (7.12\text{-}5)$$

Hence

$$G_{n_c}(\lambda \, \Delta f) = G_n[(K - \lambda) \, \Delta f] + G_n[(K + \lambda) \, \Delta f] \quad (7.12\text{-}6)$$

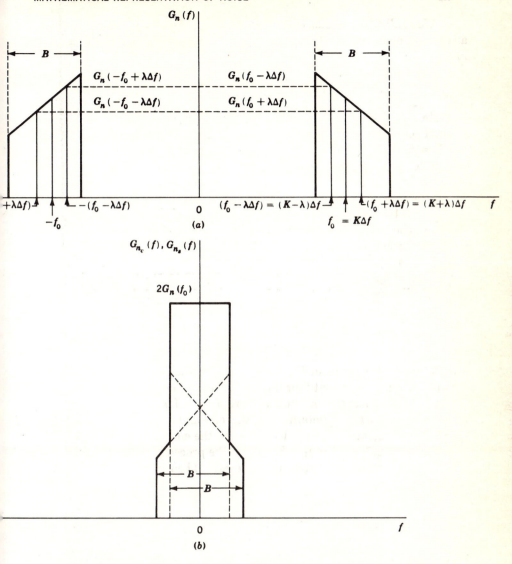

Fig. 7.12-1   (a) Power spectrum of bandlimited noise.   (b) Power spectral density of n_c and n_s.

We now set $K \Delta f = f_0$ and replace $\lambda \Delta f$ by a continuous frequency variable f, and we have, from Eq. (7.12-6),

$$G_{n_c}(f) = G_n(f_0 - f) + G_n(f_0 + f) \tag{7.12-7}$$

In a similar manner we may deduce an identical result for $G_{n_s}(f)$, namely,

$$G_{n_s}(f) = G_n(f_0 - f) + G_n(f_0 + f) \tag{7.12-8}$$

Expressed in words, Eqs. (7.12-7) and (7.12-8) say, that to find the power spectral density of $n_c(t)$ or of $n_s(t)$ at a frequency f, add the power spectral densities of $n(t)$ at the frequencies $f_0 - f$ and $f_0 + f$. In view of this result, and in view of the symmetry of a two-sided power spectral density plot as in Fig. 7.12-1, it may readily be verified that the plot of $G_{n_c}(f)$ or $G_{n_s}(f)$ may be constructed from the plot of $G_n(f)$ in the following manner:

1. Displace the positive-frequency portion of the plot of $G_n(f)$ to the left by amount f_0 so that the portion of the plot originally located at f_0 is now coincident with the ordinate.
2. Displace the negative-frequency portion of the plot of $G_n(f)$ to the right by an amount f_0.
3. Add the two displaced plots. The result of applying this procedure to the plot of Fig. 7.12-1a is shown in Fig. 7.12-1b.

A case of special interest is considered in the following example.

Example 7.12-1   White noise with power spectral density $\eta/2$ is filtered by a rectangular bandpass filter with $H(f) = 1$, centered at f_0 and having a bandwidth B. Find the power spectral density of $n_c(t)$ and $n_s(t)$. Calculate the power in $n_c(t)$, $n_s(t)$, and $n(t)$.

Solution   Since the filter is rectangular with $H(f) = 1$, the power spectral density of the output noise $n(t)$ is

$$G_n(f) = \begin{cases} \dfrac{\eta}{2} & f_0 - \dfrac{B}{2} \leq |f| \leq f_0 + \dfrac{B}{2} \\ 0 & \text{elsewhere} \end{cases} \qquad (7.12\text{-}9)$$

Hence, $G_n(f_0 + f) = G_n(f_0 - f)$, and the spectral density of $n_c(t)$ and $n_s(t)$ is

$$G_{n_c}(f) = G_{n_s}(f) = G_n(f_0 - f) + G_n(f_0 + f)$$
$$= \frac{\eta}{2} + \frac{\eta}{2} = \eta \qquad |f| \leq \frac{B}{2} \quad (7.12\text{-}10)$$

Note the extremely important result that the magnitude of $G_{n_c}(f) = G_{n_s}(f)$ is *twice* the magnitude of $G_n(f_0 + f)$.

The power (variance) of $n_c(t)$, and of $n_s(t)$, is

$$\sigma_{n_c}^2 = \sigma_{n_s}^2 = \int_{-B/2}^{B/2} G_{n_c}(f)\, df = \eta B \qquad (7.12\text{-}11)$$

The power (variance) of $n(t)$ is

$$\sigma_n^2 = \int_{-f_0-B/2}^{-f_0+B/2} G_n(f)\, df + \int_{f_0-B/2}^{f_0+B/2} G_n(f)\, df = 2\frac{\eta}{2}B = \eta B \quad (7.12\text{-}12)$$

Thus, the power of $n_c(t)$, $n_s(t)$, and $n(t)$ is equal.

7.13 PROBABILITY DENSITY OF $n_c(t)$, $n_s(t)$, AND THEIR TIME DERIVATIVES

We have noted that $n_c(t)$ and $n_s(t)$ are gaussian random processes with mean values of zero. If the noise $n(t)$ has a power spectral density $\eta/2$ over a bandwidth B, then, as noted in the preceding example, $\sigma_{n_c}^2 = \sigma_{n_s}^2 = \eta B$. Using Eq. (2.12-1) with $m = 0$, we find that the probability densities of the random variables n_c and n_s [that is, $n_c(t)$ and $n_s(t)$ at any fixed time] are given by

$$f(n_c) = \frac{1}{\sqrt{2\pi\eta B}} e^{-n_c^2/2\eta B} \tag{7.13-1a}$$

$$f(n_s) = \frac{1}{\sqrt{2\pi\eta B}} e^{-n_s^2/2\eta B} \tag{7.13-1b}$$

Since $n_c(t)$ and $n_s(t)$ are gaussian, the time derivatives $\dot{n}_c(t)$ and $\dot{n}_s(t)$ are also gaussian, because the operation of differentiation is an operation performed by a linear filter (Eq. 7.9-11), and from Sec. 7.3 we know that filtering gaussian noise does not change its probability density. To write the probability densities of $\dot{n}_c(t)$ and $\dot{n}_s(t)$, we need first to evaluate their variances $\sigma_{\dot{n}_c}^2$ and $\sigma_{\dot{n}_s}^2$. Noting that differentiation is equivalent to multiplying each spectral component by $j\omega$, we find

$$G_{\dot{n}_c}(f) = |j\omega|^2 G_{n_c}(f) = 4\pi^2 f^2 G_{n_c}(f) \tag{7.13-2}$$

so that, using Eq. (7.12-10), we find

$$\sigma_{\dot{n}_c}^2 = \int_{-B/2}^{B/2} G_{\dot{n}_c}(f)\, df = \int_{-B/2}^{B/2} 4\pi^2 f^2 \eta\, df = \frac{\pi^2}{3}\eta B^3 \tag{7.13-3}$$

with an identical result for $\sigma_{\dot{n}_s}^2$. Hence we find

$$f(\dot{n}_c) = \frac{\exp\{-\dot{n}_c^2/[(2\pi^2/3)\eta B^3]\}}{\sqrt{(2\pi^3/3)\eta B^3}} \tag{7.13-4}$$

with a similar expression for $f(\dot{n}_s)$. Assuming that the four random variables n_c, n_s, $\dot{n}_c$, and $\dot{n}_s$ are independent, the joint distribution function for the four variables is the product of the individual densities. Hence from Eqs. (7.13-1) and (7.13-3) we find

$$f(n_c, n_s, \dot{n}_c, \dot{n}_s) =$$
$$\frac{\exp[-(n_c^2 + n_s^2)/2\eta B]\, \exp[-(\dot{n}_c^2 + \dot{n}_s^2)/(2\pi^2/3)\eta B^3]}{[(2\pi^2/\sqrt{3})\eta B^2]^2} \tag{7.13-5}$$

We shall have occasion to use Eq. (7.13-5) in Chap. 10 in connection with an analysis of threshold effects in frequency modulation.

In arriving at Eq. (7.13-5), we assumed that the four random variables involved were independent. That such is indeed the case may be verified in the manner indicated in Prob. 7.13-1.

7.14 SHOT NOISE

We shall often need to know the power spectral density of a noise waveform such as indicated in Fig. 7.14-1. This noise consists of pulses all of identical *shape* but of *random amplitudes* and *random times of occurrences*. We make no assumptions concerning the pulse shape. We do assume, however, that the random process is stationary in that the average repetition rate and the average power of the waveform are both constant.

The pulse train of Fig. 7.14-1 is representative of a type of noise called *shot noise*. The term shot noise comes from consideration of the nature of current flow in a circuit which includes a thermionic diode or a semiconductor junction. In such a circuit each charge carrier contributes to current flow during the course of the time it is crossing from cathode to anode, or crossing the junction. Therefore the current in the circuit consists of a succession of current pulses, each pulse corresponding to the flight of a carrier across the thermionic diode or junction. While the average number of carrier crossings per unit time may be constant, the noise associated with the current results from the randomness of the time each carrier is shot from the cathode to anode, or shot across the junction.

Let us initially construct the noise waveform $p(t)$ shown in Fig. 7.14-2. Here we see a *single* pulse which is regularly repeated every

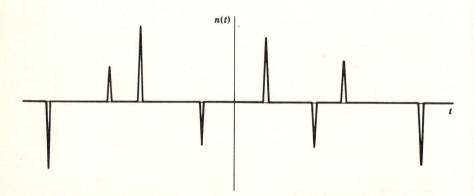

Fig. 7.14-1 Shot noise. Pulses of random amplitude and time of occurrence.

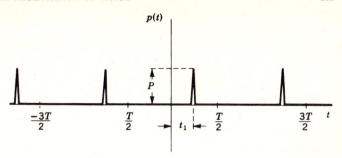

Fig. 7.14-2 A periodic pulse waveform. The amplitude P and the time t_1 are random variables.

T sec. The pulse amplitude P is a random variable, as is the time of occurrence t_1 of each pulse within its interval. Since the waveform is periodic with a period of T sec, we may expand $p(t)$ in a Fourier series as

$$p(t) = \sum_{k=-\infty}^{\infty} D_k e^{jk2\pi t/T} = \sum_{k=-\infty}^{\infty} D_k e^{jk2\pi \Delta f t} \qquad (7.14\text{-}1)$$

Here $\Delta f = 1/T$ and $D_k = |D_k| e^{j\theta_k}$. Both $|D_k|$ and θ_k are random variables. From the time-displacement theorem (Prob. 1.10-3) we have the result that θ_k is linearly related to the time of occurrence of the pulse within its interval T. When this time of occurrence varies from $t = -T/2$ to $t = +T/2$, θ_k changes by 2π. We assume that the pulse may occur at any place with equal likelihood. Therefore θ_k is a random variable whose probability density is uniform over all angles. The random variable $|D_k|$ relates only to the amplitude of the pulse. We assume that the amplitude and the time of occurrence of the pulse are independent. Therefore the random variables $|D_k|$ and θ_k are independent.

The power of each spectral term in the series of Eq. (7.14-1) is $\overline{|D_k|^2} = \overline{D_k D_k^*}$. As in Sec. 7.2 we introduce the power spectral density $G_p(k\,\Delta f)$ associated with the periodic pulse train, and write

$$G_p(k\,\Delta f)\,\Delta f = \overline{D_k D_k^*} \qquad (7.14\text{-}2)$$

Now let the average time of separation of pulses in $n(t)$ in Fig. 7.14-1 be T_s. Then the number of pulses in the larger interval T is T/T_s. If we add $T/T_s \equiv n$ waveforms, each similar to $p(t)$, we shall have a periodic waveform $n_T(t)$ in which the average pulse spacing is the same as in $n(t)$. This periodic waveform consists of pulses of random amplitudes and times of occurrence with average pulse spacing as in $n(t)$. Hence $n_T(t)$ may represent $n(t)$ in an interval T, in the same manner as is indicated in Fig. 7.2-1 for gaussian-type noise.

We have

$$n_T(t) = \sum_{\lambda=1}^{n} p_\lambda(t) = \sum_{\lambda=1}^{n} \sum_{k=-\infty}^{\infty} D_{k\lambda} e^{jk2\pi \Delta ft} \tag{7.14-3}$$

Since $n_T(t)$ is periodic with period T sec we may write

$$n_T(t) = \sum_{k=-\infty}^{\infty} F_k e^{jk2\pi \Delta ft} \tag{7.14-4}$$

The power spectral density of $n_T(t)$ associated with the frequency interval $k \Delta f$ is $G_{n_T}(k \Delta f)$ given by

$$G_{n_T}(k \Delta f) \Delta f = \overline{F_k F_k^*} \tag{7.14-5}$$

To calculate $\overline{F_k F_k^*}$, we equate coefficients of like spectral terms in Eqs. (7.14-3) and (7.14-4). Then

$$F_k = (D_{k1} + D_{k2} + \cdots + D_{kn}) \tag{7.14-6}$$

so that

$$\overline{F_k F_k^*} = \overline{(D_{k1} + D_{k2} + \cdots + D_{kn})(D_{k1}^* + D_{k2}^* + \cdots + D_{kn}^*)} \tag{7.14-7}$$

In performing the multiplication indicated in Eq. (7.14-7), we find that products of terms which have *unlike subscripts* give no net contribution. Consider, for example, a combination such as results from the cross multiplication of the first two terms in F with the first two terms in F^*. The terms with unlike subscripts are

$$\overline{D_{k1}D_{k2}^* + D_{k1}^* D_{k2}} = \overline{|D_{k1}|e^{j\theta_{k1}}|D_{k2}|e^{-j\theta_{k2}} + |D_{k1}|e^{-j\theta_{k1}}|D_{k2}|e^{j\theta_{k2}}}$$
$$= 2\overline{|D_{k1}|\,|D_{k2}|\cos(\theta_{k1} - \theta_{k2})} \tag{7.14-8}$$

Since the angles θ_k and the amplitudes $|D_k|$ are independent, we have

$$2\overline{|D_{k1}|\,|D_{k2}|\cos(\theta_{k1} - \theta_{k2})} = 2\overline{|D_{k1}|\,|D_{k2}|}\;\overline{\cos(\theta_{k1} - \theta_{k2})} \tag{7.14-9}$$

Furthermore, since both θ_{k1} and θ_{k2} have uniform probability densities, so also does $\theta_{k1} - \theta_{k2}$ (Sec. 7.8 and Prob. 7.8-3). Hence the mean value of $\cos(\theta_{k1} - \theta_{k2}) = 0$, and Eq. (7.14-8) becomes

$$\overline{D_{k1}D_{k2}^* + D_{k1}^* D_{k2}} = 0 \tag{7.14-10}$$

with a like result for any other product of terms with unlike subscripts.

The multiplication of terms of like subscript yields products $\overline{D_{k\lambda} D_{k\lambda}^*}$. There are n such products, and since the expected value of D_k is independent of λ, we drop the λ subscript. We have, from Eqs. (7.14-5) and (7.14-7),

$$G_{n_T}(k\,\Delta f)\,\Delta f = n\overline{|D_k|^2} \tag{7.14-11}$$

We employ Eq. (1.2-2) to evaluate the Fourier coefficient D_k.

$$D_k = \frac{1}{T}\int_{-T/2}^{T/2} p(t)e^{-jk2\pi\,\Delta ft}\,dt \tag{7.14-12}$$

Note that $p(t)$ consists of a *single* pulse in the region of integration. Combining Eqs. (7.14-11) and (7.14-12) and using $n = T/T_s$ and $\Delta f = 1/T$, we find

$$G_{n_T}(k\,\Delta f) = \frac{1}{T_s}\overline{\left|\int_{-T/2}^{T/2} p(t)e^{-jk2\pi\,\Delta ft}\,dt\right|^2} \tag{7.14-13}$$

Finally, we let the period of repetition T become infinite, so that $k\,\Delta f$ is replaced by the continuous variable f. The periodic shot noise waveform $n_T(t)$ becomes the waveform $n(t)$. The spectral density of $n(t)$ is then

$$G_n(f) = \frac{1}{T_s}\overline{\left|\int_{-\infty}^{\infty} p(t)e^{-j2\pi ft}\,dt\right|^2}$$

$$= \frac{1}{T_s}\overline{|P(f)|^2} \qquad\qquad -\infty < f < \infty \tag{7.14-14}$$

where $P(f)$ is the Fourier transform of the *single* pulse $p(t)$.

We shall have occasion to use Eq. (7.14-14) for the special case where $p(t)$ is an impulse of strength I. For this case, $|P(f)| = I$, and

$$G_n(f) = \frac{I^2}{T_s} \qquad -\infty < f < \infty \tag{7.14-15}$$

Equation (7.14-15) shows that when the shot noise consists of a randomly spaced train of impulses, the noise-power spectral density is white.

PROBLEMS

7.2-1. (a) A symmetrical square wave makes excursions between $+V$ and $-V$ volt, where V is a random variable uniformly distributed between 1 and 2V. It has a fundamental frequency 10^3 Hz. Make a plot of the two-sided power spectral density of the waveform.

(b) What fraction of the normalized power of the waveform is contained in the frequency range -3×10^3 to $+3 \times 10^3$ Hz?

7.2-2. Noise $n(t)$ has the power spectral density shown.

(a) Find the normalized power P of the noise in terms of η and f_M. Find P for $\eta = 1\ \mu V^2/Hz$ and $f_M = 10$ kHz.

(b) Find and plot the autocorrelation function $R_n(\tau)$ of the noise.

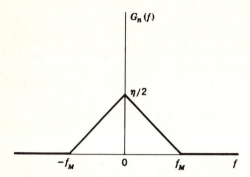

Fig. P7.2-2

7.3-1. Gaussian noise $n(t)$ of zero mean has a power spectral density

$$G_n(f) = 2 \,\mu\text{V}^2/\text{Hz} \qquad |f| \leq 1 \text{ kHz}$$
$$= 0 \qquad \qquad \text{elsewhere}$$

(a) What is the normalized power of the noise?

(b) Write the probability density function $f(n)$ of the noise.

(c) The noise $n(t)$ is passed through a filter. The power output of the filter is one-half the power of $n(t)$. Write the probability density function for the output noise of the filter.

7.4-1. (a) Two gaussian noise spectral components $n_k(t)$ and $n_l(t)$ are approximated as in Eqs. (7.4-10) and (7.4-11). The variance of a_k is $\sigma_k^2 = 1$, and the variance of a_l is $\sigma_l^2 = 2$. What is the normalized power of the sum $n_s(t) = n_k(t) + n_l(t)$?

(b) Write the probability density function for the noise waveform $n_s(t)$.

7.4-2. (a) Gaussian noise in a very narrow spectral range is represented approximately by Eq. (7.4-1). The normalized power of the noise is 0.01 μW. Write the expression for the probability density functions of the coefficients a_k and b_k.

(b) The narrowband noise in (a) is approximated by the representation of Eq. (7.4-2). Write the expression for the probability density functions of c_k and of θ_k.

7.6-1. White noise with two-sided power spectral density $\eta/2$ is passed through a low-pass RC network with time constant $\tau = RC$ and thereafter through an ideal amplifier with voltage gain of 10.

(a) Write the expression for the autocorrelation function $R_n(\tau)$ of the white noise.

(b) Write the expression for the power spectral density of the noise at the output of the amplifier.

(c) Write the expression for the autocorrelation of the output noise in (b).

7.7-1. Consider the noise waveforms $n_1(t), n_2(t), n_3(t), \ldots, n_N(t)$ where

$$E[n_j(t)] = 0 \qquad j = 1, 2, \ldots, N$$

and

$$E[n_j(t)n_k(t)] = \begin{cases} 1 & j = k \\ \frac{1}{2} & |j - k| = 1 \\ 0 & |j - k| > 1 \end{cases}$$

(a) Calculate the power dissipated in a 1-ohm resistor by the noise

$$n(t) = \sum_{j=1}^{N} n_j(t)$$

(b) Assuming that each of the $n_j(t)$ has a gaussian probability density, write the probability density for $n(t)$.

7.8-1. Noise $n(t)$ amplitude modulates a carrier having a random phase

$$v(t) = n(t) \cos (\omega_0 t + \varphi)$$

Consider that $E[n^2(t)] = \sigma^2$ and that the probability density of the random variable φ is $1/2\pi$ from $-\pi$ to π. Assume that $n(t)$ and φ are independent. Show that the power dissipated by $v(t)$ is $E(v^2) = \sigma^2/2$. Is $v(t)$ stationary?

7.8-2. $n(t) = n_1 \cos (\omega_1 t + \varphi_1) + n_2 \cos (k\omega_1 t + \varphi_2)$, where n_1, n_2, φ_1, and φ_2 are uncorrelated. Show that if $v(t) = n(t) \cos (\omega_0 t + \theta)$, where θ is a random variable uniformly distributed from $-\pi$ to π, then $E(v^2) = E(n^2)/2 = \frac{1}{4}[E(n_1^2) + E(n_2^2)]$.

7.8-3. The angles θ_k and θ_l are random variables with probability densities which are uniform over all angles. Starting with the convolution integral of Eq. (2.16-4) show formally that $\theta_k + \theta_l$ and $\theta_k - \theta_l$ also have uniform probability densities. (*Hint:* Let the probability density be represented by the radius vector r in a cylindrical coordinate system. Then the uniform density over all angles is represented by the radius of a circle of length $1/2\pi$.)

7.8-4. The two-sided power spectral density of noise $n(t)$ is shown in Fig. P7.8-4.
   (a) Plot the power spectral density of the product $n(t) \cos 2\pi f_1 t$.
   (b) Calculate the normalized power of the product in the frequency range $-(f_2 - f_1)$ to $(f_2 - f_1)$.
   (c) Repeat parts (a) and (b) for the product $n(t) \cos 2\pi[(f_2 + f_1)/2]t$.

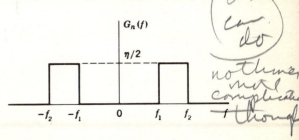

Fig. P7.8-4

7.8-5. A noise waveform $n(t)$ has the bandlimited power spectral density shown in Fig. P7.8-5, with $a = 10^{-6}$ V²/Hz and $f_M = 10^4$ Hz. Plot the power spectral density of $n(t) \sin 2\pi \times 10^6 t$ and find its normalized power over all frequencies.

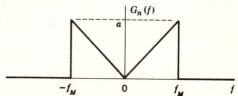

Fig. P7.8-5

7.9-1. A white-noise current source having a power spectral density $G_n(f) = \eta/2$ is filtered by a narrowband, parallel, single-tuned circuit having a resonant frequency at f_0 and a 3-dB bandwidth $B \ll f_0$.

    (a) Find the output voltage noise power.

    (b) Compare your result with Eq. (7.9-6).

7.10-1. Calculate the noise bandwidth of a single-tuned RLC filter centered at f_0 and having a 3-dB bandwidth $B \ll f_0$.

7.10-2. Calculate the noise bandwidth of the gaussian filter $|H(\omega)|^2 = e^{-\omega^2}$.

7-11-1. Show that if a_n and b_n are gaussian, $n_c(t)$ and $n_s(t)$ are gaussian.

7-11-2. Show that $E(n_c^2) = E(n_s^2)$.

7-11-3. Show that the autocorrelation functions of $n_c(t)$ and $n_s(t)$ are the same.

7-11-4. Show that $E[n_c(t)n_s(t)] = 0$.

7-11-5. Show that $E[n_c(t)n_s(t + \tau)]$ is not always equal to zero. Prove that it is equal to zero when $G_n(f)$ is an *even* function with respect to f_0.

7-11-6. Verify that, with $n_c(t)$ and $n_s(t)$ given as in Eqs. (7.11-5) and (7.11-6), $n(t)$ given by Eq. (7.11-2) is identical with $n(t)$ given by Eq. (7.11-1).

7.12-1. Using Eq. (7.11-2), show that when $E[n_c(t_1)n_s(t_2)] = 0$ [that is, $G_n(f)$ is ar even function of f, with respect to f_0],

$$R_{nn}(\tau) = R_{n_c n_c}(\tau) \cos \omega_0 \tau = R_{n_s n_s}(\tau) \cos \omega_0 \tau$$

7.12-2. Noise $n(t)$ has the power spectral density shown. We write

$$n(t) = n_c(t) \cos 2\pi f_0 t - n_s(t) \sin 2\pi f_0 t$$

Make plots of the power spectral densities of $n_c(t)$ and $n_s(t)$ for the cases

    (a) $f_0 = f_1$

    (b) $f_0 = f_2$

    (c) $f_0 = \frac{1}{2}(f_2 + f_1)$

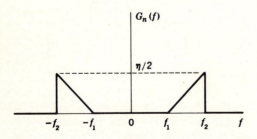

Fig. P7.12-2

7.12-3. (a) If $G_n(f) = \alpha^2/(\alpha^2 + \omega^2)$, show that $R_n(\tau) = K e^{-\alpha|\tau|}$. Find K.

    (b) If

$$G_n(f) = \frac{\alpha^2/2}{\alpha^2 + (\omega - \omega_0)^2} + \frac{\alpha^2/2}{\alpha^2 + (\omega + \omega_0)^2}$$

show that $R_n(\tau) = Ke^{-\alpha|\tau|} \cos \omega_0\tau$. Here $G_n(f)$ is the spectral density of noise which has been filtered by a narrowband (*symmetrical*) single-tuned filter.

7.13-1. When are $n_c(t)$, $n_s(t)$, $\dot{n}_c(t)$, and $\dot{n}_s(t)$ uncorrelated?

7.14-1. Consider a train of rectangular pulses. The kth pulse has a width τ and a height A_k. A_k is a random variable which can have the values 1, 2, 3, . . . , 10 with equal probability. Assuming statistical independence between amplitudes, find $G_n(f)$. Assume that the average time separating the pulses is T_s. Use Eq. (7.14-14).

7.14-2. A pulse train consists of pulses having a height of 1 volt and a width which can be either τ_1 or τ_2 with equal probability. Assume statistical independence between pulse widths. If the mean time between pulses is T_s, find $G_n(f)$, using Eq. (7.14-14).

REFERENCES

1. Davenport, W., and W. Root: "Random Signals and Noise," McGraw-Hill Book Company, New York, 1958.
   Papoulis, A.: "Probability, Random Variables, and Stochastic Processes," McGraw-Hill Book Company, New York, 1965.

8
Noise in Amplitude-modulation Systems

In Chap. 3 we described a number of different amplitude-modulation communication systems. In the present chapter we shall compare the performance of these systems under the circumstances that the received signal is corrupted by noise.

8.1 AMPLITUDE-MODULATION RECEIVER

A system for processing an amplitude-modulated carrier and recovering the baseband modulating system is shown in Fig. 8.1-1. We assume that the signal has suffered great attenuation during the course of its transmission over the communication channel and hence is in need of amplification. The input to the system might be a signal furnished by a receiving antenna which receives its signal from a transmitting antenna. The carrier of the received signal is called a *radio-frequency* (RF) carrier, and its frequency is the *radio* frequency f_{rf}. The input signal is amplified in an RF amplifier and then passed on to a *mixer*. In the mixer the modulated RF carrier is mixed (i.e., multiplied) with a sinusoidal waveform generated by a local oscillator which operates at a frequency f_{osc}. The process of mixing is also called *heterodyning*, and since, as is to be explained, the heterodyning local-oscillator frequency f_{osc} is selected to be *above* the radio frequency f_{rf}, the system is often referred to as a *superheterodyne* system.

The process of mixing generates sum and difference frequencies. Thus the mixer output consists of a carrier of frequency $f_{osc} + f_{rf}$ and a carrier $f_{osc} - f_{rf}$. Each carrier is modulated by the baseband signal to the same extent as was the input RF carrier. The sum frequency is rejected by a filter. This filter is not shown explicitly in Fig. 8.1-1 and may be considered to be part of the mixer. The difference-frequency carrier is called the *intermediate frequency* (IF) carrier, that is, $f_{if} = f_{osc} - f_{rf}$. The modulated IF carrier is applied to an IF amplifier. The process just described, in which a modulated RF carrier is replaced by a modulated IF carrier, is called *conversion*. The combination of the mixer and local oscillator is called a *converter*.

The IF amplifier output is passed, through an IF carrier filter, to the demodulator in which the baseband signal is recovered, and finally through a baseband filter. The baseband filter may include an amplifier,

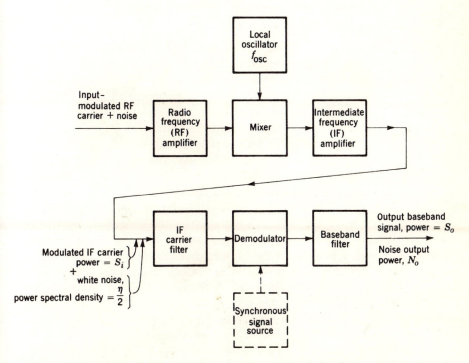

Fig. 8.1-1   A receiving system for an amplitude-modulated signal.

not explicitly indicated in Fig. 8.1-1. If synchronous demodulation is used, a synchronous signal source will be required.

The only absolutely essential operation performed by the receiver is the process of frequency translation back to baseband. This process is, of course, the inverse of the operation of modulation in which the baseband signal is frequency-translated to a carrier frequency. The process of frequency translation is performed in the system of Fig. 8.1-1 in part by the converter and in part by the demodulator. For this reason the converter is sometimes referred to as the *first detector*, while the demodulator is then called the *second detector*. The only other components of the system are the linear amplifiers and filters, none of which would be essential if the signal were strong enough and there were no need for multiplexing.

It is apparent that there is no essential need for an initial conversion before demodulation. The modulated RF carrier may be applied directly to the demodulator. However, the superheterodyne principle, which is rather universally incorporated into receivers, has great merit, as is discussed in the next section.

8.2 ADVANTAGE OF THE SUPERHETERODYNE PRINCIPLE: SINGLE CHANNEL

A signal furnished by an antenna to a receiver may have a power as low as some tens of picowatts, while the required output signal may be of the order of tens of watts. Thus the magnitude of the required gain is very large. In addition, to minimize the noise power presented to the demodulator, filters are used which are no wider than is necessary to accommodate the baseband signal. Such filters should be rather flat-topped and have sharp skirts. It is more convenient to provide gain and sharp flat-topped filters at low frequencies than at high. By way of example, in commercial FM broadcasting, the RF carrier frequency is in the range of 100 MHz, while at the FM receiver the IF frequency is 10.7 MHz.

Thus, in Fig. 8.1-1 the largest part, by far, of the required gain is provided by the IF amplifier, and the critical filtering done by the IF filter. While Fig. 8.1-1 suggests a separate amplifier and filter, actually in physical receivers these two usually form an integral unit. For example, the IF amplifier may consist of a number of amplifier stages, each one contributing to the filtering. Some filtering will also be incorporated in the RF amplifier. But this filtering is not critical. It serves principally to limit the total noise power input to the mixer and thereby avoids *overloading* the mixer with a noise waveform of excessive amplitude.

RF amplification is employed whenever the incoming signal is very small. This is because of the fact that RF amplifiers, such as masers, are *low-noise* devices; i.e., an RF amplifier can be designed to provide relatively high gain while generating relatively little noise. When RF amplification is not employed, the signal is applied directly to the mixer. The mixer provides relatively little gain and generates a relatively large noise power. Calculations showing typical values of gain and noise power generation in RF, mixer, and IF amplifiers are presented in Sec. 14.14.

MULTIPLEXING

An even greater merit of the superheterodyne principle becomes apparent when we consider that we shall want to tune the receiver to one or another of a number of different signals, each using a different RF carrier. If we were not to take advantage of the superheterodyne principle, we would require a receiver in which many stages of RF amplification were employed, each stage requiring tuning. Such tuned-radio-frequency (TRF) receivers were, as a matter of fact, commonly employed during the early days of radio communication. It is difficult enough to operate at the higher radio frequencies; it is even more difficult to *gang-tune* the individual stages over a wide band, maintaining at the same time a reasonably sharp flat-topped filter characteristic of constant bandwidth.

In a *superhet* receiver, however, we need but change the frequency of the local oscillator to go from one RF carrier frequency to another. Whenever f_{osc} is set so that $f_{osc} - f_{rf} = f_{if}$, the mixer will convert the input modulated RF carrier to a modulated carrier at the IF frequency, and the signal will proceed through the demodulator to the output. Of course, it is necessary to gang the tuning of the RF amplifier to the frequency control of the local oscillator. But again this ganging is not critical, since only one or two RF amplifiers and filters are employed.

Finally, we may note the reason for selecting f_{osc} higher than f_{rf}. With this higher selection the fractional change in f_{osc} required to accommodate a given range of RF frequencies is smaller than would be the case for the alternative selection.

8.3 SINGLE-SIDEBAND SUPPRESSED CARRIER (SSB-SC)

The receiver of Fig. 8.1-1 is suitable for the reception and demodulation of all types of amplitude-modulated signals, single sideband or double sideband, with and without carrier. The only essential changes required to accommodate one type of signal or another are in the demodulator and in the bandwidth of the IF carrier filter. Hence, from this point, our interest will focus on the section of receiver beginning with the IF filter and through to the output.

The signal input to the IF filter is an amplitude-modulated IF carrier. The normalized power (power dissipated in a 1-ohm resistor) of this signal is S_i.   The signal arrives with noise.   Added, is the noise generated in the RF amplifier and amplified in the RF amplifier and IF amplifiers.   The IF amplifiers and mixer are also sources of noise, i.e., thermal noise, shot noise, etc., but this noise, lacking the gain of the RF amplifier, represents a second-order effect.   (See Sec. 14.11.)   We shall assume that the noise is gaussian, white, and of two-sided power spectral density $\eta/2$.   The IF filter is assumed rectangular and of bandwidth no wider than is necessary to accommodate the signal.   The output baseband signal has a power S_o and is accompanied by noise of total power N_o.

CALCULATION OF SIGNAL POWER

With a single-sideband suppressed-carrier signal, the demodulator is a multiplier as shown in Fig. 8.3-1a.   The carrier is $A \cos 2\pi f_c t$.   For synchronous demodulation the demodulator must be furnished with a synchronous local carrier $\cos 2\pi f_c t$.   We assume that the upper sideband is being used; hence the carrier filter has a bandpass, as shown in Fig. 8.3-1b, that extends from f_c to $f_c + f_M$, where f_M is the baseband bandwidth.   The bandwidth of the baseband filter extends from zero to f_M as shown in Fig. 8.3-1c.

Let us assume that the baseband signal is a sinusoid of angular frequency f_m ($f_m \leq f_M$).   The carrier frequency is f_c, and, since we have assumed that the upper sideband is being used, the received signal is

$$s_i(t) = A \cos [2\pi(f_c + f_m)t] \tag{8.3-1}$$

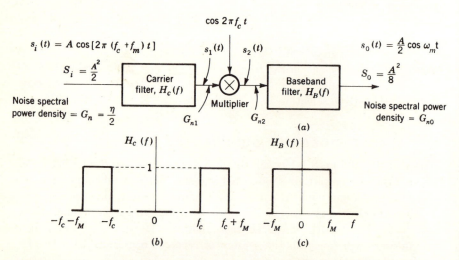

Fig. 8.3-1   (a) A synchronous demodulator operating on a single-sideband single-tone signal.   (b) The bandpass range of the carrier filter.   (c) The passband of the low-pass baseband filter.

The output of the multiplier is

$$s_2(t) = s_i(t) \cos \omega_c t = \frac{A}{2} \cos [2\pi(2f_c + f_m)t] + \frac{A}{2} \cos 2\pi f_m t \qquad (8.3\text{-}2)$$

Only the difference-frequency term will pass through the baseband filter. Therefore the output signal is

$$s_o(t) = \frac{A}{2} \cos 2\pi f_m t \qquad (8.3\text{-}3)$$

which is the modulating signal amplified by $\frac{1}{2}$.

The input signal power is

$$S_i = \frac{A^2}{2} \qquad (8.3\text{-}4)$$

while the output signal power is

$$S_o = \frac{1}{2}\left(\frac{A}{2}\right)^2 = \frac{A^2}{8} = \frac{S_i}{4} \qquad (8.3\text{-}5)$$

Thus

$$\frac{S_o}{S_i} = \frac{1}{4} \qquad (8.3\text{-}6)$$

We may readily see that, even though Eq. (8.3-6) was deduced on the assumption of a sinusoidal baseband signal, the result is entirely general. For suppose that the baseband signal were quite arbitrary in waveshape. Then the single-sideband signal generated by this baseband signal may be resolved into a series of harmonically related spectral components. The input power is the sum of the powers in these individual components. Next, we note, as was discussed in Chap. 3, that superposition applies to the multiplication process being used for demodulation. Therefore, the output signal power generated by the simultaneous application at the input of many spectral components is simply the sum of the output powers that would result from each spectral component individually. Hence S_i and S_o in Eqs. (8.3-4) and (8.3-5) are properly the *total* powers, independently of whether a single or many spectral components are involved.

CALCULATION OF NOISE POWER

We now calculate the output noise N_o. For this purpose, we recall from Sec. 7.8 that when a noise spectral component at a frequency f is multiplied by $\cos 2\pi f_c t$, the original noise component is replaced by two components, one at frequency $f_c + f$ and one at frequency $f_c - f$, each new component having one-fourth the power of the original.

The input noise is white and of spectral density $\eta/2$. The noise input to the multiplier has a spectral density G_{n1} as shown in Fig. 8.3-2a. The density of the noise after multiplication by $\cos 2\pi f_c t$ is G_{n2} as is shown

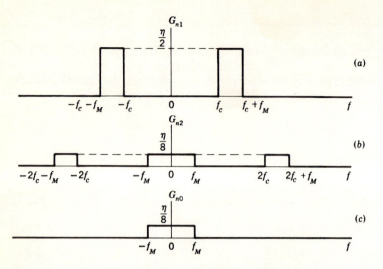

Fig. 8.3-2 Spectral densities of noises in SSB demodulator. (*a*) Density G_{n1} of noise input to multiplier. (*b*) Density G_{n2} of noise output of multiplier. (*c*) Density G_{no} of noise output of baseband filter.

in Fig. 8.3-2*b*. Finally the noise transmitted by the baseband filter is of density G_{no} as in Fig. 8.3-2*c*. The total noise output is the area under the plot in Fig. 8.3-2*c*. We have, then, that

$$N_o = 2f_M \frac{\eta}{8} = \frac{\eta f_M}{4} \tag{8.3-7}$$

USE OF QUADRATURE NOISE COMPONENTS

It is of interest to calculate the output noise power N_o in an alternative manner using the transformation of Eq. (7.11-2):

$$n(t) = n_c(t) \cos 2\pi f_c t - n_s(t) \sin 2\pi f_c t \tag{8.3-8}$$

We now apply Eq. (8.3-8) to the noise output of the IF filter so that $n(t)$ has the spectral density G_{n1} as in Fig. 8.3-2*a*. The spectral densities of $n_c(t)$ and $n_s(t)$ are [see Eqs. (7.12-7) and (7.12-8)]:

$$G_{n_c}(f) = G_{n_s}(f) = G_{n1}(f_c - f) + G_{n1}(f_c + f) \tag{8.3-9}$$

We observe that for $0 \leq f \leq f_M$, $G_{n1}(f_c + f) = \eta/2$, while $G_{n1}(f_c - f) = 0$, so that $G_{n_c}(f)$ and $G_{n_s}(f)$ are as shown in Fig. 8.3-3.

Multiplying $n(t)$ by $\cos 2\pi f_c t$ yields

$$n(t) \cos 2\pi f_c t = n_c(t) \cos^2 2\pi f_c t - n_s(t) \sin 2\pi f_c t \cos 2\pi f_c t$$
$$= \tfrac{1}{2} n_c(t) + \tfrac{1}{2} n_c(t) \cos 4\pi f_c t - \tfrac{1}{2} n_s(t) \sin 4\pi f_c t \tag{8.3-10}$$

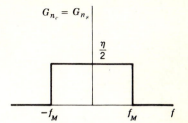

Fig. 8.3-3 Power spectral densities of G_{n_c} and G_{n_s}.

The spectra of the second and third terms in Eq. (8.3-10) extend over the range $2f_c - f_M$ to $2f_c + f_M$ and are outside the baseband filter. The output noise is, therefore,

$$n_o(t) = \tfrac{1}{2} n_c(t) \tag{8.3-11}$$

The spectral density of $n_o(t)$ is then $G_{no} = \tfrac{1}{4} G_{n_c} = \tfrac{1}{4}(\eta/2) = \eta/8$. Hence as before, as shown in Fig. 8.3-2c, the spectral density G_{no} is $\eta/8$ over the range $-f_M$ to f_M, and the total noise is again $N_o = \eta f_M/4$.

CALCULATION OF SIGNAL-TO-NOISE RATIO (SNR)

Finally we may calculate, using Eqs. (8.3-6) and (8.3-7), the *signal-to-noise ratio* at the output, S_o/N_o. We have

$$\frac{S_o}{N_o} = \frac{S_i/4}{\eta f_M/4} = \frac{S_i}{\eta f_M} \tag{8.3-12}$$

The importance of S_o/N_o is that it serves as a *figure of merit* of the performance of a communication system. Certainly, as S_o/N_o increases, it becomes easier to distinguish and to reproduce the modulating signal without error or confusion. If a system of communication allows the use of more than a single type of demodulator (say, synchronous or nonsynchronous), that ratio S_o/N_o will serve as a figure of merit with which to compare demodulators.

We observe from Eq. (8.3-12) that to increase the output signal-to-noise power ratio, we can increase the transmitted signal power, restrict the baseband frequency range, or make the receiver quieter.

8.4 DOUBLE-SIDEBAND SUPPRESSED CARRIER (DSB-SC)

When a baseband signal of frequency range f_M is transmitted over a DSB-SC system, the bandwidth of the carrier filter must be $2f_M$ rather than f_M. Thus, input noise in the frequency range $f_c - f_M$ to $f_c + f_M$ will contribute to the output noise, rather than only in the range f_c to $f_c + f_M$ as in the SSB case.

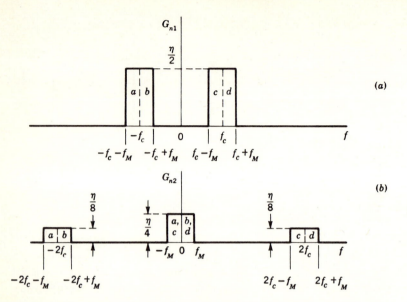

Fig. 8.4-1 Spectral densities of noise in DSB demodulation. (a) Density G_{n1} of noise at output of IF filter. (b) Density G_{n2} of noise output of baseband filter.

CALCULATION OF NOISE POWER

This situation is illustrated in Fig. 8.4-1a, which shows the spectral density $G_{n1}(f)$ of the white input noise after the IF filter. This noise is multiplied by $\cos \omega_c t$. The multiplication results in a frequency shift by $\pm f_c$ and a reduction of power in the power spectral density of the noise by a factor of 4. Thus, the noise in region d of Fig. 8.4-1a shifts to regions d shown in Fig. 8.4-1b. Similarly regions a, b, and c of Fig. 8.4-1a are translated by $\pm f_c$ and are also attenuated by 4 as shown in Fig. 8.4-1b. Note that the noise-power spectral density in the region between $-f_M$ and $+f_M$ is $\eta/4$, while the noise density in the SSB case, as shown in Fig. 8.3.2c is only $\eta/8$. Hence the output noise power is twice as large as the output noise power for SSB given in Eq. (8.3-5). The output noise for DSB after baseband filtering is therefore

$$N_o = \frac{\eta}{4}(2f_M) = \frac{\eta f_M}{2} \tag{8.4-1}$$

CALCULATION OF SIGNAL POWER

We might imagine that for equal received powers, the ratio S_o/N_o for DSB would be only half the corresponding ratio for SSB. We shall now see that such is not the case, and that the ratio S_o/N_o is the *same* in the two cases. Again, without loss in generality, let us assume a sinusoidal baseband signal of frequency $f_m \leq f_M$. To keep the received power the same as in the SSB case, that is, $S_i = A^2/2$, we write

$$s_i(t) = \sqrt{2}\, A \cos 2\pi f_m t \cos 2\pi f_c t$$

$$= \frac{A}{\sqrt{2}} \cos [2\pi(f_c + f_m)t] + \frac{A}{\sqrt{2}} \cos [2\pi(f_c - f_m)t] \qquad (8.4\text{-}2)$$

The received power is then

$$S_i = \frac{1}{2}\left(\frac{A}{\sqrt{2}}\right)^2 + \frac{1}{2}\left(\frac{A}{\sqrt{2}}\right)^2 = \frac{A^2}{2} \qquad (8.4\text{-}3)$$

as in Eq. (8.3-4).

In the demodulator (multiplier), $s_i(t)$ is multiplied by $\cos \omega_c t$. The upper-sideband term in Eq. (8.4-2) yields a signal within the passband of the baseband filter given by

$$s_o'(t) = \frac{A}{2\sqrt{2}} \cos 2\pi f_m t \qquad (8.4\text{-}4)$$

The lower-sideband term of Eq. (8.4-2) yields

$$s_o''(t) = \frac{A}{2\sqrt{2}} \cos 2\pi f_m t \qquad (8.4\text{-}5)$$

Observe, most particularly in Eqs. (8.4-4) and (8.4-5), that $s_o'(t)$ and $s_o''(t)$ are in phase and that hence the output signal is

$$s_o(t) = s'(t) + s''(t) = \frac{A}{\sqrt{2}} \cos 2\pi f_m t \qquad (8.4\text{-}6)$$

which has a power

$$S_o = \frac{A^2}{4} = \frac{S_i}{2} \qquad (8.4\text{-}7)$$

rather than $S_o = A^2/8 = S_i/4$ as in Eq. (8.3-5) for the SSB case. Thus we see that when a received signal of fixed power is split into two sideband components each of half power, as in DSB, rather than being left in a single sideband, the output signal power increases by a factor of 2. This increase results from the fact that the contributions from each sideband yield output signals which are in phase. A doubling in amplitude causes a fourfold increase in power. This fourfold increase, due to the inphase addition of s_o' and s_o'', is in part undone by the need to split the input power into two half-power sidebands. Thus the overall improvement in output signal power is by a factor of 2.

On the other hand, the noise outputs due to noise spectral components symmetrically located with respect to the carrier are uncorrelated with one another. The two resultant noise spectral components in the output, although of the same frequency, are uncorrelated. Hence the combination of the two yields a power which is the sum of the two powers individually, not larger than the sum, as is the case with the signal.

CALCULATION OF SIGNAL-TO-NOISE RATIO

Returning now to the calculation of signal-to-noise ratio, we find from Eqs. (8.4-1) and (8.4-7) that

$$\frac{S_o}{N_o} = \frac{S_i}{\eta f_M} \tag{8.4-8}$$

exactly as for SSB-SC.

ARBITRARY MODULATING SIGNAL

In the discussion in the present section concerning DSB we have assumed that the baseband signal waveform is sinusoidal. As pointed out in Sec. 8.3, this assumption causes no loss of generality because of the linearity of the demodulation. Nonetheless it is often convenient to have an expression for the power of a DSB signal in terms of the arbitrary waveform $m(t)$ of the baseband modulating signal. Hence let the received signal be

$$s_i(t) = m(t) \cos 2\pi f_c t \tag{8.4-9}$$

The power of $s_i(t)$ is

$$S_i \equiv \overline{s_i^2(t)} = \overline{m^2(t) \cos^2 2\pi f_c t} = \tfrac{1}{2}\overline{m^2(t)} + \tfrac{1}{2}\overline{m^2(t) \cos (4\pi f_c t)} \tag{8.4-10}$$

Now $m(t)$ can always be represented as a sum of sinusoidal spectral components. [Of interest, albeit of no special relevance in the present discussion, is the fact that if $m(t)$ is bandlimited to f_M, $m^2(t)$ is bandlimited to $2f_M$. See Prob. 8.4-1.] Hence $m^2(t) \cos 4\pi f_c t$ consists of a sum of sinusoidal waveforms in the frequency range $2f_c \pm 2f_M$. The average value of such a sum is zero, and we therefore have

$$S_i \equiv \overline{s_i^2(t)} = \tfrac{1}{2}\overline{m^2(t)} \tag{8.4-11}$$

When the signal $s_i(t)$ in Eq. (8.4-9) is demodulated by multiplication by $\cos 2\pi f_c t$, and the product passed through the baseband filter, the output is $s_o(t) = m(t)/2$. The output signal power is

$$S_o = \frac{\overline{m^2(t)}}{4} \tag{8.4-12}$$

so that, from Eqs. (8.4-11) and (8.4-12),

$$S_o = \frac{S_i}{2} \tag{8.4-13}$$

that is, the same result as given in Eq. (8.4-7) for an assumed single sinusoidal modulating signal.

USE OF QUADRATURE NOISE COMPONENTS TO CALCULATE N_o

It is again interesting to calculate N_o using the transformation of Eq. (7.11-2):

$$n(t) = n_c(t) \cos 2\pi f_c t - n_s(t) \sin 2\pi f_c t \qquad (8.4\text{-}14)$$

The power spectral density of $n_c(t)$ and $n_s(t)$ are [see Eqs. (7.12-7) and (7.12-8)]

$$G_{n_c}(f) = G_{n_s}(f) = G_{n1}(f_c + f) + G_{n1}(f_c - f) \qquad (8.4\text{-}15)$$

In the frequency range $|f| \leq f_M$, $G_{n1}(f_c + f) = G_{n1}(f_c - f) = \eta/2$. Thus

$$G_{n_c}(f) = G_{n_s}(f) = \eta \qquad |f| \leq f_M \qquad (8.4\text{-}16)$$

(This result was also derived in Example 7.12-1.)

The result of multiplying $n(t)$ by $\cos 2\pi f_c t$ yields

$$n(t) \cos 2\pi f_c t = \tfrac{1}{2} n_c(t) + \tfrac{1}{2} n_c(t) \cos 4\pi f_c t$$
$$- \tfrac{1}{2} n_s(t) \sin 4\pi f_c t \qquad (8.4\text{-}17)$$

Baseband filtering eliminates the second and third terms, leaving

$$n_o(t) = \tfrac{1}{2} n_c(t) \qquad (8.4\text{-}18)$$

The power spectral density of $n_o(t)$ is then

$$G_{no}(f) = \frac{1}{4} G_{n_c}(f) = \frac{\eta}{4} \qquad -f_M \leq f \leq f_M \qquad (8.4\text{-}19)$$

The output noise power N_o is, therefore,

$$N_o = \frac{\eta}{4} 2f_M = \frac{\eta f_M}{2} \qquad (8.4\text{-}20)$$

This result is, of course, identical with Eq. (8.4-1), which was obtained by considering directly the effect on a noise spectral component of a multiplication by $\cos 2\pi f_c t$.

8.5 DOUBLE SIDEBAND WITH CARRIER (DSB)

Let us now consider the case where a carrier accompanies the double-sideband signal. Demodulation is achieved synchronously as in SSB-SC and DSB-SC. The carrier is used as a *transmitted reference* to obtain the reference signal $\cos \omega_c t$ (see Prob. 8.5-1). We note that the carrier increases the total input-signal power but makes no contribution to the output-signal power. Equation (8.4-8) applies directly to this case, provided only that we replace S_i by $S_i{}^{(SB)}$, where $S_i{}^{(SB)}$ is the power in the sidebands alone. Then

$$\frac{S_o}{N_o} = \frac{S_i^{(SB)}}{\eta f_M}$$ (8.5-1)

Suppose that the received signal is

$$s_i(t) = A[1 + m(t)] \cos 2\pi f_c t$$ (8.5-2)

$$= A \cos 2\pi f_c t + A m(t) \cos 2\pi f_c t$$

where $m(t)$ is the baseband signal which amplitude-modulates the carrier $A \cos 2\pi f_c t$. The carrier power is $A^2/2$. The sidebands are contained in the term $A m(t) \cos 2\pi f_c t$. The power associated with this term is $(A^2/2)\overline{m^2(t)}$, where $\overline{m^2(t)}$ is the time average of the square of the modulating waveform. We then have that the total input power S_i is given by

$$S_i = \frac{A^2}{2} + S_i^{(SB)} = \frac{A^2}{2} [1 + \overline{m^2(t)}]$$ (8.5-3)

Eliminating A^2, we have

$$S_i^{(SB)} = \frac{\overline{m^2(t)}}{1 + \overline{m^2(t)}} S_i$$ (8.5-4)

or, with Eq. (8.5-1),

$$\frac{S_o}{N_o} = \frac{\overline{m^2(t)}}{1 + \overline{m^2(t)}} \frac{S_i}{\eta f_M}$$ (8.5-5)

In terms of the carrier power $P_c \equiv A^2/2$, we get, from Eqs. (8.5-3) and (8.5-5), that

$$\frac{S_o}{N_o} = \overline{m^2(t)} \frac{P_c}{\eta f_M}$$ (8.5-6)

If the modulation is sinusoidal, with $m(t) = m \cos 2\pi f_m t$ (m a constant), then

$$s_i(t) = A(1 + m \cos 2\pi f_m t) \cos 2\pi f_c t$$ (8.5-7)

In this case $\overline{m^2(t)} = m^2/2$ and

$$\frac{S_o}{N_o} = \frac{m^2}{2 + m^2} \frac{S_i}{\eta f_M}$$ (8.5-8)

When the carrier is transmitted only to synchronize the local demodulator waveform $\cos 2\pi f_c t$, relatively little carrier power need be transmitted. In this case $m \gg 1$, $m^2/(2 + m^2) \cong 1$, and the signal-to-noise ratio is not greatly reduced by the presence of the carrier. On the other hand, when envelope demodulation is used (Sec. 3.4), it is required that $m \leq 1$.

When $m = 1$, the carrier is 100 percent modulated. In this case $m^2/(2 + m^2) = \frac{1}{3}$, so that of the power transmitted, only one-third is in the sidebands which contribute to signal power output.

A FIGURE OF MERIT

We observe that in each demodulation system considered so far, the ratio $S_i/\eta f_M$ appeared in the expression for output signal-to-noise ratio (SNR) [see Eqs. (8.3-12), (8.4-8), and (8.5-5)]. This ratio is the output signal power S_i divided by the product ηf_M. To give the product ηf_M some physical significance, we consider it to be the noise power N_M at the input, measured in a frequency band equal to the *baseband frequency*. Thus

$$N_M \equiv \frac{\eta}{2} 2f_M = \eta f_M \tag{8.5-9}$$

The ratio $S_i/\eta f_M$ is, therefore, often referred to as the *input signal-to-noise ratio* S_i/N_M. It needs to be kept in mind that N_M is the noise power transmitted through the IF filter only when the IF filter bandwidth is f_M. Thus N_M is the true input noise power only in the case of single sideband.

For the purpose of comparing systems, we introduce the *figure of merit* γ, defined by

$$\gamma \equiv \frac{S_o/N_o}{S_i/N_M} \tag{8.5-10}$$

The results given above may now be summarized as follows:

$$\gamma = \begin{cases} 1 & \text{SSB-SC} & (8.5\text{-}11) \\[4pt] 1 & \text{DSB-SC} & (8.5\text{-}12) \\[10pt] \dfrac{\overline{m^2(t)}}{1 + \overline{m^2(t)}} & \text{DSB} & (8.5\text{-}13) \\[10pt] \dfrac{m^2}{2 + m^2} & \text{DSB with sinusoidal modulation} & (8.5\text{-}14) \end{cases}$$

A point of interest in connection with *double-sideband* synchronous demodulation is that, for the purpose of suppressing output-noise power, the carrier filter of Fig. 8.3-1 is not necessary. A noise spectral component at the input which lies outside the range $f_c \pm f_M$ will, after multiplication in the demodulator, lie outside the passband of the baseband filter. On the other hand, if the carrier filter is eliminated, the magnitude of the noise signal which reaches the modulator may be large enough to overload the active devices used in the demodulator. Hence, such carrier filters are normally included, but the purpose is *overload suppression* rather than *noise suppression*. In single sideband, of course, the situation is different, and the carrier filter does indeed suppress noise.

8.6 SQUARE-LAW DEMODULATOR

We saw in Sec. 3.6 that a double-sideband signal with carrier may be demodulated by passing the signal through a network whose input-output characteristic is not linear. Such nonlinear demodulation has the advantage, over the linear synchronous demodulation methods, that a synchronous local carrier need not be obtained. This eliminates the rather costly synchronizing circuits. In this section we discuss the performance and determine the output SNR of a nonlinear demodulator which uses a network whose output signal y (voltage or current) is related to the input signal x (voltage or current) by

$$y = \lambda x^2 \tag{8.6-1}$$

in which λ is a constant. As shown in Fig. 8.6-1, this nonlinear network, which constitutes the demodulator, is preceded by a bandpass IF filter of bandwidth $2f_M$ and is followed by a baseband low-pass filter of bandwidth f_M.

We discuss this square-law demodulator in part for its own intrinsic interest, but also because it exhibits an important characteristic which is not displayed by the linear synchronous demodulators. We have previously adopted the quantity $\gamma \equiv (S_o/N_o)/(S_i/N_M)$ as a figure of merit for the performance of demodulators in the presence of noise. We observed [Eqs. (8.5-11) to (8.5-14)] that this figure of merit is not a function of the input signal-to-noise ratio S_i/N_M. Therefore, if the input S_i/N_M decreases, say, by a factor α, the output S_o/N_o will also decrease by α. The nonlinear demodulator also has a range where the figure of merit γ is independent of S_i/N_M. However, as S_i/N_M decreases, there is a point, a

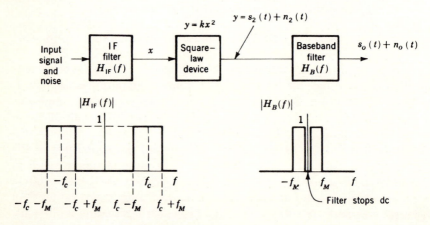

Fig. 8.6-1 The square-law AM demodulator.

threshold, at which the output S_o/N_o decreases more rapidly than does the input S_i/N_M. This threshold often marks the limit of usefulness of the demodulator.

ANALYSIS

We assume that we have a carrier of amplitude A and angular frequency ω_c, amplitude-modulated by a baseband signal $m(t)$. Then the received signal is $s_i(t) = A[1 + m(t)] \cos \omega_c t$. We assume that input noise $n(t)$, having a power spectral density $\eta/2$ for $f_c - f_M < |f| < f_c + f_M$ after IF filtering, has been added to the signal. The input to the demodulator is therefore

$$x(t) = A[1 + m(t)] \cos \omega_c t + n(t) \tag{8.6-2}$$

and the output is, from Eq. (8.6-1),

$$y(t) = \lambda\{A[1 + m(t)] \cos \omega_c t + n(t)\}^2 \tag{8.6-3}$$

We now expand Eq. (8.6-3) and drop those terms whose power spectral density falls outside the baseband filter. The baseband filter is designed to pass the modulation and suppress as much noise as possible. Thus, the maximum frequency passed is f_M. On the low-frequency end, the filter extends only low enough to pass the minimum signal frequency, often 100 to 300 Hz. However, we assume for simplicity that the filter passes all low-frequency components with the exception of dc (zero frequency). Thus we drop dc terms.

The spectrum of the baseband signal $m(t)$ extends to f_M. Then, as shown in Prob. 8.4-1, the spectrum of $m^2(t)$ extends to $2f_M$. We therefore drop terms of the form $m(t) \cos 2\omega_c t$ and $m^2(t) \cos 2\omega_c t$, since the spectrum of these terms extends, respectively, over the range $2f_c \pm f_M$ and $2f_c \pm 2f_M$. The signal that then remains, at the input to the baseband filter, is

$$s_2(t) = \lambda A^2 m(t) \left[1 + \frac{m(t)}{2} \right] \tag{8.6-4}$$

The noise, after squaring, is

$$n_2(t) = 2\lambda A n(t)[1 + m(t)] \cos \omega_c t + \lambda n^2(t) \tag{8.6-5}$$

We make the simplification of assuming $|m(t)| \ll 1$. It was noted in Sec. 3.6 that this restriction is required in order to avoid significant signal distortion. Then

$$s_2(t) \approx \lambda A^2 m(t) \tag{8.6-6}$$

and

$$n_2(t) \approx 2\lambda A n(t) \cos \omega_c t + \lambda n^2(t) \tag{8.6-7}$$

The signal $s_2(t)$ is directly proportional to $m(t)$ and therefore is frequency-limited to f_M. The baseband filter therefore passes the entire signal $s_2(t)$ and $s_o(t) = s_2(t)$. The output signal power is

$$S_o = \lambda^2 A^4 \overline{m^2(t)} \tag{8.6-8}$$

Since $n(t)$ has a power spectral density $\eta/2$, the spectral density of $n(t) \cos \omega_c t$ is $\eta/4$ between $-f_M$ and $+f_M$. (See Fig. 8.4-1b.) Therefore the noise power N_o', due to the term $2\lambda A n(t) \cos \omega_c t$ in Eq. (8.6-7), is

$$N_o' = 4\lambda^2 A^2 \frac{\eta}{4} 2f_M = 2\lambda^2 A^2 \eta f_M \tag{8.6-9}$$

CALCULATION OF N_o''

We turn now to the calculation of the noise power N_o'' which results from the term $\lambda n^2(t)$. For this purpose, as shown in Fig. 8.6-2, we divide the spectral range of the noise $(f_c - f_M$ to $f_c + f_M)$ into $2K + 1$ intervals, each Δf wide. Then, as discussed in Sec. 7.2, we represent the power in each interval by a spectral line of power $\eta \Delta f/2$ located at the center of the interval. We represent the noise $n(t)$ in the manner of Eq. (7.2-2) so that

$$n(t) = \sum_{k=-K}^{+K} c_k \cos\left[(2\pi f_c + k\,\Delta f)t + \theta_k\right] \tag{8.6-10}$$

From Eq. (7.4-6), with $G_n(k\,\Delta f) = \eta/2$, we have

$$\overline{c_k^2} = 2\eta\,\Delta f \tag{8.6-11}$$

Let us now single out two *particular* spectral components in Eq. (8.6-10), separated by a frequency $\rho\,\Delta f$. Calling the sum of this specific set of spectral components $n_{k,\rho}(t)$, we have

$$
\begin{aligned}
n_{k,\rho}(t) = {} & c_k \cos\left[(2\pi f_c + k\,\Delta f)t + \theta_k\right] \\
& + c_{k+\rho} \cos\left\{[2\pi f_c + (k+\rho)\,\Delta f]t + \theta_{k+\rho}\right\}
\end{aligned}
\tag{8.6-12}
$$

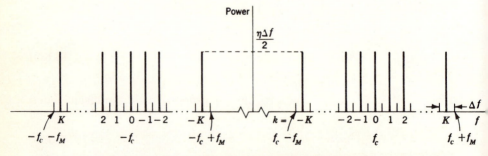

Fig. 8.6-2 The spectral range $|f - f_c| \le f_M$ of the noise $n(t)$ of power spectral density $\eta/2$ is divided into intervals Δf. The power in each interval is represented approximately by a single spectral line of power $\eta\,\Delta f/2$.

In the course of forming the product $n(t) \times n(t) = n^2(t)$, we shall have to form the product $n_{k,\rho}(t) \times n_{k,\rho}(t) = n_{k,\rho}^2(t)$. Such a product will give rise to a spectral component at the sum frequency $2f_c + (2k + \rho)\,\Delta f$ and to a component at the difference frequency $\rho\,\Delta f$. The sum frequency component is of no interest to us since it lies outside the baseband frequency range 0 to f_M. The difference frequency component, as is readily verified from Eq. (8.6-12), is

$$n_\rho(t) = c_k c_{k+\rho} \cos (2\pi\rho\,\Delta ft + \theta_{k+\rho} - \theta_k) \tag{8.6-13}$$

The power P_ρ associated with $n_\rho(t)$ may readily be deduced on the basis of the discussion in Sec. 7.8 concerning noise-noise mixing. We note that $n_\rho(t)$ in Eq. (8.6-13) is larger by a factor of 2 than the difference frequency term in Eq. (7.8-8). This factor of 2 results from the fact that $n_\rho(t)$ arises from the multiplication of the *sum* of two noise spectral components by itself, while the difference term in Eq. (7.8-8) arises from the product of one noise spectral component with another. Applying the result given in Eq. (7.8-8) to the present case, we find, since $\overline{c_k^2} = \overline{c_{k+\rho}^2}$ and using Eq. (8.6-11), that

$$P_\rho \equiv \overline{n_\rho^2(t)} = \tfrac{1}{2}\,\overline{c_k^2}\,\overline{c_{k+\rho}^2} = 2(\eta\,\Delta f)^2 \tag{8.6-14}$$

Different particular sets of spectral components in Eq. (8.6-10) are separated by the frequency difference $\rho\,\Delta f$ as in Eq. (8.6-13) and have power given by Eq. (8.6-14). These spectral components, albeit at the same frequency, are uncorrelated. Hence the total power at frequency $\rho\,\Delta f$ is simply the sum of the individual powers. Therefore, to find the total power associated with the frequency $\rho\,\Delta f$, we need now only calculate the number of pairs of spectral components in Eq. (8.6-10) which are separated by $\rho\,\Delta f$.

Now there are $2K + 1$ spectral components in $n(t)$, as given by Eq. (8.6-10), spaced by intervals Δf. Hence the number of pairs p of components separated by a frequency $\rho\,\Delta f$ is $p = 2K + 1 - \rho$. Since, in the limit, as $\Delta f \to 0$, $K \to \infty$, we shall ignore the unity in comparison with K and write

$$p \approx 2K - \rho \tag{8.6-15}$$

We have further, as is apparent in Fig. 8.6-2, that $f_M = (K + \tfrac{1}{2})\,\Delta f$. For the reason just indicated we write

$$f_M \approx K\,\Delta f \tag{8.6-16}$$

The power in the frequency interval Δf at $\rho\,\Delta f$ is the product of the number of component pairs p and the power P_ρ associated with the difference-frequency component $n_\rho(t)$ resulting from each pair. Hence, if $G_{n^2}(\rho\,\Delta f)$ is the two-sided power spectral density, at the frequency $\rho\,\Delta f$,

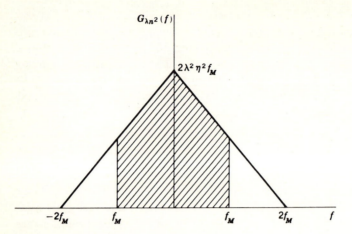

Fig. 8.6-3 Plot of power spectral density $G_{\lambda n^2}(f)$ in baseband region.

then, from Eqs. (8.6-14) and (8.6-15),

$$2G_{n^2}(\rho\,\Delta f)\,\Delta f = (2K - \rho)2(\eta\,\Delta f)^2 \tag{8.6-17}$$

Using Eq. (8.6-16) and replacing $\rho\,\Delta f$ by a continuous variable f, as $\Delta f \to 0$, we find from Eq. (8.6-17) that $G_{n^2}(f) = \eta^2(2f_M - f)$ so that the power spectral density of $\lambda n^2(t)$ is

$$G_{\lambda n^2}(f) = \lambda^2\eta^2(2f_M - f) \tag{8.6-18}$$

as plotted in Fig. 8.6-3.   The baseband filter has a bandwidth that extends only to f_M.   Therefore the noise power output N_o'' is given by the area of the shaded region in the triangular plot of Fig. 8.6-3.   We find that

$$N_o'' = 3\lambda^2\eta^2 f_M^2 \tag{8.6-19}$$

SIGNAL-TO-NOISE RATIO; THRESHOLD

The total output-noise power is, from Eqs. (8.6-9) and (8.6-19),

$$N_o = N_o' + N_o'' = 2\lambda^2\eta f_M A^2 + 3\lambda^2\eta^2 f_M^2 \tag{8.6-20}$$

The output signal-to-noise ratio is, from Eqs. (8.6-8) and (8.6-20),

$$\frac{S_o}{N_o} = \frac{A^4\overline{m^2(t)}}{2\eta f_M A^2 + 3\eta^2 f_M^2} \tag{8.6-21}$$

We now rewrite Eq. (8.6-21) using as before the symbols $P_c \equiv A^2/2$ for the carrier power and $N_M \equiv \eta f_M$ for the input-noise power in the frequency range f_M.   We find that

$$\frac{S_o}{N_o} = \overline{m^2(t)} \frac{P_c}{N_M} \frac{1}{1 + \frac{3}{4}(N_M/P_c)} \tag{8.6-22}$$

A comparison of Eqs. (8.6-22) and (8.5-6) is of interest.   In both cases the input signal is an amplitude-modulated carrier with both sidebands present.   The equations are identical except for the additional factor $1/[1 + \frac{3}{4}(N_M/P_c)]$ which appears in the case of the square-law demodulator.   This factor has its origin in the noise-times-noise term $\lambda n^2(t)$ in Eq. (8.6-5).   When the carrier power P_c is very much larger than N_M, the extra factor may be ignored, and the square-law demodulator performs as well as the linear synchronous demodulator.   Otherwise, however, the square-law demodulator is at a disadvantage.

Equation (8.6-22) is plotted (solid plot) in Fig. 8.6-4 with the vari-

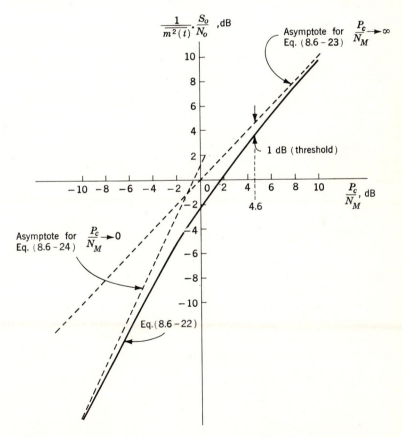

Fig. 8.6-4 Performance of a square-law demodulator illustrating the phenomenon of threshold.

ables expressed in terms of their decibel (dB) equivalents, i.e., the abscissa is marked off in units of $10 \log (P_c/N_M)$. Above threshold, when P_c/N_M is very large, Eq. (8.6-22) becomes

$$\frac{S_o}{N_o} = \overline{m^2(t)} \, \frac{P_c}{N_M} \tag{8.6-23}$$

Below threshold, when $P_c/N_M \ll 1$, Eq. (8.6-22) becomes

$$\frac{S_o}{N_o} = \frac{4}{3} \, \overline{m^2(t)} \left(\frac{P_c}{N_M}\right)^2 \tag{8.6-24}$$

For comparison, Eqs. (8.6-23) and (8.6-24) have also been plotted in Fig. 8.6-4. We observe the occurrence of a threshold in that, as P_c/N_M decreases, the demodulator performance curve falls progressively further away from the straight-line plot corresponding to P_c/N_M very large. The *threshold* point is chosen arbitrarily to be the point at which the performance curve falls away by 1 dB as shown. On this basis it turns out that the threshold occurs when $P_c/N_M = 4.6$ dB or when $P_c = 2.9 N_M$.

8.7 THE ENVELOPE DEMODULATOR

We again consider an AM signal with modulation $|m(t)| < 1$. To demodulate this DSB signal we shall use a network which accepts the modulated carrier and provides an output which follows the waveform of the *envelope* of the carrier. The diode demodulator of Sec. 3.6 is a physical circuit which performs the required operation to a good approximation. As usual, as in Fig. 8.3-1, the demodulator is preceded by a bandpass filter with center frequency f_c and bandwidth $2f_M$, and is followed by a low-pass baseband filter of bandwidth f_M.

It is convenient in the present discussion to use the noise representation given in Eq. (7.11-2):

$$n(t) = n_c(t) \cos \omega_c t - n_s(t) \sin \omega_c t \tag{8.7-1}$$

If the noise $n(t)$ has a power spectral density $\eta/2$ in the range $|f - f_c| \leq f_M$ and is zero elsewhere as shown in Fig. 8.4-1, then, as explained in Sec. 7.12, both $n_c(t)$ and $n_s(t)$ have the spectral density η in the frequency range $-f_M$ to f_M.

At the demodulator input, the input signal plus noise is

$$s_1(t) + n_1(t) = A[1 + m(t)] \cos \omega_c t + n_c(t) \cos \omega_c t - n_s(t) \sin \omega_c t \tag{8.7-2a}$$

$$= \{A[1 + m(t)] + n_c(t)\} \cos \omega_c t - n_s(t) \sin \omega_c t \tag{8.7-2b}$$

where A is the carrier amplitude and $m(t)$ the modulation. In a phasor diagram, the first term of Eq. (8.7-2b) would be represented by a phasor of amplitude $A[1 + m(t)] + n_c(t)$, while the second term would be represented by a phasor perpendicular to the first and of amplitude $n_s(t)$. The phasor sum of the two terms is then represented by a phasor of amplitude equal to the square root of the sum of the squares of the amplitudes of the two terms. Thus, the output signal plus noise just prior to baseband filtering is the envelope (phasor sum):

$$s_2(t) + n_2(t) = \{(A[1 + m(t)] + n_c(t))^2 + n_s^2(t)\}^{1/2} \qquad (8.7\text{-}3a)$$

$$= \{A^2[1 + m(t)]^2 + 2A[1 + m(t)]n_c(t) + n_c^2(t) + n_s^2(t)\}^{1/2} \qquad (8.7\text{-}3b)$$

We should now like to make the simplification in Eq. (8.7-3b) that would be allowed if we might assume that both $|n_c(t)|$ and $|n_s(t)|$ were much smaller than the carrier amplitude A. The difficulty is that n_c and n_s are noise "waveforms" for which an explicit time function may not be written and which are described only in terms of the statistical distributions of their instantaneous amplitudes. No matter how large A and how small the values of the standard deviation of $n_c(t)$ or $n_s(t)$, there is always a finite probability that $|n_c(t)|$, $|n_s(t)|$, or both, will be comparable to, or even larger than, A. On the other hand, if the standard deviations of $n_c(t)$ and $n_s(t)$ are much smaller than A, the *likelihood* that n_c or n_s will approach or exceed A is rather small. For example, since n_c and n_s have gaussian distributions, the probability that $n_c(t)$ is greater than twice its standard deviation is only 0.045, and only 0.00006 that it exceeds 4 times its standard deviation. Hence, if $\sqrt{n_c^2(t)} \ll A$ and we assume that $|n_c(t)| \ll A$, the assumption is *usually* valid.

Assuming then that $|n_c(t)| \ll A$ and $|n_s(t)| \ll A$, the "noise-noise" terms $n_c^2(t)$ and $n_s^2(t)$ may be dropped, leaving us with the approximation

$$s_2(t) + n_2(t) \approx \{A^2[1 + m(t)]^2 + 2A[1 + m(t)]n_c(t)\}^{1/2} \qquad (8.7\text{-}4a)$$

$$= A[1 + m(t)] \left\{1 + \frac{2n_c(t)}{A[1 + m(t)]}\right\}^{1/2} \qquad (8.7\text{-}4b)$$

Using now the further approximation that $(1 + x)^{1/2} \approx 1 + x/2$ for small x, we have finally that

$$s_2(t) + n_2(t) \approx A[1 + m(t)] + n_c(t) \qquad (8.7\text{-}5)$$

The output-signal power measured after the baseband filter, and neglecting dc terms, is $S_o = A^2\overline{m^2(t)}$. Since the spectral density of $n_c(t) = \eta$, the output-noise power after baseband filtering is $N_o = 2\eta f_M$. Again using the symbol $N_M(\equiv \eta f_M)$ to stand for the noise power at the input in the baseband range f_M, and using Eq. (8.5-3), we find that

$$\gamma \equiv \frac{S_o/N_o}{S_i/N_M} = \frac{\overline{m^2(t)}}{1 + \overline{m^2(t)}} \qquad (8.7\text{-}6)$$

The result is the same as given in Eq. (8.5-13) for synchronous demodulation. To make a comparison with the square-law demodulator, we assume $\overline{m^2(t)} \ll 1$. In this case, as before, $S_i \cong P_c$, and Eq. (8.7-6) reduces to Eq. (8.5-6). Hence we have the important result that *above threshold* the synchronous demodulator, the square-law demodulator, and the envelope demodulator all perform equally well, provided $\overline{m^2(t)} \ll 1$.

THRESHOLD

Like the square-law demodulator, the envelope demodulator exhibits a threshold. As the input signal-to-noise ratio decreases, a point is reached where the signal-to-noise ratio at the output decreases more rapidly than at the input. The calculation of signal-to-noise ratio is quite complex, and we shall therefore be content to simply state the result[1] that for $S_i/N_M \ll 1$, and $\overline{m^2(t)} \ll 1$

$$\frac{S_o}{N_o} = \frac{\overline{m^2(t)}}{1.1} \left(\frac{S_i}{N_M} \right)^2 \tag{8.7-7}$$

Equation (8.7-7) obviously indicates a poorer performance than indicated by Eq. (8.7-6), which applies above threshold.

Since both square-law demodulation and envelope demodulation exhibit a threshold, a comparison is of interest. We had assumed in square-law demodulation that $\overline{m^2(t)} \ll 1$. Then, as noted above, $S_i \cong A^2/2 = P_c$ the carrier power, and Eq. (8.7-7) becomes

$$\frac{S_o}{N_o} = \frac{1}{1.1} \overline{m^2(t)} \left(\frac{P_c}{N_M} \right)^2 \tag{8.7-8}$$

which is to be compared with Eq. (8.6-24) giving S_o/N_o below threshold for the square-law demodulator.

The comparison indicates that, below threshold, the square-law demodulator performs better than the envelope detector. Actually this advantage of the square-law demodulator is of dubious value. Generally, when a demodulator is operated below threshold to any appreciable extent, the performance may be so poor as to be nearly useless. What is of greater significance is that the comparison suggests that the threshold in square-law demodulation is lower than the threshold in envelope demodulation. Therefore a square-law demodulator will operate above threshold on a weaker signal than will an envelope demodulator.

In summary, on strong signals all demodulators work equally well except that the square-law demodulator requires that $\overline{m^2(t)} \ll 1$ to avoid baseband-signal distortion. On weak signals, synchronous demodulation does best since it exhibits no threshold. When synchronous demodulation is not feasible, square-law demodulation does better than envelope demodulation. It is also interesting to note that voice signals require a 40-dB output signal-to-noise ratio for high quality. In this case both the linear-envelope detector and the square-law detector operate above threshold.

PROBLEMS

8.1-1. (a) A superheterodyne receiver using an IF frequency of 455 kHz is tuned to 650 kHz. It is found that the receiver picks up a transmission from a transmitter whose carrier frequency is 1560 kHz. Suggest a reason for this undesired reception and suggest a remedy. (These frequencies, 650 kHz and 1560 kHz, are referred to as *image frequencies*. Why?)

8.3-1. Let $g(t)$ be a waveform characterized by a power spectral density $G(f)$. Assume $G(f) = 0$ for $|f| \geq f_1$. Show that the time-average value of $g(t) \cos 2\pi f_c t$ is zero if $f_c > f_1$.

8.3-2. As noted in Sec. 3.10, if $m(t)$ is an arbitrary baseband waveform, a received SSB signal may be written

$$s_i(t) = m(t) \cos 2\pi f_c t + \hat{m}(t) \sin 2\pi f_c t \qquad \text{See Carlson}$$

Here $\hat{m}(t)$ is derived from $m(t)$ by shifting by 90° the phase of every spectral component in $m(t)$.

(a) Show that $m(t)$ and $\hat{m}(t)$ have the same power spectral densities and that $\overline{m^2(t)} = \overline{\hat{m}^2(t)}$.

(b) Show that if $m(t)$ has a spectrum which extends from zero frequency to a maximum frequency f_M, then $m^2(t)$, $\hat{m}^2(t)$, and $m(t)\hat{m}(t)$ all have spectra which extend from zero frequency to $2f_M$.

(c) Show that the normalized power S_i of $s_i(t)$ is $\overline{m^2(t)} = \overline{\hat{m}^2(t)}$. (*Hint:* Use the results of Prob. 8.3-1.)

(d) Calculate the normalized power S_o of the demodulated SSB signal, i.e., the signal $s_i(t)$ multiplied by $\cos 2\pi f_c t$ and then passed through a baseband filter. Show that $S_o = \overline{m^2(t)}/4$ and hence that $S_o/S_i = \frac{1}{4}$. [*Note:* This problem establishes more generally the result given in Eq. (8.3-6) which was derived on the basis of the assumption that the modulating waveform is a sinusoid.]

8.3-3. Prove that Eq. (8.3-11) is correct by sketching the power spectral density of Eq. (8.3-10).

8.3-4. A baseband signal $m(t)$ is transmitted using SSB as in Prob. 8.3-2. Assume that the power spectral density of $m(t)$ is

$$G_m(f) = \begin{cases} \dfrac{\eta_m}{2} \dfrac{|f|}{f_M} & |f| < f_M \\ 0 & |f| > f_M \end{cases}$$

Find:

(a) The input signal power.

(b) The output signal power.

(c) If white gaussian noise with power spectral density $\eta/2$ is added to the SSB signal, find the output SNR. The baseband filter cuts off at $f = f_M$.

8.3-5. A received SSB signal has a spectrum which extends over the range from $f_c = 1$ MHz to $f_c + f_M = 1.003$ MHz. The signal is accompanied by noise with uniform power spectral density 10^{-9} watt/Hz.

    (a) The noise $n(t)$ is expressed as $n(t) = n_c(t) \cos 2\pi f_c t - n_s(t) \sin 2\pi f_c t$. Find the power spectral densities of the quadrature components $n_c(t)$ and $n_s(t)$ of the noise in the spectral range $f_c \le f \le f_c + f_M$.

    (b) The signal plus its accompanying noise is multiplied by a local carrier $\cos 2\pi f_c t$. Plot the power spectral density of the noise at the output of the multiplier.

    (c) The signal plus noise, after multiplication, is passed through a baseband filter and an amplifier which provides a voltage gain of 10. Plot the power spectral density of the noise at the amplifier output, and calculate the total noise output power.

8.4-1. Show that $m^2(t)$ in Eq. (8.4-10) is bandlimited to $2f_M$.

8.4-2. Repeat Prob. 8.3-4 if DSB rather than SSB modulation is employed.

8.4-3. A carrier of amplitude 10 mV at f_c is 50 percent amplitude-modulated by a sinusoidal waveform of frequency 750 Hz. It is accompanied by thermal noise of two-sided power spectral density $\eta/2 = 10^{-3}$ watt/Hz. The signal plus noise is passed through the filter shown. The signal is demodulated by multiplication with a local carrier of amplitude 1 volt.

    (a) Find the output signal power.
    (b) Find the output noise power.

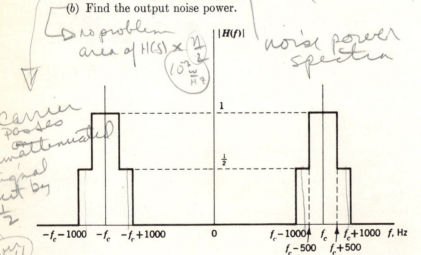

Fig. P8.4-3

8.5-1. The signal $[\epsilon + m(t)] \cos \omega_c t$ is synchronously detected. The reference signal $\cos \omega_c t$, used to multiply the incoming signal, is obtained by passing the input signal through a narrowband filter of bandwidth B, as shown in Fig. P8.5-1.

    (a) Calculate S_i.

    (b) Calculate $v_R(t)$ due to the input signal alone. Calculate the noise power accompanying $v_R(t)$.

    (c) Comment on the effect of the noisy reference on the output SNR.

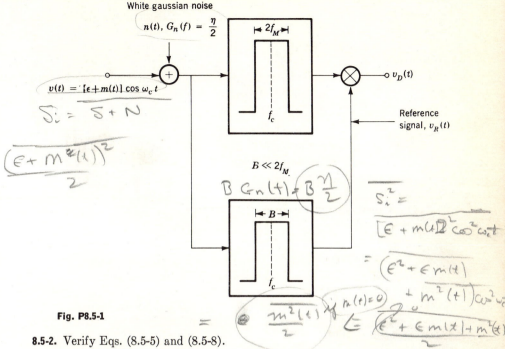

White gaussian noise

$n(t)$, $G_n(f) = \dfrac{\eta}{2}$

$v(t) = [\epsilon + m(t)] \cos \omega_c t$

$S_i = S + N$

$\dfrac{(\epsilon + m^2(t))^2}{2}$

$2f_M$

f_c

$v_D(t)$

Reference signal, $v_R(t)$

$B \ll 2f_M$

$B\, G_n(t) = B\dfrac{\eta}{2}$

$\leftarrow B \rightarrow$

f_c

$S_i^2 =$

$[E + m(t)]^2 \cos^2 \omega_c t$

$= \dfrac{(\epsilon^2 + \epsilon m(t)}{2}$

$+ m^2(t)) \cos^2 \omega_c t$

if $m(t) = 0$

$\leq (\epsilon^2 + \epsilon m(t) + m^2(t))$

$\dfrac{m^2(t)}{2}$

Fig. P8.5-1

8.5-2. Verify Eqs. (8.5-5) and (8.5-8).

8.5-3. (a) Show that the output SNR of a DSB-SC signal which is synchronously detected is independent of the IF bandwidth; i.e., Eq. (8.4-8) is independent of the IF bandwidth.

   (b) Show that the output SNR of an SSB signal which is synchronously detected is dependent on the IF bandwidth. To do this, consider an IF bandwidth which extends from $f_0 - B$ to $f_0 + f_M$, where $f_M > B > 0$. Calculate the output SNR using this bandwidth.

8.5-4. In the received amplitude-modulated signal $s_i(t) = A[1 + m(t)] \cos 2\pi f_c t$, $m(t)$ has the power spectral density $G_m(f)$ specified in Prob. 8.3-4. The received signal is accompanied by noise of power spectral density $\eta/2$. Calculate the output signal-to-noise ratio.

8.6-1. Verify Eq. (8.6-4) by showing graphically that all the neglected terms have spectra falling outside the range $|f| \leq f_M$.

8.6-2. Given $2K + 1$ spectral components of a noise waveform spaced by intervals Δf. Show that the number of pairs of components separated by a frequency $\rho \, \Delta f$ is $p = 2K + 1 - \rho$ [i.e., verify the discussion leading to Eq. (8.6-15)].

8.6-3. In a DSB transmission a carried of frequency 1 MHz and of amplitude 2 volts is amplitude-modulated to the extent of 10 percent by a sinusoidal baseband signal of frequency 5 kHz. The signal is corrupted by white noise of two-sided power spectral density 10^{-6} watt/Hz. The demodulator is a device whose input-output characteristic is given by $v_o = 3v_i^2$, where v_o and v_i are respectively the output and input voltage. The IF filter, before the demodulator, has a rectangular transfer characteristic of unity gain and 10-kHz bandwidth. By error, the IF filter is tuned so that its center frequency is at 1.002 MHz.

$2 \cos(2\pi\, 10^6\, t)\ \sin 2\pi\, 5\times10^3 t$

(a) Calculate the signal waveform at the demodulator output and calculate its normalized power.

(b) Calculate the noise power at the demodulator output and the signal-to-noise ratio.

8.6-4. Plot $(1/\overline{m^2})(S_o/N_o)$ versus P_c/N_M in Eq. (8.6-22) and show that the 1-dB threshold occurs when $P_c/N_M = 4.6$ dB.

8.6-5. Assume that $\overline{m^2(t)} = 0.1$ and that $S_o/N_o = 30$ dB. Find P_c/N_M in Eq. (8.6-22). Are we above threshold?

8.7-1. A baseband signal $m(t)$ is superimposed as an amplitude modulation on a carrier in a DSB transmission from a transmitting station. The instantaneous amplitude of $m(t)$ has a probability density which falls off linearly from a maximum at $m = 0$ to zero at $|m| = 0.1$ volt. The spectrum of $m(t)$ extends over a frequency range from zero to 10 kHz. The level of the modulating waveform applied to the modulator is adjusted to provide for maximum allowable modulation. It is required that under this condition the total power supplied by the transmitter to its antenna be 10 kw. Assume that the antenna appears to the transmitter as a resistive load of 72 ohms.

(a) Write an expression for the voltage at the input to the antenna.

(b) At a receiver, tuned to pick up the transmission, the level of the carrier at the input to the diode demodulator is 3 volts. What is the maximum allowable power spectral density at the demodulator input if the signal-to-noise ratio at the receiver output is to be at least 20 dB?

8.7-2. Plot $(1/\overline{m^2})(S_o/N_o)$ versus S_i/N_M for the envelope demodulator. Assume that $\overline{m^2} \ll 1$, and find the intersection of the above-threshold and the below-threshold asymptotes.

REFERENCE

1. Davenport, W., and W. Root: "Random Signals and Noise," McGraw-Hill Book Company, New York, 1958.

9

Noise in Frequency-modulation Systems

In this chapter we discuss the performance of frequency-modulation systems in the presence of additive noise. We shall show how, in an FM system, an improvement in output signal-to-noise power ratio can be made through the sacrifice of bandwidth.

9.1 AN FM DEMODULATOR

The receiving system of Fig. 8.1-1 may be used with an AM or FM signal. When used to recover a frequency-modulated signal, the AM demodulator is replaced by an FM demodulator such as the limiter-discriminator shown in Fig. 9.1-1.

THE LIMITER

In an FM system, the baseband signal varies only the frequency of the carrier. Hence any amplitude variation of the carrier must be due to noise alone. The *limiter* is used to suppress such amplitude-variation noise. In a limiter, diodes, transistors, or other devices are used to con-

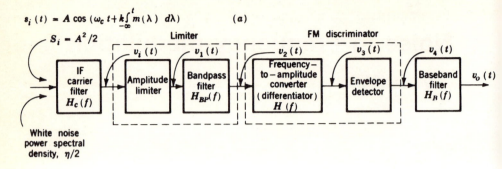

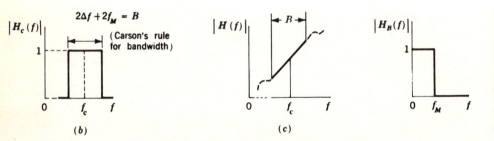

Fig. 9.1-1 A limiter-discriminator used to demodulate an FM signal.

struct a circuit in which the output voltage v_1 is related to the input voltage v_i in the manner shown in Fig. 9.1-2a.   The output follows the input only over a limited range.   A cycle of the carrier is shown in Fig. 9.1-2b and the output waveform in Fig. 9.1-2c.   In limiter operation the carrier amplitude is very large in comparison with the limited range of the limiter, actually much larger than is indicated in Fig. 9.1-2.   As a consequence,

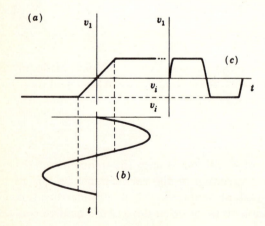

Fig. 9.1-2 (a) A limiter input-output characteristic. (b) A cycle of the input carrier. (c) The output waveform.

the output waveform is a *square* wave. Thus, the output has a wave-shape which is nearly entirely independent of modest changes in carrier amplitude. The bandpass filter, following the limiter, selects the fundamental frequency component of the square wave. Therefore the filter output is again sinusoidal. It has an amplitude which is very nearly independent of the input-carrier amplitude, but does, of course, have the same instantaneous frequency as does the input carrier. In a physical circuit the limiter and filter generally form an integral unit so that actually there is no point in the limiter-filter combination where the square-wave waveform may be observed.

In this text we shall assume that the discriminator is always preceded by an ideal *hard* limiter, i.e., that the limiter output is a perfect square wave, no matter how small $v_i(t)$ may be.

THE DISCRIMINATOR

The discriminator consists also of two component parts. The first of these is a network which, over the range of excursion of the instantaneous frequency, exhibits a transfer characteristic $H(f)$ such that $|H(f)|$ varies linearly with frequency. When the constant-amplitude FM signal passes through this network, it will appear at the output with an amplitude variation (i.e., an envelope) which varies with time precisely as does the instantaneous frequency of the carrier. The baseband signal is now recovered by passing this amplitude-modulated waveform through an envelope demodulator such as the diode detector of Fig. 3.4-2. The input to the envelope detector is frequency-modulated as well as amplitude-modulated, but the detector does not respond to the frequency modulation.

MATHEMATICAL REPRESENTATION OF THE OPERATION OF THE LIMITER-DISCRIMINATOR

The frequency-to-amplitude converter necessary to obtain frequency demodulation need have an $|H(f)|$ which varies linearly with ω over only a limited range, and its slope may be positive or negative. Further, the phase variation of $H(f)$ is of no consequence. However, as a matter of mathematical convenience we shall assume that $H(j\omega)$ is given by

$$H(j\omega) = j\sigma\omega \qquad (9.1\text{-}1)$$

where σ is a constant. The advantage of such a selection of $H(j\omega)$ is that (see Fig. 9.1-1) the output of the converter $v_3(t)$ is related to the input $v_2(t)$ by the equation

$$v_3(t) = \sigma \frac{d}{dt} v_2(t) \qquad (9.1\text{-}2)$$

Equation (9.1-2) results from the fact that a multiplication by $j\omega$ in the frequency domain is equivalent to differentiation in the time domain; i.e.,

$$\sigma \frac{d}{dt} \Leftrightarrow j\sigma\omega \qquad (9.1\text{-}3)$$

Now suppose that the voltage $v_2(t)$ applied to the converter is

$$v_2(t) = A_L \cos[\omega_c t + \phi(t)] \qquad (9.1\text{-}4)$$

Here A_L is the *limited* amplitude of the carrier so that A_L is fixed and independent of the input amplitude, and $\omega_c t + \phi(t)$ is the instantaneous phase. Then from Eq. (9.1-2)

$$v_3(t) = -\sigma A_L \left[\omega_c + \frac{d}{dt}\phi(t) \right] \sin[\omega_c t + \phi(t)] \qquad (9.1\text{-}5)$$

The output of the envelope detector is, using $\alpha \equiv \sigma A_L$,

$$v_4(t) = \sigma A_L \left[\omega_c + \frac{d}{dt}\phi(t) \right] = \alpha\omega_c + \alpha\frac{d}{dt}\phi(t) \qquad (9.1\text{-}6)$$

Thus, in summary, we see that if the input waveform to the discriminator is given by Eq. (9.1-4), the discriminator output is calculated from Eq. (9.1-6). Note that the discriminator output is proportional to the frequency of the input, $d\phi/dt$ (see Sec. 4.2).

9.2 CALCULATION OF OUTPUT SIGNAL AND NOISE POWERS

Let us now consider that the input signal to the IF carrier filter of Fig. 9.1-1 is

$$s_i(t) = A \cos\left[\omega_c t + k \int_{-\infty}^{t} m(\lambda)\, d\lambda \right] \qquad (9.2\text{-}1)$$

where $m(t)$ is the frequency-modulating baseband waveform. We assume that the signal is embedded in additive white gaussian noise of power spectral density $\eta/2$. The IF carrier filter has a bandwidth $B = 2\Delta f + 2f_M$ (Carson's rule for bandwidth; see Sec. 4.7). This filter passes the signal with negligible distortion and eliminates all noise outside the bandwidth B. The signal with its accompanying noise is ideally limited, discriminated, and after passing through the baseband filter, appears at the output as a signal $s_o(t)$ and a noise waveform $n_o(t)$.

It can be shown that, when the signal-to-noise ratio is high, the noise does not affect the output-signal power. We shall accept this result without giving a proof.[1] In calculating the output-signal power we shall therefore ignore the noise. When the signal $s_i(t)$ in Eq. (9.2-1) arrives at the output of the limiter shown in Fig. 9.1-1, the signal is $s_2(t)$ [corresponding to $v_2(t)$] given by

$$s_2(t) = A_L \cos \left[\omega_c t + k \int_{-\infty}^{t} m(\lambda) \, d\lambda \right] \tag{9.2-2}$$

Using the result of Eq. (9.1-6) and setting

$$\phi(t) = k \int_{-\infty}^{t} m(\lambda) \, d\lambda \tag{9.2-3}$$

we find for the output of the discriminator

$$s_4(t) = \alpha \omega_c + \alpha k m(t) \tag{9.2-4}$$

The baseband filter rejects the dc component and passes the signal component without distortion. Thus, the output signal is $s_o(t) = \alpha k m(t)$, and the output-signal power is

$$S_o = \alpha^2 k^2 \overline{m^2(t)} \tag{9.2-5}$$

OUTPUT-NOISE POWER

Let us now calculate the noise output of the FM discriminator which results from the presence at the input of white noise having a power spectral density $\eta/2$. To facilitate the computation, we set the modulation $m(t) = 0$. It can be shown, although the proof is complex and will not be undertaken here, that the noise output is approximately independent of $m(t)$.[1] The carrier and noise pass through the IF filter $H_c(\omega)$, which filters the noise. The resulting noise is represented as in Eq. (7.11-2). Thus the carrier and noise at the limiter input are

$$\begin{aligned} v_i(t) &= A \cos \omega_c t + n_c(t) \cos \omega_c t - n_s(t) \sin \omega_c t \\ &= [A + n_c(t)] \cos \omega_c t - n_s(t) \sin \omega_c t \end{aligned} \tag{9.2-6}$$

A phasor diagram of the signal and noise is shown in Fig. 9.2-1. Note that the phasor representing $n_c(t) \cos \omega_c t$ is in phase with the carrier phasor $A \cos \omega_c t$. The phasor representing $n_s(t) \sin \omega_c t$ has an amplitude $n_s(t)$ and is in phase-quadrature with the other two terms. The envelope $R(t)$ is easily computed using Fig. 9.2-1 and is

$$R(t) = \sqrt{[A + n_c(t)]^2 + [n_s(t)]^2} \tag{9.2-7}$$

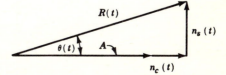

Fig. 9.2-1 A phasor diagram of the terms in Eq. (9.2-6).

Similarly, the phase $\theta(t)$ is

$$\theta(t) = \tan^{-1} \frac{n_s(t)}{A + n_c(t)} \qquad (9.2\text{-}8)$$

Thus, the signal and noise forming $v_i(t)$ can be written as

$$v_i(t) = R(t) \cos [\omega_c t + \theta(t)] \qquad (9.2\text{-}9)$$

We ignore the time-varying envelope $R(t)$, since all time variations are removed by the limiter. The output of the limiter-bandpass filter is therefore

$$v_2(t) = A_L \cos [\omega_c t + \theta(t)] \qquad (9.2\text{-}10)$$

where again A_L is determined by the limiter, and is a constant.

Let us assume that we are operating under the condition of high-input SNR. Then the noise power is much smaller than the carrier power. In this case we assume that most of the time $|n_c(t)| \ll A$ and $|n_s(t)| \ll A$, on the basis of the justification presented in Sec. 8.7. With these assumptions and using the approximation $\tan \theta \approx \theta$ for small θ, we have, from Eq. (9.2-8),

$$\theta(t) \approx \frac{n_s(t)}{A} \qquad (9.2\text{-}11)$$

Thus, $v_2(t)$ is approximately

$$v_2(t) = A_L \cos \left[\omega_c t + \frac{n_s(t)}{A} \right] \qquad (9.2\text{-}12)$$

Comparing Eq. (9.2-12) with Eq. (9.1-4) we see that $n_s(t)/A$ is $\phi(t)$. Then, from Eq. (9.1-6) we find

$$v_4(t) = \alpha \left[\omega_c + \frac{1}{A} \frac{d}{dt} n_s(t) \right] \qquad (9.2\text{-}13)$$

If we drop the dc term in Eq. (9.2-13), the noise $n_4(t)$ at the input to the baseband filter is

$$n_4(t) = \frac{\alpha}{A} \frac{d}{dt} n_s(t) \qquad (9.2\text{-}14)$$

The spectral density of $n_s(t)$ is η over the frequency range $|f| \leq B/2$ (see Sec. 7.12). The differentiation is equivalent to passing $n_s(t)$ through a network whose transfer function is $H(j\omega) = j\omega$. Hence, as shown in Fig. 9.2-2a, the operation performed on $n_s(t)$ is equivalent to passing $n_s(t)$ through a network with $H(j\omega) = j\alpha\omega/A$. Then $|H(j\omega)|^2 = \alpha^2\omega^2/A^2$. Therefore the spectral density of $n_4(t)$ is $G_{n_4}(f)$ given by

$$G_{n_4}(f) = \frac{\alpha^2\omega^2}{A^2} \eta \qquad |f| \leq \frac{B}{2} \qquad (9.2\text{-}15)$$

This spectral density is plotted in Fig. 9.2-2b.

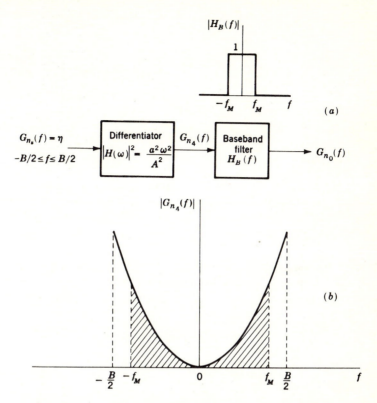

Fig. 9.2-2 (*a*) Indicating the operations performed by the discriminator and baseband filter on the noise output of the limiter. (*b*) The variation with frequency of the power spectral density at the output of an FM demodulator.

Since the baseband filter passes frequencies just up to f_M, only the shaded area in Fig. 9.2-2*b* contributes to the output-noise power. This output power can therefore be calculated by computing the shaded area. The output-noise power N_o is

$$N_o = \int_{-f_M}^{f_M} G_{n_4}(f)\,df = \frac{\alpha^2 \eta}{A^2} \int_{-f_M}^{f_M} 4\pi^2 f^2 \, df = \frac{8\pi^2}{3} \frac{\alpha^2 \eta}{A^2} f_M^3 \qquad (9.2\text{-}16)$$

OUTPUT SNR

The output signal-to-noise ratio can now be computed using Eqs. (9.2-5) and (9.2-16). We find

$$\frac{S_o}{N_o} = \frac{\alpha^2 k^2 \overline{m^2(t)}}{(8\pi^2/3)(\alpha^2 \eta/A^2) f_M^3} = \frac{3}{4\pi^2} \frac{k^2 \overline{m^2(t)}}{f_M^2} \frac{A^2/2}{\eta f_M} \qquad (9.2\text{-}17)$$

Now let us consider that the modulating signal $m(t)$ is sinusoidal and produces a frequency deviation Δf. Then the input signal $s_i(t)$ given in Eq. (9.2-1) may be written [see Eq. (4.4-4)]

$$s_i(t) = A \cos\left(\omega_c t + \frac{\Delta f}{f_m} \sin 2\pi f_m t\right) \tag{9.2-18}$$

where f_m is the modulating frequency. Comparing Eq. (9.2-1) with (9.2-18), we have, after differentiating the argument, that

$$km(t) = 2\pi \, \Delta f \cos 2\pi f_m t \tag{9.2-19}$$

Hence

$$k^2 \overline{m^2(t)} = \frac{4\pi^2(\Delta f)^2}{2} = 2\pi^2(\Delta f)^2 \tag{9.2-20}$$

Substituting this result into Eq. (9.2-17) yields

$$\frac{S_o}{N_o} = \frac{3}{2}\left(\frac{\Delta f}{f_M}\right)^2 \frac{A^2/2}{\eta f_M} = \frac{3}{2}\beta^2 \frac{S_i}{N_M} \tag{9.2-21}$$

where $\beta \equiv \Delta f/f_M$ is the modulation index, $S_i = A^2/2$ is the input-signal power, and $N_M \equiv \eta f_M$ is the noise power at the input in the baseband bandwidth f_M. We also have that

$$\gamma_{\text{FM}} \equiv \frac{S_o/N_o}{S_i/N_M} \equiv \frac{3}{2}\beta^2 \tag{9.2-22}$$

which is to be compared with values of γ given in Eqs. (8.5-11) and (8.5-12) for AM systems.

It is to be noted that in Eqs. (9.2-21) and (9.2-22) the symbol β is being used in a sense which is somewhat different from the sense in which we have used it previously in Chap. 4. Previously, we used β to represent the modulation index associated with a sinusoidal modulating waveform of frequency f_m, that is, $\beta = \Delta f/f_m$, where Δf is the frequency deviation produced by the sinusoidal waveform. In the present instance, $\beta(\equiv \Delta f/f_M)$ characterizes not a particular sinusoidal modulating waveform but rather a particular set of specifications associated with an FM system. Thus Eq. (9.2-22) has the following interpretation. Suppose that in an FM system whose baseband frequency range is f_M we consider only sinusoidal modulation. Suppose we assume that independently of the frequency of the modulation its amplitude (and, hence, the frequency deviation Δf) is kept fixed. Then, independently of the frequency of the modulation (up to f_M) the performance criterion γ will be fixed at $3\beta^2/2$.

9.3 COMPARISON OF FM AND AM

A comparison of the performances of FM and conventional AM (double sideband with carrier) is of interest. Equation (9.2-22) applies for sinusoidal frequency modulation of the carrier with modulation index β.   Let us compare this result with the corresponding result for sinusoidal amplitude modulation with 100 percent modulation.   We find for γ_{AM} from Eq. (8.5-14) that $\gamma_{AM} = \frac{1}{3}$, so that from Eq. (9.2-22)

$$\frac{\gamma_{FM}}{\gamma_{AM}} = \frac{9}{2}\beta^2 \tag{9.3-1}$$

In the comparison of FM and AM leading to Eq. (9.3-1), we have assumed in the two cases equal input-noise-power spectral density $\eta/2$, equal baseband bandwidth f_M, and equal input-signal power S_i.

Several authors prefer to make the comparison not on the basis of equal signal power but rather on the basis of *equal signal power* measured when the modulation $m(t) = 0$.   In this case, as is easily verified (Prob. 9.3-2), we find that Eq. (9.3-1) is replaced by

$$\frac{\gamma_{FM}}{\gamma_{AM}} = 3\beta^2 \tag{9.3-2}$$

A comparison on this basis is not completely fair, however, since the total *transmitted* powers are unequal when the unmodulated carrier powers are the same.

It is clear from either Eq. (9.3-1) or Eq. (9.3-2) that FM offers the possibility of improved signal-to-noise ratio over AM.   The improvement begins when $9\beta^2/2 \approx 1$ (Eq. 9.3-1) or when $3\beta^2 \approx 1$ (Eq. 9.3-2) corresponding to $\beta \cong \sqrt{2}/3 \cong 0.5$ or $\beta \cong 1/\sqrt{3} \cong 0.6$.   As β increases, the improvement becomes more pronounced, but this improvement is achieved at the expense of requiring greater bandwidth.   To see the relationship between improvement and bandwidth sacrifice, let us assume that β is large enough so that Carson's rule for bandwidth formula [Eq. (4.7-2)],

$$B_{FM} = 2(\beta + 1)f_M \tag{9.3-3}$$

may be approximated by $B_{FM} \approx 2\beta f_M$.   The bandwidth of the AM system is $B_{AM} = 2f_M$ so that Eq. (9.3-1) may be rewritten

$$\frac{\gamma_{FM}}{\gamma_{AM}} = \frac{9}{2}\left(\frac{B_{FM}}{B_{AM}}\right)^2 \tag{9.3-4}$$

Accordingly, each increase in bandwidth by a factor of 2 increases γ_{FM}/γ_{AM} by a factor of 4 (6 dB).

We observe then a characteristic of FM which is not shared by AM. The FM system allows us to sacrifice bandwidth for the sake of improving signal-to-noise ratio.   The improvement begins to make itself apparent when $\beta \cong 0.5$ or 0.6.   This value of β is roughly the value of β which establishes the demarcation between "narrowband" and "wideband" FM.   Thus, signal-to-noise improvement is a feature of wideband FM not shared by narrowband FM. Equations (9.3-1) and (9.3-2) suggest a continuous improvement in performance with increased β (and correspondingly increased bandwidth). Such is indeed the case as long as the noise power admitted by the carrier filter continues to be small in comparison with the signal power. It will be recalled that in deriving Eq. (9.2-22) we made this assumption of relatively small noise (see Eq. 9.2-11). When the bandwidth becomes so large that the noise power is not relatively small, the performance of the FM system degrades rapidly, i.e., the system exhibits a *threshold*. We shall see in Sec. 10.1 that when the input-noise power is not small in comparison with the input-signal power, the system performance may be improved by *restricting* the bandwidth, by *reducing* the modulation index.

We shall now comment briefly on what might appear to be an anomalous situation. We refer to the fact that the performance of the FM system is improved as the bandwidth increases. We might, on intuitive grounds, have expected the opposite effect, since widening the bandwidth admits more noise into the system. In the AM system the carrier filter bandwidth extends over the range $f_c \pm f_M$, while in FM the bandwidth extends over the range $f_c \pm \beta f_M$. The fact is, however, that noise in the range $f_c \pm \beta f_M$ but outside the range $f_c \pm f_M$ does not appear at the output of the FM system. To see this clearly, we refer to Eq. (9.2-14), which relates the output noise to the quadrature component of the input noise. This equation indicates that each spectral component of input noise gives rise to an output spectral component of the same frequency. Consider, for example, the pair of noise spectral components at frequencies $f_c \pm f_n$ with $f_M < f_n < \beta f_M$. Such noise components give rise to a baseband noise component at a frequency f_n which will not pass through the baseband filter whose cutoff is at f_M. Each such pair of noise spectral components gives rise to a baseband noise component which is independent of all other noise components. That is, the noise spectral components are not correlated.

On the other hand, let the carrier be FM-modulated by a sinusoidal baseband signal of frequency $f_s > f_M/2$. Then, as discussed in Sec. 4.5, the carrier is accompanied by sidebands separated from the carrier by $\pm f_s$, $\pm 2f_s$, etc. Except for the sidebands at $\pm f_s$, all other sidebands are separated from the carrier by more than the baseband bandwidth. However, these sidebands are correlated in such a way that the sidebands at, say, $\pm 2f_s$, do not give rise to a baseband signal at frequency $2f_s$ but serve rather to increase the amplitude of the baseband signal at f_s, as do the higher-order sidebands. Thus, the apparent anomaly is resolved by recognizing that noise outside the range $f_c \pm f_M$ does not pass through the system, while signal sidebands outside this range do contribute to the eventual output signal of the demodulator.

9.4 PREEMPHASIS AND DEEMPHASIS, SINGLE CHANNEL

Suppose that we undertake to transmit a baseband signal using FM modulation and naturally require the best possible signal-to-noise ratio for a given carrier power, noise spectral density, IF bandwidth, and baseband bandwidth. Then clearly, before applying the baseband signal to the FM modulator, we shall raise the level of the modulating baseband signal to the maximum extent possible in order to modulate the carrier as vigorously as possible. How shall we know that we have reached the maximum allowable level of the modulating signal? One way of making such a determination is to demodulate the modulated signal and measure the distortion. The distortion occurs because eventually the frequency deviation exceeds the specified IF bandwidth. The modulating signal level may be raised only until the distortion exceeds a specified value.

When, however, the baseband signal happens to be an audio signal, it turns out that something further can be done. An audio signal usually has the characteristic that its power spectral density is relatively high in the low-frequency range and falls off rapidly at higher frequencies. For example, speech has little power spectral density above about 3 kHz. And while music, of course, extends farther into the high-frequency range, the feature still persists that most of its power is in the low-frequency region. As a consequence, when we examine the spectrum of the sidebands associated with a carrier which is frequency-modulated by an audio signal, we find that the power spectral density of the sidebands is greatest near the carrier and relatively small near the limits of the allowable frequency band allocated to the transmission. The manner in which we may take advantage of these spectral features, which are characteristic of audio signals, in order to improve the performance of an FM system, is shown in Fig. 9.4-1.

We observe in Fig. 9.4-1 that, at the transmitting end, the baseband signal $m(t)$ is not applied directly to the FM modulator but is first passed through a filter of transfer characteristic $H_p(\omega)$, so that the modulating signal is $m_p(t)$. The modulated carrier is transmitted across a communication channel, during which process, as usual, noise is added to the signal.

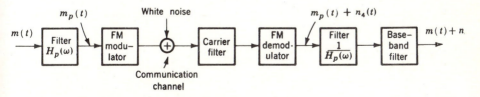

Fig. 9.4-1 Preemphasis and deemphasis in an FM system.

The receiver is a conventional discriminator except that a filter has been introduced before the baseband filter. The transfer characteristic of this filter is the reciprocal of the characteristic of the transmitter filter. This receiver filter of transfer characteristic $1/H_p(\omega)$ may equally well be placed after the baseband filter since both filters are linear. We observe that any modification introduced into the baseband signal by the first filter, prior to modulation, is presently undone by the second filter which follows the discriminator. Hence, the output signal at the receiver is exactly the same as it would be if the filters had been omitted entirely. The noise, however, passes through only the receiver filter, and this filter may then be used to suppress the noise to some extent.

The selection of the transfer characteristic $H_p(\omega)$ is based on the following considerations. We note that at the output of the demodulator the spectral density of the noise, given by $G_{n_4}(f)$ in Eq. (9.2-15) and shown in Fig. 9.2-2b, increases with the square of the frequency. Hence the receiver filter will be most effective in suppressing noise if the response of the filter falls off with increasing frequency, that is, if the filter transmission is lowest where the spectral density of the noise is highest. In such a case the transmitter filter must exhibit a rising response with increasing frequency. Let us assume initially that we design the transmitter filter so that it serves only to increase the spectral density of the higher-frequency components of the signal $m(t)$. Such a filter must necessarily increase the power in the modulating signal, thereby increasing the distortion above its specified maximum value. We have, however, noted above that the spectral density of the modulated carrier is relatively small near the edges of the allowed frequency band. We may then expect that such a filter may possibly raise the signal spectral density only near the edges of the allowed frequency band and cause only a small increase in distortion. In this case we might expect that if the modulating signal power is lowered to decrease the distortion to the allowed value, we end up with a net advantage, i.e., the improvement due to raising the spectral density near the edges of the allowable band outweighs the disadvantage due to the need to lower the level of the modulating signal. We shall see that such is indeed the case. The premodulation filtering in the transmitter, to raise the power spectral density of the baseband signal in its upper-frequency range, is called *preemphasis* (or *predistortion*). The filtering at the receiver to undo the signal preemphasis and to suppress noise is called *deemphasis*.

SNR IMPROVEMENT USING PREEMPHASIS

We recall from Sec. 4.13 that the bandwidth occupied by the output of an FM modulator is fixed if the normalized power (or mean square frequency deviation) of the modulating signal is kept fixed (see Prob. 9.4-3).

Hence, referring to Fig. 9.4-1, we require that the normalized power of the baseband signal $m(t)$ must be the same as the normalized power of the preemphasized signal $m_p(t)$. We begin by expressing the normalized power of a signal in terms of its spectral density. If then $G_m(f)$ is the power spectral density of $m(t)$, the density of $m_p(t)$ is $|H_p(f)|^2 G_m(f)$, and we require that

$$P_m = \int_{-f_M}^{f_M} G_m(f)\, df = \int_{-f_M}^{f_M} |H_p(f)|^2 G_m(f)\, df \tag{9.4-1}$$

where f_M is the maximum frequency of the modulating signal.

In the absence of deemphasis the output noise is given by N_o in Eq. (9.2-16). With the deemphasis filter, the output noise is

$$N_{od} = \left(\frac{\alpha}{A}\right)^2 4\pi^2 \eta \int_{-f_M}^{f_M} f^2 \left|\frac{1}{H_p(f)}\right|^2 df \tag{9.4-2}$$

The ratio of the noise output without deemphasis to the noise output with deemphasis is $N_o/N_{od} \equiv \Re$. From Eqs. (9.2-16) and (9.4-2) we have

$$\Re = \frac{(\alpha/A)^2 (4\pi^2 \eta) \int_{-f_M}^{f_M} f^2\, df}{(\alpha/A)^2 (4\pi^2 \eta) \int_{-f_M}^{f_M} f^2/|H_p(f)|^2\, df} = \frac{f_M^3/3}{\int_0^{f_M} f^2\, df/|H_p(f)|^2} \tag{9.4-3}$$

Since the signal itself is unaffected in the overall process, this quantity $\Re$ is the ratio by which preemphasis-deemphasis improves the signal-to-noise ratio. We are at liberty to select $H_p(f)$ in Eq. (9.4-3) arbitrarily, provided only that $H_p(f)$ satisfies the constraint imposed by Eq. (9.4-1).

In the next section we discuss the application of preemphasis-deemphasis to commercial FM broadcasting.

9.5 PREEMPHASIS AND DEEMPHASIS IN COMMERCIAL FM BROADCASTING

We have noted that the spectral density of the noise at the output of an FM demodulator increases with the square of the frequency. Hence a deemphasis network at the receiver will be most effective in suppressing noise if its response falls with increasing frequency. In commercial FM the deemphasis is performed by the simple low-pass resistance-capacitance network of Fig. 9.5-1a. This network has a transfer function $H_d(f)$ given by

$$H_d(f) = \frac{1}{1 + jf/f_1} \tag{9.5-1}$$

where $f_1 = 1/2\pi RC$. At the transmitter we require an inverse network. A simple network which may be adjusted to provide the required response is shown in Fig. 9.5-1b. Let us assume that $r \ll R$, and that in the base-

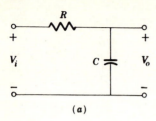

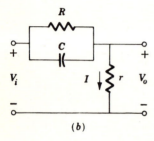

Fig. 9.5-1 (a) Deemphasis network and (b) preemphasis network used in commercial television.

band audio-frequency range r is also very small in comparison with the reactance of the capacitance C. In this case we can approximately compute the current $I(f)$ by neglecting the presence of r. We find, for the audio range,

$$I(f) = V_i(f) \left(\frac{1}{R} + j\omega C \right)$$

(9.5-2)

The output voltage $V_o(f) = rI(f)$, and the transfer function $H_p(f) \equiv V_o(f)/I(f)$ is

$$H_p(f) = \frac{r}{R} (1 + j\omega CR) = \frac{r}{R} \left(1 + j \frac{f}{f_1} \right)$$

(9.5-3)

where, as before, $f_1 = 1/2\pi RC$. Hence $H_p(f)$ has a frequency dependence inverse to $H_d(f)$ as required, in order that no net distortion be introduced into the signal. Thus, $H_p(f)H_d(f) = r/R = $ constant.

Normalized logarithmic plots of H_p and H_d are shown in Fig. 9.5-2. The transfer function H_d actually has a second breakpoint, as shown, at $f_2 = 1/2\pi rC$. Since $r \ll R$, $f_2 \gg f_1$, and it is easily arranged that this second breakpoint lie well outside the baseband spectral range and hence be irrelevant.

The improvement in signal-to-noise ratio which results from preemphasis depends on the frequency dependence of the power spectral density of the baseband signal. Let us assume that the spectral density of a typical audio signal, say music, may reasonably be represented as

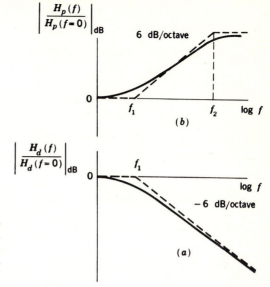

Fig. 9.5-2 Normalized logarithmic plots of the frequency characteristics of (a) the deemphasis network and (b) the preemphasis network.

having a frequency dependence given by

$$G_m(f) = \begin{cases} G_0 \dfrac{1}{1 + (f/f')^2} & |f| \le f_M \\ 0 & \text{elsewhere} \end{cases} \tag{9.5-4}$$

where G_0 is the spectral density at low frequencies, while f' is the frequency at which $G(f)$ has fallen by 3 dB from its low-frequency value. Let us further assume that we have adjusted the preemphasis network so that $f_1 = f'$.

The baseband signal to be preemphasized is transmitted through the network of Fig. 9.5-1b and also through an adjustable gain amplifier. The amplifier will make up for the attenuation produced by the preemphasis network, and its gain is adjusted to the point where the full allowable bandwidth is occupied by the modulated carrier. Hence, the baseband signal $m(t)$ passes through a network whose transfer function is

$$H_p(f) = K\left(1 + j\frac{f}{f_1}\right) \tag{9.5-5}$$

where K is the product of the amplifier gain and the ratio r/R [see Eq. (9.5-3)]. The coefficient K is adjusted so that the constraint of Eq. (9.4-1) is satisfied. Using Eq. (9.5-4) with $f' = f_1$ and using Eq. (9.5-5), we find by substitution into Eq. (9.4-1) that K is determined by the condition that

$$P_m = \int_{-f_M}^{f_M} \frac{G_0\,df}{1 + (f/f_1)^2} = \int_{-f_M}^{f_M} K^2 G_0\,df \tag{9.5-6}$$

Integrating and solving for K^2, we find

$$K^2 = \frac{f_1}{f_M} \tan^{-1} \frac{f_M}{f_1} \tag{9.5-7}$$

Using this value of K, and substituting $H_p(f)$ in Eq. (9.5-5) into Eq. (9.4-3), we compute $\Re$, the improvement which results from preemphasis and subsequent deemphasis. We find that

$$\Re = \frac{\tan^{-1}(f_M/f_1)}{3(f_1/f_M)[1 - (f_1/f_M)\tan^{-1}(f_M/f_1)]} \tag{9.5-8}$$

If the power in the baseband signal is principally confined to the lower frequencies, so that $f_M/f_1 \gg 1$, then $\Re$ becomes approximately

$$\Re \cong \frac{\pi}{6} \frac{f_M}{f_1} \tag{9.5-9}$$

In commercial FM broadcasting $f_1 = 2.1$ kHz, while f_M may reasonably be taken as $f_M = 15$ kHz. In this case $f_M/f_1 = 7.5$, and from Eq. (9.5-9) $\Re \cong 4.7$ corresponding to 6.7 dB improvement in the output SNR when using preemphasis. Since output signal-to-noise ratios are typically 40 to 50 dB, this represents an improvement of approximately 15 percent; a significant improvement.

Preemphasis is particularly effective in FM systems which are used for transmission of audio signals. This effectiveness results from the fact that the spectral density of the audio signal is smallest precisely where the spectral density of the noise is greatest. For, as noted, the audio signal spectral density falls off with increasing frequency while the noise spectral density increases with the square of the frequency. The advantage of using preemphasis is less pronounced in an AM system, such as DSB with carrier, which is employed in commercial AM broadcasting. For in AM the noise spectral density is constant and does not increase with frequency. Preemphasis is also frequently employed in connection with phonographic recording to suppress needle scratch noise.

STEREO FM

As noted in Sec. 9.2, and as shown in Fig. 9.2-2, the noise output of the discriminator in an FM receiver has a parabolic power spectral density in the range $-B/2$ to $B/2$. In commercial FM the bandwidth $B \simeq 200$ kHz so that $B/2 \simeq 100$ kHz. In monophonic FM, only the noise in the spectral range of the baseband filter is passed on to the output. In stereo (Sec. 4.14), the situation is different. The difference signal $L(t) - R(t)$ as received occupies the frequency range from 23 to 53 kHz. When this difference signal is translated to baseband, so also is the noise in this difference-signal frequency interval. The noise in this difference-signal frequency interval is substantially larger than in the baseband frequency range, because the difference-signal frequency range is twice as large as the baseband range and also because of the parabolic form of the noise power spectral density. As a result, commercial stereo FM is noisier than monophonic FM.

For the sake of the advantage gained thereby, as well as for the sake of compatibility, preemphasis is incorporated in stereo FM just as in monophonic FM. In stereo FM, at the transmitter, the sum signal $L(t) + R(t)$ and the difference signal $L(t) - R(t)$ are passed through identical preemphasis filters. At the receiver the sum and difference signals are passed through identical deemphasis filters before the sum and difference signals are combined to generate the individual signals $L(t)$ and $R(t)$. The preemphasis and deemphasis filters are as shown in Fig. 9.5-1, with transfer characteristics as indicated in Fig. 9.5-2. As in monophonic FM, $f_1 = 2.1$ kHz (corresponding to a time constant of 75 μsec).

The overall result, taking the preemphasis and deemphasis into account, and assuming a baseband frequency range extending from nominally zero to 15 kHz, is that commercial stereo FM yields a signal-to-noise ratio about 22 dB poorer than monophonic FM. (See Prob. 9.5-2.) This disadvantage turns out to be tolerable simply because of the high power with which commercial broadcasts are transmitted.

9.6 PHASE MODULATION IN MULTIPLEXING

One method which is used to multiplex voice channels prior to long distance transmission is shown in Fig. 9.6-1. The speech signals $m_1(t)$, $m_2(t)$, etc., are each bandlimited to 3 kHz and are the baseband signals used to generate a sequence of SSB-SC signals, whose carrier frequencies f_1, f_2, etc., are separated by 4 kHz. The outputs of the SSB modulators are added, and this sum of signals forms a composite baseband signal that is used to angle-modulate (FM, PM, or some combination of the two) a carrier of frequency f_c. If 1000 speech signals are multiplexed in this way, the baseband signal nominally extends over a bandwidth 1000 $\times$ 4 kHz = 4 MHz. The angle modulator excites an antenna whose signal is directed at a receiving antenna. Not indicated in the figure are the repeater stations which may be interposed between initial transmitter and final receiver in a system which connects very distant points. For example, much of the telephone communication between the United States and Europe presently employs a system as shown in Fig. 9.6-1, and uses a satellite repeater.

At the receiver, an angle demodulator recovers the composite baseband signal. The individual baseband signals $m_1(t)$, $m_2(t)$, etc., are

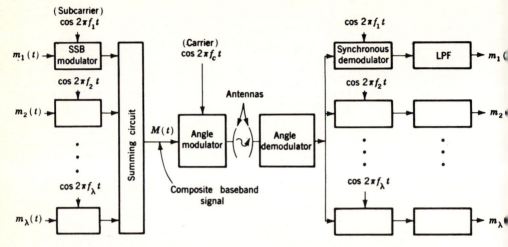

Fig. 9.6-1 A system of frequency division multiplexing.

recovered by synchronous demodulation and by passing the output of each synchronous demodulator through a low-pass filter of bandwidth 3 kHz.

Suppose now that the angle modulator at the transmitter is an FM modulator, that is, the instantaneous frequency of the modulator is directly proportional to the instantaneous magnitude of the composite baseband signal $M(t)$. Then the angle demodulator would be a frequency demodulator, such as the discriminator discussed in Sec. 9.1. However, we have noted in that section that, with white noise present at the input to the discriminator, the spectral density of the noise output is quadratic. As a consequence a channel associated with a higher subcarrier frequency will be *noisier* than a channel associated with a lower subcarrier frequency. That such is the case is easily seen in Fig. 9.6-2. Here we have plotted the positive half of the parabolic spectral density $G_{n_o}(f)$ of the output noise of the discriminator. Two subcarrier frequencies f_l and f_h are indicated, together with the frequency band occupied by the associated SSB signals (the upper sidebands are arbitrarily chosen). The noise power in each of the individual channels is proportional to its respective shaded area. If then the transmitter power is raised to a level where the signal-to-noise ratio is acceptable in the highest channel, the signal-to-noise ratio in the lower channels will be better than is required.

This variation in noise power from channel to channel may be corrected, and the entire system rendered thereby more efficient, by transmitting a carrier which is *phase-modulated* by the composite signal, rather than frequency-modulated. For under these circumstances we

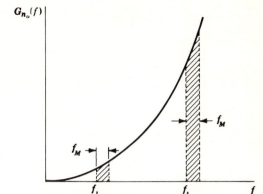

Fig. 9.6-2 To illustrate that in the multiplex system of Fig. 9.6-1, using FM, channels associated with high carrier frequencies are noisier than those associated with lower frequencies.

shall be required to use at the receiver a *phase demodulator*. A phase demodulator is a device whose instantaneous output signal is proportional to the instantaneous phase of its input signal. The student may readily verify that a discriminator followed by an integrator constitutes a phase demodulator. The important point, for our present purpose, about a phase demodulator is that, with such a demodulator, a uniform (white) noise spectral density at the input gives rise to a uniform output-noise spectral density. That such is the case is readily seen from Eq. (9.2-11). In Eq. (9.2-11), $\theta(t) = n_s(t)/A$ is the phase-modulation noise. Since $\theta(t)$ and $n_s(t)$ are directly related, the form of the power spectral density of each is identical. Hence, in summary, we have the result that if the composite signal $M(t)$ in Fig. 9.6-1 is transmitted (and necessarily received) as a phase-modulated signal, the noise power in each channel will be the same.

9.7 COMPARISON BETWEEN FM AND PM IN MULTIPLEXING

We noted in the preceding section that phase modulation offers some advantage over frequency modulation in a multichannel FDM system. We compare the two systems now, to see more quantitatively the advantage that accrues from the use of phase modulation.

A frequency-modulation system is shown in Fig. 9.7-1a and a phase-modulation system is indicated in Fig. 9.7-1b. We have chosen to construct the phase modulator as a frequency modulator preceded by a differentiator. That such an arrangement does indeed constitute a phase modulator was pointed out in Sec. 4.3. Similarly we have constructed a phase demodulator as a discriminator followed by an integrator. We may now, if we please, view the system of Fig. 9.7-1b as a frequency-

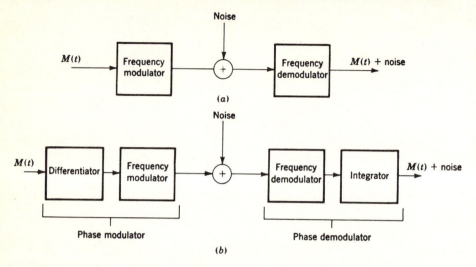

Fig. 9.7-1   Comparison of an FM system in (a) with a phase-modulation system in (b).

modulation system in which preemphasis and deemphasis have been incorporated. The differentiator is the preemphasis network, and the integrator is the deemphasis network. The combination of preemphasis and deemphasis leaves the signal unaltered; hence the output signal is the same in Fig. 9.7-1a and b. However, the noise power per channel is the same for each channel in (b), while in (a) the noise is most pronounced in the top channel. Assuming that both channels (a) and (b), are constrained to use the same bandwidth, we shall now compute the improvement afforded by channel (b) in the signal-to-noise ratio of the top channel.

 If there are N channels, the frequency range of the topmost channel of the composite signal $M(t)$ extends from $(N - 1)f_M$ to Nf_M, where f_M is the frequency range of an individual component signal. Applying Eq. (9.2-16) to the present case, we find that, in the absence of deemphasis, the noise output of the top channel is

$$N_{o,\text{top}} = 2\frac{\alpha^2\eta}{A^2}\int_{(N-1)f_M}^{Nf_M} 4\pi^2 f^2\, df \tag{9.7-1}$$

$$\simeq \frac{8\pi^2\alpha^2\eta N^2 f_M^3}{A^2}$$

since $N \gg 1$. The preemphasis (differentiator) circuit has a transfer function whose magnitude squared is

$$|H_p(f)|^2 = 4\pi^2\tau^2 f^2 \tag{9.7-2}$$

in which τ^2 is a constant.   For the deemphasis filter we have

$$|H_d(f)|^2 = \frac{1}{|H_p(f)|^2} = \frac{1}{4\pi^2\tau^2f^2} \tag{9.7-3}$$

Applying Eq. (9.4-1) to the composite signal $M(t)$, we find that the condition of equal bandwidth requires that

$$\int_{-Nf_M}^{+Nf_M} G_M(f)\ df = \int_{-Nf_M}^{+Nf_M} |H_p(f)|^2 G_M(f)\ df \tag{9.7-4}$$

In Eq. (9.7-4) $G_M(f)$ is the power spectral density of the composite signal. Assuming that $G_M(f)$ is constant, we find by substituting Eq. (9.7-2) into Eq. (9.7-4) that

$$\tau^2 = \frac{3}{4\pi^2 N^2 f_M^2} \tag{9.7-5}$$

In the presence of a deemphasis filter with transfer function $H_d(f)$ the output-noise power of the top channel becomes, from Eq. (9.7-1),

$$N_{od,\text{top}} = 2\,\frac{\alpha^2\eta}{A^2} \int_{(N-1)f_M}^{Nf_M} \frac{4\pi^2 f^2\ df}{|H_p(f)|^2} \tag{9.7-6}$$

The improvement factor $\mathfrak{R}$ for the top channel is calculated from Eqs. (9.7-1), (9.7-3), (9.7-5), and (9.7-6).   We find

$$\mathfrak{R}_{\text{top}} \equiv \frac{N_{o,\text{top}}}{N_{od,\text{top}}} = 3\ (= 4.8\text{ dB}) \tag{9.7-7}$$

9.8   EFFECT OF TRANSMITTER NOISE

The discussion of the preceding section is somewhat unrealistic in that it was assumed that the only noise with which we have to contend is noise introduced in the communications channel.   Actually, some noise is introduced by the transmitter itself.   For example, the frequency modulator may have some "jitter."   That is, there is some random variation of frequency even in the absence of a modulating signal.   Let us represent this jitter as due to a noise voltage present at the input to the frequency modulator, and let us assume for explicitness that the spectral density of this noise voltage is uniform.   If we now also assume that the spectral density of the composite baseband signal is also uniform, then the situation with the PM system can be represented as shown in Fig. 9.8-1a.   In this figure the composite input signal is $M(t)$ of spectral density $G_M(f)$, and the output of the amplifier-differentiator is $M'(t)$ of spectral density $G_{M'}(f)$.   The transmitter noise $n(t)$, with spectral density $G_n(f)$, is placed at the junction between the amplifier-differentiator and the frequency modulator.   The positive-frequency portion of the assumed *flat* power spectral density of $G_M(f)$ is shown in Fig. 9.8-1b,

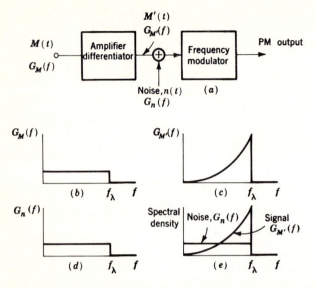

Fig. 9.8-1 (a) A PM system in which noise is introduced before transmission. (b) The spectral density of the system. (c) The spectral density of the signal after differentiation. (d) The spectral density of the noise. (e) Comparison of spectral densities of signal and noise at input to modulator.

while the parabolic density of $G_{M'}(f)$ appears in Fig. 9.8-1c. The noise spectral density is shown in Fig. 9.8-1d, while finally in Fig. 9.8-1e we compare the spectral densities of the signal and noise at the input to the frequency modulator.

Observe that at the lower frequencies the noise exceeds the signal. To keep the signal level above the noise at low frequencies, it is necessary to modify the transfer characteristic of the network which precedes the frequency modulator. A suitable network is one whose transfer characteristic increases with frequency to emphasize the high frequencies without suppressing the low frequencies. A network similar to the pre-emphasis circuit of Fig. 9.5-1b is suitable. In practice when such a pre-emphasis circuit is employed, together with its reciprocal deemphasis circuit, the 4.8-dB advantage quoted above for PM over FM is not realized. In a 1000-channel voice multiplex system the advantage is more nearly 3 or 4 dB.

PROBLEMS

9.1-1. In an FM receiver we find it advantageous to use an amplitude limiter. In the "amplitude-modulation" receivers discussed in Chap. 8 we did not find it advantageous to use a "frequency limiter." Discuss.

9.1-2. The AM-FM signal $[1 + m(t)] \cos(\omega_c t + \beta \sin \omega_M t)$, $|m(t)| < 1$ is ideally limited and then bandpass-filtered. If the bandwidth of the bandpass filter is selected to pass the fundamental frequency and to reject frequencies of the order of $2\omega_c$, $3\omega_c$, etc., and in addition if the filter bandwidth is also chosen to pass 98 percent of the signal energy contained in the fundamental and its sidebands, find:

(a) The bandwidth of the filter.

(b) The minimum ratio of ω_c/ω_M for a given β.

9.1-3. (a) The signal $v(t) = 2(1 + 0.1 \cos 2\pi \times 10^3 t) \cos 2\pi \times 10^6 t$ is passed through a filter which removes the lower sideband. Write an expression for the filter output $v_o(t)$ in a form which makes explicitly evident that $v_o(t)$ is an angle-modulated waveform with varying amplitude, that is

$$v_o(t) = A(t) \cos [2\pi \times 10^6 t + \theta(t)]\}$$

Find $A(t)$ and $\theta(t)$ and make reasonable approximations to simplify $v_o(t)$.

(b) The waveform $v_0(t)$ is applied to a network whose transfer function is $H(f) = j\sigma\omega$ (with $\sigma = 1/2\pi$ sec) in the frequency range $f_c = 10^6 \pm 10^3$ Hz and $H(f) = 0$ otherwise. The output of this network is then applied to an envelope demodulator. Write an expression for the output of the demodulator.

9.1-4. We note from Eq. (9.1-6) that, since $\alpha = \sigma A_L$, the output of the limiter-discriminator is $v_4 = \sigma A_L \, d\phi/dt$. Hence the more severe the limiting, i.e., the smaller A_L, the smaller will be the signal output of the limiter-discriminator. A student commented that limiting should be just barely adequate to remove amplitude variations in order to keep A_L as large as possible. Is this comment valid? Discuss.

9.2-1. For Eq. (9.2-21) make a plot of $(S_o/N_o)_{dB}$ against $(S_i/N_M)_{dB}$ for values of $\beta = 1, 5, 10, 100$. On the same set of axes include the corresponding plot for synchronous demodulation in a linear system (i.e., SSB or DSB-SC).

9.2-2. Commercial FM transmission is allocated a bandwidth $B = 200$ kHz. Assume a receiver with a rectangular IF filter of corresponding bandwidth. If the discriminator output is applied directly to a baseband filter with $f_M = 15$ kHz, what fraction of the noise output of the discriminator passes through the filter?

9.2-3. Consider two frequency-modulated signals, the first where

$$km(t) = \beta\omega_M \cos \omega_M t$$

and the second where $km(t)$ is equal to a gaussian signal, which is white with power spectral density $\eta_m/2$, $|f| \le f_M$, and zero elsewhere. If the output SNR of each system is the same, calculate the ratio of the IF bandwidths needed to pass 98 percent of the signal energy in each case.

9.2-4. If the input SNR $S_i/N_M = 30$ dB, calculate S_o/N_o when

$$km(t) = \beta\omega_M \cos \omega_M t$$

and $\beta = 1, 5, 10, 100$.

9.2-5. In an FM system the baseband bandwidth is f_M. The modulation is sinusoidal and the bandwidth B is to be kept constant. Write an expression for γ_{FM}, defined in Eq. (9.2-22) as a function of the modulation index, in terms of B and f_M.

9.2-6. The input SNR $S_i/N_M = 30$ dB, and an $S_o/N_o = 48$ dB is required. Find the mean-squared frequency deviation in terms of the frequency range, f_M.

9.3-1. Verify Eq. (9.3-1).

9.3-2. Derive Eq. (9.3-2).

9.3-3. A baseband signal $m(t)$ whose spectrum extends to $f_M = 5$ kHz is transmitted by FM. The rms frequency deviation produced by $m(t)$ is 100 kHz. The same signal $m(t)$ is also transmitted by SSB. If the output signal-to-noise ratios for FM and SSB are to be the same, compare the input signal-to-noise ratios for the two types of transmission.

9.4-1. The power spectral density of a modulating signal $m(t)$ is given by

$$G_m(f) = \frac{G_0}{1 + (f/f_1)^2}$$

where $f_1 \ll f_M$. If a preemphasis circuit is to be used, where $|H_p(f)|^2 = K^2 f^2$, find K^2 if the preemphasis is not to increase the bandwidth.

9.4-2. If $G_m(f) = \eta_m/2$, $|f| \le f_M$ and zero elsewhere,

(a) Find $\int_{-f_M}^{f_M} |H_p(f)|^2 \, df$.

(b) If $|H_p(f)|^2 = k_1^2 + k_2^2 f^2$, find k_1 and k_2 to maximize $\Re$, subject to the constraint of part a of this problem.

9.4-3. Consider that a carrier is FM-modulated by a baseband waveform $m(t)$. Let the bandwidth occupied by the FM waveform be defined as in Eq. (4.14-1):

$$B = 2 \left[\frac{\int_{-\infty}^{\infty} \nu^2 G(\nu) \, d\nu}{\int_{-\infty}^{\infty} G(\nu) \, d\nu} \right]^{1/2}$$

(a) Show that, in this case, the bandwidth is proportional to the normalized power of $m(t)$, i.e., $B^2 = c^2 \overline{m^2(t)}$, with c^2 a constant.

(b) Let $m(t)$ have a normalized power of 2 volts^2 and suppose that the sensitivity of the FM modulator is such that a change by 1 volt in $m(t)$ produces a frequency change of 10 kHz. Find the bandwidth B.

9.5-1. Preemphasis is to be used in conjunction with a DSB-SC system. The spectral density of the baseband signal, of spectral range f_M, is as given in Eq. (9.5-4), and the preemphasis filter has the transfer function given in Eq. (9.5-5). Assume $f_1 = f'$. Impose the constraint that the preemphasis is not to increase the transmitted power. Calculate the preemphasis improvement $\Re$ and compare with Eq. (9.5-9).

9.5-2. (a) In a stereo FM system the baseband spectral range is $f_M = 15$ kHz. Assume that preemphasis is not used. Find the ratio of the noise output power to the noise power for monophonic transmission. See Fig. 4.23-1.

(b) Taking preemphasis into account as described in Sec. 9.5, show that stereo FM is noisier than monophonic FM.

9.5-3. Plot Eqs. (9.5-8) and (9.5-9) to determine when Eq. (9.5-9) can be employed.

9.5-4. If, in Eq. (9.5-8), $f_1 = 2.1$ kHz and $f_M = 15$ kHz, calculate $\mathcal{R}$. Compare your result with that given in the text which was obtained using Eq. (9.5-9).

9.7-1. (a) One thousand signals are multiplexed by the system of Fig. 9.6-1. The angle modulator is a frequency modulator, i.e., the composite baseband signal $M(t)$ frequency modulates the carrier. The lowest SSB carrier frequency is 10 kHz and each baseband signal is allocated 4 kHz. Assume that each baseband signal has the same normalized power. The channel with the least noise turns out to have a signal-to-noise ratio of 80 dB. If a signal-to-noise ratio lower than 20 dB were not acceptable, how many channels of the system would be usable.

(b) Assume instead that $M(t)$ phase-modulates the carrier. Assume also that the total transmitted power is the same as in part (a). Calculate the signal-to-noise ratio in any channel.

9.7-2. A 4-MHz TV signal, and one thousand 4-kHz audio signals, are multiplexed onto a single FM carrier (the audio signals are SSB-modulated to obtain this goal, the TV signal is left at baseband and is therefore channel 1). The power spectral density of the composite signal is 3 W/Hz over its entire spectral range.

(a) Find the spectral range of the composite signal.

(b) Calculate the output SNR for channel 1, the TV signal, in terms of the input SNR.

(c) Calculate the output SNR for the top channel.

REFERENCE

1. Sakrison, D.: "Communication Theory," John Wiley & Sons, Inc., New York, 1968.

10

Threshold in Frequency Modulation

We have seen (Secs. 8.6 and 8.7) that the nonlinear AM demodulators, such as the square-law demodulator and the envelope demodulator, exhibit a threshold. The FM discriminator also exhibits a threshold, and in this chapter we shall discuss the mechanism which is responsible for such an FM threshold. We shall also discuss two demodulator circuits which have lower thresholds than the discriminator. These circuits are the *phase-locked loop* (PLL) and the *frequency demodulator with feedback* (FMFB).

10.1 THRESHOLD IN FREQUENCY MODULATION

In a communication system in which the modulation is linear and demodulation is accomplished by coherent detection (for example, SSB and DSB-SC), we have the result that [see Eqs. (8.5-11) and (8.5-12)]

$$\frac{S_o}{N_o} = \frac{S_i}{N_M} \tag{10.1-1}$$

or equivalently

$$10 \log \frac{S_o}{N_o} \equiv \left[\frac{S_o}{N_o}\right]_{\text{dB}} = \left[\frac{S_i}{N_M}\right]_{\text{dB}} \equiv 10 \log \frac{S_i}{N_M} \qquad (10.1\text{-}2)$$

In FM we have [see Eq. (9.2-21)]

$$\frac{S_o}{N_o} = \frac{3}{2} \beta^2 \frac{S_i}{N_M} \qquad (10.1\text{-}3)$$

or equivalently

$$\left[\frac{S_o}{N_o}\right]_{\text{dB}} = \left[\frac{S_i}{N_M}\right]_{\text{dB}} + 10 \log \frac{3}{2} \beta^2 \qquad (10.1\text{-}4)$$

In Fig. 10.1-1 we have plotted Eq. (10.1-2) in a coordinate system in which the coordinate axes are marked off in decibels. The plot is a straight line passing through the origin. We have also plotted Eq. (10.1-4) for two values of $\beta(> \sqrt{2/3})$. These plots, as indicated by the dashed extensions,

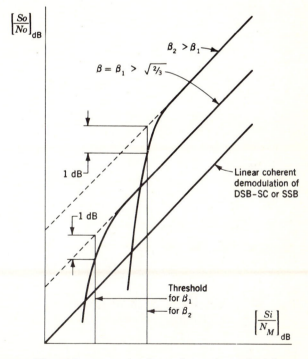

Fig. 10.1-1 Plots of output signal-to-noise ratio against input signal-to-noise ratio for linear modulation and demodulation and also for an FM system. Illustrating the phenomenon of threshold in FM.

are also straight lines.   In the FM case a plot for a value of β is raised by the amount $10 \log (3\beta^2/2)$ above the plot for a linear coherent modulation system.   The quantity $10 \log (3\beta^2/2)$ expresses in decibels precisely the improvement afforded by the FM system in return for a sacrifice of bandwidth.

Experimentally it is determined, however, that the FM system exhibits a threshold.   Thus, as indicated by the solid-line plots, for each value of β, as S_i/N_M decreases, a point is reached where S_o/N_o falls off much more sharply than S_i/N_M.   The threshold value of S_i/N_M is arbitrarily taken to be the value at which S_o/N_o falls 1 dB below the dashed extension.   We note that for larger β the threshold S_i/N_M is also higher. Suppose, then, that we are operating with a modulation index β_1 above the threshold for β_1 but below the threshold for β_2.   Suppose, further, that at this input S_i/N_M we should increase β from β_1 to β_2 hoping thereby to improve the output-signal-to-noise ratio S_o/N_o by a sacrifice of bandwidth.   We would find, however, as is apparent from Fig. 10.1-1, that such a sacrifice of bandwidth would actually decrease the output SNR. Similarly we see that, if we are operating sufficiently below threshold for any value of β, we do better with a linear coherent system than with an FM system.

The onset of threshold may be observed by examining the noise output of an FM discriminator on a cathode-ray oscilloscope.   At high input-signal-to-noise ratios (S_i/N_M), the noise displays the usual random-variation characteristic of thermal noise generally.   A typical output-noise waveform has the appearance shown in Fig. 10.1-2.   We can deter-

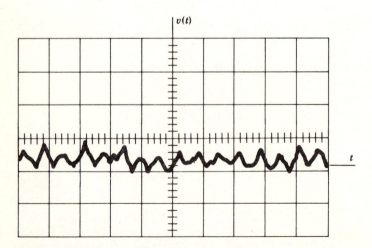

Fig. 10.1-2   Thermal noise at discriminator output.

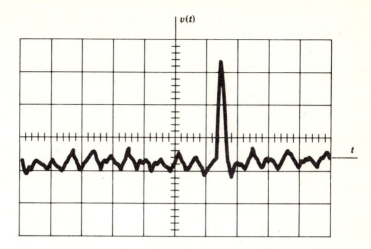

Fig. 10.1-3 A spike superimposed on a background of smooth (thermal) noise.

mine experimentally that the instantaneous noise amplitude has a gaussian probability density as is to be anticipated from the discussion of Sec. 7.3.   As S_i/N_M decreases, a point is reached where the character of the noise waveform changes markedly.   The noise now has the appearance indicated in Fig. 10.1-3.   Here we observe, superimposed on the background thermal-type noise (usually referred to as *smooth* noise), a pulse-type noise waveform.   Because of its appearance this new component of noise is often referred to as *spike* noise or *impulse* noise.   If we were to listen to the discriminator, we would hear a clicking sound on each occasion that we observed a spike.

The appearance of these spikes denotes the onset of threshold.   In succeeding sections we shall discuss the origin of these spikes.   We shall show that although the frequency of occurrence of a spike is small, the noise energy associated with a spike is very large compared with the energy of the smooth noise occurring during a comparable time interval. Hence the *spike* noise greatly increases the total noise output and thereby causes a threshold.

10.2 OCCURRENCE OF SPIKES

An FM demodulator consisting of an IF filter, limiter-discriminator, and a baseband filter is shown in Fig. 10.2-1.   The IF filter has a bandwidth B, centered at the IF frequency f_c.   The baseband filter extends from $-f_M$ to $+f_M$ but does not pass dc.   In order to simplify our discussion, we assume that the input signal is an unmodulated carrier; that is, $m(t) = 0$,

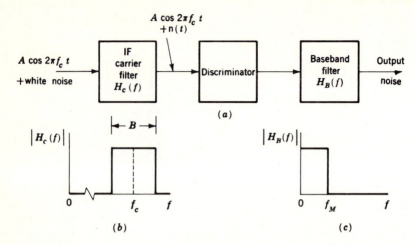

Fig. 10.2-1 (*a*) An FM discriminator and associated filters. (*b*) The bandpass range of the carrier filter. (*c*) The passband of the baseband filter.

accompanied by white noise. The white noise is filtered and shaped by the IF filter.

In the present and in succeeding sections we shall have occasion to use the quadrature component representation of noise as expressed in Eq. (7.11-2). It will be convenient to use a notation different from the notation of Eq. (7.11-2). We shall replace $n_c(t)$ by $x(t)$ and $n_s(t)$ by $y(t)$. In this alternative notation, which is as commonly found in the literature as is the original notation, Eq. (7.11-2) reads, with $2\pi f_c$ replaced by ω_c,

$$n(t) = x(t) \cos \omega_c t - y(t) \sin \omega_c t \qquad (10.2\text{-}1)$$

The notation of Eq. (10.2-1) is especially appropriate when $n(t)$ is to be represented as a combination of phasors in a coordinate system rotating counterclockwise with angular frequency ω_c. In such a coordinate system $n(t)$ is represented by the phasor sum of $x(t)$ in the horizontal direction (i.e., the x direction) and $y(t)$ in the vertical direction (i.e., the y direction). In terms of this new notation, the carrier and noise output of the IF filter, which is the input $v_i(t)$ to the demodulator, is given by

$$v_i(t) = A \cos \omega_c t + x(t) \cos \omega_c t - y(t) \sin \omega_c t \qquad (10.2\text{-}2)$$

This equation can be rewritten in phasor notation as

$$v_i(t) = \text{Re} \left\{ [A + x(t)]e^{j\omega_c t} + y(t)e^{j[\omega_c t + (\pi/2)]} \right\}$$
$$= \text{Re} \left\{ \underbrace{e^{j\omega_c t}[A + x(t) + jy(t)]}_{\text{phasors}} \right\} \qquad (10.2\text{-}3)$$

The phasor $A + x(t)$ lies along the horizontal axis and $y(t)$ lies along the vertical axis.

It is convenient, when discussing the occurrence of spikes, to talk about the *amplitude* of the noise $r(t) = \sqrt{x^2 + y^2}$ and its random phase angle $\varphi(t) = \tan^{-1} y(t)/x(t)$. [It was shown in Sec. 2.14 that the density of $r(t)$ is Rayleigh and that the density of $\varphi(t)$ is $1/2\pi$ from $-\pi$ to $+\pi$.]

The phasor diagram for $v_i(t)$ is shown in Fig. 10.2-2. Here we see that the phasor sum of signal A and noise $r(t)$ is defined as $R(t)$. The angle that $R(t)$ makes with the horizontal axis is called $\theta(t)$. Then

$$A + x + jy = A + r(t)e^{j\varphi(t)} = R(t)e^{j\theta(t)} \tag{10.2-4}$$

and that $v_i(t)$ in Eq. (10.2-2) is

$$v_i(t) = \text{Re } [e^{j\omega_c t}R(t)e^{j\theta(t)}] = R(t) \cos [\omega_c t + \theta(t)] \tag{10.2-5}$$

As was discussed in Sec. 9.2, the output of the discriminator is proportional to $\dot{\theta}(t) = d\theta/dt$.

Let us now consider Fig. 10.2-2a to see how $\theta(t)$, and hence $\dot{\theta}(t)$, are affected as the noise $r(t)$ and $\varphi(t)$ vary. If the noise power is small in comparison with the carrier power, we expect that $r(t) \ll A$ most of the time, and that the end point of the resultant phasor $R(t)$ will never wander far from the end point of the carrier phasor. Thus, as $\varphi(t)$ changes, the angle $\theta(t)$ remains small.

If, however, the ratio $S_i/\eta f_M$ decreases, the likelihood of $r(t)$ being much less than A also decreases. When $r(t)$ becomes comparable in magnitude to the carrier amplitude A, the locus of the end point of the resultant phasor $R(t)$ moves away from the end point of the carrier phasor and

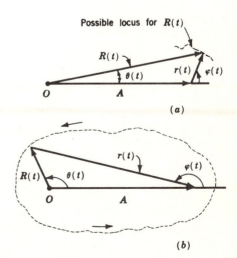

Fig. 10.2-2 (a) A noise phasor $r(t)$ is added to a carrier phasor of amplitude A. The sum is the resultant phasor $R(t)$. (b) A locus for the end point of $R(t)$ which will give rise to a spike.

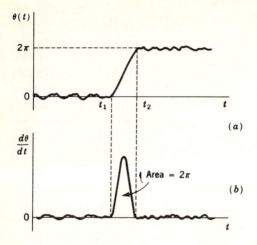

Fig. 10.2-3 (*a*) A plot of $\theta(t)$ for a case in which the end point of $R(t)$ in Fig. 10.2-2*b* executes a rotation around the origin. (*b*) A plot of $d\theta/dt$ as a function of time.

may, as shown in Fig. 10.2-2*b*, even rotate about the origin.   The variation of the phase angle $\theta(t)$, at and near the time of occurrence of this event, is shown in Fig. 10.2-3*a*.   If the rotation of the end point of $R(t)$ around the origin occurs in the interval between t_1 and t_2, the angle $\theta(t)$ changes by 2π rad during this time interval.   Preceding t_1 and following t_2, $r(t) \ll A$, and the usual small random variations of $\theta(t)$ occur.

We are interested in the discriminator output, which is proportional to the instantaneous frequency $d\theta/dt$, and we have therefore plotted $d\theta/dt$ in Fig. 10.2-3*b*.   Notice that when $\theta(t)$ changes by 2π rad, $d\theta/dt$ appears as a sharp *spike* or *impulse* with area 2π.   To show that the area under the spike waveform is indeed 2π, we simply integrate $d\theta/dt$ over the time interval t_1 to t_2 during which $\theta(t)$ changes by 2π.   Thus

$$\text{Area} = \int_{t_1}^{t_2} \frac{d\theta}{dt}\, dt = \theta \Big|_{t_1}^{t_2} = 2\pi \tag{10.2-6}$$

The waveform shown in Fig. 10.2-3*b* is the *spike* noise referred to in Sec. 10.1 and is the waveform which is presented to the input of the baseband filter.   Since these spikes occur only rarely and are impulse-like in character, they represent a *shot* noise phenomenon.   The noise power at the baseband filter output is calculated in Sec. 10.4, using the results given in Sec. 7.14.

10.3   SPIKE CHARACTERISTICS

When the noise amplitude $r(t)$ is comparable to the carrier amplitude A, the resultant phasor $R(t)$ may clearly execute all sorts of random, wide-ranging excursions which will cause $\theta(t)$, and hence the frequency $d\theta/dt$,

to experience large changes.   Why then do we single out for special consideration the excursion which carries the resultant in a complete rotation
around the origin?   The reason for our special concern may be seen by
comparing the noise outputs for the two cases shown in Fig. 10.3-1a.
The path of $R(t)$, marked *spike path*, carries the end point of $R(t)$ completely around the origin as in Fig. 10.2-2b and results in a waveform for
$\theta(t)$ as shown in Fig. 10.2-3a and in a waveform for the discriminator
output as shown in Fig. 10.2-3b.   The second path, marked *triplet path*,
departs from the spike path only slightly, but, most importantly, it does

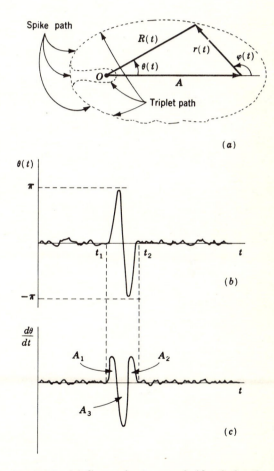

Fig. 10.3-1   (a) Comparing two nearly identical paths
for the end point of $R(t)$.   One path encircles the
origin, the other does not.   (b) and (c) $\theta(t)$ and $d\theta/dt$
for the path which does not encircle the origin.

not encircle the origin. For this path, $\theta(t)$ and the discriminator output appear as shown in Fig. 10.3-1b and c. Notice that the angle $\theta(t)$ increases nearly to π, reverses to nearly $-\pi$, and then returns to a small value. The waveform of $\theta(t)$ displays, not a simple jump of 2π, but rather a pulse doublet. The waveform of $d\theta/dt$ displays not a single pulse but rather three pulses, two of positive and one of negative polarity. It is most important to recognize that the total algebraic area under the waveform in Fig. 10.3-1c is zero. That is, the sum of the positive areas $A_1 + A_2$ is equal to the negative area A_3. This can be seen by applying the calculation of Eq. (10.2-6) to the determination of the total area and noting that the total change in θ between t_1 and t_2 is zero.

We can now illustrate how, in an FM system, the noise energy present at the output of the baseband filter shown in Fig. 10.2-1 is much larger when a spike is generated than when a triplet occurs. We have already noted that if a carrier, accompanied by noise, is passed through a bandpass filter of bandwidth B, the spectral components of the noise output of the discriminator will extend over the frequency range from $-B/2$ to $B/2$. Of this noise, only the part which lies in the range $-f_M$ to f_M will appear at the output of the baseband filter. In Fig. 10.3-2a and b we have sketched the magnitudes of the Fourier transforms: $F_S(j\omega)$, the spike waveform shown in Fig. 10.2-3b, and $F_T(j\omega)$, the triplet waveform shown in Fig. 10.3-1c. We have confined the spectral range to the interval $-B/2$ to $+B/2$ as required, and have indicated that for the triplet $F_T(j\omega) = 0$ at $f = 0$, while for the spike $F_S(j\omega) \neq 0$ at $f = 0$. These features of the transforms, at $f = 0$, are verified in the following way. The Fourier transform of a function $f(t)$ is

$$F(j\omega) = \int_{-\infty}^{\infty} f(t)e^{-j\omega t}\, dt \tag{10.3-1}$$

so that

$$[F(j\omega)]_{\omega=0} = \int_{-\infty}^{\infty} f(t)\, dt = \text{area under } f(t) \tag{10.3-2}$$

Thus, the value of the transform at $f = 0$ is equal to the area under $f(t)$. As already noted, the spike has a finite area, while the triplet area is zero.

Since the passband of the baseband filter extends only to $f_M \ll B/2$, only the energy represented by the shaded areas in Fig. 10.3-2a and b will appear at the filter output. It is apparent from this figure that a spike will contribute much more noise than the triplet. More generally, when the resultant phasor executes any path that carries it around the origin, the corresponding noise energy output will be much larger than any other excursion of comparable wide range which does not succeed in causing an encirclement of the origin.

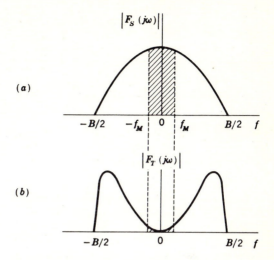

Fig. 10.3-2 Qualitative plots of the Fourier transform of a pulse-type waveform with spectral components limited to range $-B/2$ to $+B/2$. (*a*) For case where waveform has dc component. (*b*) For case where dc component is zero.

SPIKE DURATION

We may make an order-of-magnitude estimate of the duration of the spike. For this purpose we use the general principle, discussed in Sec. 1.14, that when a pulse is made up of spectral components, confined mainly to a frequency range 0 to f_{max}, the time duration of the pulse is of the order of $1/f_{max}$. The spike output of the discriminator has a spectrum that extends over the range 0 to $B/2$, where B is the bandpass of the carrier filter. Hence, we estimate that the spike duration is approximately $2/B$ sec.

10.4 CALCULATION OF THRESHOLD IN AN FM DISCRIMINATOR

To calculate the spike noise contribution at the output of an FM discriminator, we need to know the power spectral density of the spike noise. The spike noise is similar to the shot noise discussed in Sec. 7.14. We may therefore use Eq. (7.14-14), which states that the power spectral density is given by

$$G_S(f) = \frac{1}{T_s} \overline{|P(j\omega)|^2} \tag{10.4-1}$$

where $P(j\omega)$ is the Fourier transform of the spike, and T_s is the mean time interval between spikes. From the discussion of Sec. 10.3 we know that $|P(j\omega)|$ will have the general form shown in Fig. 10.3-2a. For the purpose of calculating noise power at the output of the baseband filter, we need know $G_S(f)$ only in the range $-f_M$ to $+f_M$. In this range $|P(j\omega)|$ does not change very much, provided $B/2 \gg f_M$. We may therefore assume that $|P(j\omega)|$ is constant and equal to $|P(0)|$, that is, to the value of the Fourier transform at $f = 0$.

Let $\theta(t)$ in Fig. 10.3-1 make a rotation by 2π so that a spike is formed. Then the waveform of the output-noise voltage spike from the discriminator is

$$[n(t)]_{\text{spike}} = \alpha \frac{d\theta}{dt} \tag{10.4-2}$$

where α is a constant, i.e., the discriminator constant. From Eq. (10.3-2) $|P(0)|$ is equal to the area of $[n(t)]_{\text{spike}}$. Since α is a constant and the area under the waveform $d\theta/dt$ is 2π, we have

$$|P(0)| = 2\pi\alpha \tag{10.4-3}$$

Since each spike has the same area, 2π, the average value of $|P(0)|$ is also given by Eq. (10.4-3), and from Eq. (10.4-1)

$$G_S(f) = \frac{4\pi^2\alpha^2}{T_s} \tag{10.4-4}$$

Thus the total output-noise power due to the spikes is simply

$$N_S = \frac{4\pi^2\alpha^2}{T_s} 2f_M \tag{10.4-5}$$

The expression for N_S given in Eq. (10.4-5) is not very useful until we learn how T_s depends on the parameters of the system. It is proven in Sec. 10.5 that, for an unmodulated carrier,

$$\frac{1}{T_s} = \frac{B}{2\sqrt{3}} \text{ erfc} \sqrt{\frac{f_M}{B} \frac{S_i}{N_M}} \tag{10.4-6}$$

where erfc is the complimentary error function, i.e.,

$$\text{erfc } x \equiv \frac{2}{\sqrt{\pi}} \int_x^\infty e^{-\lambda^2} d\lambda = 1 - \frac{2}{\sqrt{\pi}} \int_0^x e^{-\lambda^2} d\lambda \tag{10.4-7}$$

From Eqs. (10.4-5) and (10.4-6) we have

$$N_S = \frac{4\pi^2\alpha^2 B f_M}{\sqrt{3}} \text{ erfc} \sqrt{\frac{f_M}{B} \frac{S_i}{N_M}} \tag{10.4-8}$$

The total output-noise power is $N_o = N_G + N_S$, where N_G is the gaussian (smooth) noise given by Eq. (9.2-16), and the spike noise N_S is given by

Eq. (10.4-8). The signal power is given by Eq. (9.2-5). Combining these equations, we have, with $S_i = A^2/2$, and $N_M = \eta f_M$,

$$\frac{S_o}{N_o} = \frac{[3k^2\overline{m^2(t)}/4\pi^2 f_M^2](S_i/N_M)}{1 + (\sqrt{3}B/f_M)(S_i/N_M)\ \mathrm{erfc}\ \sqrt{(f_M/B)(S_i/N_M)}} \qquad (10.4\text{-}9)$$

The threshold effect is evident in Eq. (10.4-9). At very high S_i/N_M the complementary error function approaches zero. The right-hand member of the denominator of Eq. (10.4-9) becomes zero, and S_o/N_o varies linearly with S_i/N_M. As S_i/N_M decreases, the denominator increases above unity, and S_o/N_o decreases at a more rapid rate than does S_i/N_M.

We shall see in Sec. 10.6 that the spike output of a discriminator is very much larger when the input carrier is modulated than when the carrier is not modulated. In deriving Eq. (10.4-8), we have used Eq. (10.4-6) which applies only in the absence of modulation, or when the carrier is frequency-modulated to a very slight extent. Thus Eq. (10.4-9) is valid only when the bandwidth occupied by the modulated signal is very small in comparison with the bandwidth B of the IF carrier filter.

10.5 CALCULATION OF MEAN TIME BETWEEN SPIKES

In this section we shall calculate the mean time between spikes. We shall derive Eq. (10.4-6), which was used in the preceding section in our calculation of S_o/N_o near threshold.

A phasor diagram of the carrier and noise at the input to the discriminator is shown in Fig. 10.5-1. The input noise is expressed in terms

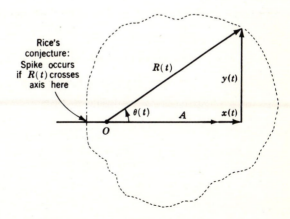

Fig. 10.5-1 Phasor diagram using quadrature noise components.

of the quadrature components $x(t)$ and $y(t)$. In Secs. 10.2 and 10.3 we showed that a spike resulted in the discriminator output when $\theta(t)$ rotated 2π rad. If the rotation was counterclockwise so that $\theta(t)$ *increased* by 2π the spike was positive, and if the rotation was clockwise so that $\theta(t)$ *decreased* by 2π the spike was negative.

We shall assume that if the end point of $R(t)$ has crossed the horizontal axis so that $\theta(t)$ *passes through* π rad, then $\theta(t)$ will continue to increase (or decrease) giving rise to a positive (or negative) spike. Thus we assume that to ensure a spike output, it is not actually necessary to observe a complete rotation of $R(t)$, but it is adequate that there be guaranteed a rotation of at least π rad. If, then, at an instant of time $t = t_1$,

$$x(t_1) < -A \tag{10.5-1a}$$

$$y(t_1) = 0 \tag{10.5-1b}$$

and

$$\dot{y}(t_1) < 0 \tag{10.5-1c}$$

then $\theta(t)$ has *increased* through π rad, and a positive spike will result. Similarly if, at an instant of time $t = t_2$,

$$x(t_2) < -A \tag{10.5-2a}$$

$$y(t_2) = 0 \tag{10.5-2b}$$

and

$$\dot{y}(t_2) > 0 \tag{10.5-2c}$$

then $\theta(t)$ has *decreased* through π rad, and a negative spike results. This hypothesis is due to Rice[1] and has been verified experimentally in a variety of different circumstances.[2] We shall employ these results, Eqs. (10.5-1) and (10.5-2), to determine the probability of occurrence of a spike.

Let us consider an interval of time Δt and calculate the probability P_- of a negative spike occurring during this time interval. This probability is the probability that the conditions given by Eq. (10.5-2) are satisfied at some instant t_2 within the time interval Δt. These conditions are illustrated in Fig. 10.5-2. In this figure we see $R(t)$ and $\theta(t)$ just after $\theta(t)$ has *decreased* through π, which is equivalent to $y(t)$ *increasing* through zero. It is this *increase* which results in $\dot{y}(t_2) > 0$.

Thus, we write

$$P_- = P\left[x(t_2) < -A,\ y(t_2) = 0,\ \frac{dy}{dt}\bigg|_{t_2} > 0 \right] \tag{10.5-3}$$

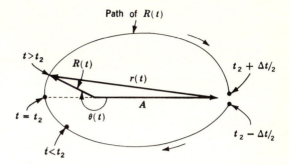

Fig. 10.5-2   Locus of $R(t)$ and $\theta(t)$ to cause a negative spike.

In Sec. 7.13 we determined the joint probability density of x, $\dot{x}$, y, and $\dot{y}$. In the present instance we need only the joint density of x, y, and $\dot{y}$. Using Eq. (7.13-5), we have, after replacing n_c and n_s by x and y, respectively

$$f(x,y,\dot{y}) = \overbrace{\frac{e^{-x^2/2\eta B}}{\sqrt{2\pi\eta B}}}^{f(x)} \; \overbrace{\frac{e^{-y^2/2\eta B}}{\sqrt{2\pi\eta B}}}^{f(y)} \; \overbrace{\frac{e^{-\dot{y}^2/(2\pi^2 B^3\eta/3)}}{\sqrt{2\pi^3 B^3\eta/3}}}^{f(\dot{y})} \tag{10.5-4}$$

and

$$P_- = \int_{-\infty}^{-A} dx \int_0^{\infty} d\dot{y} \int_{-\Delta y/2}^{\Delta y/2} dy \; f(x,y,\dot{y}) \tag{10.5-5}$$

Before integrating Eq. (10.5-5), let us investigate the limits on the integrals to verify that Eq. (10.5-5) really represents Eq. (10.5-3).   First we see that $\int_{-\infty}^{-A} dx\, f(x)$ will yield the probability that $x(t_2) < -A$. Next, we note that $\int_{-\Delta y/2}^{\Delta y/2} dy\, f(y) = f(0)\,\Delta y$ gives the probability of $y(t_2)$ passing through 0.   Finally we have $\int_0^{\infty} d\dot{y}\, f(\dot{y})$, which is the probability of $\dot{y}(t_2)$ being greater than zero.   These are exactly the conditions of Eq. (10.5-3).

We now evaluate Eq. (10.5-5).   The first integral that we determine is

$$\int_{-\Delta y/2}^{\Delta y/2} dy\, f(y) = f(0)\,\Delta y \tag{10.5-6}$$

Referring to Eq. (10.5-4), we see that $f(0) = 1/\sqrt{2\pi\eta B}$.   We next consider the factor Δy.   This is the differential change in y occurring during a differential change in time Δt.   Thus, we write $\Delta y = (\Delta y/\Delta t)\,\Delta t$. Since Δy and Δt are differential quantities, $\Delta y/\Delta t$ can be approximated by dy/dt. Equation (10.5-5) now becomes

$$P_- = \int_{-\infty}^{-A} dx\, f(x) \int_0^\infty d\dot{y}\, f(\dot{y}) \frac{\Delta t}{\sqrt{2\pi B}}\, \dot{y} \tag{10.5-7}$$

The next integral to be evaluated is

$$\int_0^\infty d\dot{y}\, f(\dot{y})\dot{y} = \int_0^\infty d\dot{y}\, \dot{y}\, \frac{e^{-\dot{y}^2/(2\pi^2 B^3\eta/3)}}{\sqrt{2\pi^3 B^3\eta/3}} \tag{10.5-8}$$

This integral is easily evaluated after changing the variable of integration to

$$\lambda = \frac{\dot{y}^2}{2\pi^2 B^3\eta/3} \tag{10.5-9}$$

Then

$$d\lambda = \frac{\dot{y}\, d\dot{y}}{\pi^2 B^3\eta/3} \tag{10.5-10}$$

Thus, Eq. (10.5-8) becomes

$$\sqrt{\frac{\pi B^3\eta}{6}} \int_0^\infty d\lambda\, e^{-\lambda} = \sqrt{\frac{\pi B^3\eta}{6}} \tag{10.5-11}$$

Equation (10.5-7) now becomes

$$\begin{aligned} P_- &= \int_{-\infty}^{-A} dx\, f(x) \left(\sqrt{\frac{\pi B^3\eta}{6}} \frac{\Delta t}{\sqrt{2\pi\eta B}} \right) \\ &= \frac{B\,\Delta t}{\sqrt{12}} \int_{-\infty}^{-A} dx\, \frac{e^{-x^2/2\eta B}}{\sqrt{2\pi\eta B}} \end{aligned} \tag{10.5-12}$$

Again, we make a change in variable of integration. This time we set

$$\lambda = \frac{-x}{\sqrt{2\eta B}} \tag{10.5-13}$$

Then

$$P_- = \frac{B\,\Delta t}{4\sqrt{3}} \int_{\frac{A}{\sqrt{2\eta B}}}^\infty \frac{2}{\sqrt{\pi}}\, d\lambda\, e^{-\lambda^2} = \frac{B\,\Delta t}{4\sqrt{3}} \operatorname{erfc} \sqrt{\frac{A^2}{2\eta B}} \tag{10.5-14}$$

The input-signal power is $S_i = A^2/2$, while the input noise in the baseband bandwidth is $N_M = \eta f_M$. Hence we may rewrite Eq. (10.5-14) in the form

$$P_- = \left(\frac{B}{4\sqrt{3}} \operatorname{erfc} \sqrt{\frac{f_M}{B} \frac{S_i}{N_M}} \right) \Delta t \tag{10.5-15}$$

In one second there are $1/\Delta t$ intervals in which a spike might occur.

Since P_- is the probability of occurrence of a negative spike within any such individual interval, the expected number of pulses in a second, N_-, is

$$N_- = \frac{P_-}{\Delta t} \tag{10.5-16}$$

Hence, N_- is found from Eqs. (10.5-15) and (10.5-16) to be

$$N_- = \frac{B}{4\sqrt{3}} \operatorname{erfc} \sqrt{\frac{f_M}{B} \frac{S_i}{N_M}} \tag{10.5-17}$$

From the symmetry displayed in Fig. 10.5-2 we see that the average number of positive spikes N_+ is the same as the average number of negative spikes N_-. Thus, the total number of spikes occurring per second in the presence of a carrier alone (no modulation) is, on the average,

$$N_c = N_- + N_+ = 2N_- = 2N_+ \tag{10.5-18}$$

The average time between spikes T_s is, therefore,

$$T_s = \frac{1}{N_c} = \frac{2\sqrt{3}}{B \operatorname{erfc} \sqrt{(f_M/B)(S_i/N_M)}} \tag{10.5-19}$$

10.6 EFFECT OF MODULATION

When the noise spectral components swing the resultant phasor in Fig. 10.2-2 around the origin in the counterclockwise direction, a positive spike occurs. Similarly, a clockwise rotation produces a negative spike. When the carrier frequency is at the center frequency of the carrier filter, the noise spectral components are symmetrically located in frequency above and below the carrier. In this case, because of the symmetry, positive and negative spikes are equally likely.

Now, however, let us assume that we have *offset* the carrier so that it lies just within the passband of the carrier filter at the lower-frequency limit of the passband. Let us consider a new coordinate system in which the phasor for this offset carrier is at rest. This offset phasor is accompanied by noise spectral components almost all of which are at higher frequency than the carrier and extend over a frequency range, with respect to the carrier, of B rather than $B/2$ as was the case when the carrier was centrally located. In our new coordinate system all the noise spectral components rotate counterclockwise. Hence only positive-frequency spikes can occur. Further, since the noise causing the spike has a spectral range B rather than $B/2$ (see Prob. 10.6-1), the spikes will be narrower than in the case of the symmetrically located carrier. Since the spike area is fixed, the spike amplitude must be larger. Finally, because of the higher relative frequencies of the noise components with respect to the offset carrier, whatever is going to happen will happen more frequently.   Hence the rate of occurrence of spikes will increase.

Suppose we use a cathode-ray oscilloscope to examine the output waveform of the discriminator of Fig. 10.2-1.   Let us assume that polarities have been adjusted so that a frequency increase in the carrier produces a positive deflection on the scope.   In the case where the carrier is centrally located in the carrier filter passband, the scope trace will display the output noise with an equal frequency of occurrence of positive and negative spikes.   As the carrier frequency is lowered, the trace will move downward, and the number of positive and negative spikes will become asymmetrical.   The positive spikes will become more frequent and larger in amplitude, while the negative spikes will become smaller and occur less frequently.   Eventually, only positive spikes will appear, and the frequency of occurrence of these positive spikes will be greater than the frequency of positive and negative spikes together in the symmetrical case.

When the carrier is modulated, we may view the modulation as simply a continuously varying frequency offset.   When the baseband signal appears at the output of the discriminator, negative spikes are encountered most frequently at the positive extremity of the signal, and positive spikes at the negative extremity.   The characteristic appearance of a recovered sinusoidal modulation is shown in Fig. 10.6-1.   The important fact to note is not so much the spike polarity; rather, it is the unfortunate fact that in the region of threshold the noise power increases further when the carrier is modulated.

When the frequency of the input is offset by an amount $\delta f = km(t)/2\pi$ from the frequency f_c carrier, the number of spikes increases by an amount δN.   This increase δN (in the presence of a carrier which is frequency-modulated) is related to δf by

$$\delta N = |\delta f| e^{-(f_M/B)(S_i/N_M)} \qquad (10.6-1)$$

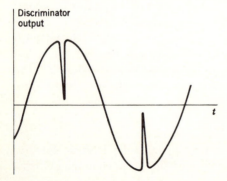

Fig. 10.6-1 Spikes present on sinusoidal output signal.

The polarity of these *additional* spikes as observed at the discriminator output is always in the direction *opposite* to the polarity of the output signal produced by the offset δf. The derivation of Eq. (10.6-1) is an extension of the derivation of Eq. (10.5-18) and will not be given here. (See Prob. 10.6-2.)

Suppose that the input signal is sinusoidally modulated at a frequency f_m ($\le f_M$). Then the input is

$$v(t) = A \cos\left(\omega_c t + \frac{\Delta f}{f_m} \sin 2\pi f_m t\right) \qquad (10.6\text{-}2)$$

where Δf is the frequency deviation. Then

$$\delta f = \Delta f \cos 2\pi f_m t \qquad (10.6\text{-}3)$$

and the average $\overline{|\delta f|} = (2/\pi)\,\Delta f$. The average value of δN is, therefore,

$$\overline{\delta N} = \frac{2\Delta f}{\pi}\,e^{-(f_M/B)(S_i/N_M)} \qquad (10.6\text{-}4)$$

The total number of spikes per second is $N = N_c + \overline{\delta N}$, the latter two terms given, respectively, by Eqs. (10.5-19) and (10.6-4). It may, however, be verified (Prob. 10.6-4) that near and below threshold $\overline{\delta N} \gg N_c$. Therefore the total average number of spikes is $N \cong \overline{\delta N}$. The average time between spikes T_s is, then,

$$T_s = \frac{1}{N} \approx \frac{1}{\overline{\delta N}} \qquad (10.6\text{-}5)$$

We may now calculate the output SNR of the FM discriminator when demodulating a sinusoidally modulated carrier where

$$\delta f = km(t)/2\pi = \Delta f \cos \omega_m t$$

We use Eqs. (9.2-16), (9.2-21), (10.4-5), (10.6-4), and (10.6-5) to find

$$\frac{S_o}{N_o} = \frac{(\tfrac{3}{2})\beta^2 (S_i/N_M)}{1 + (12\beta/\pi)(S_i/N_M)\exp\{-\tfrac{1}{2}[1/(\beta+1)](S_i/N_M)\}} \qquad (10.6\text{-}6)$$

where $\beta = \Delta f/f_M$.

Equation (10.6-6) is plotted in Fig. 10.6-2 for $\beta = 3$ and $\beta = 12$. Note that, when we are operating at a value of S_i/N_M which is above threshold for both values of β, the higher β, corresponding to a wider bandwidth, results in an increase in output SNR. On the other hand, at $S_i/N_M = 20$ dB, the output SNR is higher for $\beta = 3$ than for $\beta = 12$.

If the input signal is modulated by a sample function of a gaussian random process, we have

$$v(t) = A \cos\left[\omega_c t + k \int_{-\infty}^{t} m(\lambda)\,d\lambda\right] + n(t) \qquad (10.6\text{-}7)$$

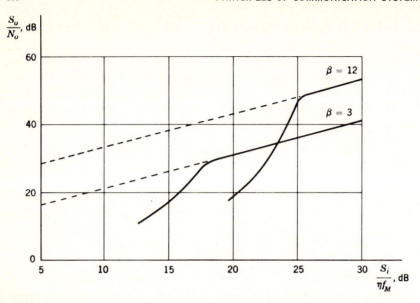

Fig. 10.6-2 Output SNR of an FM discriminator when demodulating an FM signal which is sinusoidally modulated.

where $km(t)$ is the instantaneous frequency deviation. Hence the rms frequency deviation produced is

$$(\Delta f)_{\text{rms}} = \sqrt{\overline{k^2 m^2(t)}}/2\pi \tag{10.6-8}$$

We begin our calculation of the output SNR by obtaining an expression for the output-signal power. The output-signal power is found from Eqs. (9.2-5) and (10.6-8):

$$S_o = \alpha^2 k^2 \overline{m^2(t)} = \alpha^2 4\pi^2 (\Delta f_{\text{rms}})^2 \tag{10.6-9}$$

The FM noise, neglecting spikes, is given by Eq. (9.2-16):

$$N_G = \alpha^2 \frac{8\pi^2}{3} \frac{\eta f_M{}^3}{A^2} \tag{10.6-10}$$

The spike noise is found by combining Eqs. (10.4-5), (10.6-1), and (10.6-5), with the result

$$N_S = \alpha^2 (8\pi^2 f_M) \overline{|\delta f|} \exp\left[-(f_m/B)(S_i/\eta f_M)\right] \tag{10.6-11}$$

Here $\delta f = km(t)/2\pi$ is the instantaneous frequency deviation and thus $\overline{|\delta f|}$ is the average value of the magnitude of the instantaneous frequency deviation. Noting that $m(t)$ is gaussian, we have that δf is gaussian with an average value of 0 and a standard deviation $(\Delta f)_{\rm rms}$.   Hence

$$\overline{|\delta f|} = \int_{-\infty}^{\infty} \frac{|\delta f| \exp{[-(\delta f)^2/2(\Delta f_{\rm rms})^2]}}{\sqrt{2\pi}(\Delta f_{\rm rms})} \, d\,(\delta f) \qquad (10.6\text{-}12)$$

This integral can be evaluated by integrating from 0 to infinity and doubling the result.   Thus

$$\overline{|\delta f|} = \sqrt{\frac{2}{\pi}} \,(\Delta f_{\rm rms}) \int_0^{\infty} \frac{\delta f}{(\Delta f_{\rm rms})^2} \exp{[-(\delta f)^2/2(\Delta f_{\rm rms})^2]} \, d\,(\delta f)$$

$$(10.6\text{-}13)$$

This integral is directly integrable [change variables to $x = (\delta f)^2/2(\Delta f_{\rm rms})^2$]. The result is

$$\overline{|\delta f|} = \sqrt{\frac{2}{\pi}} \,(\Delta f_{\rm rms}) \qquad (10.6\text{-}14)$$

Substituting Eq. (10.6-14) into Eq. (10.6-11) yields

$$N_S = \alpha^2(8\pi^2 f_M)\left(\sqrt{\frac{2}{\pi}}\,(\Delta f_{\rm rms})\right)\exp{[-(f_M/B)(S_i/\eta f_M)]} \qquad (10.6\text{-}15)$$

The output SNR is found by combining Eqs. (10.6-9) to (10.6-11). The final result is

$$\frac{S_o}{N_o} = \frac{3(\Delta f_{\rm rms}/f_M)^2 S_i/\eta f_M}{1 + 6\sqrt{2/\pi}\,(\Delta f_{\rm rms}/f_M)(S_i/\eta f_M)\exp{[-(f_M/B)(S_i/\eta f_M)]}}$$

$$(10.6\text{-}16)$$

Equation (10.6-13) is a function of the input SNR, $S_i/\eta f_M$, and the rms modulation index $\Delta f_{\rm rms}/f_M$ [note that using Eq. (4.13-5), we have $B/f_M = 4.6\Delta f_{\rm rms}/f_M$].

10.7 THE PHASE-LOCKED LOOP

When it is necessary to modulate a carrier, we usually employ either amplitude modulation or frequency modulation.   When noise is a problem (as is often the case), we would be inclined to use FM and allow a sacrifice of bandwidth for the sake of improved output signal-to-noise ratio.   In particular we should like to be able to avail ourselves of the advantage of FM when the input signal-to-noise ratio is low.   It is therefore most disconcerting to find that there is an FM threshold which precludes such use of FM.   For this reason a great deal of effort has gone into studies of methods to lower the FM threshold.

In the following sections we discuss several types of FM demodulators which, while they offer no improvement over the conventional dis-

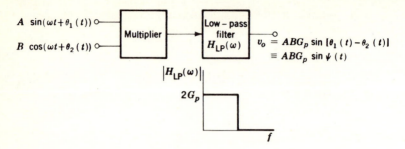

Fig. 10.7-1   A phase comparator consisting of a multiplier and a filter.

criminator above threshold, do provide threshold improvement.   One such demodulator is the *phase-locked loop*.[3]

The phase-locked loop (PLL) is a feedback system which may be used to extract a baseband signal from a frequency-modulated carrier. The basic component building blocks of a phase-locked loop are a *phase comparator* and a *voltage-controllable oscillator* (abbreviated VCO).   Before discussing the PLL, we shall discuss the phase comparator and the VCO.

PHASE COMPARATOR

A phase comparator is a device with two input ports and a single output port.   If periodic signals of identical frequency but with a timing difference (i.e., a phase difference if the signals are sinusoidal) are applied to the inputs, the output is a voltage which depends on the timing difference. One way in which a phase comparator may be constructed is by the combination of a multiplier and a filter as shown in Fig. 10.7-1.   Here we assume two sinusoidal input voltages of amplitudes A and B, frequency ω, and with time-varying phases $\theta_1(t)$ and $\theta_2(t)$.   The student may easily verify (Prob. 10.7-1) that the output of the multiplier consists of the term

$$v_o = \frac{AB}{2} \sin [\theta_1(t) - \theta_2(t)] \equiv \frac{AB}{2} \sin \psi(t) \tag{10.7-1}$$

plus other terms whose spectral components cluster around 2ω.   Hence, if $\theta_1(t)$ and $\theta_2(t)$ have bandwidths less than 2ω, a low-pass filter may separate and pass only the term $(AB/2) \sin \psi(t)$.   In Fig. 10.7-1 we have assumed an amplifier with a gain $2G_p$ in the filter stage of the phase comparator so that the output voltage is $v_o = ABG_p \sin \psi(t)$.   The comparator output is plotted in Fig. 10.7-2a as a function of ψ.

Suppose, however, that we convert the input sinusoids to square waves before application to the multiplier.   Such conversion may be achieved by hard limiting and amplifying the input signals.   The student may easily verify that if the square waves are clipped so that they have

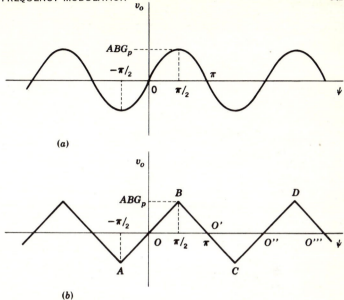

Fig. 10.7-2 (*a*) Phase characteristic of the comparator of Fig. 10.7-1 with sinusoidal inputs. (*b*) Phase characteristic with square-wave inputs.

peak amplitudes A and B, the comparator output varies with ψ in the manner shown in Fig. 10.7-2*b*. The plot in Fig. 10.7-2*b* displays the same periodicity as does the plot in Fig. 10.7-2*a* except that the sinusoidal variation has been replaced by a piecewise-linear variation. Thus

$$
v_o = \begin{cases} ABG_p \dfrac{\psi}{\pi/2} & |\psi| \leq \dfrac{\pi}{2} \\[2mm] -ABG_p \dfrac{\psi - \pi}{\pi/2} & \dfrac{\pi}{2} \leq \psi \leq \dfrac{3\pi}{2} \\[2mm] \text{etc.} \end{cases} \tag{10.7-2}
$$

VOLTAGE-CONTROLLED OSCILLATOR

A voltage-controlled oscillator is a source of a periodic signal whose frequency may be determined by a voltage applied to the VCO from an external source. Any frequency modulator may serve as a VCO. In practice, for simplicity, the VCO's employed in phase-locked loops are of the parameter-variation type described in Sec. 4.15.

Suppose that a VCO generates a sinusoidal waveform of amplitude B and, in the absence of a frequency-controlling voltage, operates at an angular frequency ω_c. Let the frequency sensitivity of the VCO be G_0 rad/(sec)(volt). G_0 is the change in the instantaneous angular frequency

ω_i produced by a change in the frequency-controlling voltage v; that is, $G_0 = d\omega_i/dv$. In the present discussion, G_0 plays the same role as the constant k in, for example, Eq. (9.2-1). Thus the oscillator signal v_{osc} is

$$v_{\text{osc}} = B \cos \left(\omega_c t + G_0 \int_{-\infty}^{t} v(\lambda) \, d\lambda \right) \tag{10.7-3}$$

In this case, the instantaneous angular frequency is

$$\omega_i(t) = \omega_c + G_0 v(t) \tag{10.7-4}$$

as required.

PHASE-LOCKED LOOP DEMODULATOR

The manner in which a phase comparator and VCO are connected to form a PLL is shown in Fig. 10.7-3. The controlling voltage for the VCO is v_o, taken from the output of the comparator, and the output of the VCO furnishes one input of the comparator. The PLL of Fig. 10.7-3 is called a *first-order* loop for reasons that are explained below.

We discuss qualitatively how the PLL may be used to recover the baseband signal from an FM modulated carrier. Figure 10.7-3 shows a frequency-modulated carrier $A \sin [\omega_c t + \varphi(t)]$ applied to one input of the phase comparator. The carrier frequency is ω_c, and if the modulating baseband signal is $m(t)$, then $\varphi(t) = k \int_{-\infty}^{t} m(\lambda) \, d\lambda$, with k a constant. Now let us assume that initially $\varphi(t) = 0$, and that we have adjusted the VCO so that when its input voltage $v_o = 0$, its frequency is precisely ω_c, the carrier frequency. Let us further adjust the VCO output signal to have a 90° phase shift relative to the carrier. This phase shift is required so that the comparator output shall be zero when $v_o(t) = 0$. Then the situation we have established is certainly a state of equilibrium. The two inputs to the comparator differ in phase by 90°; the comparator output, which is the VCO input, is zero. Therefore this initial setting of the VCO will not be disturbed.

Now let the frequency of the input signal make an abrupt change ω

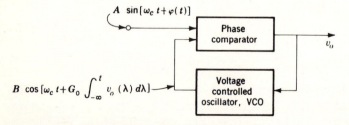

Fig. 10.7-3   A (first-order) phase-locked loop.

at time $t = 0$. Then, beginning at $t = 0$, $\varphi(t) = \omega t$, since $d\varphi/dt = \omega$. That is, the abrupt frequency change causes the phase $\varphi(t)$ to begin to increase linearly with time. The phase difference at the comparator input will generate a positive output v_o, which will in turn increase the frequency of the VCO. A new equilibrium point will eventually be established in the system, when the frequency of the VCO has been increased to equal the frequency of the input signal. When equilibrium is established, the input signal and the VCO output will be of identical frequency but no longer different in phase by 90°. For if the VCO is to operate at a frequency other than its initial frequency ω_c, there must be an output v_o, and hence a departure from the 90° phase difference at the input which yields $v_o = 0$.

The VCO output given in Fig. 10.7-3 is the same as Eq. (10.7-3) with v replaced by v_o. If the input and VCO frequencies are to be the same at equilibrium, we require that

$$\frac{d\varphi(t)}{dt} = \frac{d}{dt} G_0 \int_{-\infty}^{t} v_o(\lambda) \, d\lambda \tag{10.7-5}$$

Letting $d\varphi/dt = \omega$, we have

$$v_o(t) = \frac{\omega}{G_0} \tag{10.7-6}$$

Thus the output voltage is proportional to the frequency change as required in an FM demodulator. We see then that if the input-carrier frequency changes continuously, and at a rate which is slow in comparison with the time required for the PLL to establish a new equilibrium operating point, the PLL output is continuously proportional to the *frequency variation* of the carrier.

We readily note that the PLL is a feedback control system in which the error signal is the phase difference between the modulated carrier and the VCO signal. Initially, with an unmodulated carrier, the operating point of the phase comparator may be adjusted to be at the origin in Fig. 10.7-2a or b. In the presence of modulation the operating point will move up and down along the portion of the plot between $-\pi/2$ and $\pi/2$.

10.8   ANALYSIS OF THE PHASE-LOCKED LOOP

We now present an elementary analysis of the PLL of Fig. 10.7-3. We assume for simplicity that the phase characteristic of the phase comparator is *piecewise-linear* as shown in Fig. 10.7-2b. The results will apply equally well to a PLL using a comparator with the characteristic shown in Fig.

10.7-2a. From Fig. 10.7-3 we have that the phase angle difference $\psi(t)$ is

$$\psi(t) \equiv \varphi(t) - G_0 \int_{-\infty}^{t} v_o(\lambda)\, d\lambda \tag{10.8-1}$$

Differentiating Eq. (10.8-1) and transposing, we have

$$\frac{d\psi}{dt} + G_0 v_o = \frac{d\varphi}{dt} \tag{10.8-2}$$

The output voltage can be eliminated using Eq. (10.7-2). The result is

$$\frac{d\psi}{dt} + \frac{\psi}{\tau} = \frac{d\varphi}{dt} \qquad \text{for } |\psi| \leq \frac{\pi}{2} \tag{10.8-3}$$

where

$$\tau \equiv \frac{\pi}{2ABG_pG_0} \tag{10.8-4}$$

Since ψ and v_o are directly related by Eq. (10.7-2), Eq. (10.8-3) can also be written as

$$\frac{dv_o}{dt} + \frac{v_o}{\tau} = \frac{1}{G_0\tau}\frac{d\varphi}{dt} \qquad \text{for } |\psi| \leq \frac{\pi}{2} \tag{10.8-5}$$

Equation (10.8-5) shows explicitly the relationship between the frequency offset $d\varphi/dt$ and the output voltage $v_o(t)$. Equation (10.8-3) or (10.8-5) is the differential equation of the PLL. Its solution, subject to the appropriate initial conditions, describes the PLL performance. Parenthetically, we note that with the phase comparator having the sinusoidal characteristic shown in Fig. 10.7-2a, the equation corresponding to Eq. (10.8-3) is

$$\frac{d\psi}{dt} + \frac{\sin \psi}{\tau'} = \frac{d\varphi(t)}{dt} \tag{10.8-6}$$

with $\tau' = 1/ABG_pG_0$. For small ψ, where $\sin \psi \approx \psi$, Eq. (10.8-6) would be identical with Eq. (10.8-3), except that the time constants would be different, this difference reflecting only the fact that in Fig. 10.7-2 the slopes at the origin are different in parts (a) and (b).

Suppose that the PLL is operating initially with $\psi = 0$, that is, the VCO and input-carrier frequency are identical and exactly 90° out of phase. Now let us again ask the question raised before; that is, what happens if the angular frequency of the carrier is abruptly changed by an amount ω $(d\varphi/dt = \omega)$? Subject to these initial conditions, the solution of Eq. (10.8-3) is

$$\psi = \omega\tau(1 - e^{-t/\tau}) \qquad |\psi| \leq \frac{\pi}{2} \tag{10.8-7}$$

Then if $\omega\tau \leq \pi/2$, ψ will approach its new steady-state value with a time constant τ. The new equilibrium steady state is at

$$\psi = \psi_e = \omega\tau \leq \frac{\pi}{2} \tag{10.8-8}$$

or correspondingly, using Eqs. (10.7-2) and (10.8-4), the steady-state output voltage v_o is given by

$$v_o = \frac{\omega}{G_0} \tag{10.8-9}$$

OPERATING RANGE

We note that as long as the phase comparator operates in the range AOB in Fig. 10.7-2b, that is, $|\psi| \leq \pi/2$, the steady-state operating point is given by $\psi_e = \omega\tau$. The range R of the PLL is the angular frequency change ω of the input carrier which will just carry the steady-state operating point from O to B or from O to A. This range R is given by

$$R = \omega_{\max} = \frac{\psi_{\max}}{\tau} = \frac{\pi/2}{\tau} = \frac{\pi}{2\tau} \tag{10.8-10}$$

For the phase comparator of Fig. 10.7-2a we find R by using Eq. (10.8-6). Set $d\varphi(t)/dt = \omega$ and $d\psi/dt = 0$ (steady state). Then, when $\psi = \pi/2$, $\sin\psi = 1$. Thus $R = \omega_{\max} = 1/\tau'$. We observe that the behavior of the PLL is determined entirely by the single parameter τ (or τ'). This parameter depends on the gains G_0 and G_p and also on the amplitudes A and B of the input and VCO signals.

As τ becomes progressively smaller, the range of the PLL becomes progressively larger, and the speed with which the PLL responds to a frequency change ω increases. The reason the range becomes larger is, as a matter of fact, precisely because the loop does respond faster. A *fast-acting* loop will not permit the phase $\varphi(t)$ to depart appreciably from the phase of the VCO. In the limit, when $\tau \to 0$, ψ will remain close to $\psi = 0$ and hence close to the origin 0 in Fig. 10.7-2b, no matter how large ω should become.

BANDWIDTH

Since the PLL response is given by a linear differential equation, it has the characteristics of a filter. To see that such is the case, let $\Psi(s) \equiv \mathcal{L}[\psi(t)]$ and $\Omega(s) \equiv \mathcal{L}(d\varphi/dt) \equiv \mathcal{L}[\omega(t)]$. Then from Eq. (10.8-3) we have

$$\Psi(s) = \frac{\Omega(s)}{s + 1/\tau} \tag{10.8-11}$$

The transfer function of the PLL, $\psi(s)/\Omega(s)$, which relates the output signal to the input signal, i.e., the change in the input angular frequency, has a transfer function

$$H(s) = \frac{\Psi(s)}{\Omega(s)} = \frac{1}{s + 1/\tau} \tag{10.8-12}$$

The transfer function has a single pole at $s = -1/\tau$, and the PLL is therefore referred to as a *first-order* PLL.

Suppose that the angular frequency variation of the input signal were sinusoidal: $\omega(t) = \Delta\omega_m \cos \omega_m t$, with $\Delta\omega_m$ the amplitude, and ω_m the angular frequency of the sinusoidal variation of the instantaneous frequency. Then for a fixed frequency deviation $\Delta\omega_m$, an increase in the modulating frequency ω_m results in a decrease in $\psi(t)$ and hence $v_o(t)$. The angular frequency $(\omega_m)_{3\,dB}$ at which the response will have fallen by 3 dB is seen from Eq. (10.8-12) to be

$$(\omega_m)_{3\,dB} = \frac{1}{\tau} \tag{10.8-13}$$

A PLL will consequently introduce distortion between the original modulating signal and the signal recovered at the PLL output. The distortion will be the same as would be introduced if the modulating signal had been passed through a low-pass resistance-capacitance network of time constant τ. The distortion may be decreased by making the PLL 3-dB cutoff frequency much higher than the highest-frequency spectral component of the modulating signal.

A point worthy of note is to be seen in Eqs. (10.8-10) and (10.8-13). We observe that a single time constant τ determines both the range of the PLL and its frequency response. We are therefore not at liberty, in this first-order loop, to vary these parameters independently. We shall see in Sec. 10.11 how this limitation can be remedied.

As long as the operating point of the comparator remains in the range AOB in Fig. 10.7-2b, the PLL demodulator accomplishes no function which is not performed at least equally well by a conventional discriminator. The PLL displays its special merit, however, precisely when it is operated in such a manner that its operating point ranges outside the limits A and B in Fig. 10.7-2b. In the next section we discuss such operation.

10.9   STABLE AND UNSTABLE OPERATING POINTS

We have seen that in the range AOB in Fig. 10.7-2b, the differential equation for the PLL is

$$\frac{d\psi}{dt} + \frac{\psi}{\tau} = \frac{d\varphi}{dt} \tag{10.9-1}$$

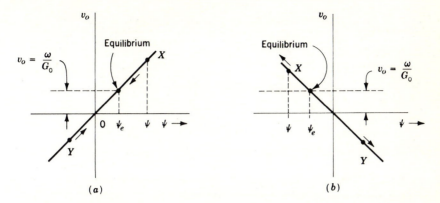

Fig. 10.9-1 (*a*) Illustrating that an equilibrium point is stable if the operating point of a PLL is on a positive-slope portion of the phase comparator. (*b*) Illustrating the instability that results when the operating point is on a negative-slope portion of the comparator.

This equation applies equally well for operation over any positive-slope region in Fig. 10.7-2*b*, provided that ψ measures the departure of the phase from the phase at which the phase characteristic crosses the axis $v_o = 0$. Thus, for example, the equation applies to the region $CO''D$ if ψ is measured from O''. Similarly, by retracing the derivation, we find that in a negative-slope region, as, say, $BO'C$, if ψ is measured from O', the equation is (see Prob. 10.9-1)

$$\frac{d\psi}{dt} - \frac{\psi}{\tau} = \frac{d\varphi}{dt} \tag{10.9-2}$$

In Fig. 10.9-1*a* we have drawn a portion of the phase-comparator characteristic in which the slope is positive. Let us assume that the input-carrier angular frequency has been offset by an amount ω. Then, as given in Eq. (10.8-9), the new *steady-state* equilibrium is, as shown, at $v_o = \omega/G_0$, and the corresponding equilibrium value of ψ is $\psi_e = \omega\tau$ as in Eq. (10.8-8). Suppose, however, that because of the past history of the PLL, the operating point happens to find itself initially at point X where the phase angle is ψ. We have then from Eq. (10.9-1) that

$$\frac{d\psi}{dt} = \frac{d\varphi}{dt} - \frac{\psi}{\tau} \tag{10.9-3a}$$

$$= \omega - \frac{\psi}{\tau} \tag{10.9-3b}$$

$$= \frac{\psi_e}{\tau} - \frac{\psi}{\tau} = \frac{1}{\tau}(\psi_e - \psi) \tag{10.9-3c}$$

Now if, as in the case of point X, $\psi > \psi_e$, then $d\psi/dt$ is negative. Hence the instantaneous operating point X moves to the left along the comparator-phase characteristic. The point X will, as we have already seen, approach the equilibrium point asymptotically, with time constant τ. Similarly if the operating point is at Y, then $\psi < \psi_e$, and $d\psi/dt$ is positive so that again Y approaches the equilibrium point. The equilibrium point is a *stable* point.

Let us now consider the negative-slope portion of the phase characteristic such as $BO'C$ in Fig. 10.7-2b. Here we use Eq. (10.9-2) instead of Eq. (10.9-1). Repeating the above discussion using Fig. 10.9-1b, we find that initial operating points such as X and Y result in ψ moving *away* from, not *toward*, the equilibrium point ψ_e. Thus in Fig. 10.9-1b the equilibrium point is *unstable*. In the next section we apply these simple notions to explain how the PLL acts to suppress spikes.

10.10 SPIKE SUPPRESSION

Let us assume that the input to a PLL is an unmodulated carrier, offset from ω_c by an amount ω. Let the steady-state output voltage of the PLL corresponding to this offset be $v_o = V_o = \omega/G_0$ as indicated by the dashed horizontal line in Fig. 10.10-1. That is, the offset ω is so large that the horizontal line ZZ' does not intersect the phase characteristic. There are no *actual* equilibrium points. However, two extended equilibrium points have been indicated at points P_2 and P_3. The point P_2 is the intersection of AOB with ZZ'. An initial operating point P_1 on AOB will move as though headed for equilibrium at P_2 for, until point B is reached, the PLL does not know that AOB does not continue to P_2. Similarly the *equilibrium* point for operation along $BO'C$ is at P_3.

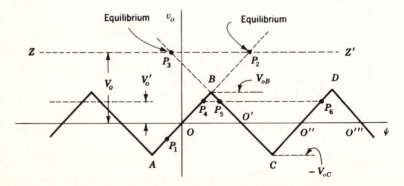

Fig. 10.10-1 Illustrating the PLL response for various assumed fixed-frequency offsets of the carrier signal.

If the PLL is initially at the origin with $\psi = 0$, and at time $t = 0$ a phase offset ω is added to ω_c, the operating point will move from 0 to B. The operating point cannot remain at B, since B is not an equilibrium point. However, the point B is also on the negative-slope segment $BO'C$. Hence it will recede from the equilibrium point P_3 and proceed toward C. Continuing in the same way, we see that from C the operating point will go to D, i.e., toward the equilibrium point, and so on. If the output voltage corresponding to point B is $v_o = v_{oB}$, then the output voltage will oscillate with excursion from $+v_{oB}$ to $-v_{oB}$. The waveform of this oscillation is easily calculated. Note that ψ constantly increases and does not oscillate.

Suppose that the offset were such to establish equilibrium at the level V_o' as indicated in Fig. 10.10-1; then if the operating point is initially on AOB, it will settle at P_4. If the operating point is initially on $BO'C$, it will eventually settle at P_4 or P_6, depending on whether the initial point is above or below P_5. If the operating point is initially constrained to be precisely at P_5, it will remain there only as long as no disturbance displaces it, even ever so slightly. If a disturbance does move the operating point from P_5, the operating point will end up again at P_4 or P_6, depending on the direction of the disturbance.

After these preliminaries we may turn our attention to the matter of spikes. For this purpose we shall inquire into the response of the PLL to an "artificial" spike. Thus we consider that the input carrier is at the carrier angular frequency ω_c except for a short time T during which the angular frequency is offset by an amount ω. Such artificial spikes are shown by the solid-line plot in Fig. 10.10-2 in which the offset angular frequency ω is plotted as a function of time. In a physical situation the duration T of the spike will be determined by the bandwidth of the IF carrier filter. The spike amplitude ω_S will then be determined by the condition that the area under the spike be 2π, that is, $\omega_S T = 2\pi$.

We have seen [Eq. (10.8-10)] that the angular frequency corresponding to a steady-state operating point at B in Fig. 10.10-1 is $\omega_{\max} = \pi/2\tau$. The dashed waveforms in Fig. 10.10-2 indicate the PLL responses for three relative values of $\omega_{\max}$ and ω_S.

Case 1. The steady-state output voltage v_o corresponding to ω is $v_o = \omega/G_0$ [see Eq. (10.8-9)]. In Fig. 10.10-2a we assume $\omega_{\max} > \omega_S$. The output voltage of the PLL responds to the spike by rising asymptotically with time constant τ to the steady-state level $v_o = \omega_S/G_0$. At this level the operating point of the PLL has not reached the point B in Fig. 10.10-1 since we assume that $\omega_{\max} > \omega_S$. At the termination of the spike the PLL returns to its initial level. The output-voltage waveform v_o has the form of the dashed plot. This output waveform is a replica of the input-

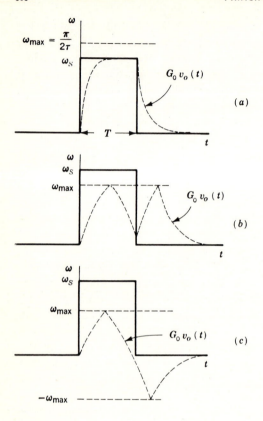

Fig. 10.10-2 Various possible responses of a PLL to an "artificial" spike. The different cases correspond to different selections of the time constant of the PLL.

spike waveform except rounded because of the time constant of the PLL. In this case, a spike input produces a spike output.

Case 2. Here in Fig. 10.10-2b $\omega_{max} < \omega_S$. Hence the operating point reaches the limit of its range (point B in Fig. 10.10-1) before it has been able to respond fully to the spike amplitude. After reaching point B, the operating point starts down along $BO'C$. However, the spike ends before point O' is reached. When the spike has terminated, ω is again equal to zero, and O' is an *unstable* equilibrium point. Thus, the operating point retraces its path back over B and into the positive-slope region, finally returning to the starting point at O. An input spike now yields two spikes at the output.

  Note that the change from case 1 to case 2 is made by increasing τ, thereby lowering ω_{max} and also slowing the response of the PLL.

Case 3. Here in Fig. 10.10-2c ω_{max} has been reduced still further and has been set at such a level that the operating point reaches the limit of the range in a time slightly less than one-half the spike duration.   Therefore when the spike terminates, the operating point has passed O' and hence continues to C and finally settles at O''.   In this case a spike input has yielded an output which consists of a positive spike followed by a negative spike.   Such an output waveform as indicated by the dashed plot in Fig. 10.10-2c is called a *doublet*.   We may reasonably expect that the total area under a doublet will be near zero.   Therefore the doublet will yield very little energy in the baseband in comparison with the energy yielded by the spike itself.   Here, then, qualitatively is the mechanism by which the PLL suppresses the noise of spikes.   The PLL changes spikes into doublets of relatively small energy.

If all the spikes encountered in a physical situation were of precisely the same waveform, then, as the preceding discussion indicates, it would be possible to adjust the time constant τ of the PLL so that each spike might be replaced by a doublet.   Observe that only a small range of τ is suitable.   The time constant τ must be large enough to avoid the responses indicated in Fig. 10.10-2a and b.   On the other hand, increasing τ results in decreasing the PLL bandwidth [Eq. (10.8-13)].   If the PLL is being used to demodulate a carrier frequency modulated by a baseband signal of bandwidth f_M, then we require that $(1/\tau) \gg f_M$.

Unfortunately all spikes are not identical.   Even in the absence of modulation there will be some variation in spike waveform, and with modulation the spikes will vary considerably during the course of the modulation cycle.   Still, as a matter of practice it turns out to be possible to adjust a PLL to convert many of the spikes into doublets and thereby effect a net improvement in performance.

10.11   SECOND-ORDER PHASE-LOCKED LOOP

We now discuss the second-order PLL, and we shall explain, qualitatively, how the second-order loop manages to suppress spikes more effectively than does the first-order loop.

In Fig. 10.11-1 the PLL has been modified by the inclusion of a filter.   This filter is not to be compared with the filter described in Sec. 10.7 as having been incorporated in the phase comparator.   The phase-comparator filter was used to suppress the carrier frequency and its harmonics and hence performs no filtering in the passband of the PLL. Hence, while the phase-comparator filter has absolutely no influence on the PLL performance, the filter introduced in Fig. 10.11-1 is deliberately introduced to have such influence.

It will be more convenient, in our present discussion, to refer not

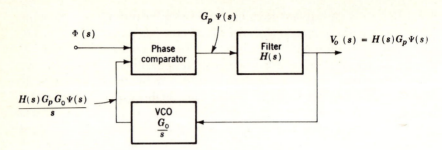

Fig. 10.11-1 A phase-locked loop which includes a filter.

to the angle-modulated carrier waveforms encountered in the PLL but rather directly to the phases of these waveforms. Thus, in Fig. 10.11-1 the signal input is $\varphi(t)$ with Laplace transform $\Phi(s)$. The output voltage of the phase comparator is $G_p\Psi(s)$, where $\psi(t)$ is the phase-angle difference at the phase-comparator input, and G_p is a constant of proportionality. The VCO provides a carrier having a phase proportional to the *integral* of its input voltage $v_o(t)$. Hence the VCO has been characterized as having a transform G_0/s.

With $H(s) \equiv 1$, the PLL of Fig. 10.11-1 is identical with the loop of Fig. 10.7-3. We have, in this case

$$\Psi(s) \equiv \Phi(s) - \frac{G_pG_0\Psi(s)}{s} \qquad (10.11\text{-}1)$$

Using $s\Phi(s) = \mathcal{L}[d\varphi(t)/dt] \equiv \Omega(s)$ and $1/\tau = G_0G_p$, we find from Eq. (10.11-1) that

$$\Psi(s) = \frac{s\Phi(s)}{s + G_pG_0} = \frac{\Omega(s)}{s + G_0G_p} = \frac{\Omega(s)}{s + 1/\tau} \qquad |\psi(t)| \leq \frac{\pi}{2} \quad (10.11\text{-}2)$$

exactly as in Eq. (10.8-11).

Now let us assume a *proportional-plus-integral* filter with $H(s)$ given by

$$H(s) = 1 + \frac{K}{s} \qquad (10.11\text{-}3)$$

where K is a constant. We now find that the phase-angle difference $\psi(t)$ has a transform

$$\Psi(s) = \frac{s^2\Phi(s)}{s^2 + s/\tau + K/\tau} = \frac{s\Omega(s)}{s^2 + s/\tau + K/\tau} \qquad |\psi(t)| \leq \frac{\pi}{2} \quad (10.11\text{-}4)$$

while the output voltage $v_o(t)$ has a transform

$$V_o(s) = G_pH(s)\Psi(s) = \frac{s(s + K)G_p\Phi(s)}{s^2 + s/\tau + K/\tau} = \frac{(s + K)G_p\Omega(s)}{s^2 + s/\tau + K/\tau} \qquad (10.11\text{-}5)$$

We observe that the expressions for $\Psi(s)$ and $V_o(s)$ have denominators which, being quadratic in s, give rise to two poles. Hence the term *second-order* PLL.

We inquire now, as we did for the first-order loop, about the steady-state response of the second-order loop to an abrupt change of magnitude ω in the angular frequency of the input carrier. To make this calculation, we use the *final-value theorem* which states that if $F(s) = \mathcal{L}[f(t)]$, then

$$\lim_{s \to 0} [sF(s)] = \lim_{t \to \infty} f(t) \tag{10.11-6}$$

Applying this theorem to $\Psi(s)$ and $V_o(s)$ in Eqs. (10.11-4) and (10.11-5) with $\Omega(s) = \omega/s$, we find

$$\psi(\infty) = 0 \tag{10.11-7}$$

and

$$v_o(\infty) = G_p\omega\tau \tag{10.11-8}$$

Thus, we observe from Eq. (10.11-8) that the steady-state response at the output of the second-order loop is the same as in the first-order loop. However, we see from Eq. (10.11-7) that the phase difference $\psi(t)$, unlike the situation obtained in the first-order loop, does not respond with a steady-state displacement from its initial position of equilibrium. Instead, while it may make a transient excursion, it thereafter settles right back to its starting point. Another way of emphasizing this same point is to recognize that the transfer function that relates $\Psi(s)$ to the input signal $\Omega(s)$ in Eq. (10.11-4) is the transfer function of a bandpass filter, while the transfer function that relates $V_o(s)$ to $\Omega(s)$ in Eq. (10.11-5) is the transfer function of a low-pass filter. Therefore, by appropriately selecting the cutoff frequencies of these filters, it is possible to arrange that the output of the PLL respond properly to the modulating signal, while the operating point of the phase comparator responds hardly at all and always hovers close to its initial point of equilibrium.

The relevance of these considerations to spike suppression may now be seen. Consider the first-order loop and let the input be a step in frequency which carries the phase-comparator operating point toward A in Fig. 10.10-1 and holds it there at point P_1. If, now, a spike develops because of noise, this spike will be in the direction to drive the operating point in the opposite direction toward B. The spike will not be suppressed unless the operating point does indeed get to point B and beyond toward C, as explained in Sec. 10.10. However, the initial displacement of the operating point away from 0 toward A due to the input signal makes such spike suppression less likely than if the initial operating point were at point O. On the other hand, in the second-order loop, after some initial response to the input-frequency step, the phase comparator settles back at point O.   Thus if an input spike occurs, that is, $\varphi(t)$ changes by 2π rad, the distance moved to point B and hence the time required to reach point B is less than in the first-order loop.   As a consequence such a spike has a greater likelihood of suppression.

10.12   OUTPUT SNR OF A PHASE-LOCKED LOOP[4]

The input signal to an FM demodulator is

$$v_i(t) = R(t) \cos[\omega_c t + \varphi(t)] \tag{10.12-1}$$

Here $R(t)$ is the envelope of the waveform which results from the superposition of the carrier of amplitude A and the noise $n(t)$.   Similarly $\varphi(t) = \varphi_s(t) + \varphi_n(t)$, where $\varphi_s(t)$ is the angular modulation due to the signal, and $\varphi_n(t)$ due to the noise.   Let us ignore, for the present, the effect of the noise on the envelope and assume that $R(t) = A$.

With this assumption, an FM discriminator, when presented with the waveform of Eq. (10.12-1), will yield an output voltage

$$v_o(t) = \alpha \frac{d\varphi(t)}{dt} \tag{10.12-2}$$

where α is a constant.   The student may verify from Eq. (10.8-5) that the output voltage of a first-order PLL is related to $d\varphi(t)/dt$ as indicated in Fig. 10.12-1 (see Prob. 10.12-1).   The time constant of the RC circuit is equal to the time constant τ of the PLL.   If we neglect the effect of this low-pass RC circuit, then

$$v_o(t) = \frac{1}{G_0} \frac{d\varphi(t)}{dt} \tag{10.12-3}$$

just as for the discriminator except with α replaced by $1/G_0$.   Now both the discriminator and the PLL will be followed by a low-pass baseband filter of cutoff frequency f_M.   If, as is always the case, we select the 3-dB frequency of the RC circuit, which represents the phase-locked loop, to be much larger than f_M, then, as far as the baseband filter output is concerned, the RC circuit is indeed of no consequence.   Above thresh-

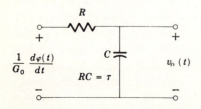

Fig. 10.12-1   A representation of the relationship between the output voltage v_o of a first-order PLL and the frequency modulation $d\varphi/dt$ of an input carrier.

old, where spikes are extremely rare, the discriminator and the PLL, when combined with the baseband filter, process the input waveform, signal plus noise, in exactly the same way.   The output SNR does not depend on the constant α nor on the constant $1/G_0$.   Therefore, above threshold, both discriminator and PLL yield the same output SNR.

THE EFFECT OF LIMITING

Now let us take account of the fact that the envelope of $v_i(t)$ in Eq. (10.12-1) actually varies because of noise.   A discriminator is always accompanied in practice by a limiter.   Hence the result stated in Eq. (10.12-2) continues to apply, since the limiter removes all amplitude variation.   In the case of the PLL we note from Fig. 10.12-1 that the carrier amplitude enters only through the time constant τ, for as is to be seen in Eq. (10.8-4) τ depends on A.   Still, if the 3-dB frequency of the RC filter in Fig. 10.12-1 is very much larger than the baseband filter cutoff frequency, the small variations in τ resulting from the noise can have little effect.   At least such will be the case above threshold where the noise power is very small in comparison with the carrier power. Hence, we have the result that above threshold the PLL does not require a limiter, and the result of Eq. (10.12-3) is valid.

Above threshold it is immaterial whether we use a limiter.   We shall now show qualitatively that below threshold it is advantageous *not* to use a limiter.   As explained in Sec. 10.10 in connection with Fig. 10.10-2, it is necessary to restrict the range $\omega_{\max}$ ($= \pi/2\tau$) of the PLL in order to achieve spike suppression.   Such restriction of the range has the disadvantage that it also restricts the ability of the PLL to follow the frequency deviations of the carrier due to signal modulation.   It would therefore be of advantage if we could keep the range large in the absence of a spike and yet be able to restrict the range during the course of the formation of a spike.   We may now see that if the PLL is operated without a limiter, the PLL will automatically restrict its own range during a spike.

Referring to Fig. 10.2-2b, we see that when the noise $r(t)$ is comparable to A, that is, when a spike occurs, then as the resultant $R(t)$ executes a rotation by 2π, its average magnitude will be smaller than A. [It is of course possible to conceive that $r(t)$ is so large that $R(t)$ is larger than A, but such a situation would have a small probability of occurrence and is therefore of no interest in the present discussion.]   Furthermore, since the range $\omega_{\max} = \pi/2\tau$ and τ is inversely proportional to the carrier envelope [see Eq. (10.8-4)], the range of the PLL automatically reduces during a spike.   This advantageous operating characteristic of the PLL would be lost if the envelope amplitude were kept constant by the use of a limiter.

OUTPUT SNR

We have seen that the number of spikes present at the PLL output is fewer than at the output of the FM discriminator, and threshold *extension* results. The output-noise power due to spikes present at the PLL output is N_S and is found by replacing $1/T_s$ by $N \equiv N_c + \delta N$, the average number of spikes per second occurring at the PLL output, in Eq. (10.4-5). The result is

$$N_S = \frac{8\pi^2 f_M N}{G_0^2} \tag{10.12-4}$$

Note that α is replaced by $1/G_0$ in this equation. Since the PLL suppresses spikes, N is smaller than for the discriminator. Thus, N_S is smaller in the PLL.

From Eqs. (9.2-5) and (9.2-16) with α replaced by $1/G_0$, and using Eq. (10.12-4), we have

$$\frac{S_o}{N_o} = \frac{\overline{k^2\, m^2(t)}}{(8\pi^2/3)\eta f_M^3/A^2 + 8\pi^2 f_M N} \tag{10.12-5}$$

If we simplify and assume sinusoidal modulation, $km(t) = \Delta\omega \cos \omega_m t$, then Eq. (10.12-5) becomes

$$\frac{S_o}{N_o} = \frac{\tfrac{3}{2}\beta^2\, S_i/N_M}{1 + 6(N/f_M)(S_i/N_M)} \tag{10.12-6}$$

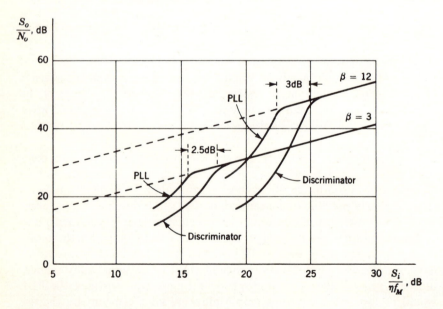

Fig. 10.12-2   Threshold extension of a PLL.

While an analytic expression for N of a PLL has not been determined, computer simulation of the PLL has resulted in the determination of N and of threshold.[5]   The results obtained are shown in Fig. 10.12-2. We see from this figure that with $\beta = 12$, the PLL extends threshold by 3 dB, while with $\beta = 3$, threshold extension of 2.5 dB results.

10.13   THE FM DEMODULATOR USING FEEDBACK

An FM demodulator using feedback (FMFB) is shown in Fig. 10.13-1. This demodulator, like the PLL, decreases the S_i/N_M at threshold.   The construction of the FMFB and the PLL are similar inasmuch as a VCO is frequency-modulated by the output signal $v_o(t)$ of the FMFB demodulator.   Also, as in the PLL, the input $v_i(t)$ of carrier frequency f_c ($= \omega_c/2\pi$) is multiplied with the output v_{osc} of the VCO.   In the present case, however, the frequency of the VCO is offset from f_c by an amount f_0.   The FMFB includes in its forward transmission path a bandpass filter and also a limiter-discriminator, neither of which is encountered in the PLL. The bandpass filter, following the multiplier, is centered at the offset frequency f_0, and hence passes the difference-frequency output of the multiplier.

We show now that the FMFB recovers the baseband signal from an FM modulated carrier, and when operating above threshold yields the same signal-to-noise output as does a simple limiter-discriminator. Let the input signal and noise $v_i(t)$ be

$$v_i(t) = R(t) \sin \left[\omega_c t + \varphi_s(t) + \varphi_n(t) \right] \tag{10.13-1}$$

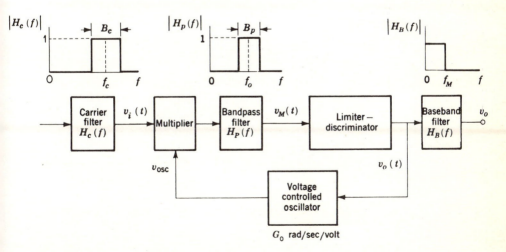

Fig. 10.13-1   The FM demodulator using feedback.

where $R(t)$ is the envelope of the carrier signal and the noise, $\varphi_s(t)$ is the angular modulation due to the signal, and $\varphi_n(t)$ is due to the noise.   If $v_i(t)$ was the input to a limiter-discriminator whose baseband output was α times the departure of the instantaneous angular frequency from the carrier frequency ω_c, the output voltage of the discriminator $v_o(t)$ would be

$$v_o(t) = \alpha \frac{d}{dt} [\varphi_s(t) + \varphi_n(t)] \tag{10.13-2}$$

Returning to the FMFB shown in Fig. 10.13-1, we represent the VCO output as

$$v_{\text{osc}} = B \cos \left[(\omega_c - \omega_0)t + G_0 \int_{-\infty}^{t} v_o(\lambda) \, d\lambda \right] \tag{10.13-3}$$

where B is the VCO amplitude and G_0, as introduced in Eq. (10.7-3), is the change in angular frequency of the VCO per unit change in $v_o(t)$. We neglect temporarily the effect of the bandpass filter following the multiplier except to take account of the fact that it passes only the difference-frequency component of the multiplier.   On this basis, the product signal $v_M(t)$, which is equal to the low-pass component of $v_i v_{\text{osc}}$, is

$$v_M(t) = \frac{AB}{2} \cos \left[\omega_0 t + \varphi_s(t) + \varphi_n(t) - G_0 \int_{-\infty}^{t} v_o(\lambda) \, d\lambda \right] \tag{10.13-4}$$

This product signal is applied to the limiter-discriminator.   The output of the discriminator $v_o(t)$ is then

$$v_o(t) = \alpha \left[\frac{d\varphi_s(t)}{dt} + \frac{d\varphi_n(t)}{dt} - G_0 v_o(t) \right] \tag{10.13-5}$$

Solving for $v_o(t)$ yields

$$v_o(t) = \frac{\alpha}{1 + \alpha G_0} \frac{d}{dt} (\varphi_s + \varphi_n) \tag{10.13-6}$$

We observe from Eq. (10.13-6), and from comparing this equation with Eq. (10.13-2), that the FMFB does indeed demodulate.   We see further that the only difference between the output of the discriminator and the FMFB is that the amplitude of output of the FMFB is smaller by the factor $(1 + \alpha G_0)$.   Since both signal and noise have been reduced by the same factor $1/(1 + \alpha G_0)$, the signal-to-noise ratio is the same for the FMFB and the FM discriminator.

We turn our attention now to the bandpass filter and ask how narrow it can be made to pass the signal without undue distortion. Using Eq. (10.13-6), we may rewrite Eq. (10.13-4) as

$$v_M(t) = \frac{AB}{2} \cos \left[\omega_0 t + \frac{1}{1 + \alpha G_0} (\varphi_s + \varphi_n) \right] \tag{10.13-7}$$

We observe, in comparing Eq. (10.13-7) with Eq. (10.13-1), that the feedback has suppressed the frequency deviation, produced by the signal φ_s, by the factor $1/(1 + \alpha G_0)$. Consider, for example, that the modulation is sinusoidal. Then $\varphi_s(t) = \beta \sin \omega_m t$, and the phase of the signal present in the multiplied signal is $\varphi_s(t)/(1 + \alpha G_0) = [\beta/(1 + \alpha G_0)] \sin \omega_m t$. Hence, if the bandwidth of the carrier filter is B_c and the bandwidth of the bandpass filter preceding the discriminator is B_p, then

$$B_p = 2\left(\frac{\beta}{1 + \alpha G_0} + 1\right) f_m \tag{10.13-8}$$

and

$$B_c = 2(\beta + 1)f_m \tag{10.13-9}$$

so that

$$B_p = \frac{[\beta/(1 + \alpha G_0)] + 1}{\beta + 1} B_c \tag{10.13-10}$$

If $\beta = 9$ and $\alpha G_0 = 8$, then $B_p = \frac{1}{5}B_c$.

Note that the bandwidths B_p and B_c given by Eqs. (10.13-8) and (10.13-9) employ Carson's rule for the bandwidth of *rectangular* filters. In practice the carrier filter with bandwidth B_c is *almost* rectangular. However, the bandpass filter with bandwidth B_p is usually a single-tuned filter, and to pass 98 percent of the signal energy, B_p must be greater than the value given in Eq. (10.13-8). The reason for employing the single-tuned circuit is related to the fact that the FMFB is a feedback system. If a single-tuned filter is used, the system is stable; if, however, a more sophisticated filter containing many poles is employed, the FMFB may oscillate.

10.14 THRESHOLD EXTENSION USING THE FMFB[6]

Let us consider the operation of the FMFB under two conditions of G_0. First let the VCO sensitivity $G_0 = 0$. Then, referring to Fig. 10.13-1, we find there is no feedback. The modulation index is not reduced, and $B_p = B_c$ [see Eq. (10.13-10)]. The VCO now serves only to shift the carrier frequency ω_c to an IF frequency ω_0. Therefore no threshold extension occurs.

Our second condition for G_0 is $G_0 \to \infty$. Let the output voltage of the VCO be written as

$$v_{osc} = B \cos\left[(\omega_c - \omega_0)t + \varphi_{osc}(t)\right] \tag{10.14-1}$$

where, comparing Eq. (10.14-1) with Eq. (10.13-3), we see that

$$\varphi_{osc}(t) = G_0 \int_{-\infty}^{t} v_o(\lambda)\, d\lambda \tag{10.14-2}$$

Since G_0 is extremely large, $v_o(t)$ must be infinitesimally small for $\varphi_{\text{osc}}(t)$ to remain finite.

Refer now to Eq. (10.13-5) and replace $G_0 v_o(t)$ by $\dot{\varphi}_{\text{osc}}(t)$. We have then

$$v_o(t) = \frac{1}{G_0} \frac{d\varphi_{\text{osc}}(t)}{dt} = \alpha \left(\frac{d\varphi_s}{dt} + \frac{d\varphi_n}{dt} - \frac{d\varphi_{\text{osc}}}{dt} \right) \tag{10.14-3}$$

For very large G_0 we can assume $v_o(t) = 0$. Equation (10.14-3) now becomes

$$\frac{d\varphi_{\text{osc}}}{dt} \approx \frac{d\varphi_s}{dt} + \frac{d\varphi_n}{dt} \tag{10.14-4}$$

Equation (10.14-4) shows that when the amount of feedback becomes very large, the frequency of the VCO approaches the sum of the frequency of the input signal and the frequency of the input noise. Note that the output of the multiplier becomes a very narrowband FM signal.

Thus, whenever $\varphi_n(t)$ rotates by 2π, resulting in a discriminator spike, we see from Eq. (10.14-4) that φ_{osc} also rotates by 2π, resulting in a spike in the FMFB. Thus there is no threshold extension when $G_0 \to \infty$.

We now show that if G_0 is neither zero nor very large, threshold extension results. We assume first that the modulation is sinusoidal and that the demodulation is performed by a conventional limiter-discriminator circuit. From Eqs. (10.5-18), (10.5-19), and (10.6-4) we find that the total average number of spikes per second N is

$$N_{\text{discr}} = N_c + \overline{\delta N}$$

$$= \frac{B}{2\sqrt{3}} \operatorname{erfc} \sqrt{\frac{f_M}{B} \frac{S_i}{N_M}} + \frac{2\Delta f}{\pi} \exp\left[-(f_M/B)(S_i/N_M) \right] \tag{10.14-5}$$

In Eq. (10.14-5), B is the bandwidth of the IF filter and Δf is the frequency deviation of the sinusoidally modulated carrier.

In the case of the FMFB, N is also given by Eq. (10.14-5), provided that B is replaced by B_p (see Fig. 10.13-1) and Δf is multiplied by the factor $1/(1 + \alpha G_0)$. Hence,

$$N_{\text{FMFB}} = \frac{B_p}{2\sqrt{3}} \operatorname{erfc} \sqrt{\frac{f_M}{B_p} \frac{S_i}{N_M}}$$

$$+ \frac{2\Delta f}{\pi(1 + \alpha G_0)} \exp\left[-(f_M/B_p)(S_i/N_M) \right] \tag{10.14-6}$$

Both N_{discr} and N_{FMFB} depend on the ratio S_i/N_M. It can, however, be shown that S_i/N_M is the same whether measured at the output of the IF filter or at the output of the filter of bandwidth B_p. However, since $B_p < B$ and since $1 + \alpha G_0 > 1$, we see in comparing Eqs. (10.14-5) and (10.14-6) that $N_{\text{FMFB}} < N_{\text{discr}}$ and that hence the threshold is extended.

PROBLEMS

10.2-1. (a) A carrier, $A \cos 2\pi f_c t$, is accompanied by noise having quadrature components $x(t)$ and $y(t)$. Assume that for an interval from $t = 0$ to $t = 1/f$, $x(t)$ and $y(t)$ may be approximated by

$$x(t) = \frac{A}{2} [\sin 2\pi ft]$$

$$y(t) = \frac{A}{2} [\cos 2\pi ft - 1]$$

Draw the path of the resultant phasor $R(t)$ in the phasor diagram of Fig. 10.2-2. Make a qualitative plot of $\theta(t)$ and of $d\theta/dt$.

(b) Repeat (a) if f is replaced by $2f$ in the expression for $y(t)$.

(c) Repeat (a) if the functions describing $x(t)$ and $y(t)$ are interchanged.

(d) Repeat (c) if the amplitude A in the expression for $x(t)$ and $y(t)$ is replaced by $3A$.

10.2-2. Consider a random pulse train where each pulse has a duration $\tau = 2/B$ and an amplitude πB. If the average time between pulses is T, and if the pulse train is filtered by a rectangular low-pass filter of bandwidth f_M,

(a) Show that the power measured at the filter output is $(4\pi^2/T)(2f_M)$, if $f_M \ll B/2$.

(b) Obtain an expression for the power measured at the filter output when f_M is equal to $B/2$.

10.3-1. A random pulse train consists of pulses given by the equation $p(t) = \sin Bt$, $0 \leq t \leq 2\pi/B$, separated by the average time interval T.

(a) Find $|P(f)|^2$.

(b) If the pulse train is filtered by a low-pass filter with a cutoff frequency $f_M \ll B$, show that the power measured at the filter output is very small in comparison with the power of the pulse train.

10.4-1. Threshold is defined as the value of $S_i/\eta f_M$ which results in S_o/N_o decreasing by 1 dB from the value

$$\frac{S_o}{N_o} = \frac{3k^2 \overline{m^2}}{4\pi^2 f_M^2} \frac{S_i}{\eta f_M}$$

(a) Show that when the carrier is unmodulated, threshold is reached when

$$0.26 = \sqrt{3} \frac{B}{f_M} \frac{S_i}{\eta f_M} \text{ erfc} \sqrt{\frac{f_M}{B} \frac{S_i}{\eta f_M}}$$

(b) Plot $S_i/\eta f_M$ at threshold as a function of B/f_M.

10.6-1. Consider that the signal and noise at the input to an FM system are

$$v_i(t) = \cos(\omega_c t + \Omega t) + x(t) \cos \omega_c t - y(t) \sin \omega_c t$$

and that

$$G_x(f) = G_y(f) = \eta \qquad -\frac{B}{2} \leq f \leq +\frac{B}{2}$$

(a) Show that $v_i(t)$ can also be written as

$$v_i(t) = \cos(\omega_c t + \Omega t) + x'(t) \cos(\omega_c t + \Omega t) + y'(t) \sin(\omega_c t + \Omega t)$$

Find $x'(t)$ and $y'(t)$.

(b) Find $G_{x'}(f) = G_{y'}(f)$. Show that if $\Omega = B/2$, the spectral density of x' (and y') extends to B Hz. Comment on the significance of this result with regard to the duration of a spike.

(c) Find $E[x'(t)y'(t)]$, i.e., show that since x and y are uncorrelated, x' and y' are also uncorrelated.

10.6-2. Using the results of Prob. 10.6-1 and Eqs. (10.5-1) and (10.5-2),

(a) Show that a negative spike occurs when $x' < -1$, y' goes through 0, and $\dot{y}' > 0$.

(b) Show also that a positive spike occurs when $x' < -1$, y' goes through 0, and $\dot{y}' < 0$.

(c) Using Eq. (10.5-4) with x replaced by x', and y by y', derive Eq. (10.6-1), where $\delta f = \Omega/2\pi$.

10.6-3. An FM carrier is modulated by the signal $m(t)$, which is gaussian, and bandlimited to f_M Hz. The instantaneous frequency deviation produced by $m(t)$ is $km(t)$. If the mean value of the square of $km(t)$ is $\overline{(\Delta\omega)^2}$, find $\overline{|\delta f|}$.

10.6-4. An FM carrier is sinusoidally modulated as in Eq. (10.6-2). Show, using Eqs. (10.6-4) and (10.5-19), that $\overline{\delta N} \gg N_c$.

10.6-5. Verify Eq. (10.6-6).

10.6-6. (a) Plot the ratio $S_i/\eta f_M$ at threshold as a function of β [use Eq. (10.6-6)].

(b) Compare your results with those found in Prob. 10.4-1b.

10.6-7. The FM signal $v(t) = \cos\left[\omega_c t + k \int_{-\infty}^{t} m(\lambda)\, d\lambda\right]$ is demodulated using an FM discriminator. If the signal $m(t)$ is gaussian and is bandlimited to f_M Hz, and $S_i/\eta f_M = 20$ dB, find the maximum rms deviation possible so that the discriminator is operating at threshold.

10.7-1. Show that if the two inputs to a phase comparator are $V_1(t) = A \cos[\omega_c t + \theta_1(t)]$ and $V_2(t) = B \cos[\omega_c t + \theta_2(t)]$, the output of the phase comparator is $v_o(t) = (AB/2) \cos[\theta_1(t) - \theta_2(t)] + (AB/2) \cos[2\omega_c t + \theta_1(t) + \theta_2(t)]$.

10.7-2. Show that Eq. (10.7-2) describes the output of a phase comparator when its input consists of "square" waves rather than sine waves.

10.8-1. A first-order PLL, with a phase-comparator characteristic as shown in Fig. 10.7-2b, is initially adjusted so that in the presence of a carrier alone $\psi = 0$. The VCO has the property that when its input is changed by 1V, its frequency changes by 10^5 Hz. The time constant of the PLL is $\tau = 10^{-4}$ sec.

(a) The carrier is frequency-modulated by a sinusoidal waveform. What is the maximum allowable peak frequency deviation of the carrier if the PLL is to recover the modulating waveform without distortion?

(b) With the frequency deviation at its peak as in (a), make a plot of the PLL output voltage v_0 as a function of the frequency of the modulating waveform.

10.8-2. Derive Eq. (10.8-6).

10.9-1. Derive Eq. (10.9-2).

10.9-2. A first-order PLL has a phase-comparator characteristic as in Fig. 10.7-2b. The input carrier is unmodulated and the PLL is adjusted so that it is in equilibrium with $\psi = 0$. By the application of external constraints, the operating point is forced to $\psi = 3\pi/4$. At the time $t = 0$, these constraints are removed.

(a) Write the differential equation for the phase angle ψ.

(b) If $\tau = 10^{-3}$ sec, calculate the time at which the operating point reaches B in Fig. 10.7-2b. Calculate $d\psi/dt$ at the time the constraints are released and at the time point B is reached.

10.10-1. A first-order PLL has a phase-comparator characteristic as in Fig. 10.7-2b. The PLL is adjusted so that, in the presence of a carrier alone, $\psi = 0$. The time constant of the PLL is $\tau = 10^{-4}$ sec. At time $t = 0$ the carrier is abruptly offset by 5 kHz. Plot ψ as a function of time up to the point where ψ attains the value $\psi = 3\pi/2$.

10.10-2. A first-order PLL has a phase characteristic as in Fig. 10.7-2b. The PLL is adjusted so that in the presence of a carrier alone $\psi = -\pi/4$. The time constant of the PLL is $\tau = 5 \times 10^{-5}$ sec and the sensitivity of the VCO is such that a 1V input changes its frequency by 10 kHz. At time $t = 0$ the carrier is abruptly offset by 7.5 kHz (in the direction to increase ψ for the interval $0 \leq t \leq 10^{-4}$ sec). Draw a plot of the output voltage waveform of the PLL.

10.10-3. A first-order PLL has a phase characteristic as in Fig. 10.7-2b. In the presence of a carrier alone $\psi = 0$. The time constant is $\tau = 10^{-3}$ sec and $G_0 = 1$ kHz/volt. At time $t = 0$ the carrier is offset by 0.5 kHz for the interval $0 \leq t \leq T$, i.e., the frequency of the carrier is modulated by a pulse of duration T.

(a) How long must the pulse last (say $T = T_0$) if at the end of the pulse $\psi = \pi$ (in which case the subsequent behavior of the PLL is indeterminate).

(b) Assume $T = 0.9T_0$. Draw the waveform at the output of the PLL.

(c) Assume $T = 1.1T_0$. Draw the waveform at the output of the PLL.

10.10-4. A first-order PLL is used to demodulate the signal $\cos [\omega_c t + \Omega t - \varphi_s(t)]$. The VCO is initially operating at a frequency ω_c. Assume $\Omega = \pi/4\tau$.

(a) Find $v_0(t)$ when $\varphi_s(t) = 0$.

(b) If after the PLL has reached equilibrium $\varphi_s(t) = -\lambda t, 0 \leq t \leq 2\pi/\lambda$ and zero elsewhere, find the minimum value of the product $\lambda\tau$ to *avoid* a spike.

(c) If $\varphi_s(t) = +\lambda t, 0 \leq t \leq 2\pi/\lambda$ and zero elsewhere, find the minimum value of the $\lambda\tau$ product to *avoid* a spike.

10.11-1. Find the differential equation of the piecewise-linear second-order PLL in the region $\pi/2 \leq \psi \leq 3\pi/2$.

10.11-2. Find the differential equation of the second-order PLL if the phase comparator has a sinusoidal characteristic.

10.12-1. Show that Fig. 10.12-1 represents the first-order PLL.

10.12-2. Plot N/f_M versus $S_i/\eta f_M$ at threshold. This is called the *threshold hyperbola*. If $S_i/\eta f_M = 20$ dB and $f_M = 5$ kHz, find N to produce threshold.

10.14-1. Show that the ratio $S_i/\eta f_M$ is unchanged when measured after the carrier filter or the bandpass filter.

10.14-2. Show that for the FMFB, $\delta N \gg N_c$. Use Eq. (10.14-6) and plot $N_c/\delta N$ as a function of $S_i/\eta B_p$ with $[(1 + \alpha G_0)B_p]/\Delta f = 4$ (this corresponds to choosing $1 + \alpha G_0 = \beta$, which represents reasonably good design).

10.14-3. Find the threshold extension possible for $\beta = 3$ and 5, if $1 + \alpha G_0 = \beta$.

REFERENCES

1. Rice, S. O.: "Time-series Analysis," chap. 25, John Wiley & Sons, Inc., New York, 1963.
2. Schilling, D. L., E. Nelson, and K. Clarke: Discriminator Response to an FM Signal in a Fading Channel, *IEEE Trans. Commun. Tech.*, April, 1967.
   Schilling, D. L., E. Hoffman, and E. Nelson: Error Rates For Digital Signals Demodulated by an FM Discriminator, *IEEE Trans. Commun. Tech.*, August, 1967.
   Nelson, E., and D. L. Schilling: The Response of an FM Discriminator to a Digital FM Signal in Randomly Fading Channels, *IEEE Trans. Commun. Tech.*, August, 1968.
3. Schilling, D. L.: The Response of an APC System to FM Signals and Noise, *Proc. IEEE*, October, 1963.
4. Schilling, D. L., and J. Billig: Threshold Extension Capability of the PLL and the FMFB, *Proc. IEEE*, May, 1964.
5. Osborne, P., and D. L. Schilling: Threshold Response of a Phase Locked Loop, *Proc. Int. Conf. Commun.*, 1968.
6. Hoffman, E., and D. L. Schilling: Threshold of the FMFB, *Proc. Int. Conf. Commun.*, 1969.

11
Data Transmission

A pulse-code modulation system using binary encoding transmits a sequence of binary digits, that is, 1s and 0s. These digits may be represented in a number of ways. For example, a 1 may be represented by a voltage V held for a time T, while a zero is represented by a voltage $-V$ held for an equal time. In general the binary digits are encoded so that a 1 is represented by a signal $s_1(t)$ and a 0 by a signal $s_2(t)$, where $s_1(t)$ and $s_2(t)$ each have a duration T. The resulting signal may be transmitted directly or, as is more usually the case, used to modulate a carrier as in PSK, DPSK, or FSK (see Secs. 11.7 to 11.10). The received signal is corrupted by noise, and hence there is a finite probability that the receiver will make an error in determining, within each time interval, whether a 1 or a 0 was transmitted.

In this chapter we make calculations of such error probabilities and discuss methods to minimize them. The discussion will lead us to the concept of the *matched filter*.

11.1   A BASEBAND SIGNAL RECEIVER

Consider that a binary-encoded PCM baseband signal consists of a time sequence of voltage levels $+V$ or $-V$. If there is a guard interval between the bits, the signal forms a sequence of positive and negative pulses. In either case there is no particular interest in preserving the waveform of the signal after reception. We are interested only in knowing within each bit interval whether the transmitted voltage was $+V$ or $-V$. With noise present, the received signal and noise together will yield sample values generally different from $\pm V$. In this case, what deduction shall we make from the sample value concerning the transmitted bit?

Suppose, as is always the case with the types of noise we consider, that the noise voltage has a probability density which is entirely symmetrical with respect to zero volts. Then the probability that the noise has increased the sample value is the same as the probability that the noise has decreased the sample value. It then seems entirely reasonable that we can do no better than to assume that if the sample value is positive the transmitted level was $+V$, and if the sample value is negative the transmitted level was $-V$. It is, of course, possible that at the sampling time the noise voltage may be of magnitude larger than V and of a polarity opposite to the polarity assigned to the transmitted bit. In this case an error will be made as indicated in Fig. 11.1-1. Here the transmitted bit is represented by the voltage $+V$ which is sustained over an interval T from t_1 to t_2. Noise has been superimposed on the level $+V$ so that the voltage v represents the received signal and noise. If now the sampling should happen to take place at a time $t = t_1 + \Delta t$, an error will have been made.

We can reduce the probability of error by processing the received signal plus noise in such a manner that we are then able to find a sample time where the sample voltage due to the signal is emphasized relative to the sample voltage due to the noise. Such a processer (receiver) is shown

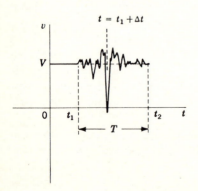

Fig. 11.1-1 Illustration that noise may cause an error in the determination of a transmitted voltage level.

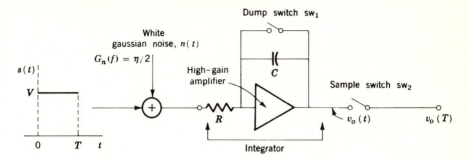

Fig. 11.1-2   A receiver for a binary coded baseband PCM signal.

in Fig. 11.1-2.   The signal input during a bit interval is indicated.   As a matter of convenience we have set $t = 0$ at the beginning of the interval. The waveform of the signal $s(t)$ before $t = 0$ and after $t = T$ has not been indicated since, as will appear, the operation of the receiver during each bit interval is independent of the waveform during past and future bit intervals.

The signal $s(t)$ with added white gaussian noise $n(t)$ of power spectral density $\eta/2$ is presented to an integrator.   At time $t = 0+$ we required that capacitor C be uncharged.   Such a discharged condition may be ensured by a brief closing of switch SW_1 at time $t = 0-$, thus relieving C of any charge it may have acquired during the previous interval.   The sample is taken at the output of the integrator by closing this sampling switch SW_2.   This sample is taken at the end of the bit interval, at $t = T$. The signal processing indicated in Fig. 11.1-2 is described by the phrase *integrate and dump*, the term *dump* referring to the abrupt discharge of the capacitor after each sampling.

PEAK SIGNAL TO RMS NOISE OUTPUT VOLTAGE RATIO

The integrator yields an output which is the integral of its input multiplied by $1/RC$.   Using $\tau = RC$, we have

$$v_o(T) = \frac{1}{\tau} \int_0^T [s(t) + n(t)]\, dt = \frac{1}{\tau} \int_0^T s(t)\, dt + \frac{1}{\tau} \int_0^T n(t)\, dt$$

$$\text{(11.1-1)}$$

The sample voltage due to the signal is

$$s_o(T) = \frac{1}{\tau} \int_0^T V\, dt = \frac{VT}{\tau} \tag{11.1-2}$$

The sample voltage due to the noise is

$$n_o(T) = \frac{1}{\tau} \int_0^T n(t)\, dt \tag{11.1-3}$$

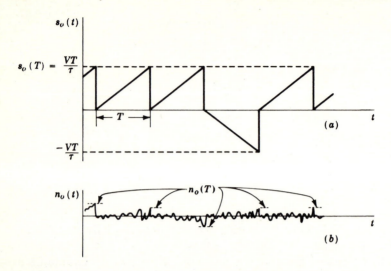

Fig. 11.1-3   (a) The signal output and (b) the noise output of the integrator of Fig. 11.1-2.

This noise-sampling voltage $n_o(T)$ is a gaussian random variable in contrast with $n(t)$, which is a gaussian random process.

The variance of $n_o(T)$ was found in Sec. 7.9 [see Eq. (7.9-17)] to be

$$\sigma_o^2 = \overline{n_o^2(T)} = \frac{\eta T}{2\tau^2} \tag{11.1-4}$$

and, as noted in Sec 7.3, $n_o(T)$ has a gaussian probability density.

The output of the integrator, before the sampling switch, is $v_o(t) = s_o(t) + n_o(t)$. As shown in Fig. 11.1-3a, the signal output $s_o(t)$ is a ramp, in each bit interval, of duration T. At the end of the interval the ramp attains the voltage $s_o(T)$ which is $+VT/\tau$ or $-VT/\tau$, depending on whether the bit is a 1 or a 0. At the end of each interval the switch SW_1 in Fig. 11.1-2 closes momentarily to discharge the capacitor so that $s_o(t)$ drops to zero. The noise $n_o(t)$, shown in Fig. 11.1-3b, also starts each interval with $n_o(0) = 0$ and has the random value $n_o(T)$ at the end of each interval. The sampling switch SW_2 closes briefly just before the closing of SW_1 and hence reads the voltage

$$v_o(T) = s_o(T) + n_o(T) \tag{11.1-5}$$

We would naturally like the signal voltage to be as large as possible in comparison with the noise voltage. Hence a figure of merit of interest is the signal-to-noise ratio

$$\frac{[s_o(T)]^2}{[n_o(T)]^2} = \frac{2}{\eta} V^2 T \tag{11.1-6}$$

This result is calculated from Eqs. (11.1-2) and (11.1-4). Note that the signal-to-noise ratio increases with increasing bit duration T and that it depends on V^2T which is the normalized energy of the bit signal. Therefore, a bit represented by a narrow, high amplitude signal and one by a wide, low amplitude signal are equally effective, provided V^2T is kept constant.

It is instructive to note that the integrator filters the signal and the noise such that the signal voltage increases linearly with time, while the standard deviation (rms value) of the noise increases more slowly, as $\sqrt{T}$. Thus, the integrator enhances the signal relative to the noise, and this enhancement increases with time as shown in Eq. (11.1-6).

11.2 PROBABILITY OF ERROR

Since the function of a PCM receiver is to distinguish the bit 1 from the bit 0 in the presence of noise, a most important characteristic is the probability that an error will be made in such a determination. We now calculate this error probability P_e for the integrate-and-dump receiver of Fig. 11.1-2.

We have seen that the probability density of the noise sample $n_o(T)$ is gaussian and hence appears as in Fig. 11.2-1. The density is therefore given by

$$f[n_o(T)] = \frac{e^{-n_o{}^2(T)/2\sigma_o{}^2}}{\sqrt{2\pi\sigma_o^2}} \qquad (11.2-1)$$

where σ_o^2, the variance, is $\sigma_o^2 \equiv \overline{n_o^2(T)}$ given by Eq. (11.1-4). Suppose, then, that during some bit interval the input-signal voltage is held at, say, $-V$. Then, at the sample time, the signal sample voltage is $s_o(T) = -VT/\tau$, while the noise sample is $n_o(T)$. If $n_o(T)$ is posi-

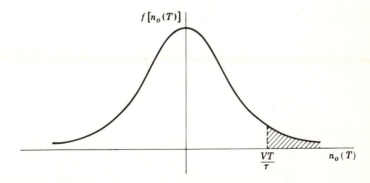

$f[n_o(T)]$

$\dfrac{VT}{\tau}$ $n_o(T)$

Fig. 11.2-1 The gaussian probability density of the noise sample $n_o(T)$.

tive and larger in magnitude than VT/τ, the total sample voltage $v_o(T) = s_o(T) + n_o(T)$ will be positive.   Such a positive sample voltage will result in an error, since as noted earlier, we have instructed the receiver to interpet such a positive sample voltage to mean that the signal voltage was $+V$ during the bit interval.   The probability of such a mis-interpretation, that is, the probability that $n_o(T) > VT/\tau$, is given by the area of the shaded region in Fig. 11.2-1.   The probability of error is, using Eq. (11.2-1),

$$P_e = \int_{VT/\tau}^{\infty} f[n_o(T)]\, dn_o(T) = \int_{VT/\tau}^{\infty} \frac{e^{-n_o^2(T)/2\sigma_o^2}}{\sqrt{2\pi\sigma_o^2}}\, dn_o(T) \qquad (11.2\text{-}2)$$

Defining $x \equiv n_o(T)/\sqrt{2}\,\sigma_o$, and using Eq. (11.1-4), Eq. (11.2-2) may be rewritten as

$$P_e = \frac{1}{2}\frac{2}{\sqrt{\pi}} \int_{x=V\sqrt{T/\eta}}^{\infty} e^{-x^2}\, dx$$

$$= \frac{1}{2}\operatorname{erfc}\left(V\sqrt{\frac{T}{\eta}}\right) = \frac{1}{2}\operatorname{erfc}\left(\frac{V^2T}{\eta}\right)^{\!\frac{1}{2}} = \frac{1}{2}\operatorname{erfc}\left(\frac{E_s}{\eta}\right)^{\!\frac{1}{2}} \qquad (11.2\text{-}3)$$

in which $E_s = V^2T$ is the signal energy of a bit.

If the signal voltage were held instead at $+V$ during some bit interval, then it is clear from the symmetry of the situation that the probability of error would again be given by P_e in Eq. (11.2-3).   Hence Eq. (11.2-3) gives P_e quite generally.

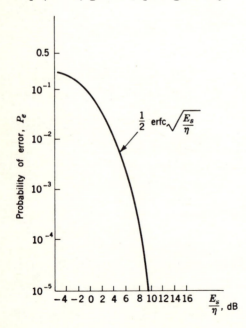

Fig. 11.2-2   Variation of P_e versus E_s/η.

The probability of error P_e, as given in Eq. (11.2-3), is plotted in Fig. 11.2-2. Note that P_e decreases rapidly as E_s/η increases. The maximum value of P_e is $\frac{1}{2}$. Thus, even if the signal is entirely lost in the noise so that any determination of the receiver is a sheer guess, the receiver cannot be wrong more than half the time on the average.

11.3 THE OPTIMUM FILTER

In the receiver system of Fig. 11.1-2, the signal was passed through a filter (i.e., the integrator), so that at the sampling time the signal voltage might be emphasized in comparison with the noise voltage. We are naturally led to ask whether the integrator is the optimum filter for the purpose of minimizing the probability of error. We shall find that for the received signal contemplated in the system of Fig. 11.1-2 the integrator is indeed the optimum filter. However, before returning specifically to the integrator receiver, we shall discuss optimum filters more generally.

We assume that the received signal is a binary PCM waveform. One binary bit is represented by a signal waveform $s_1(t)$ which persists for time T, while the other bit is represented by the waveform $s_2(t)$ which also lasts for an interval T. For example, in the case of transmission at baseband, as shown in Fig. 11.1-2, $s_1(t) = +V$, while $s_2(t) = -V$; for PSK signalling, $s_1(t) = A \cos \omega_0 t$ and $s_2(t) = -A \cos \omega_0 t$; while for FSK, $s_1(t) = A \cos (\omega_0 + \Omega)t$ and $s_2(t) = A \cos (\omega_0 - \Omega)t$.

As shown in Fig. 11.3-1 the input, which is $s_1(t)$ or $s_2(t)$, is corrupted by the addition of noise $n(t)$. The noise is gaussian and has a spectral density $G(f)$. [In most cases of interest the noise is white, so that $G(f) = \eta/2$. However, we shall assume the more general possibility, since it introduces no complication to do so.] The signal and noise are filtered and then sampled at the end of each bit interval. The output sample is either $v_o(T) = s_{o1}(T) + n_o(T)$ or $v_o(T) = s_{o2}(T) + n_o(T)$. We assume that immediately after each sample, every energy-storing element in the filter has been discharged.

In the absence of noise the output sample would be $v_o(T) = s_{o1}(T)$ or $s_{o2}(T)$. With noise, we shall assume that $s_1(t)$ has been transmitted

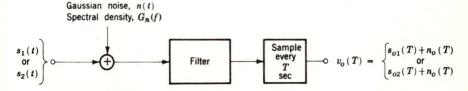

Fig. 11.3-1 A receiver for binary coded PCM.

if $v_o(T)$ is *closer* to $s_{o1}(T)$ than to $s_{o2}(T)$. Similarly, we assume $s_2(t)$ has been transmitted if $v_o(T)$ is *closer* to $s_{o2}(T)$. The decision boundary is therefore midway between $s_{o1}(T)$ and $s_{o2}(T)$. For example, in the baseband system of Fig. 11.1-2, where $s_{o1}(T) = VT/\tau$ and $s_{o2}(T) = -VT/\tau$, the decision boundary is $v_o(T) = 0$. In general, we shall take the decision boundary to be

$$v_o(T) = \frac{s_{o1}(T) + s_{o2}(T)}{2} \tag{11.3-1}$$

The probability of error for this general case may be deduced as an extension of the considerations used in the baseband case. Suppose that $s_{o1}(T) > s_{o2}(T)$ and that $s_2(t)$ was transmitted. If, at the sampling time, the noise $n_o(T)$ is positive and larger in magnitude than the voltage difference $\frac{1}{2}[s_{o1}(T) + s_{o2}(T)] - s_{o2}(T)$, an error will have been made. That is, an error [we decide that $s_1(t)$ is transmitted rather than $s_2(t)$] will result if

$$n_o(T) \geq \frac{s_{o1}(T) - s_{o2}(T)}{2} \tag{11.3-2}$$

Hence the probability of error is

$$P_e = \int_{[s_{o1}(T)-s_{o2}(T)]/2}^{\infty} \frac{e^{-n_o^2(T)/2\sigma_o^2}}{\sqrt{2\pi\sigma_o^2}} \, dn_o(T) \tag{11.3-3}$$

If we make the substitution $x \equiv n_o(T)/\sqrt{2}\,\sigma_o$, Eq. (11.3-3) becomes

$$P_e = \frac{1}{2}\frac{2}{\sqrt{\pi}} \int_{[s_{o1}(T)-s_{o2}(T)]/2\sqrt{2}\sigma_o}^{\infty} e^{-x^2} \, dx \tag{11.3-4a}$$

$$P_e = \frac{1}{2}\operatorname{erfc}\left[\frac{s_{o1}(T) - s_{o2}(T)}{2\sqrt{2}\,\sigma_o}\right] \tag{11.3-4b}$$

Note that for the case $s_{o1}(T) = VT/\tau$ and $s_{o2}(T) = -VT/\tau$, and, using Eq. (11.1-4), Eq. (11.3-4b) reduces to Eq. (11.2-3) as expected.

The complementary error function is a monotonically decreasing function of its argument. (See Fig. 11.2-2.) Hence, as is to be anticipated, P_e decreases as the difference $s_{o1}(T) - s_{o2}(T)$ becomes larger and as the rms noise voltage σ_o becomes smaller. The optimum filter, then, is the filter which maximizes the ratio

$$\gamma = \frac{s_{o1}(T) - s_{o2}(T)}{\sigma_o} \tag{11.3-5}$$

We now calculate the transfer function $H(f)$ of this optimum filter. As a matter of mathematical convenience we shall actually maximize γ^2 rather than γ.

CALCULATION[1] OF THE OPTIMUM-FILTER TRANSFER FUNCTION $H(f)$

The fundamental requirement we make of a binary-encoded PCM receiver is that it distinguishes the voltages $s_1(t) + n(t)$ and $s_2(t) + n(t)$. We have seen that the ability of the receiver to do so depends on how large a particular receiver can make γ. It is important to note that γ is proportional not to $s_1(t)$ nor to $s_2(t)$, but rather to the *difference* between them. For example, in the baseband system we represented the signals by voltage levels $+V$ and $-V$. But clearly, if our only interest was in distinguishing levels, we would do just as well to use $+2$ volts and 0 volt, or $+8$ volts and $+6$ volts, etc. (The $+V$ and $-V$ levels, however, have the advantage of requiring the least average power to be transmitted.) Hence, while $s_1(t)$ or $s_2(t)$ is the received signal, the signal which is to be compared with the noise, i.e., the signal which is relevant in all our error-probability calculations, is the difference signal

$$p(t) \equiv s_1(t) - s_2(t) \tag{11.3-6}$$

Thus, for the purpose of calculating the minimum error probability, we shall assume that the input signal is $p(t)$. The corresponding *output signal* of the filter is then

$$p_o(t) \equiv s_{o1}(t) - s_{o2}(t) \tag{11.3-7}$$

We shall let $P(f)$ and $P_o(f)$ be the Fourier transforms, respectively, of $p(t)$ and $p_o(t)$.

If $H(f)$ is the transfer function of the filter,

$$P_o(f) = H(f)P(f) \tag{11.3-8}$$

and

$$p_o(T) = \int_{-\infty}^{\infty} P_o(f)e^{j2\pi fT}\,df = \int_{-\infty}^{\infty} H(f)P(f)e^{j2\pi fT}\,df \tag{11.3-9}$$

The output noise $n_o(t)$ has a power spectral density $G_{n_o}(f)$ which is related to the power spectral density of the input noise $G_n(f)$ by

$$G_{n_o}(f) = |H(f)|^2 G_n(f) \tag{11.3-10}$$

Using Parseval's theorem (Eq. 1.12-5), we find that the normalized output noise power, i.e., the noise variance σ_o^2, is

$$\sigma_o^2 = \int_{-\infty}^{\infty} G_{n_o}(f)\,df = \int_{-\infty}^{\infty} |H(f)|^2 G_n(f)\,df \tag{11.3-11}$$

From Eqs. (11.3-9) and (11.3-11) we now find that

$$\gamma^2 = \frac{p_o^2(T)}{\sigma_o^2} = \frac{\left|\int_{-\infty}^{\infty} H(f)P(f)e^{j2\pi Tf}\,df\right|^2}{\int_{-\infty}^{\infty} |H(f)|^2 G_n(f)\,df} \tag{11.3-12}$$

Equation (11.3-12) is unaltered by the inclusion or deletion of the absolute value sign in the numerator since the quantity within the magnitude sign $p_o(T)$ is a positive real number. The sign has been included, however, in order to allow further development of the equation through the use of the *Schwarz inequality*.

The *Schwarz inequality* states that given arbitrary complex functions $X(f)$ and $Y(f)$ of a common variable f, then

$$\left| \int_{-\infty}^{\infty} X(f)Y(f)\,df \right|^2 \leq \int_{-\infty}^{\infty} |X(f)|^2\,df \int_{-\infty}^{\infty} |Y(f)|^2\,df \qquad (11.3\text{-}13)$$

The equal sign applies when

$$X(f) = KY^*(f) \qquad (11.3\text{-}14)$$

where K is an arbitrary constant and $Y^*(f)$ is the complex conjugate of $Y(f)$.

We now apply the Schwarz inequality to Eq. (11.3-12) by making the identification

$$X(f) \equiv \sqrt{G_n(f)}\, H(f) \qquad (11.3\text{-}15)$$

and

$$Y(f) \equiv \frac{1}{\sqrt{G_n(f)}}\, P(f)e^{j2\pi Tf} \qquad (11.3\text{-}16)$$

Using Eqs. (11.3-15) and (11.3-16) and using the Schwarz inequality, Eq. (11.3-13), we may rewrite Eq. (11.3-12) as

$$\frac{p_o^2(T)}{\sigma_o^2} = \frac{\left| \int_{-\infty}^{\infty} X(f)Y(f)\,df \right|^2}{\int_{-\infty}^{\infty} |X(f)|^2\,df} \leq \int_{-\infty}^{\infty} |Y(f)|^2\,df \qquad (11.3\text{-}17)$$

or, using Eq. (11.3-16),

$$\frac{p_o^2(T)}{\sigma_o^2} \leq \int_{-\infty}^{\infty} |Y(f)|^2\,df = \int_{-\infty}^{\infty} \frac{|P(f)|^2}{G_n(f)}\,df \qquad (11.3\text{-}18)$$

The ratio $p_o^2(T)/\sigma_o^2$ will attain its maximum value when the equal sign in Eq. (11.3-18) may be employed as is the case when $X(f) = KY^*(f)$. We then find from Eqs. (11.3-15) and (11.3-16) that the optimum filter which yields such a maximum ratio $p_o^2(T)/\sigma_o^2$ has a transfer function

$$H(f) = K \frac{P^*(f)}{G_n(f)}\, e^{-j2\pi fT} \qquad (11.3\text{-}19)$$

Correspondingly, the maximum ratio is, from Eq. (11.3-18),

$$\left[\frac{p_o^2(T)}{\sigma_o^2} \right]_{\max} = \int_{-\infty}^{\infty} \frac{|P(f)|^2}{G_n(f)}\,df \qquad (11.3\text{-}20)$$

In succeeding sections we shall have occasion to apply Eqs. (11.13-19) and (11.13-20) to a number of cases of interest.

11.4 WHITE NOISE; THE MATCHED FILTER

An optimum filter which yields a maximum ratio $p_o^2(T)/\sigma_o^2$ is called a *matched filter* when the input noise is *white*. In this case $G_n(f) = \eta/2$, and Eq. (11.3-19) becomes

$$H(f) = K \frac{P^*(f)}{\eta/2} e^{-j2\pi f T} \tag{11.4-1}$$

The impulsive response of this filter, i.e., the response of the filter to a unit strength impulse applied at $t = 0$, is

$$h(t) = \mathcal{F}^{-1}[H(f)] = \frac{2K}{\eta} \int_{-\infty}^{\infty} P^*(f)e^{-j2\pi f T}e^{j2\pi f t}\, df \tag{11.4-2a}$$

$$= \frac{2K}{\eta} \int_{-\infty}^{\infty} P^*(f)e^{j2\pi f(t-T)}\, df \tag{11.4-2b}$$

A physically realizable filter will have an impulse response which is real, i.e., not complex. Therefore $h(t) = h^*(t)$. Replacing the right-hand member of Eq. (11.4-2b) by its complex conjugate, an operation which leaves the equation unaltered, we have

$$h(t) = \frac{2K}{\eta} \int_{-\infty}^{\infty} P(f)e^{j2\pi f(T-t)}\, df \tag{11.4-3a}$$

$$= \frac{2K}{\eta} p(T - t) \tag{11.4-3b}$$

Finally, since $p(t) \equiv s_1(t) - s_2(t)$ [see Eq. (11.3-6)], we have

$$h(t) = \frac{2K}{\eta} [s_1(T - t) - s_2(T - t)] \tag{11.4-4}$$

The significance of these results for the matched filter may be more readily appreciated by applying them to a specific example. Consider then, as in Fig. 11.4-1a, that $s_1(t)$ is a triangular waveform of duration T, while $s_2(t)$, as shown in Fig. 11.4-1b, is of identical form except of reversed polarity. Then $p(t)$ is as shown in Fig. 11.4-1c, and $p(-t)$ appears in Fig. 11.4-1d. The waveform $p(-t)$ is the waveform $p(t)$ rotated around the axis $t = 0$. Finally, the waveform $p(T - t)$ called for as the impulse response of the filter in Eq. (11.4-3b) is this rotated waveform $p(-t)$ translated in the positive t direction by amount T. This last translation ensures that $h(t) = 0$ for $t < 0$ as is required for a *causal* filter.

In general, the impulsive response of the matched filter consists of $p(t)$ rotated about $t = 0$ and then delayed long enough (i.e., a time T) to

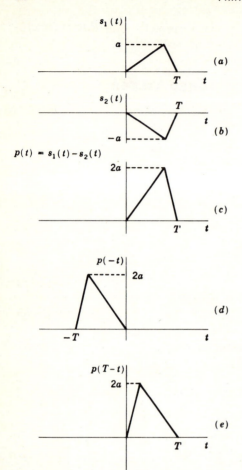

Fig. 11.4-1 The signals (a) $s_1(t)$, (b) $s_2(t)$, and (c) $p(t) = s_1(t) - s_2(t)$. (d) $p(t)$ rotated about the axis $t = 0$. (e) The waveform in (d) translated to the right by amount T.

make the filter realizable. We may note, in passing, that any additional delay that a filter might introduce would in no way interfere with the performance of the filter, for both signal and noise would be delayed by the same amount, and at the sampling time (which would need similarly to be delayed) the ratio of signal to noise would remain unaltered.

11.5 PROBABILITY OF ERROR OF THE MATCHED FILTER

The probability of error which results when employing a matched filter, may be found by evaluating the maximum signal-to-noise ratio $[p_o^2(T)/\sigma_o^2]_{\max}$ given by Eq. (11.3-20). With $G_n(f) = \eta/2$, Eq. (11.3-20) becomes

$$\left[\frac{p_o^2(T)}{\sigma_o^2}\right]_{\max} = \frac{2}{\eta}\int_{-\infty}^{\infty} |P(f)|^2\, df \tag{11.5-1}$$

From Parseval's theorem we have

$$\int_{-\infty}^{\infty} |P(f)|^2 \, df = \int_{-\infty}^{\infty} p^2(t) \, dt = \int_{0}^{T} p^2(t) \, dt \qquad (11.5\text{-}2)$$

In the last integral in Eq. (11.5-2), the limits take account of the fact that $p(t)$ persists for only a time T. With $p(t) = s_1(t) - s_2(t)$, and using Eq. (11.5-2), we may write Eq. (11.5-1) as

$$\left[\frac{p_o^2(T)}{\sigma_o^2} \right]_{\max} = \frac{2}{\eta} \int_0^T [s_1(t) - s_2(t)]^2 \, dt \qquad (11.5\text{-}3a)$$

$$= \frac{2}{\eta} \left[\int_0^T s_1^2(t) \, dt + \int_0^T s_2^2(t) \, dt - 2 \int_0^T s_1(t) s_2(t) \, dt \right] \qquad (11.5\text{-}3b)$$

$$= \frac{2}{\eta} (E_{s1} + E_{s2} - 2E_{s12}) \qquad (11.5\text{-}3c)$$

Here E_{s1} and E_{s2} are the energies, respectively, in $s_1(t)$ and $s_2(t)$, while E_{s12} is the joint energy in the product waveform $\sqrt{s_1(t)s_2(t)}$.

Suppose that we have selected $s_1(t)$, and let $s_1(t)$ have an energy E_{s1}. Then it can be shown that if $s_2(t)$ is to have the *same energy*, the optimum choice of $s_2(t)$ is

$$s_2(t) = -s_1(t) \qquad (11.5\text{-}4)$$

The choice is optimum in that it yields a maximum output signal $p_o^2(T)$ for a given signal energy. Letting $s_2(t) = -s_1(t)$, we find

$$E_{s1} = E_{s2} = -E_{s12} \equiv E_s$$

and Eq. (11.5-3c) becomes

$$\left[\frac{p_o^2(T)}{\sigma_o^2} \right]_{\max} = \frac{8E_s}{\eta} \qquad (11.5\text{-}5)$$

Rewriting Eq. (11.3-4b) using $p_o(T) = s_{o1}(T) - s_{o2}(T)$, we have

$$P_e = \frac{1}{2} \, \text{erfc} \left[\frac{p_o(T)}{2\sqrt{2}\,\sigma_o} \right] = \frac{1}{2} \, \text{erfc} \left[\frac{p_o^2(T)}{8\sigma_o^2} \right]^{1/2} \qquad (11.5\text{-}6)$$

Combining Eq. (11.5-6) with (11.5-5), we find that the minimum error probability $(P_e)_{\min}$ corresponding to a maximum value of $p_o^2(T)/\sigma_o^2$ is

$$(P_e)_{\min} = \frac{1}{2} \, \text{erfc} \left\{ \frac{1}{8} \left[\frac{p_o^2(T)}{\sigma_o^2} \right]_{\max} \right\}^{1/2} \qquad (11.5\text{-}7)$$

$$= \frac{1}{2} \, \text{erfc} \left(\frac{E_s}{\eta} \right)^{1/2} \qquad (11.5\text{-}8)$$

We note that Eq. (11.5-8) establishes more generally the idea that the error probability depends only on the signal energy and not on the signal waveshape. Previously we had established this point only for signals which had constant voltage levels.

We note also that Eq. (11.5-8) gives $(P_e)_{min}$ for the case of the matched filter and when $s_1(t) = -s_2(t)$. In Sec. 11.2 we considered the case when $s_1(t) = +V$ and $s_2(t) = -V$ and the filter employed was an integrator. There we found [Eq. (11.2-3)] that the result for P_e was identical with $(P_e)_{min}$ given in Eq. (11.5-8). This agreement leads us to suspect that for an input signal where $s_1(t) = +V$ and $s_2(t) = -V$, the integrator is the matched filter. Such is indeed the case. For when we have

$$s_1(t) = V \qquad 0 \le t \le T \tag{11.5-9a}$$

and

$$s_2(t) = -V \qquad 0 \le t \le T \tag{11.5-9b}$$

the impulse response of the matched filter is, from Eq. (11.4-4),

$$h(t) = \frac{2K}{\eta} [s_1(T - t) - s_2(T - t)] \tag{11.5-10}$$

The quantity $s_1(T - t) - s_2(T - t)$ is a pulse of amplitude $2V$ extending from $t = 0$ to $t = T$ and may be rewritten, with $u(t)$ the unit step,

$$h(t) = \frac{2K}{\eta} (2V)[u(t) - u(t - T)] \tag{11.5-11}$$

The constant factor of proportionality $4KV/\eta$ in the expression for $h(t)$ (that is, the gain of the filter) and has no effect on the probability of error since the gain affects signal and noise alike. We may therefore select the coefficient K in Eq. (11.5-11) so that $4KV/\eta = 1$. Then the inverse transform of $h(t)$, that is, the transfer function of the filter, becomes, with s the Laplace transform variable,

$$H(s) = \frac{1}{s} - \frac{e^{-sT}}{s} \tag{11.5-12}$$

The first term in Eq. (11.5-12) represents an integration beginning at $t = 0$, while the second term represents an integration with reversed polarity beginning at $t = T$. The overall response of the matched filter is an integration from $t = 0$ to $t = T$ and a zero response thereafter. In a physical system, as already described, we achieve the effect of a zero response after $t = T$ by sampling at $t = T$, so that so far as the determination of one bit is concerned we ignore the response after $t = T$.

11.6 COHERENT RECEPTION:CORRELATION

We discuss now an alternative type of receiving system which, as we shall see, is identical in performance with the matched filter receiver. Again, as shown in Fig. 11.6-1, the input is a binary PCM waveform $s_1(t)$ or

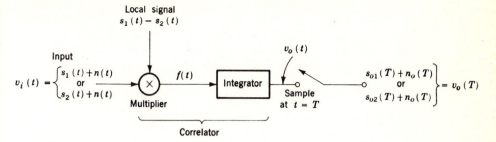

Fig. 11.6-1 A coherent system of signal reception.

$s_2(t)$ corrupted by noise $n(t)$. The bit length is T. The received signal
plus noise $v_i(t)$ is multiplied by a locally generated waveform $s_1(t) - s_2(t)$.
The output of the multiplier is passed through an integrator whose output
is sampled at $t = T$. As before, immediately after each sampling, at the
beginning of each new bit interval, all energy-storing elements in the
integrator are discharged. This type of receiver is called a *correlator*,
since we are *correlating* the received signal and noise with the waveform
$s_1(t) - s_2(t)$.

The output signal and noise of the correlator shown in Fig. 11.6-1 are

$$s_o(T) = \frac{1}{\tau} \int_0^T s_i(t)[s_1(t) - s_2(t)] \, dt \qquad (11.6\text{-}1)$$

$$n_o(T) = \frac{1}{\tau} \int_0^T n(t)[s_1(t) - s_2(t)] \, dt \qquad (11.6\text{-}2)$$

where $s_i(t)$ is either $s_1(t)$ or $s_2(t)$, and where τ is the constant of the integrator (i.e., the integrator output is $1/\tau$ times the integral of its input).
We now compare these outputs with the matched filter outputs.

If $h(t)$ is the impulsive response of the matched filter, then the output of the matched filter $v_o(t)$ can be found using the convolution integral
(see Sec. 1.11). We have

$$v_o(t) = \int_{-\infty}^{\infty} v_i(\lambda)h(t - \lambda) \, d\lambda = \int_0^T v_i(\lambda)h(t - \lambda) \, d\lambda \qquad (11.6\text{-}3)$$

The limits on the integral have been changed to 0 and T since we are
interested in the filter response to a bit which extends only over that
interval. Using Eq. (11.4-4) which gives $h(t)$ for the matched filter,
we have

$$h(t) = \frac{2K}{\eta} [s_1(T - t) - s_2(T - t)] \qquad (11.6\text{-}4)$$

so that

$$h(t - \lambda) = \frac{2K}{\eta} [s_1(T - t + \lambda) - s_2(T - t + \lambda)] \qquad (11.6\text{-}5)$$

Substituting Eq. (11.6-5) into (11.6-3), we have

$$v_o(t) = \frac{2K}{\eta} \int_0^T v_i(\lambda)[s_1(T - t + \lambda) - s_2(T - t + \lambda)] \, d\lambda \qquad (11.6\text{-}6)$$

Since $v_i(\lambda) = s_i(\lambda) + n(\lambda)$, and $v_o(t) = s_o(t) + n_o(t)$, setting $t = T$ yields

$$s_o(T) = \frac{2K}{\eta} \int_0^T s_i(\lambda)[s_1(\lambda) - s_2(\lambda)] \, d\lambda \qquad (11.6\text{-}7)$$

where $s_i(\lambda)$ is equal to $s_1(\lambda)$ or $s_2(\lambda)$. Similarly we find that

$$n_o(T) = \frac{2K}{\eta} \int_0^T n(\lambda)[s_1(\lambda) - s_2(\lambda)] \, d\lambda \qquad (11.6\text{-}8)$$

Thus $s_o(T)$ and $n_o(T)$, as calculated from Eqs. (11.6-1) and (11.6-2) for the coherent receiver, and as calculated from Eqs. (11.6-7) and (11.6-8) for the matched filter receiver, are identical. Hence the performances of the two systems are identical.

The *matched filter* and the *correlator* are not simply two distinct, independent techniques which happen to yield the same result. In fact they are two techniques of synthesizing the optimum filter $h(t)$.

11.7 PHASE-SHIFT KEYING

An important application of the coherent reception system of Sec. 11.6 is its use in phase-shift keying (PSK). Here the input signal is

$$s_1(t) = A \cos \omega_0 t \qquad (11.7\text{-}1)$$

or

$$s_2(t) = -A \cos \omega_0 t \qquad (11.7\text{-}2)$$

At the receiver a coherent local signal $s_1(t) - s_2(t) = 2A \cos \omega_0 t$ needs to be provided for the multiplier (see Fig. 11.6-1).

Since, in PSK, $s_1(t) = -s_2(t)$, Eq. (11.5-8) gives the error probability. Then, in PSK, as in baseband transmission,

$$P_e = \frac{1}{2} \operatorname{erfc} \sqrt{\frac{E_s}{\eta}} \qquad (11.7\text{-}3)$$

If a bit duration extends for a time T, which encompasses a whole number of cycles, then the signal energy is $E_s = A^2 T / 2$ so that from Eq. (11.7-3) the error probability is

$$P_e = \frac{1}{2} \operatorname{erfc} \sqrt{\frac{A^2 T}{2\eta}} \qquad (11.7\text{-}4)$$

IMPERFECT SYNCHRONIZATION

In PSK, where the signals $s_1(t)$ and $s_2(t)$ are as given in Eqs. (11.7-1) and (11.7-2), the required local waveform for the correlator in Fig. 11.6-1 is $s_1(t) - s_2(t) = 2A \cos \omega_0 t$. Then, when $s_1(t)$ is received, the output of the matched filter at the sampling time $t = T$ is $s_{o1}(T) = cA^2T$, where c is some constant depending on the gain of the integrator. Similarly, $s_{o2}(T) = -cA^2T$. The quantity $p_o(T)$ in Eq. (11.5-6), being $p_o(T) \equiv s_{o1}(T) - s_{o2}(T)$, is given by $p_o(T) = 2cA^2T$.

Suppose now that the local signal used at the correlator were not $2A \cos \omega_0 t$ as required, but rather $2A \cos (\omega_0 t + \phi)$, where ϕ is some fixed phase offset. Then, as is easily verified, the quantity $p_o(T)$ would become $2cA^2T \cos \phi$, that is, $p_o(T)$ would be reduced, being multiplied by the factor $\cos \phi$. In this case Eq. (11.5-8) or (11.7-3) would be replaced by

$$P_e = \frac{1}{2} \operatorname{erfc} \sqrt{\frac{E_s}{\eta} \cos^2 \phi} \tag{11.7-5}$$

while Eq. (11.7-4) would read

$$P_e = \frac{1}{2} \operatorname{erfc} \sqrt{\frac{A^2T \cos^2 \phi}{2\eta}} \tag{11.7-6}$$

The phase shift ϕ increases the probability of error P_e. Typical error probabilities in a communication system range from 10^{-4} to 10^{-7}. In this range, if $\phi = 25°$, the probability of error is increased by a factor of 10 as compared with the result obtained for $\phi = 0$.

11.8   FREQUENCY-SHIFT KEYING

In frequency-shift keying (FSK) the received signal is either

$$s_1(t) = A \cos (\omega_0 + \Omega)t \tag{11.8-1}$$

or

$$s_2(t) = A \cos (\omega_0 - \Omega)t \tag{11.8-2}$$

As was explained in Sec. 11.6, one way of synthesizing the matched filter is to construct the correlation receiver system of Fig. 11.6-1. This receiver will give precisely the same performance as a matched filter, provided that the local waveform is $s_1(t) - s_2(t)$. In FSK the required local waveform is

$$s_1(t) - s_2(t) = A \cos (\omega_0 + \Omega)t - A \cos (\omega_0 - \Omega)t \tag{11.8-3}$$

PROBABILITY OF ERROR

In Sec. 11.5 we calculated the probability of error for the matched filter and arrived at the result $P_e = \frac{1}{2} \operatorname{erfc} \sqrt{E_s/\eta}$ given in Eq. (11.5-8). The derivation was general, and would apply in the present case, except for the fact that we had assumed there that $s_1(t) = -s_2(t)$. This assumption is obviously not valid for FSK.

To calculate the probability of error for FSK, we return to a point in the derivation of Sec. 11.5 just before the introduction of the assumption $s_1(t) = -s_2(t)$. We start with Eq. (11.5-3a), which reads

$$\left[\frac{p_o^2(T)}{\sigma_o^2} \right]_{\max} = \frac{2}{\eta} \int_0^T [s_1(t) - s_2(t)]^2 \, dt \tag{11.8-4}$$

Substituting $s_1(t)$ and $s_2(t)$ as given in Eqs. (11.8-1) and (11.8-2) into Eq. (11.8-4) and performing the indicated integration, we find that

$$\left[\frac{p_o^2(T)}{\sigma_o^2} \right]_{\max} = \frac{2A^2T}{\eta} \left[1 - \frac{\sin 2\Omega T}{2\Omega T} + \frac{1}{2} \frac{\sin [2(\omega_0 + \Omega)T]}{2(\omega_0 + \Omega)T} \right. $$
$$\left. - \frac{1}{2} \frac{\sin [2(\omega_0 - \Omega)T]}{2(\omega_0 - \Omega)T} - \frac{\sin 2\omega_0 T}{2\omega_0 T} \right] \tag{11.8-5}$$

If we assume that the offset angular frequency Ω is very small in comparison with the carrier angular frequency ω_0 (a situation usually encountered in physical systems), then the last three terms in Eq. (11.8-5) each have the form $(\sin 2\omega_0 T)/2\omega_0 T$. This ratio approaches zero as $\omega_0 T$ increases. We further assume, as is generally the case, that $\omega_0 T \gg 1$. We may therefore neglect these last three terms. We are left with

$$\left[\frac{p_o^2(T)}{\sigma_o^2} \right]_{\max} = \frac{2A^2T}{\eta} \left(1 - \frac{\sin 2\Omega T}{2\Omega T} \right) \tag{11.8-6}$$

The quantity $[p_o^2(T)/\sigma_o^2]_{\max}$ in Eq. (11.8-6) attains its largest value when Ω is selected so that $2\Omega T = 3\pi/2$. For this value of Ω we find

$$\left[\frac{p_o^2(T)}{\sigma_o^2} \right]_{\max} = 2.42 \frac{A^2T}{\eta} = 4.84 \frac{(A^2/2)T}{\eta} \tag{11.8-7}$$

The probability of error, calculated using Eq. (11.5-6) with $[p_o^2(T)/\sigma_o^2]_{\max}$ as given in Eq. (11.8-7), is found to be

$$P_e = \frac{1}{2} \operatorname{erfc} \left\{ \frac{1}{8} \left[\frac{p_o^2(T)}{\sigma_o^2} \right]_{\max} \right\}^{1/2} \simeq \frac{1}{2} \operatorname{erfc} \left(0.6 \frac{E_s}{\eta} \right)^{1/2} \tag{11.8-8}$$

where the signal energy is $E_s = A^2T/2$.

Comparing the probability of error obtained for FSK [Eq. (11.8-8)] with the probability of error obtained for PSK [Eq. (11.7-3)], we see that equal probability of error in each system can be achieved if the signal energy in the PSK signal is 0.6 times as large as the signal energy in FSK. As a result, a 2-dB increase in the transmitted signal power is required for FSK. Why is FSK inferior to PSK? The answer is that in PSK $s_1(t) = -s_2(t)$, while in FSK this condition is not satisfied. Thus, although an optimum filter is used in each case, PSK results in considerable improvement compared with FSK.

11.9 NONCOHERENT DETECTION OF FSK

Since FSK can be thought of as the transmission of the output of either of two signal sources, the first at the frequency $\omega_1 = \omega_0 + \Omega$, and the second at the frequency $\omega_2 = \omega_0 - \Omega$, we should think that a reasonable detection system would consist of two bandpass filters, with center frequencies at ω_1 and ω_2.   The bandwidth of each filter would be adjusted to yield a maximum output when the appropriate signal is received.   Thus, when filter H_1 with center frequency ω_1 has a larger output than filter H_2 with center frequency ω_2, we would decide that $s_1(t)$ was transmitted.   Similarly we would decide that $s_2(t)$ was transmitted when the output of H_2 was greater than the output of H_1.

When using a filter receiver, we make no use of the *phase* of the incoming signal.   Such reception is, therefore, called *noncoherent* detection.   The *coherent* matched-filter detector, on the other hand, uses synchronization techniques to determine the phase of the incoming signal. Since some valuable information concerning the signal is not used, the probability of detecting the signal is reduced.   The probability of error of noncoherent FSK is found to be[2]

$$P_e = \tfrac{1}{2}e^{-E_s/2\eta} \tag{11.9-1}$$

This probability of error is compared in Sec. 11.11 with the other systems discussed.

11.10 DIFFERENTIAL PSK

The operation of differential phase-shift keying (DPSK) was explained in Sec. 6.18.   We shall now show the *suboptimum* nature of DPSK by considering the system in terms of the phasor diagram shown in Fig. 11.10-1.   In Fig. 11.10-1a we see that when no noise is present the received phase is either at angle 0 or π.   From this we draw a decision boundary at angle $\pi/2$ and decide that a 1 was sent if the phase difference between two consecutive bits differs by less than $\pi/2$, or we decide that a 0 was sent if the phase difference between two consecutive bits differs by more than $\pi/2$.

Figure 11.10-1b shows three consecutive received bits.   Each bit was transmitted as a 1, but because of noise each is perturbed from the horizontal axis as shown.   The DPSK receiver compares bit 2 with bit 1, reads an angle θ_1 which is less than 90°, and decides that bit 2 is a 1.   The DPSK receiver then compares bit 3 with bit 2, reads an angle θ_2 which is greater than 90°, and decides that a 0 was transmitted.

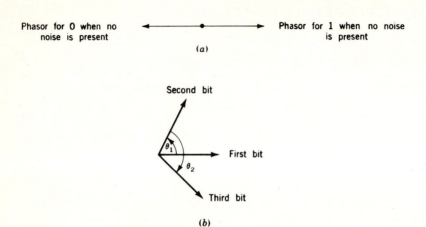

Fig. 11.10-1 Reception of DPSK. (*a*) Phasors when no noise is present. (*b*) Phasors when noise is present.

The error was due to the fact that the DPSK receiver uses only the previous bit as a reference. This method of operation is analogous to employing *poor* synchronization. If all the previous positive bits were somehow averaged by employing good synchronization, and this average were used as a reference, then the DPSK receiver would have a *stable* reference, and the error described above would not have occurred. As a matter of fact we would then have a PSK system, not a DPSK system. Thus, DPSK is *suboptimum* and results in a higher probability of error than in PSK where we have a *stable* reference phase (when perfect synchronization is assumed).

The calculation of probability of error is complicated and will not be given here.[2] The result is

$$P_e = \tfrac{1}{2}e^{-E_s/\eta} \tag{11.10-1}$$

The probability of error of DPSK and PSK are compared in Sec. 11.11.

11.11 COMPARISON OF DATA TRANSMISSION SYSTEMS

To compare the probability of error of the various data transmission systems discussed above, we refer to Fig. 11.11-1. Here we see that at small error probabilities, an increase of 1 dB in signal energy would result in DPSK performing as well as PSK. Approximately a 2-dB increase in energy is needed to have FSK yield the same probability of error, and approximately a 4-dB increase is required for noncoherent FSK.

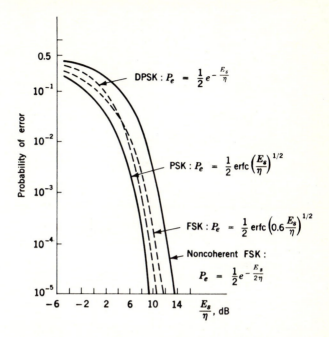

Fig. 11.11-1 Probability of error of data transmission systems.

Error probabilities of 10^{-5} are typical, and it is standard procedure to assume this probability of error and find the decibel difference in signal energy between systems, rather than to pick E_s/η and find the difference in P_e. The reader may well ask the significance of 2 dB. Here again, as in our discussion of threshold extension (Chap. 10), even 1 dB is extremely costly. One major communications company estimates an annual savings of many millions of dollars for every 1 dB that it can save.

11.12 FOUR-PHASE PSK

In previous sections of this chapter we dealt exclusively with binary communication systems, that is, systems in which, in any interval $0 \leq t \leq T$, one of two possible messages was transmitted. However, data transmission systems allowing many possible messages in the interval T (so-called M-ary systems, the number of messages being M) are also possible and are widely used. As an example of such a M-ary system we consider 4-phase PSK, which, because of its relative simplicity, is very popular.

In 4-phase PSK one of four possible waveforms is transmitted during each interval T. These waveforms are $s_1(t) = A \cos \omega_0 t$, $s_2(t) = -A \sin \omega_0 t$, $s_3(t) = -A \cos \omega_0 t$, and $s_4(t) = A \sin \omega_0 t$. These four waveforms are

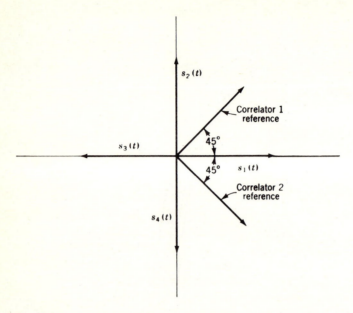

Fig. 11.12-1 A phasor diagram representation of the signals and local references in 4-phase PSK.

represented in the phasor diagram of Fig. 11.12-1. The receiver system is shown in Fig. 11.12-2. Observe that two correlators are required and that the local reference waveforms, as indicated also in Fig. 11.12-1, are $A \cos (\omega_0 t + 45°)$ and $A \cos (\omega_0 t - 45°)$.

Suppose, now, that, in the absence of the noise, the signal $s_1(t)$ is received. Let us use the symbol V_o to represent the corresponding output of correlator 1, i.e., $V_o \equiv v_{o1}(T)$ when $s_1(t)$ is received. Then, as is readily verified, the output of the two correlators corresponding to each of the four possible signals is as given in Table 11.12-1.

Thus the transmitted signal may be recognized from a determination of the outputs of *both* correlators. In the presence of noise, of course,

Table 11.12-1

Output	Signal			
	$s_1(t)$	$s_2(t)$	$s_3(t)$	$s_4(t)$
$v_{o1}(T)$	$+V_o$	$+V_o$	$-V_o$	$-V_o$
$v_{o2}(T)$	$+V_o$	$-V_o$	$-V_o$	$+V_o$

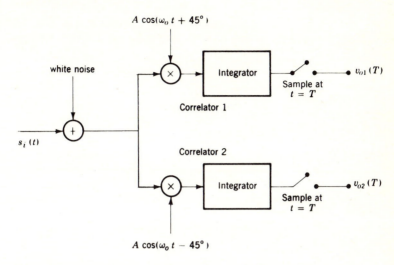

Fig. 11.12-2   A correlation receiver for 4-phase PSK.

there will be some probability that an error will be made by one or both correlators.

We note from Fig. 11.12-1, that the reference waveform of correlator 1 is at an angle $\phi = 45°$ to the axes of orientation of all of the four possible signals.   Hence, from Eq. (11.7-6), since $(\cos 45°)^2 = \frac{1}{2}$, the probability that correlator 1 will make an error is

$$P'_{e1} = \frac{1}{2}\,\text{erfc}\,\sqrt{\frac{A^2T}{4\eta}} \tag{11.12-1}$$

The probability P'_{e2} that correlator 2 will make an error is similarly given by the expression in Eq. (11.12-1).   The probability P_c that the 4-phase receiver will correctly identify the transmitted signal is equal to the product of the probabilities that both correlator 1 and correlator 2 have yielded correct results.   Thus, using $P'_e = P'_{e1} = P'_{e2}$

$$P_c = (1 - P'_e)(1 - P'_e) = 1 - 2P'_e + P'^2_e \tag{11.12-2}$$

If, as is normally the case, $P'_e \ll 1$, the last term in Eq. (11.12-2) may be neglected.   Finally, then, the probability of error of the system is

$$P_e(\text{4-phase PSK}) = 1 - P_c \simeq 2P'_e = \text{erfc}\,\sqrt{\frac{A^2T}{4\eta}} \tag{11.12-3}$$

Comparing Eq. (11.12-3) with Eq. (11.7-4) we see that the error probability in 4-phase PSK is greater than in 2-phase (binary) PSK.   This disadvantage is the price we pay for the advantage of being able to select from among more possible messages during any interval T.

The derivation above has assumed that the noise at the output of the two correlators is uncorrelated. If, for example, the two noise outputs were perfectly correlated, then both correlators would always be both right or both wrong. We have however assumed that the correlators, when they make errors, do so independently. It can be shown (Prob. 11.12-3) that when the period of integration T extends one or more *complete* cycles of the carrier waveform (a rather common arrangement) the noise outputs of the correlators, at the sampling times, are statistically independent of one another.

PROBLEMS

11.1-1. (a) Find the power spectral density $G_n(f)$ of noise $n(t)$ which has an autocorrelation function

$$R_n(\tau) = \sigma^2 e^{-|\tau/\tau_0|}$$

(b) The noise in (a) is applied to an integrator at $t = 0$. Find the mean square value $\overline{[n_0(T)]^2}$ of the noise output of the integrator at $t = T$.

(c) The noise in (a) accompanies a signal which consists of either the voltage $+V$ or the voltage $-V$ sustained for a time T. At time $t = T$ find the ratio of the integrator output due to the signal to the rms noise voltage.

11.1-2. A received signal $s(t) = \pm V$ is held for an interval T. The signal is accompanied by white gaussian noise of power spectral density $\eta/2$. The received signal is to be processed as in Fig. 11.1-2. However, as an approximation to the required integrator we use a low-pass RC circuit of 3-dB bandwidth f_c. Calculate the value of f_c for which the signal-to-noise voltage, at the sampling time, will be a maximum. For this value of f_c calculate the signal-to-rms noise ratio and compare with Eq. (11.1-6) which applies when an integrator is used. Show that for the RC network the signal-to-noise ratio is about 1 dB smaller than for the integrator.

11.2-1. A signal which can assume one of the voltages $+V$ or $-V$ is transmitted. Consider that the probability of transmitting $+V$ is $\frac{3}{4}$ while the probability of transmitting $-V$ is $\frac{1}{4}$. The signal is accompanied by white gaussian noise.

(a) Assume that the threshold voltage for decision between the two possible signals is V_t rather than zero volts. Write an expression for the probability that an error in decision will be made. An integrate and dump receiver is used as in Fig. 11.1-2.

(b) Find V_t such that the probability of error is a minimum and calculate the corresponding probability of error.

11.2-2. A signal, which can take on the voltages $+V$, 0, $-V$ with equal likelihood, is transmitted. When received, it is embedded in white gaussian noise. The receiver integrates the signal and noise for T sec.

Write an expression for the threshold voltages $\pm V_t$ so that the probability of error is independent of which signal is transmitted.

11.2-3. A received signal is either $+2V$ or $-2V$ held for a time T. The signal is corrupted by white gaussian noise of power spectral density 10^{-4} volt²/Hz. If the signal is processed by an integrate and dump receiver, what is the minimum time T during which a signal must be sustained if the probability of error is not to exceed 10^{-4}?

11.3-1. A transmitter transmits the signals $\pm V$ with equal probability; the channel noise has the power spectral density $G_n(f) = G_0/[1 + (f/f_1)^2]$.

    (a) Find the transfer function $H(f)$ of the *optimum filter*, and comment on its realizability.

    (b) Find the average probability of error when using the optimum filter.

    (c) An integrator is employed rather than the optimum filter. Find its P_e and compare with (b).

11.5-1. A signal is either $s_1(t) = A \cos 2\pi f_0 t$ or $s_2(t) = 0$ for an interval $T = n/f_0$ with n an integer. The signal is corrupted by white noise with $G_n(f) = \eta/2$. Find the transfer function of the matched filter for this signal. Write an expression for the probability of error P_e.

11.5-2. Repeat Prob. 11.5-1 if the signal is $s(t) = \pm A(1 - \cos 2\pi f_0 t)$.

11.6-1. Compare the outputs of the MF and the correlator, when the input signal is either $\pm V$, as a function of time t for $0 \leq t \leq T$. Assume white gaussian noise. Are the outputs the same for all t, or just when $t = T$?

11.6-2. A signal is $s(t) = \pm 2(t/T)$ for $0 \leq t \leq T$. The signal is corrupted by white gaussian noise of power spectral density 10^{-6} volt²/Hz.

    (a) Draw the signal waveform at the output of a matched filter receiver.

    (b) If the probability of error P_e is to be no larger than 10^{-4}, find the minimum allowable interval T.

11.7-1. Derive Eq. (11.7-5) for the error probability P_e in PSK when there is a phase offset in the local reference waveform.

11.8-1. Plot Eq. (11.8-8) versus E_s/η.

11.8-2. If the frequency offset Ω in FSK satisfies $\Omega T = n\pi$, $s_1(t)$ and $s_2(t)$ are orthogonal.

    (a) Prove this statement.

    (b) Calculate P_e.

    (c) Plot P_e versus E_s/η and compare with the results given in Eq. (11.8-8).

11.8-3. Plot the P_e in binary FSK as a function of ΩT. Select $E_s/\eta = 15$.

11.9-1. Plot Eq. (11.9-1) versus E_s/η.

11.11-1. A PSK system suffers from imperfect synchronization. Find ϕ so that the $P_e(\text{PSK}) = P_e(\text{DPSK}) = 10^{-5}$.

11.12-1. M-ary PSK involves choosing signals of the form $(\cos \omega_0 t + \theta_i)$ for M values of i.

    (a) Show how to choose the θ_i so that the probability of error of each θ_i is the same.

    (b) Find the correlator detector.

    (c) Obtain an expression for the probability of error.

11.12-2. Verify the entries in Table 11.12-1.

11.12-3. In connection with Fig. 11.12-2 show that, if T corresponds to an integral number of cycles, the two correlators make errors independently. {*Hint:* Assume white noise. Evaluate $E[n_{o1}(T)n_{o2}(T)]$ taking into account that $E[n_1(t_1)n_2(t_2)] = 0$ except when $t_1 = t_2$.}

REFERENCES

1. Stein, S., and J. Jones: "Modern Communication Principles," McGraw-Hill Book Company, New York, 1967.
2. Schwartz, M., W. R. Bennett, and S. Stein: "Communication Systems and Techniques," McGraw-Hill Book Company, New York, 1966.

12

Noise in Pulse-code and Delta-modulation Systems

In this chapter we consider again pulse-code modulation and delta modulation. Our present interest is in the output noise and the output signal-to-noise ratio. We shall take account of quantization effects and of thermal noise. It will be shown that PCM yields a better signal-to-noise ratio performance than does DM. Still, DM has such elegant simplicity of implementation that it is the system of choice in many applications.

12.1 PCM TRANSMISSION

A binary PCM transmission system is shown in Fig. 12.1-1. The baseband signal $m(t)$ is quantized, giving rise to the quantized signal $m_q(t)$, where

$$m_q(t) = m(t) + e(t) \qquad (12.1\text{-}1)$$

The term $e(t)$ is the error signal which results from the process of quantization. The quantized signal is next sampled. Sampling takes place at the Nyquist rate. The sampling interval is $T_s = 1/2f_M$, where f_M is the frequency to which the signal $m(t)$ is bandlimited.

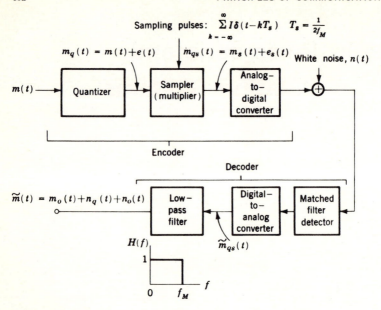

Fig. 12.1-1   A binary PCM encoder-decoder.

Sampling is accomplished by multiplying the signal $m_q(t)$ by a waveform which consists of a periodic train of pulses, the pulses being separated by the sampling interval T_s. We shall assume that the sampling pulses are narrow enough so that the sampling may be considered as instantaneous. It will be recalled (Sec. 5.1) that with such instantaneous sampling, the sampled signal may be reconstructed *exactly* by passing the sequence of samples through a low-pass filter with cutoff frequency at f_M. Now, as a matter of mathematical convenience, we shall represent each sampling pulse as an *impulse*. Such an impulse is infinitesimally narrow yet is characterized by having a finite area. The area of an impulse is called its *strength*, and an impulse of strength I is written $I\delta(t)$. The sampling-impulse train is therefore $S(t)$, given by

$$S(t) = I \sum_{k=-\infty}^{\infty} \delta(t - kT_s) \qquad T_s = \frac{1}{2f_M} \qquad (12.1\text{-}2)$$

From Eqs. (12.1-1) and (12.1-2) the quantized signal $m_q(t)$ after sampling becomes $m_{qs}(t)$, written as

$$m_{qs}(t) = m(t) \, I \sum_{k=-\infty}^{\infty} \delta(t - kT_s) + e(t) \, I \sum_{k=-\infty}^{\infty} \delta(t - kT_s)$$

$$(12.1\text{-}3a)$$

$$= m_s(t) \qquad\qquad\qquad +e_s(t) \qquad\qquad (12.1\text{-}3b)$$

The quantized strength-modulated impulse train is applied to an analog-to-digital converter. (In a physical system, the input to the A-to-D converter is a quantized amplitude-modulated *pulse* train.) The binary output of the A-to-D converter is transmitted over a communication channel and arrives at the receiver contaminated as a result of the addition of white thermal noise $n(t)$. Transmission may be direct, as indicated in Fig. 12.1-1, or the binary output signal may be used to modulate a carrier as in PSK or FSK. In any event, the received signal is detected by a matched filter to minimize errors in determining each binary bit and thereafter passed on to a D-to-A converter. The output of the D-to-A converter is called $\tilde{m}_{qs}(t)$. [In the absence of thermal noise, and assuming unity gain from the input to the A-to-D converter to the output of the D-to-A converter, we would have $\tilde{m}_{qs}(t) = m_{qs}(t)$.] Finally, the signal $\tilde{m}_{qs}(t)$ is passed through the low-pass baseband filter. At the output of this filter we find a signal $m_o(t)$ which, aside from a possible difference in amplitude, has exactly the waveform of the original baseband signal $m(t)$. This output signal, however, is accompanied by a noise waveform $n_q(t)$, which is due to the quantization, and an additional noise waveform $n_o(t)$, due to the thermal noise.

12.2 CALCULATION OF QUANTIZATION NOISE

We shall now temporarily ignore the effect of the thermal noise and shall calculate the output power due to the quantization noise in the PCM system of Fig. 12.1-1.

The sampled quantization error waveform, as given by Eq. (12.1-3), is

$$e_s(t) = e(t) \, I \sum_{k=-\infty}^{\infty} \delta(t - kT_s) \tag{12.2-1}$$

It is to be noted that if the sampling rate is selected to be the Nyquist rate for the baseband signal $m(t)$, the sampling rate will be inadequate to allow reconstruction of the error signal $e(t)$ from its samples $e_s(t)$. That such is the case is readily apparent from Fig. 12.2-1. In Fig. 12.2-1a is shown the relationship between $m_q(t)$ and $m(t)$, while in Fig. 12.2-1b is shown the error waveform $e(t)$ as a function of $m(t)$. The quantization levels are separated by amount S. We observe in Fig. 12.2-1b that $e(t)$ executes a complete cycle and exhibits an abrupt discontinuity every time $m(t)$ makes an excursion of amount S. Hence the spectral range of $e(t)$ extends far beyond the bandlimit f_M of $m(t)$.

To find the quantization noise output power N_q, we require the power spectral density of the sampled quantization error $e_s(t)$, given in Eq. (12.2-1). Since $\delta(t - kT_s) = 0$ except when $t = kT_s$, $e_s(t)$ may be

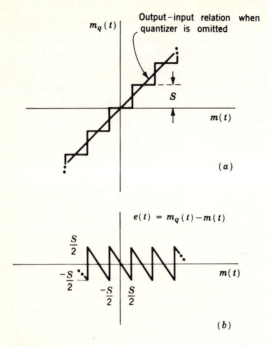

Fig. 12.2-1 (a) Plot of $m_q(t)$ as a function of $m(t)$.
(b) Plot of $e(t)$ as a function of $m(t)$.

written

$$e_s(t) = I \sum_{k=-\infty}^{\infty} e(kT_s)\delta(t - kT_s) \qquad (12.2\text{-}2)$$

The waveform of Eq. (12.2-2) consists of a sequence of impulses of area $A = e(kT_s)I$ occurring at intervals T_s. The quantity $e(kT_s)$ is the quantization error at the sampling time and is a random variable. In Sec. 7.14 we calculated the power spectral density of just such a waveform and arrived at the result given in Eq. (7.14-14). Applying Eq. (7.14-14) to the present case, we find that the power spectral density $G_{e_s}(f)$ of the sampled quantization error is (see Example 1.10-4)

$$G_{e_s}(f) = \frac{I^2}{T_s} \overline{e^2(kT_s)} \qquad (12.2\text{-}3)$$

In Sec. 6.3 [see Eq. (6.3-3)] we found that if the quantization levels are separated by amount S, then the quantization error is given by

$$\overline{e^2(t)} = \frac{S^2}{12} \qquad (12.2\text{-}4)$$

Equation (12.2-3) involves $\overline{e^2(kT_s)}$ rather than $\overline{e^2(t)}$. However, since the probability density of $e(t)$ does not depend on time, the variance of $e(t)$ is equal to the variance of $e(t = kT_s)$. Thus

$$\overline{e^2(t)} = \overline{e^2(kT_s)} \tag{12.2-5}$$

From Eqs. (12.2-3), (12.2-4), and (12.2-5) we have

$$G_{e_s}(f) = \frac{I^2}{T_s} \frac{S^2}{12} \tag{12.2-6}$$

Finally, the quantization noise N_q is, from Eq. (12.2-6),

$$N_q = \int_{-f_M}^{f_M} G_{e_s}(f) \, df = \frac{I^2}{T_s} \frac{S^2}{12} 2f_M = \frac{I^2}{T_s^2} \frac{S^2}{12} \tag{12.2-7}$$

since $2f_M = 1/T_s$.

Of more interest than the quantization noise given in Eq. (12.2-7) is the signal-to-quantization-noise ratio. To determine this ratio, we calculate, in the next section, the output-signal power expressed in terms of the quantization step S.

12.3   THE OUTPUT-SIGNAL POWER

The sampled signal which appears at the input to the baseband filter in Fig. 12.1-1 is given by $m_s(t)$ in Eq. (12.1-3) as

$$m_s(t) = m(t)I \sum_{k=-\infty}^{\infty} \delta(t - kT_s) \tag{12.3-1}$$

Since the impulses have a strength (area) I and are separated by a time T_s, the dc component of the impulse train of Eq. (12.3-1) is I/T_s. Hence the signal $m_o(t)$ at the output of the baseband filter is

$$m_o(t) = \frac{I}{T_s} m(t) \tag{12.3-2}$$

Since $T_s = 1/2f_M$, other terms in the series of Eq. (12.3-1) lie outside the passband of the filter. The normalized signal output power is, from Eq. (12.3-2),

$$\overline{m_o^2} = \frac{I^2}{T_s^2} \overline{m^2(t)} \tag{12.3-3}$$

We shall now express $\overline{m^2(t)}$ in terms of the number M of quantization levels and the step size S. A set of M quantization levels with step size S is shown in Fig. 12.3-1. We have assumed here that M is an odd integer. Hence to accommodate a presumed symmetrical signal $m(t)$ we have set one level at zero and positioned the other $M - 1$ levels

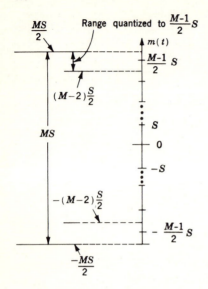

Fig. 12.3-1 Quantization levels in an M-level quantizer.

symmetrically about zero. Thus, whenever the signal $m(t)$ falls in the range $[(M-1)/2]S \pm S/2$, it will be quantized to the level $[(M-1)/2]S$, and similarly for other ranges. Hence, as may be seen in the figure, the allowable peak-to-peak excursion of $m(t)$ corresponding to our selection of quantization levels is from $-MS/2$ to $MS/2$.

We now assume that the instantaneous value of $m(t)$ may fall anywhere in its allowable range with equal likelihood. Then the probability density of the instantaneous value of m is $f(m)$ given by

$$f(m) = \frac{1}{MS} \tag{12.3-4}$$

The variance of $m(t)$, that is, $\overline{m^2(t)}$, is

$$\overline{m^2(t)} = \int_{-MS/2}^{MS/2} m^2 f(m)\ dm = \int_{-MS/2}^{MS/2} \frac{m^2\ dm}{MS} = \frac{M^2 S^2}{12} \tag{12.3-5}$$

Hence, from Eq. (12.3-3), the output-signal power is

$$S_o = \overline{m_o^2(t)} = \frac{I^2}{T_s^2} \frac{M^2 S^2}{12} \tag{12.3-6}$$

While we have derived Eq. (12.3-6) on the assumption that M is odd, it may be verified (Prob. 12.3-1) that the result is equally valid for M even.

From Eqs. (12.2-7) and (12.3-6) we find that the signal-to-quantization-noise ratio is

$$\frac{S_o}{N_q} = M^2 = (2^N)^2 = 2^{2N} \tag{12.3-7}$$

where N is the number of binary digits needed to assign individual binary-code designations to the M quantization levels.

12.4 THE EFFECT OF THERMAL NOISE

The effect of additive thermal noise is to cause the matched filter detector of Fig. 12.1-1 to make an occasional error in determining whether a binary 1 or a binary 0 was transmitted.   As we saw in Chap. 11, if the thermal noise is white and gaussian, the probability of such an error depends on the ratio E_s/η, where E_s is the signal energy transmitted during a bit and $\eta/2$ is the two-sided power spectral density of the noise. The error probability depends also on the type of modulation employed, i.e., direct transmission, PSK, FSK, etc.

Rather typically, PCM systems operate with error probabilities which are small enough so that we may ignore the likelihood that more than a single bit error will occur within a single word.   By way of example, if the error probability $P_e = 10^{-3}$ and a word has 10 bits ($N = 10$), we would expect, on the average, that 1 word would be in error for every 100 words transmitted.   A typical probability of error is 10^{-5}, for which an error would occur on the average of once every 10,000 words.

Let us assume that a code word used to identify a quantization level has N binary digits.   We assume further that the assignment of code words to levels is in the order of the numerical significance of the word. Thus, we assign 00 . . . 00 to the most negative level, 00 . . . 01 to the next-higher level until the most positive level is assigned the code word 11 . . . 11.

An error which occurs in the least significant bit of the code word corresponds to an incorrect determination by amount S in the quantized value $m_s(t)$ of the sampled signal.   An error in the next-higher significant bit corresponds to an error $2S$; in the next-higher, $4S$, etc.   Let us call the error Δm_s.   Then assuming that an error may occur with equal likelihood in any bit of the word, the variance of the error is

$$\overline{(\Delta m_s)^2} = \frac{1}{N}\ [S^2 + (2S)^2 + (4S)^2 + (8S)^2 + \cdots + (2^{N-1}S)^2]$$

$$(12.4\text{-}1)$$

The sum of the geometric progression in Eq. (12.4-1) is

$$\overline{(\Delta m_s)^2} = \frac{2^{2N} - 1}{3N}\ S^2 \cong \frac{2^{2N}}{3N}\ S^2 \tag{12.4-2}$$

for $N \geq 2$.

The preceding discussion indicates that the effect of thermal-noise errors may be taken into account by adding, at the input to the A-to-D

converter in Fig. 12.1-1, an error voltage Δm_s, and by deleting the white-noise source and the matched filter. We have assumed unity gain from the input to the A-to-D converter to the output of the D-to-A converter. Thus, the same error voltage appears at the input to the low-pass baseband filter. The result of a succession of errors is a train of impulses, each of strength $I \Delta m_s$. These impulses are of random amplitude and of random time of occurrence.

A thermal-noise error impulse occurs on each occasion when a word is in error. With P_e the probability of a bit error, the mean separation between bits which are in error is $1/P_e$ bits. With N digits per word, the mean separation between words which are in error is $1/NP_e$ words. Words are separated in time by the sampling interval T_s. Hence the mean time between words which are in error is T, given by

$$T = \frac{T_s}{NP_e} \tag{12.4-3}$$

Again using Eq. (7.14-14), we find that the power spectral density of the thermal-noise error impulse train is, using Eqs. (12.4-2) and (12.4-3),

$$G_{th}(f) = \frac{I^2\overline{(\Delta m_s)^2}}{T} = \frac{NP_eI^2\overline{(\Delta m_s)^2}}{T_s} \tag{12.4-4}$$

Using Eq. (12.4-2), we have

$$G_{th}(f) = \frac{2^{2N}S^2P_eI^2}{3T_s} \tag{12.4-5}$$

Finally, the output power due to the thermal error noise is

$$N_{th} = \int_{-f_M}^{f_M} G_{th}(f) \, df = \frac{2^{2N}S^2P_eI^2}{3T_s^2} \tag{12.4-6}$$

since $T_s = 1/2f_M$.

12.5 OUTPUT SIGNAL-TO-NOISE RATIO IN PCM

The output signal-to-noise ratio, including both quantization and thermal noise, is found by combining Eqs. (12.2-7), (12.3-6), and (12.4-6). The result is

$$\frac{S_o}{N_o} = \frac{S_o}{N_q + N_{th}} = \frac{(I^2/T_s^2)(M^2S^2/12)}{(I^2/T_s^2)(S^2/12) + (I^2/T_s^2)(P_e2^{2N}S^2/3)}$$

$$= \frac{2^{2N}}{1 + 4P_e2^{2N}} \tag{12.5-1}$$

in which we have used the fact that $M = 2^N$.

In PSK (or for direct transmission) we have, from Eq. (11.7-3), that

$$(P_e)_{\text{PSK}} = \frac{1}{2}\,\text{erfc}\,\sqrt{\frac{E_s}{\eta}} \tag{12.5-2}$$

where E_s is the signal energy of a bit and $\eta/2$ is the two-sided thermal-noise power spectral density. Also, for coherent reception of FSK we have, from Eq. (11.8-8), that

$$(P_e)_{\text{FSK}} = \frac{1}{2}\,\text{erfc}\,\sqrt{0.6\,\frac{E_s}{\eta}} \tag{12.5-3}$$

To calculate E_s, we note that if a sample is taken at intervals of T_s, and the code word of N bits occupies the entire interval between samples, then a bit has a duration T_s/N. If the received signal power is S_i, the energy E_s associated with a single bit is

$$E_s = S_i\frac{T_s}{N} = S_i\frac{1}{2f_M N} \tag{12.5-4}$$

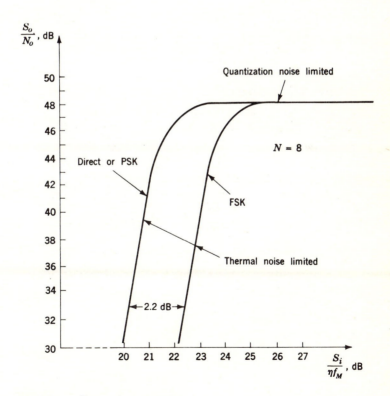

Fig. 12.5-1 Comparison of PCM transmission systems.

Combining Eqs. (12.5-1), (12.5-2), and (12.5-4), we find

$$\left(\frac{S_o}{N_o}\right)_{PSK} = \frac{2^{2N}}{1 + 2^{2N+1} \text{ erfc } \sqrt{(1/2N)(S_i/\eta f_M)}} \qquad (12.5\text{-}5)$$

Using Eq. (12.5-3) in place of (12.5-2), we have

$$\left(\frac{S_o}{N_o}\right)_{FSK} = \frac{2^{2N}}{1 + 2^{2N+1} \text{ erfc } \sqrt{(0.3/N)(S_i/\eta f_M)}} \qquad (12.5\text{-}6)$$

Equations (12.5-5) and (12.5-6) are plotted in Fig. 12.5-1 for $N = 8$. Note that for $S_i/\eta f_M \gg 1$ and $N = 8$

$$\left(\frac{S_o}{N_o}\right)_{PSK,FSK} = 10 \log (2^{16}) = 48 \text{ dB}$$

Observe that both PCM systems exhibit a threshold, the FSK threshold occurring at a $S_i/\eta f_M$ which is 2.2 dB greater than for PSK. (The threshold point is arbitrarily defined as the $S_i/\eta f_M$ value at which S_o/N_o has fallen 1 dB from the value corresponding to a large $S_i/\eta f_M$.) Experimentally, the onset of threshold in PCM is marked by an abrupt increase in a *crackling* noise analogous to the clicking noise heard below threshold in analog FM systems.

A comparison of PCM and FM is presented in Sec. 13.22.

12.6 DELTA MODULATION (DM)

The operation of a delta-modulation communication system was discussed in Sec. 6.13. At that point the discussion was entirely qualitative, and we considered neither quantization nor thermal noise. In the following sections we shall calculate the output signal-to-noise ratio of a DM system taking quantization noise and thermal noise into account.

A delta-modulation system, including a thermal-noise source, is shown in Fig. 12.6-1. The impulse generator applies to the modulator a continuous sequence of impulses $p_i(t)$ of time separation τ. The modulator output is a sequence of pulses $p_o(t)$ whose polarity depends on the polarity of the difference signal $\Delta(t) = m(t) - \tilde{m}(t)$, where $\tilde{m}(t)$ is the integrator output. We assume that the integrator has been adjusted so that its response to an input impulse of strength I is a step of size S; that is, $\tilde{m}(t) = (S/I)\int p_o(t) \, dt$.

A typical impulse train $p_o(t)$ is shown in Fig. 12.6-2a [actually, of course, the waveform $p_o(t)$ is a train of narrow pulses, each having very small energy]. Before transmission, the impulse waveform will be converted to the two-level waveform of Fig. 12.6-2b since this latter waveform has much greater power than a train of narrow pulses. This conversion is accomplished by the block in Fig. 12.6-1 marked "transmitter." The

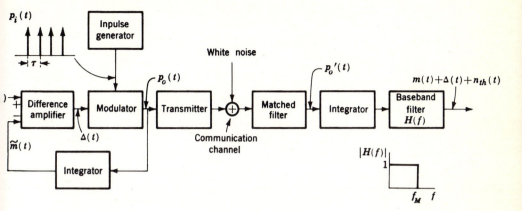

Fig. 12.6-1 A delta-modulation system.

transmitter, in principle, need be nothing more complicated than a bistable multivibrator (flip-flop). We may readily arrange that the positive impulses set the flip-flop into one of its stable states, while the negative impulses reset the flip-flop to its other stable state. The binary waveform of Fig. 12.6-2b will be transmitted directly or used to modulate a carrier as in PSK or FSK. After detection by the matched filter shown in Fig. 12.6-1, the binary waveform will be reconverted to a sequence of impulses $p_o'(t)$. In the absence of thermal noise $p_o'(t) = p_o(t)$, and the signal $\tilde{m}(t)$ is recovered at the receiver by passing $p_o'(t)$ through an integrator. We assume that transmitter and receiver integrators are identical and that the input to each consists of a train of impulses of strength $\pm I$. Hence in the absence of thermal noise, the outputs of both integrators are identical.

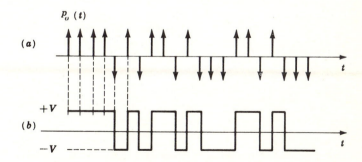

Fig. 12.6-2 (a) A typical impulse train $p_o(t)$ appearing at the modulator output in Fig. 12.6-1. (b) The two-level signal transmitted over the communication channel.

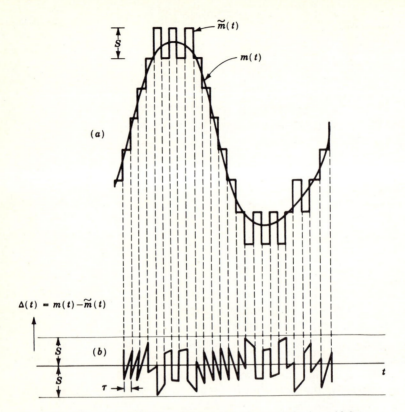

Fig. 12.7-1 The estimate $\tilde{m}(t)$ and error $\Delta(t)$ when $m(t)$ is sinusoidal.

12.7 QUANTIZATION NOISE IN DELTA MODULATION

To arrive at an estimate of the quantization noise power in delta modulation, we consider the situation represented in Fig. 12.7-1. Here, in
Fig. 12.7-1a is shown a sinusoidal input signal $m(t)$ and the waveform
$\tilde{m}(t)$ which is the delta modulator approximation to $m(t)$. In Fig. 12.7-1b
is shown the error waveform $\Delta(t)$ given by

$$\Delta(t) \equiv m(t) - \tilde{m}(t) \tag{12.7-1}$$

This error waveform is the source of the quantization noise.

We observe that, as long as slope overloading is avoided, the error
$\Delta(t)$ is always less than the step size S. (In PCM, the error is always
less than $S/2$.) We shall assume that $\Delta(t)$ takes on all values between
$-S$ and $+S$ with equal likelihood. That is, we assume that the probability density of $\Delta(t)$ is

$$f(\Delta) = \frac{1}{2S} \qquad -S \leq \Delta \leq S \tag{12.7-2}$$

The normalized power of the waveform $\Delta(t)$ is then

$$\overline{[\Delta(t)]^2} = \int_{-S}^{S} \Delta^2 f(\Delta)\, d\Delta = \int_{-S}^{S} \frac{\Delta^2}{2S}\, d\Delta = \frac{S^2}{3} \tag{12.7-3}$$

Our interest is in estimating how much of this power will pass through a baseband filter. For this purpose we need to know something about the power spectral density of $\Delta(t)$.

In Fig. 12.7-1 the period T of the sinusoidal waveform $m(t)$ has been selected so that T is an integral multiple of the step duration τ. We then observe that $\Delta(t)$ is periodic, with fundamental period T, and is, of course, rich in harmonics. Suppose, however, that the period T is changed very slightly by amount δT. Then the fundamental period of $\Delta(t)$ will not be T but will be instead $T \times \tau/\delta T$ (Prob. 12.7-3) corresponding to a fundamental frequency near zero as $\delta T \to 0$. And again, of course, $\Delta(t)$ will be rich in harmonics. Hence, in the general case, especially with $m(t)$ a random signal, it is reasonable to assume that $\Delta(t)$ has a spectrum which extends continuously over a frequency range which begins near zero.

To get some idea of the upper-frequency range of the waveform $\Delta(t)$, let us contemplate passing $\Delta(t)$ through a low-pass filter of adjustable cutoff frequency. Suppose that initially the cutoff frequency is high enough so that $\Delta(t)$ may pass with nominally no distortion. As we lower the cutoff frequency, the first type of distortion we would note is that the abrupt discontinuities in the waveform would exhibit finite rise and fall times. Such is the case since it is these abrupt changes which contribute to the high-frequency power content of the signal. To keep the distortion within reasonable limits, let us arrange that the rise time be rather smaller than the interval τ. As was discussed in Sec. 1.14, to satisfy this condition, we require that the filter cutoff frequency f_c be of the order of $f_c = 1/\tau$.

We now have made it appear reasonable, by a rather heuristic argument, that the spectrum of $\Delta(t)$ extends rather continuously from nominally zero up to $f_c = 1/\tau$. We shall assume further that over this range the spectrum is white. It has indeed been established experimentally that the spectrum of $\Delta(t)$ is approximately white over the frequency range indicated.

We may now finally calculate the quantization noise that will appear at the output of a baseband filter of cutoff frequency f_M. Since the quantization noise power in a frequency range f_c is $S^2/3$ as given by Eq. (12.7-3), the output-noise power in the baseband-frequency range f_M is

$$N_q = \frac{S^2}{3} \frac{f_M}{f_c} = \frac{S^2 \tau f_M}{3} \tag{12.7-4}$$

since $f_c = 1/\tau$. We may note also, in passing, that the two-sided power spectral density of $\Delta(t)$ is

$$G_\Delta(f) \simeq \frac{S^2/3}{2f_c} = \frac{S^2\tau}{6} \qquad -\frac{1}{\tau} \le f \le \frac{1}{\tau} \tag{12.7-5}$$

12.8 THE OUTPUT-SIGNAL POWER

In PCM, the signal power is determined by the step size and the number of quantization levels. Thus, as in Fig. 12.3-1, with step size S and M levels, the signal could make excursions only between $-MS/2$ and $+MS/2$. In delta modulation, there is no similar restriction on the amplitude of the signal waveform, because the number of levels is not fixed. On the other hand, in delta modulation there is a limitation on the *slope* of the signal waveform which must be observed if *slope overload* is to be avoided. If, however, the signal waveform changes slowly (i.e., has a frequency content in the lower-baseband range), there is nominally no limit to the signal power which may be transmitted.

Let us consider a *worst case* for delta modulation. We assume that the signal power is concentrated at the *upper end* of the baseband. Specifically let the signal be

$$m(t) = A \sin \omega_M t \tag{12.8-1}$$

with A the amplitude and $\omega_M = 2\pi f_M$, where f_M is the upper limit of the baseband frequency range. Then the signal power output power is

$$S_o = \overline{m^2(t)} = \frac{A^2}{2} \tag{12.8-2}$$

The maximum slope of $m(t)$ is $\omega_M A$. The maximum average slope of the delta modulator approximation $\tilde{m}(t)$ is S/τ, where S is the step size and τ the interval between steps. The limiting value of A just before the onset of slope overload is, therefore, given by the condition

$$\omega_M A = \frac{S}{\tau} \tag{12.8-3}$$

From Eqs. (12.8-2) and (12.8-3) we have that the maximum power which may be transmitted is

$$S_o = \frac{S^2}{2\omega_M^2 \tau^2} \tag{12.8-4}$$

The condition specified in Eq. (12.8-3) is unduly severe. A design procedure, more often employed, is to select the S/τ ratio to be equal to the rms value of the *slope* of $m(t)$. In this case the output-signal power can be increased significantly above the value given in Eq. (12.8-4).

12.9 DELTA-MODULATION OUTPUT-SIGNAL-TO-QUANTIZATION-NOISE RATIO

The output-signal-to-quantization-noise ratio for delta modulation is found by dividing Eq. (12.8-4) by Eq. (12.7-4).   The result is

$$\frac{S_o}{N_q} = \frac{3/8\pi^2}{(f_M\tau)^3} \approx \frac{3/80}{(f_M\tau)^3} \tag{12.9-1}$$

It is of interest to note that when our heuristic analysis is replaced by a rigorous analysis,[1] it is found that Eq. (12.9-1) continues to apply, except with the factor $3/80$ replaced by $3/64$, corresponding to a difference of less than 1 dB.

The dependence of S_o/N_q on the product $f_M\tau$ should be anticipated. For suppose that the signal amplitude were adjusted to the point of slope overload.   If now, say, f_M were increased by some factor, then τ would have to be reduced by this same factor in order to continue to avoid overload [see Eq. (12.8-4)].

Let us now make a comparison of the performance of PCM and DM in the matter of the ratio S_o/N_q.   We observe that the *transmitted* signals in DM (Fig. 12.6-2) and in PCM (Fig. 6.6-1) are of the same waveform, a binary pulse train.   In PCM a voltage level corresponding to a single bit persists for the time duration allocated to one bit of a code word. With sampling at the Nyquist rate, every $1/2f_M$ sec, and with N bits per code word, the bit duration is $\tau' = 1/2f_MN$.   In DM, a voltage level corresponding to a single bit is held for a duration τ which is the interval between samples.

If the communication channel is of limited bandwidth, then there is the possibility of intersymbol interference (Sec. 6.8) in either DM or PCM.   Whether such intersymbol interference occurs in DM depends on the ratio of $1/\tau$ to the bandwidth of the channel and similarly, in PCM, on the ratio of $1/\tau'$ to the channel bandwidth.   For a fixed channel bandwidth, if intersymbol interference is to be equal in the two cases, DM or PCM, we require that $\tau' = \tau$ or

$$\tau = \frac{1}{2f_MN} \tag{12.9-2}$$

Combining Eq. (12.9-1) with Eq. (12.9-2), we have, for DM,

$$\frac{S_o}{N_q} \approx 0.3N^3 \tag{12.9-3}$$

This result is now to be compared with Eq. (12.3-7) for PCM, which is

$$\frac{S_o}{N_q} = 2^{2N} \tag{12.9-4}$$

Comparing Eq. (12.9-3) with Eq. (12.9-4), we observe that for a fixed channel bandwidth the performance of DM is always poorer than PCM. By way of example, if a channel is adequate to accommodate code words in PCM with $N = 8$, Eq. (12.9-4) gives $(S_o/N_q) = 48$ dB. The same channel used for DM would, from Eq. (12.9-3), yield $(S_o/N_q) \approx 22$ dB.

12.10 DELTA PULSE-CODE MODULATION (DPCM)

In delta modulation, the approximation $\tilde{m}(t)$ is compared with the signal $m(t)$. A correction of *fixed magnitude* (of step size S) is then made in $\tilde{m}(t)$. The direction of this correction depends on whether $\tilde{m}(t)$ is greater or smaller than $m(t)$. An improved system results if the correction is not of fixed magnitude but rather increases as the error $\Delta(t) = \tilde{m}(t) - m(t)$ increases. Delta pulse-code modulation is just such a system in which, however, the correction is quantized. Delta pulse-code modulation is implemented by replacing the modulator of Fig. 12.6-1 by an M-level quantizer and sampler. The output of this quantizer-sampler is an impulse whose strength is not fixed, but is proportional to the quantized error. This quantized error sample is also applied to the transmitter. Corresponding to each quantized error sample, the transmitter impresses on the communication channel a binary waveform of N bits ($2^N = M$) which is the binary code representation of the quantized error sample.

It turns out that for $M \geq 4$ the DPCM system can result in a higher ratio S_o/N_q than ordinary PCM using the same bit duration. Thus, a combination of the simplicity of DM and the quantizing feature of PCM results in a system superior to either.

12.11 THE EFFECT OF THERMAL NOISE IN DELTA MODULATION

When thermal noise is present, the matched filter in the receiver of Fig. 12.6-1 will occasionally make an error in determining the polarity of the transmitted waveform. Whenever such an error occurs, the received impulse stream $p_o'(t)$ will exhibit an impulse of incorrect polarity. The received impulse stream is then

$$p_o'(t) = p_o(t) + p_{th}(t) \qquad\qquad (12.11\text{-}1)$$

in which $p_{th}(t)$ is the error impulse stream due to thermal noise. If the strength (area) of the individual impulses is I, then each impulse in p_{th} is of strength $2I$ and occurs only at each error. The factor of 2 results from the fact that an error *reverses* the polarity of the impulse.

The thermal-error noise appears as a stream of impulses of random time of occurrence and of strength $\pm 2I$. The average time of separation between these impulses is τ/P_e, where P_e is the bit error probability, and τ is the time allocated to a bit. If the results of Sec. 7.14 [see Eq. (7.14-14)] are used, the power spectral density of the thermal-noise impulses is

$$G_{p_{th}}(f) = \frac{P_e}{\tau}(2I)^2 = \frac{4I^2 P_e}{\tau} \tag{12.11-2}$$

Now we have already characterized the integrators (assumed identical in both transmitter and receiver) as having the property that when the integrator input is an impulse of strength I, the output is a step of amplitude S. The Fourier transform of the impulse is I, and the Fourier transform of a step of amplitude S is [using $u(t) \equiv$ unit step]

$$\mathcal{F}\{Su(t)\} = \frac{S}{j\omega} \qquad \omega \neq 0 \tag{12.11-3}$$

$$= S\pi\delta(\omega) \qquad \omega = 0$$

We may ignore the dc component in the transform since such dc components will not be transmitted through the baseband filter. Hence we may take the transfer function of the integrator to be $H_I(f)$ given by

$$H_I(f) = \frac{S}{I}\frac{1}{j\omega} \qquad \omega \neq 0 \tag{12.11-4a}$$

and

$$|H_I(f)|^2 = \left(\frac{S}{I}\right)^2 \frac{1}{\omega^2} \qquad \omega \neq 0 \tag{12.11-4b}$$

From Eqs. (12.11-2) and (12.11-4b) we find that the power spectral density of the thermal noise at the input to the baseband filter is $G_{th}(f)$ given by

$$G_{th}(f) = |H_I(f)|^2 G_{p_{th}}(f) = \frac{4S^2 P_e}{\tau\omega^2} \qquad \omega \neq 0 \tag{12.11-5}$$

It would now appear that to find the thermal-noise output, we need but to integrate $G_{th}(f)$ over the passband of the baseband filter. We have performed similar integrations on many occasions. And, in so doing, we have extended the range of integration from $-f_M$ through $f = 0$ to $+f_M$, even though we recognized that the baseband filter does not pass dc and actually has a low-frequency cutoff f_1. However, in these other cases the power spectral density of the noise near $f = 0$ is not inordinately large in comparison with the density throughout the baseband range generally. Hence, if as is normally the case, $f_1 \ll f_M$, the procedure is certainly justified as a good approximation. We observe however, that in the present case [Eq. (12.11-5)], $G_{th}(f) \to \infty$ at $\omega \to 0$, and more importantly that the integral of $G_{th}(f)$, over a range which includes $\omega = 0$, is infinite.

Let us then explicitly take account of the low-frequency cutoff f_1 of the baseband filter. The thermal-noise output is, using Eq. (12.11-5) and with $\omega = 2\pi f$,

$$N_{th} = \frac{S^2 P_e}{\pi^2 \tau}\left(\int_{-f_M}^{-f_1}\frac{df}{f^2} + \int_{f_1}^{f_M}\frac{df}{f^2}\right) \tag{12.11-6a}$$

$$= \frac{2S^2 P_e}{\pi^2 \tau}\left(\frac{1}{f_1} - \frac{1}{f_M}\right) \tag{12.11-6b}$$

$$\cong \frac{2S^2 P_e}{\pi^2 f_1 \tau} \tag{12.11-6c}$$

if $f_1 \ll f_M$. Observe that, unlike the situation encountered in all other cases, the thermal-noise output in delta modulation depends on the low-frequency cutoff rather than the high-frequency limit of the baseband range.

12.12 OUTPUT SIGNAL-TO-NOISE RATIO IN DELTA MODULATION

The output SNR is obtained by combining Eqs. (12.7-4), (12.8-4), and (12.11-6c). The result is

$$\frac{S_o}{N_o} = \frac{S_o}{N_q + N_{th}} = \frac{S^2/2\omega_M^2\tau^2}{(S^2\tau f_M/3) + (2S^2 P_e/\pi^2 f_1\tau)} \tag{12.12-1}$$

which may be rewritten

$$\frac{S_o}{N_o} = \frac{3\pi/(\omega_M\tau)^3}{1 + 24P_e/[(\omega_M\tau)(\omega_1\tau)]} \tag{12.12-2}$$

If transmission is direct or by means of PSK,

$$P_e = \frac{1}{2}\,\text{erfc}\,\sqrt{\frac{E_s}{\eta}} \tag{12.12-3}$$

where E_s, the signal energy in a bit, is related to the received signal power S_i by

$$E_s = S_i\tau \tag{12.12-4}$$

Combining Eqs. (12.12-2), (12.12-3), and (12.12-4), we have

$$\frac{S_o}{N_o} = \frac{3\pi/(\omega_M\tau)^3}{1 + \{12/[(\omega_M\tau)(\omega_1\tau)]\}\,\text{erfc}\,\sqrt{S_i\tau/\eta}} \tag{12.12-5}$$

12.13 COMPARISON OF PCM AND DM

We can now compare the output signal-to-noise ratios in PCM and DM by comparing Eqs. (12.5-5) and (12.12-5). To ensure that the communications channel bandwidth required is the same in the two cases, we use the condition, given in Eq. (12.9-2), that $\tau f_M = 1/2N$. Then Eq. (12.12-5) may be written

$$\frac{S_o}{N_o} = \frac{3N^3/\pi^2}{1 + \{12N/[\pi(\omega_1\tau)]\}\operatorname{erfc}\sqrt{(1/2N)(S_i/\eta f_M)}} \tag{12.13-1}$$

Equations (12.13-1) and (12.5-5) are compared in Fig. 12.13-1 for $N = 8$. To obtain the threshold performance of the delta-modulation system, we set $\omega_1\tau = 0.04$. The reason for this choice is as follows. Since $\tau f_M = 1/2N$,

$$\omega_1\tau = \frac{f_1}{f_M}\frac{\pi}{N} \tag{12.13-2}$$

Thus, if we consider the case of voice communication where $f_1 = 300$ Hz and $f_M = 3$ kHz, and if we choose $N = 8$, we obtain $\omega_1\tau = 0.04$.

Comparing results for PCM and DM, we see that at high-input SNR the output SNR for PCM is 26 dB greater than for DM. The threshold for DM occurs at $S_i/\eta f_M = 20.5$ dB, while for PCM threshold occurs at 22.5 dB. However, even with this threshold extension, PCM

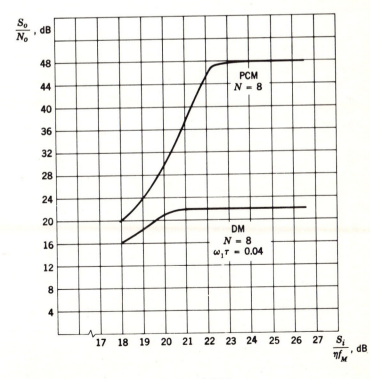

Fig. 12.13-1   A comparison of PCM and DM.

is seen to yield superior performance as compared with DM. In many applications this superior performance is outweighed by the relative simplicity of the delta modulator. As a result DM is gaining in popularity.

PROBLEMS

12.2-1. Suppose that in PCM, the sampling of the quantized waveform is not instantaneous but rather flat-topped. Let the flat-topped sampling pulses have a duration τ and an amplitude I/τ. For this case find the power spectral density $G_{e_s}(f)$ of the sampled quantization error corresponding to Eq. (12.2-3).

12.2-2. The signal $m(t)$ is sampled at Q times the Nyquist rate. Find N_q.

12.3-1. Prove that Eq. (12.3-6) is valid when M is an even integer.

12.3-2. Plot Eq. (12.3-7) as a function of N.

12.3-3. A signal $m(t)$ is not strictly bandlimited. We bandlimit $m(t)$ and then sample it. Due to the bandlimiting, distortion results even without quantization.

(a) Show that the noise caused by the distortion is $N_D = 2 \int_{f_M}^{\infty} G_m(f)\, df$,

where $G_m(f)$ is the power spectral density of $m(t)$, and f_M is the cutoff frequency of the bandlimiting filter.

(b) If

$$G_m(f) = G_o e^{-|f/f_1|}$$

find N_D.

(c) If the signal $m(t)$ is sampled at the Nyquist rate and quantized, find the total output SNR $= S_o/(N_D + N_q)$.

12.4-1. The thermal-noise error pulse train is *not* an impulse train. Each pulse has the duration of $1/2f_M$ if sampling is at the Nyquist rate. The pulse amplitude is equal to Δm_s.

(a) Find $G_{th}(f)$.

(b) Calculate N_{th}.

12.5-1. Plot S_o/N_o in Eq. (12.5-5) as a function of $S_i/\eta f_M$ for $N = 4, 8, 100$.

12.5-2. Show that threshold is defined by the equation $P_e \approx 1/[(16)2^{2N}]$. Plot P_e versus N.

12.5-3. Find the output SNR if the binary signal is transmitted using DPSK. Plot your result.

12.5-4. The signal $m(t)$ is *not* bandlimited. Its power spectral density is

$$G_m(f) = G_o e^{-|f/f_1|}$$

Using the results of Prob. 12.3-3c, obtain an expression for the ratio $S_o/N_o = S_o/(N_D + N_Q + N_{th})$.

12.5-5. A signal $m(t)$, bandlimited to 4 kHz, is sampled at *twice* the Nyquist rate and the samples transmitted by PCM. An output SNR of 47 dB is required. Find N and the minimum value of S_i/η if operation is to be above threshold.

12.5-6. Two signals $m_1(t)$ and $m_2(t)$, each bandlimited to 4 kHz, are sampled at the Nyquist rate, PCM encoded, and then time-division multiplexed. The output

SNR, including thermal-noise effects, of each demultiplexed signal is to be at least 30 dB.

(a) Sketch the entire transmitter and receiver structure.

(b) Find N and the minimum S_i/η that is required if operation is to be above threshold.

12.7-1. Let $m(t)$ be a constant M_0.

(a) Sketch the steady-state error $\Delta(t)$ for the DM shown in Fig. 12.6-1.

(b) Calculate the output quantization noise power after filtering. Assume a step size S and that the time between samples is τ. Let the filter bandwidth be f_M.

12.7-2. Let $m(t) = Kt$, where $K = S/\tau$.

(a) Sketch $\Delta(t)$ for the DM shown in Fig. 12.6-1.

(b) Sketch the filtered output quantization noise power. Let the baseband filter bandwidth be f_M.

12.7-3. Assume initially that $m(t)$ in Fig. 12.6-1 is periodic with a period T which is an integral multiple of the sampling interval τ. Now let T change, very slightly, by an amount δT. Show that the fundamental period of the error waveform $\Delta(t)$ becomes $(T\tau)/\delta T$.

12.8-1. The baseband input signal $m(t)$ in Fig. 12.6-1 is a random waveform with power spectral density which is uniform at $\eta_m/2$ up to the bandlimit f_M. To keep slope overload within tolerable limits, the design criterion to be used is that S/τ is to be equal to the rms value of the slope of $m(t)$. For this condition calculate the maximum power S_o and compare with Eq. (12.8-4).

12.9-1. A signal $m(t)$ is to be encoded using either DM or PCM. The signal-to-quantization-noise ratio $S_o/N_q \geq 30$ dB. Find the ratio of the PCM to DM bandwidths required.

12.9-2. Plot $\left(\dfrac{S_o}{N_q}\right)_{\text{PCM}}$ versus $\left(\dfrac{S_o}{N_q}\right)_{\text{DM}}$ for equal bandwidths.

12.10-1. A DPCM system is shown in Fig. P12.10-1.

(a) If $m(t) = Kt$, find $K_{\max}$ to avoid slope overloading.

(b) If a two-level quantizer is used, $S/\tau = K_{\max}$. With this value of $K_{\max}$, find τ'/τ, τ' being the time interval allowed for a bit in the transmitted PCM waveform.

(c) Sketch $\tilde{m}(t)$ and $\Delta(t)$, if $S/\tau = 2K_{\max}$.

(d) Comment on the quantization noise of the DPCM system as compared with the DM system.

12.11-1. Verify Eq. (12.11-6c).

12.12-1(a). Show that the threshold in DM occurs when $100\,P_e \approx (\omega_M\tau)^2(f_1/f_M)$.

(b) Plot P_e versus $\omega_M\tau$ for $f_1/f_M = 0.01, 0.04, 0.1$.

12.12-2. Plot Eq. (12.12-5) for $\omega_M\tau = 10$ and $\omega_1\tau = 0.04$.

12.12-3(a) Derive an expression for the output SNR when the binary signal is transmitted using DPSK.

(b) If $\omega_m\tau = 10$ and $\omega_1\tau = 0.04$, compare the thresholds obtained using PSK and DPSK.

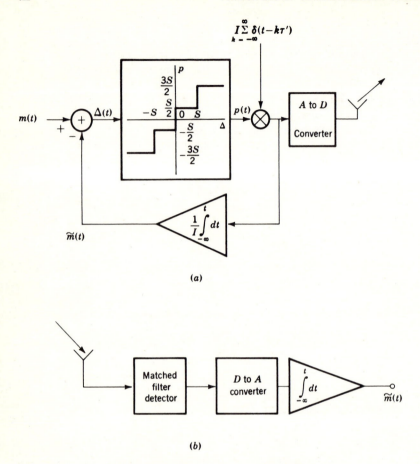

Fig. P12.10-1 A DPCM system. (*a*) Encoder. (*b*) Decoder.

12.13-1. A PCM and a DM system are both designed to yield an output SNR of 30 dB. Let $f_m = 4$ kHz and assume PCM sampling at 5 times the Nyquist rate.

  (*a*) Compare the bandwidths required for each system.

  (*b*) If $f_1/f_M = 0.04$, compare the threshold of each system.

REFERENCE

1. Van de Weg, H.: Quantizing Noise of a Single Integration Delta Modulation System with an N-digit Code, *Philips Res. Rept.* 8, pp. 367–385, 1953.

13
Information Theory and Coding

We have seen that communication systems are limited in their performance by the available signal power, the inevitable background noise, and the need to limit bandwidth. The performance of each of the systems studied is not equal; some are better than others. We are naturally led to ask whether there is some system, which we have not yet encountered, which is superior to any so far considered. More specifically, we inquire what would be the characteristics of an *ideal* system, that is, a system which is not limited by our engineering ingenuity and inventiveness but limited rather only by the *fundamental nature of the physical universe*. In the course of this inquiry we shall be led to discuss the following matters:

1. What are the performance characteristics of such an ideal system?
2. How, at least in principle, is such a system to be realized?
3. How do the performances of existing systems compare with that of the ideal system?

In response to the first of these questions we shall have occasion to introduce some elemental concepts of *information theory*,[1,2] a theory which, as employed here, is due principally to Shannon.[3]　The second question will lead us to an elementary discussion of coding.　The third question is discussed beginning with Sec. 13.19.

13.1　DISCRETE MESSAGES

In considering communication systems broadly, we may assume, without loss of generality, that the function of a communication system is to convey, from transmitter to receiver, a sequence of messages which are selected from a *finite* number of *predetermined* messages.　Thus, within some time interval one of these messages is transmitted.　During the next time interval another of these messages (or possibly the same message) is transmitted.　It should be noted that while the messages are predetermined and hence known by the receiver, the message selected for transmission during a particular interval is not known a priori by the receiver. Thus the receiver does not have the burden of extracting an arbitrary signal from a background of noise, but need only perform the operation of identifying which of a number of allowable messages was transmitted. The job of the receiver is not to answer the question "What was the message?" but rather the question "Which one?"　Rather generally, the probability that a particular message has been selected for transmission will not be the same for all messages.　In this case, we assume that the probability of occurrence of each possible message is known at the receiver.

We recognize that this transmission of known messages is actually nothing more than an extension of the concept of *quantization* discussed in Sec. 12.2.　When a signal is quantized, the receiver need determine only which of the quantized levels is encountered at each sampling time.　The discrete predetermined messages then consist of those quantization levels which may be transmitted directly or encoded into any one of a number of forms.　It is to be noted, of course, that such quantization imposes no restriction on the precision with which an arbitrary signal may be transmitted, since, in principle, the number of quantization levels may be increased without limit.

We may be rather free in our interpretation of the term "message." Thus, suppose that, during some time interval, there is generated at the transmitter one of a number of predetermined waveforms.　If the receiver correctly identifies the waveform, then the receiver has received the transmitted message.　For example, suppose that a quantization level is encoded into a binary waveform as in binary PCM.　Then we may view the quantization level as a message, but we may also view each binary digit as a message.　Thus a number of messages conveyed by a succession of binary digits may convey the single message contained in one quantization sample.

13.2 THE CONCEPT OF AMOUNT OF INFORMATION

Let us consider a communication system in which the allowable messages are m_1, m_2, . . . , with probabilities of occurrence p_1, p_2, ... Of course, $p_1 + p_2 + \cdots = 1$. Let the transmitter select message m_k, of probability p_k; let us further assume that the receiver has correctly identified the message. Then we shall say, by way of definition of the term *information*, that the system has conveyed an *amount of information* I_k given by

$$I_k \equiv \log_2 \frac{1}{p_k} \qquad\qquad (13.2\text{-}1)$$

The concept of *amount of information* is so essential to our present interest that it behooves us to examine with some care the implications of Eq. (13.2-1). We note first that while I_k is an entirely dimensionless number, by convention, the "unit" it is assigned is the *bit*. Thus, by way of example, if $p_k = \frac{1}{4}$, $I_k = \log_2 4 = 2$ bits. The unit bit is employed principally as a reminder that in Eq. (13.2-1) the base of the logarithm is 2. When the natural logarithmic base is used, the unit is the *nat*, and when the base is 10, the unit is the *Hartley* or *decit*. (The use of such units in the present case is analogous to the unit *radian* used in angle measure and *decibel* used in connection with power ratios.) The use of the base 2 is especially convenient when binary PCM is employed. For, if the 2 possible binary digits (bits) may occur with equal likelihood, each with a probability $\frac{1}{2}$, then the correct identification of the binary digit conveys an amount of information $I = \log_2 2 = 1$ bit.

In the past we have used the term bit as an abbreviation for the phrase *binary digit*. When there is an uncertainty whether the word bit is intended as an abbreviation for binary digit or as a unit of information measure, it is customary to refer to a binary digit as a *binit*. Note that if the probabilities of two possible binits are not equally likely, one binit conveys *more* and one conveys *less* than 1 *bit* of information. For example, if the binits 0 and 1 occur with probabilities of $\frac{1}{4}$ and $\frac{3}{4}$, respectively, then binit 0 conveys information in amount $\log_2 4 = 2$ bits, while binit 1 conveys information in amount $\log_2 \frac{4}{3} = 0.42$ bit.

Suppose that there are M equally likely and independent messages and that $M = 2^N$, with N an integer. In this case the information in each message is

$$I = \log_2 M = \log_2 2^N = N \text{ bits} \qquad\qquad (13.2\text{-}2a)$$

Suppose, further, that we choose to identify each message by binary PCM code words. The number of binary digits required for each of the 2^N messages is also N. Hence in this case the information in each message as measured in bits is numerically the same as the number of binits needed to encode the messages.

When $p_k = 1$, we have a trivial case since only one possible message is allowed. In this instance, since the receiver knows the message, there is really no need for transmission. We find that $I = \log_2 1 = 0$. As p_k decreases from 1 to 0, I_k increases monotonically, going from 0 to infinity. Thus, a greater amount of information has been conveyed when the receiver correctly identifies a less likely message.

When two independent messages m_k and m_l are correctly identified, the amount of information conveyed is the sum of the information associated with each of the messages individually. The individual information amounts are

$$I_k = \log_2 \frac{1}{p_k} \tag{13.2-2b}$$

$$I_l = \log_2 \frac{1}{p_l} \tag{13.2-2c}$$

Since the messages are *independent*, the probability of the composite message is $p_k p_l$ with corresponding information content

$$I_{k,l} = \log_2 \frac{1}{p_k p_l} = \log_2 \frac{1}{p_k} + \log_2 \frac{1}{p_l} = I_k + I_l \tag{13.2-3}$$

It is of interest to note that the term *information* applied to the symbol I_k in Eq. (13.2-1) is rather aptly chosen, since there is some correspondence between the properties of I_k and the meaning of the word *information* as used in everyday speech. For example, suppose that an airplane dispatcher calls the weather bureau of a distant city during daylight hours to inquire about the present weather. If he receives in response the message, "There is daylight here," he will surely judge that he has received *no* information since he was certain beforehand that such was the case. On the other hand, if he hears "It is not raining here," he will consider that he has received information since he might anticipate such a situation with less than perfect certainty. Further, suppose the dispatcher receives some weather information on two different days. Then he might indeed consider that the total information received was the sum of the information received in the individual weather reports.

Furthermore, consider that the weather bureau of a city located in the desert, where it has not rained for 25 years, is called before each flight. The call is required although everyone "knows" that the weather will be clear. However, one day the call is made and the reply is that a very heavy rainstorm is in progress and that the flight must be canceled. The information received is then very great.

13.3 AVERAGE INFORMATION, ENTROPY

Suppose we have M different messages m_1, m_2, . . . , with probabilities of occurrence p_1, p_2, ... Suppose further that during a long period of transmission a sequence of L messages has been generated. Then, if L is very large, we may expect that $p_1 L$ messages of m_1, $p_2 L$ messages of m_2, etc., will have occurred in the sequence. The total information in such a sequence will be

$$I_{\text{total}} = p_1 L \log_2 \frac{1}{p_1} + p_2 L \log_2 \frac{1}{p_2} + \cdots \tag{13.3-1}$$

The *average information* per message interval, represented by the symbol H, will then be

$$H \equiv \frac{I_{\text{total}}}{L} = p_1 \log_2 \frac{1}{p_1} + p_2 \log_2 \frac{1}{p_2} + \cdots = \sum_{k=1}^{M} p_k \log_2 \frac{1}{p_k} \tag{13.3-2}$$

This average information is also referred to by the term *entropy*.

We have seen that when there is only a single possible message ($p_k = 1$), the receipt of that message conveys no information. At the other extreme, as $p_k \rightarrow 0$, $I_k \rightarrow \infty$. However, since

$$\lim_{p \rightarrow 0} p \log \frac{1}{p} = 0 \tag{13.3-3}$$

the *average information* associated with an extremely unlikely message is *zero*.

As an example of the dependence of H on the probabilities of messages, let us consider the case of just two messages with probabilities p and $(1 - p)$. The average information per message is

$$H = p \log_2 \frac{1}{p} + (1 - p) \log_2 \frac{1}{1 - p} \tag{13.3-4}$$

A plot of H as a function of p is shown in Fig. 13.3-1. Note that $H = 0$ at $p = 0$ and at $p = 1$. The maximum value of H may be located by setting to zero dH/dp as calculated from Eq. (13.3-4). It is then found that, as indicated in the figure, the maximum occurs at $p = \frac{1}{2}$, that is, when the two messages are equally likely. The corresponding H is

$$H_{\max} = \frac{1}{2} \log_2 2 + \frac{1}{2} \log_2 2 = \log_2 2 = 1 \text{ bit/message} \tag{13.3-5}$$

When there are M messages, it may likewise be proved that H becomes a maximum when all the messages are equally likely. (The student is

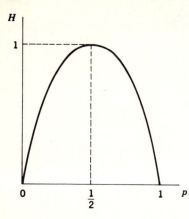

Fig. 13.3-1 Average information H, for case of two messages, plotted as a function of the probability p of one of the messages.

guided through the details of this proof in Prob. 13.3-3.) In this case each message has a probability $p = 1/M$ and

$$H_{\max} = \sum \frac{1}{M} \log_2 M = \log_2 M \tag{13.3-6}$$

since there are M terms in the summation.

13.4 INFORMATION RATE

If the *source* of the messages generates messages at the rate r messages per second, then the *information rate* is defined to be

$$R \equiv rH = \text{average number of bits of information/second} \tag{13.4-1}$$

Example 13.4-1 An analog signal is bandlimited to B Hz, sampled at the Nyquist rate, and the samples are quantized into 4 levels. The quantization levels Q_1, Q_2, Q_3, and Q_4 (messages) are assumed independent and occur with probabilities $p_1 = p_4 = \frac{1}{8}$ and $p_2 = p_3 = \frac{3}{8}$. Find the information rate of the source.

Solution The average information H is

$$H = p_1 \log_2 \frac{1}{p_1} + p_2 \log_2 \frac{1}{p_2} + p_3 \log_2 \frac{1}{p_3} + p_4 \log_2 \frac{1}{p_4}$$

$$= \frac{1}{8} \log_2 8 + \frac{3}{8} \log_2 \frac{8}{3} + \frac{3}{8} \log_2 \frac{8}{3} + \frac{1}{8} \log_2 8 \tag{13.4-2}$$

$$= 1.8 \text{ bits/message}$$

The information rate R is

$$R = rH = 2B(1.8) = 3.6B \text{ bits/sec}$$

We might choose to transmit the messages, referred to in the above example, by binary PCM. Thus we might identify each message by a binary code as indicated in the following table.

Message	Probability	Binary code
Q_1	$\dfrac{1}{8}$	0 0
Q_2	$\dfrac{3}{8}$	0 1
Q_3	$\dfrac{3}{8}$	1 0
Q_4	$\dfrac{1}{8}$	1 1

If we transmit $2B$ messages per second, then, since each message requires 2 binits, we shall be transmitting $4B$ binits per second. We have seen that a binit is *capable* of conveying 1 bit of information. Hence with $4B$ binits per second we *should* be able to transmit $4B$ bits of information per second. We note in the above illustration that we are actually transmitting only 3.6 bits of information per second. We are therefore not taking full advantage of the ability of the binary PCM to convey information. One way to rectify this situation is to select different quantization levels such that each level is equally likely. In this case we would find that the average information per message is

$$H = 4(\tfrac{1}{4} \log_2 4) = 2 \text{ bits/message}$$

and

$$R = rH = 2B(2) = 4B \text{ bits/sec}$$

Suppose, however, that it was not convenient or appropriate to change the messages. In this case we might seek an alternative *coding* scheme (rather than binary encoding) in which, on the average, the number of bits per message was fewer than 2 (ideally 1.8). Such a coding scheme is illustrated in Prob. 13.4-1.

13.5 SHANNON'S THEOREM, CHANNEL CAPACITY

The importance of the concept of information rate is that it enters into a theorem due to Shannon[3] which is fundamental to the theory of communications. This theorem is concerned with the rate of transmission of information over a communication channel. While we have used the term *communication channel* on many occasions, it is well to emphasize at this point, that the term, which is something of an abstraction, is intended to encompass all the features and component parts of the transmission system which introduce noise or limit the bandwidth. Shannon's theorem says that it is possible, in principle, to devise a means whereby a communications system will transmit information with an arbitrarily small probability of error provided that the information rate R is less than or equal to a rate C called the *channel capacity*. The technique used to approach this end is called *coding* and is discussed beginning with Sec. 13.8. To put the matter more formally, we have the following:

Theorem Given a source of M equally likely messages, with $M \gg 1$, which is generating information at a rate R. Given a channel with channel capacity C. Then, if

$$R \leq C \tag{13.5-1}$$

there exists a *coding* technique such that the output of the source may be transmitted over the channel with a probability of error of receiving the message which may be made arbitrarily small.

The important feature of the theorem is that it indicates that for $R \leq C$ transmission may be accomplished without error in the presence of noise. This result is surprising. For in our consideration of noise, say, gaussian noise, we have seen that the probability density of the noise extends to *infinity*. We should then imagine that there will be some times, however infrequent, when the noise *must* override the signal thereby resulting in errors. However, Shannon's theorem says that this need not be the case.

There is a negative statement associated with Shannon's theorem. It states the following:

Theorem Given a source of M equally likely messages, with $M \gg 1$, which is generating information at a rate R; then if

$$R > C$$

the probability of error is close to unity for every possible set of M transmitter signals.

This *negative* theorem shows that if the information rate R exceeds a specified value C, the error probability will increase toward unity as M increases, and that, generally, increasing the complexity of the coding results in an increase in the probability of error.

13.6 CAPACITY OF A GAUSSIAN CHANNEL

A theorem which is complementary to Shannon's theorem and applies to a channel in which the noise is gaussian is known as the Shannon-Hartley theorem.

Theorem The channel capacity of a white, bandlimited gaussian channel is

$$C = B \log_2 \left(1 + \frac{S}{N} \right) \text{ bits/sec} \tag{13.6-1}$$

where B is the channel bandwidth, S the signal power, and N is the total noise within the channel bandwidth, that is, $N = \eta B$, with $\eta/2$ the (two-sided) power spectral density.

This theorem, although restricted to the gaussian channel, is of fundamental importance. First, we find that channels encountered in physical systems generally are, at least approximately, gaussian. Second, it turns out that the results obtained for a gaussian channel often provide a *lower bound* on the performance of a system operating over a nongaussian channel. Thus, if a particular encoder-decoder is used with a gaussian channel and an error probability P_e results, then with a nongaussian channel another encoder-decoder can be designed so that the P_e will be smaller. We may note that channel capacity equations corresponding to Eq. (13.6-1) have been derived for a number of nongaussian channels.

The derivation of Eq. (13.6-1) for the capacity of a gaussian channel is rather formidable and will not be undertaken. However, the result may be made to appear reasonable by the following considerations. Suppose that, for the purpose of transmission over the channel, the messages are represented by fixed voltage levels. Then, as the source generates one message after another in sequence, the transmitted signal $s(t)$ takes on a waveform similar to that shown in Fig. 13.6-1.

The received signal is accompanied by noise whose root-mean-square voltage is σ. The levels have been separated by an interval $\lambda\sigma$, where λ is a number presumed large enough to allow recognition of individual levels with an acceptable probability of error. Assuming an even number of levels, the levels are located at voltages $\pm\lambda\sigma/2$, $\pm3\lambda\sigma/2$, etc. If there are to be M possible messages, then there must be M levels. We assume that the messages and hence the levels occur with equal likelihood. Then the average signal power is

$$S = \frac{2}{M} \left\{ \left(\frac{\lambda\sigma}{2} \right)^2 + \left(\frac{3\lambda\sigma}{2} \right)^2 + \cdots + \left[\frac{(M-1)\lambda\sigma}{2} \right]^2 \right\} \tag{13.6-2a}$$

$$= \frac{M^2 - 1}{12} (\lambda\sigma)^2 \tag{13.6-2b}$$

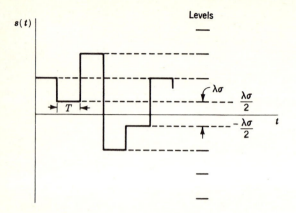

Fig. 13.6-1   A sequence of messages is represented by a
waveform $s(t)$ which assumes voltage levels corresponding to the messages.

The number of levels for a given signal power is, from Eq. (13.6-2b),

$$M = \left(1 + \frac{12S}{\lambda^2\sigma^2}\right)^{1/2} = \left(1 + \frac{12}{\lambda^2}\frac{S}{N}\right)^{1/2} \tag{13.6-3}$$

where $N = \sigma^2$ is the noise power.   Each message conveys an average
amount of information

$$H = \log_2 M = \log_2\left(1 + \frac{12}{\lambda^2}\frac{S}{N}\right)^{1/2}$$
$$= \frac{1}{2}\log_2\left(1 + \frac{12}{\lambda^2}\frac{S}{N}\right) \text{ bits/message} \tag{13.6-4}$$

To find the information rate of the signal waveform $s(t)$ of Fig. 13.6-1,
we need to estimate how many messages per unit time may be carried by
this signal.   That is, we need to estimate the interval T which should be
assigned to each message to allow the transmitted levels to be recognized
individually at the receiver, even though the bandwidth B of the channel is
limited.   Now the principal effect of limited bandwidth on the signal waveform $s(t)$ will be a *rounding* of the initially abrupt transitions from one level
to another.   We saw in Sec. 1.14 that when an abrupt step is applied to
an ideal low-pass filter of bandwidth B, the response has a 10 to 90 percent
rise time τ given by $\tau = 0.44/B$.   We may then reasonably estimate
that if we set $T = \tau$, we shall be able to distinguish levels *reliably*.   Since
our discussion is heuristic, let us, as a matter of convenience, take
$T = \tau = 0.5/B$.   The message rate is then

$$r = \frac{1}{T} = 2B \text{ messages/second} \tag{13.6-5}$$

Since the transmission of any of the M messages is equally likely,

$$H = \log_2 M$$

Thus our channel is transferring information at a rate $R = rH$. Since we have presumably taken what precautions are necessary to ensure that the channel is just able to allow the transmission with acceptable probability of error, $R \approx C$. The channel capacity is, therefore, found by combining Eqs. (13.6-4) and (13.6-5):

$$C \approx R = rH = B \log_2 \left(1 + \frac{12}{\lambda^2} \frac{S}{N}\right) \tag{13.6-6}$$

Comparing Eq. (13.6-6) with Eq. (13.6-1) of the Shannon-Hartley theorem, we observe that the results would be identical if we set $12/\lambda^2 = 1$, that is, $\lambda = 3.5$.

This heuristic discussion leading to Eq. (13.6-6) was undertaken to make the Shannon-Hartley theorem [Eq. (13.6-1)] rather intuitively acceptable. It needs to be emphasized, however, that Eq. (13.6-6) specifies the rate at which information may be transmitted with *small* error, while the Shannon-Hartley theorem contemplates that, with a sufficiently sophisticated transmission technique, transmission at channel capacity is possible with arbitrarily small error.

13.7 BANDWIDTH—S/N TRADEOFF

The Shannon-Hartley theorem, Eq. (13.6-1), indicates that a noiseless gaussian channel ($S/N = \infty$) has an infinite capacity. On the other hand, the channel capacity does not become infinite as the bandwidth becomes infinite because, with an increase in bandwidth, the noise power also increases. Thus for a fixed signal power and in the presence of white gaussian noise the channel capacity approaches an upper limit with increasing bandwidth. We now calculate that limit. Using $N = \eta B$ in Eq. (13.6-1), we have

$$C = B \log_2 \left(1 + \frac{S}{\eta B}\right) = \frac{S}{\eta} \frac{\eta B}{S} \log_2 \left(1 + \frac{S}{\eta B}\right) \tag{13.7-1a}$$

$$= \frac{S}{\eta} \log_2 \left(1 + \frac{S}{\eta B}\right)^{\eta B/S} \tag{13.7-1b}$$

We recall that $\lim_{x \to 0} (1 + x)^{1/x} = e$ (the naperian base), and identifying x as $x = S/\eta B$, we find that Eq. (13.7-1b) becomes

$$\lim_{B \to \infty} C = \frac{S}{\eta} \log_2 e = 1.44 \frac{S}{\eta} \tag{13.7-2}$$

The Shannon-Hartley principle also indicates that we may trade off bandwidth for signal-to-noise ratio and vice versa. For example, if $S/N = 7$ and $B = 4$ kHz, we find $C = 12 \times 10^3$ bits/sec. If the SNR is increased to $S/N = 15$ and B decreased to 3 kHz, the channel capacity remains the same. With a 3-kHz bandwidth the noise power will be $\frac{3}{4}$ as large as with 4 kHz. Thus the signal power will have to be increased by the factor $(3/4) \times (15/7) = 1.6$. Therefore, this 25 percent reduction in bandwidth requires a 60 percent increase in signal power.

It is of great interest, and even a little surprising, to recognize that the trade off between bandwidth and signal-to-noise ratio is not limited by a lower limit on bandwidth. Specifically, suppose that we want to transmit a signal whose spectral range extends up to a frequency f_M. Let us quantize the signal so that we may determine an information rate for the signal, and let us assume that the channel has a channel capacity greater than the signal information rate. Now, however, suppose that it turns out that the bandwidth B of the channel is less than f_M. Say, as an extreme example, that the bandwidth is 1 Hz while f_M is 1000 Hz. Then is it really possible, in principle, to receive the signal with arbitrarily small probability of error? The answer is yes.

To make the idea somewhat more acceptable that such errorless signal reception is indeed possible, consider the extreme case where there is no noise at all. Suppose that the signal with $f_M = 1000$ Hz is transmitted through a channel which can be represented as a low-pass RC circuit with cut off at 1 Hz. Then the received signal will be greatly attenuated and severely distorted. If, however, there is no noise, then we are entirely free to make up for the attenuation by the use of an amplifier and to correct the frequency distortion by the use of an equalizer. The signal is then recoverable precisely as it was transmitted.

13.8 USE OF ORTHOGONAL SIGNALS TO ATTAIN SHANNON'S LIMIT[4]

In the preceding section we saw that in the (trivial) case of no noise the channel capacity was infinite, and that no matter how restricted the bandwidth was, it was always possible to receive a signal without error, as predicted by the Shannon limit. In the present section we shall discuss one possible method of attaining performance predicted by Shannon's theorem in the presence of gaussian noise. We shall see that, to obtain errorless transmission for $R \leq C$ requires an extremely complicated communication system. In physically realizable systems, therefore, we must accept a performance less than optimum.

Orthogonal Signals. The system we are to describe involves the use of a set $s_1(t)$, $s_2(t)$, . . . of orthogonal signals. Such signals $s_i(t)$, defined as being orthogonal over the interval 0 to T, have the property that

$$\int_0^T s_i(t)s_j(t)\, dt = 0 \qquad \text{for } i \neq j \tag{13.8-1}$$

A familiar example of a set of orthogonal signals is the set of sinusoidal waveforms whose frequencies are integral multiples of some fundamental frequency, i.e., the sinusoidal waveforms which form the basis of a Fourier series expansion. Another example is pulse-position modulation. Here, the interval T is divided into M disjoint segments, each of duration T/M. A pulse occurring in any of the M time slots represents one of the M messages. Since the time slots do not overlap, Eq. (13.8-1) is satisfied.

We shall assume that the amplitudes of the signals have been adjusted so that in every case

$$\int_0^T s_i^2(t)\, dt = E_s \tag{13.8-2}$$

that is, each signal has the same energy E_s in the interval T and also the same power $S = E_s/T$.

MATCHED-FILTER RECEPTION

Let us assume that our message source generates M messages, each with equal likelihood. Let each message be represented by one of the orthogonal sets of signals $s_1(t)$, $s_2(t)$, . . . , $s_M(t)$. The message interval is T. The signals are transmitted over a communications channel where they are corrupted by additive white gaussian noise. As shown in Fig. 13.8-1, at the receiver a determination of which message has been trans-

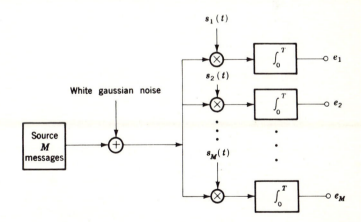

Fig. 13.8-1 The messages of a source are represented by orthogonal signals $s_1(t)$, $s_2(t)$, . . . , $s_M(t)$. Reception is accomplished by the use of correlation detectors (matched filters).

mitted is made through the use of M matched filters, that is, correlators. Each correlator consists of a multiplier followed by an integrator. The local inputs to the multipliers are, as shown, the signals $s_i(t)$. Suppose then, in the absence of noise, that the signal $s_i(t)$ is transmitted, and the output of each integrator is sampled at the end of a message interval. Then, because of the orthogonality condition of Eq. (13.8-1), all integrators will have zero output, except that the ith integrator output will be E_s, as given by Eq. (13.8-2).

We have assumed in this discussion that when the signal $s_i(t)$ is transmitted it is received without distortion. Since the signal is arbitrary and of finite duration, we have, in effect, assumed that the channel bandwidth is unlimited.

In the presence of an additive noise waveform $n(t)$, the output of the lth correlator $(l \neq i)$ will be

$$e_l = \int_0^T n(t)s_l(t) \, dt \equiv n_l \tag{13.8-3}$$

This quantity n_l is a *random variable*. In Prob. 13.8-2 the student is guided through a proof that (a) the random variable n_l is gaussian, (b) that it has a mean value zero, and (c) that it has a mean-square value, i.e., a variance σ^2 given by

$$\sigma^2 = \frac{\eta E_s}{2} \tag{13.8-4}$$

and (d) that $E(e_l e_m) = 0$; that is, the outputs of the matched filter are independent. The gaussian character and zero mean value of n_l do not depend on the form of the deterministic signal $s_l(t)$. The variance σ^2 depends only on the signal energy. Since we have selected the orthogonal signals $s_1(t), s_2(t), \ldots$, to have the same energy [Eq. (13.8-2)], the noise sample value at the output of each correlator will have the same statistical properties.

The correlator corresponding to the transmitted message $s_i(t)$ will have an output

$$e_i = \int_0^T [s_i(t) + n(t)]s_i(t) \, dt \tag{13.8-5a}$$

$$= \int_0^T s_i^2(t) \, dt + \int_0^T n(t)s_i(t) \, dt \tag{13.8-5b}$$

$$= E_s + n_i \tag{13.8-5c}$$

from Eqs. (13.8-2) and (13.8-3). Again n_i is a random variable with statistical properties identical with those specified for n_l.

CALCULATION OF ERROR PROBABILITY

To determine which message has been transmitted, we shall compare the matched-filter outputs $e_1, e_2, \ldots, e_M$. We shall decide that $s_i(t)$ has been transmitted if the corresponding output e_i is larger than the output of any other filter. We shall now calculate the probability that such a determination will lead to an error.

The probability that some arbitrarily selected output e_l is less than the output e_i is

$$P(e_l < e_i) = \frac{1}{\sqrt{2\pi\sigma^2}} \int_{-\infty}^{e_i} e^{-e_l^2/2\sigma^2} \, de_l \tag{13.8-6}$$

in which σ^2 is given by Eq. (13.8-4). Observe, in Eq. (13.8-6), that $P(e_l < e_i)$ depends only on e_i and not on e_l; that is, $P(e_l < e_i)$ does not depend on which output $e_l(l \neq i)$ has been selected for comparison with e_i.

The probability that, say, e_1 and e_2 are *both* smaller than e_i is

$$P(e_1 < e_i \text{ and } e_2 < e_i) = P(e_1 < e_i)P(e_2 < e_i) \tag{13.8-7a}$$

$$= [P(e_1 < e_i)]^2 = [P(e_2 < e_i)]^2 \tag{13.8-7b}$$

since e_1 and e_2 are independent (Prob. 13.8-2), and as we have noted, $P(e_l < e_i)$ does not depend on e_l. Hence, the probability P_L that e_i is the largest of the outputs is

$$P_L \equiv P(e_i > e_1, e_2, \ldots, e_{i-1}, e_{i+1}, \ldots, e_M)$$

$$= \left(\frac{1}{\sqrt{2\pi\sigma^2}} \int_{-\infty}^{e_i} e^{-e_l^2/2\sigma^2} \, de_l \right)^{M-1} \tag{13.8-8a}$$

$$= \left(\frac{1}{\sqrt{2\pi\sigma^2}} \int_{-\infty}^{E_s+n_i} e^{-e_l^2/2\sigma^2} \, de_l \right)^{M-1} \tag{13.8-8b}$$

from Eq. (13.8-5c). If we let $x \equiv e_l/(\sqrt{2}\,\sigma)$, Eq. (13.8-8b) becomes

$$P_L = \left(\frac{1}{\sqrt{\pi}} \int_{-\infty}^{\sqrt{E_s/\eta}+n_i/\sqrt{2}\sigma} e^{-x^2} \, dx \right)^{M-1} \tag{13.8-9}$$

where we have used Eq. (13.8-4) to show that $E_s/(\sqrt{2}\,\sigma) = \sqrt{E_s/\eta}$. Observe that P_L depends on the two deterministic parameters E_s/η and M and on the single random variable, $n_i/\sqrt{2}\,\sigma$; that is,

$$P_L = P_L \left(\frac{E_s}{\eta}, M, \frac{n_i}{\sqrt{2}\,\sigma} \right)$$

To find the probability that e_i is the largest output without reference to the noise output n_i of the ith correlator, we need but average P_L over all possible values of n_i. This average is the probability that we shall be correct in deciding that the transmitted signal corresponds to the correlator which yields the largest output. Let us call this probability P_c. The probability of an error is then $P_e = 1 - P_c$.

We have seen that n_i is a gaussian random variable with zero mean and variance σ^2. Hence the average value of P_L, considering all possible values of n_i, is

$$P_c = 1 - P_e = \frac{1}{\sqrt{2\pi\sigma^2}} \int_{-\infty}^{\infty} P_L(E_s/\eta, M, n_i/\sqrt{2}\,\sigma) e^{-n_i^2/2\sigma^2}\, dn_i$$

(13.8-10)

If we let $y \equiv n_i/(\sqrt{2}\,\sigma)$, Eq. (13.8-10) becomes, using Eq. (13.8-9)

$$1 - P_e \equiv P_c = \left(\frac{1}{\sqrt{\pi}}\right)^M \int_{-\infty}^{\infty} e^{-y^2} \left(\int_{-\infty}^{\sqrt{E_s/\eta}+y} e^{-x^2}\, dx\right)^{M-1} dy$$

(13.8-11)

The integral in Eq. (13.8-11) is rather formidable. It has been evaluated[4] by numerical integration using a digital computer, and the results for several values of M are shown in Fig. 13.8-2. Here the ordinate is the error probability P_e. We may note that it appears from Eq. (13.8-11) that P_e is a function of E_s/η and M. Since the received power is $S_i = E_s/T$, we may eliminate E_s. Hence it can be shown (Prob. 13.8-3) that P_e is a function only of the ratio S_iT/η and M. The abscissa of the plot of Fig. 13.8-2 has been marked off in units of the ratio $S_iT/(\eta \log_2 M)$. Note that $(\log_2 M)/T = R$, the rate of information transfer (since each of the M signals has equal probability of occurrence), so that $S_iT/(\eta \log_2 M) = S_i/\eta R$.

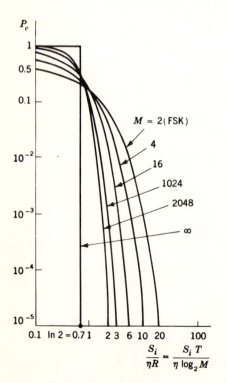

Fig. 13.8-2 Probability of error with orthogonal signals. (*Courtesy of A. J. Viterbi, "Principles of Coherent Communications," McGraw-Hill Book Co.,* 1966.)

13.9 EFFICIENCY OF ORTHOGONAL SIGNAL TRANSMISSION

We observe that for all M, P_e decreases as $S_i/\eta R$ increases. Also, we note that as $S_i/\eta R \to \infty$, $P_e \to 0$. The axis $P_e = 0$ does not appear on the plot because of the logarithmic scale of the ordinate. In the plot for $M \to \infty$, $P_e = 0$, provided that

$$\frac{S_i}{\eta R} \geq \ln 2 \tag{13.9-1}$$

and $P_e = 1$ otherwise. In general, as M increases, S_i/η must also increase to keep P_e small, otherwise P_e increases.

Some of the features in Fig. 13.8-2 are certainly to have been anticipated readily enough on a qualitative basis. Thus we note that for fixed M and R, P_e decreases as the signal power S_i goes up or as the background noise density η goes down. Similarly, we note that for fixed S_i, η, and M, the P_e decreases as we allow more time T (that is, decreasing R) for the transmission of a single message. This result is to have been anticipated as an easy generalization of our earlier discussion (Sec. 11.1) concerning baseband binary PCM transmission [see particularly Eq. (11.1-6) and the ensuing comments].

However, the real usefulness to us of the results given in Fig. 13.8-2 is that, in the first place, they allow us the satisfaction of verifying, in this one special case, the validity of Shannon's principle and of the Shannon-Hartley result for channel capacity. In the second place, they allow us a basis on which to make an engineering judgement concerning how much complexity is warranted in terms of the reduction in error which results therefrom. We consider now the first point.

SHANNON LIMIT

Let us assume that our signal source generates 4 equally likely messages $m_1(t)$, $m_2(t)$, $m_3(t)$, $m_4(t)$, and that a time $T = 1$ sec is to be allowed as the transmission interval for a message. We shall refer to these messages as *elemental* messages in the sense that they extend over the smallest time interval in which some information may be conveyed. The transmission rate is $R = (\log_2 4)/1 = 2$ bits/sec. Let us assume that $S_i/\eta = 10$ sec^{-1}. Then $S_i/\eta R = 5$, and, as indicated on the plot of Fig. 13.8-2, we may expect an error probability $P_e = 10^{-2}$. At the receiver we require 4 matched filters. We further assume that the channel bandwidth is unlimited.

Now let us join two successive intervals into an extended interval of length $2T = 2$ sec. Since 4 messages are possible in each interval T, there are $4 \times 4 = 16$ message combinations, i.e., new orthogonal messages possible in the extended interval. We now consider $2T$ as the message interval, and the transmission rate R is, of course, unaltered, for we now have $R = (\log_2 16)/2 = 2$ bits/sec as before. If we transmit using these 16 new orthogonal messages, then we shall complicate the receiver very seriously, since we now require 16 matched filters in place of the original 4. The advantage which accrues to us in return for this added complexity is to be seen by referring again to Fig. 13.8-2. For now we find, with $S_i/\eta R = 5$ again, that the error probability has been reduced to $P_e \cong 5 \times 10^{-4}$.

If we extend our message intervals to $3T$, $4T$, etc., the number of composite messages increases correspondingly to 64, 256, etc., reducing the error probability further, but correspondingly increasing the number of matched filters required. In the limit, as the number of message intervals increases without bound, $M \to \infty$, and the error probability P_e will approach zero provided that Eq. (13.9-1) is satisfied. From this condition we may calculate the maximum allowable errorless transmission rate $R_{\max}$, which, by definition, is the channel capacity C. We find

$$R_{\max} \equiv C = \frac{S_i}{\eta} \frac{1}{\ln 2} = \frac{S_i}{\eta} \log_2 e = 1.44 S_i/\eta \tag{13.9-2}$$

Equation (13.9-2) is in agreement with Eq. (13.7-2) which we deduced from the Shannon-Hartley theorem for a channel of unlimited bandwidth.

When we transmit elemental messages, then, of course, some information is received in each elemental interval T. Suppose, however, we form composite messages each of which extends over many time intervals. Then no information will be received except at the termination of the extended interval. At this termination, all the information in all the intervals will be received at once. In the limit, as we increase the number of intervals without bound, and hence let $M \to \infty$, to reduce the error probability to zero, we find that we must wait a time $T \to \infty$ before we receive any information.

In summary, we have examined one possible way of achieving errorless transmission. We have shown that such transmission is possible through the use of orthogonal signals and matched filters over a channel of unlimited bandwidth. We have also calculated the channel capacity and arrived at a result which is consistent with the Shannon-Hartley theorem.

TRADEOFFS

The plots of Fig. 13.8-2 may be used not only to illustrate some general principles as we have already done, but also to determine what advantage results in one direction as a result of a sacrifice in another direction. By way of example, suppose we are required to transmit information at some fixed information rate R, in the presence of a fixed background-noise power spectral density, with an error probability $P_e = 10^{-5}$. Suppose we use $M = 2$. Then, from Fig. 13.8-2 we find that $S_i/\eta R = 20$. Now consider what happens if M is increased to $M = 4$. Then to keep R fixed, T must be increased by the factor $(\log_2 4/\log_2 2) = 2$. Again referring to Fig. 13.8-2, we see that $S_i/\eta R$ can now be reduced to 10 and still result in $P_e = 10^{-5}$. Since η and R are fixed, S_i can be halved (reduced by 3 dB). Thus, to increase the number of messages while maintaining the same information rate R and P_e requires that the time T be increased; however, the required signal power can be decreased.

If with $M = 2$ the signal were sampled every millisecond, then with $M = 4$ we would transmit a 2-msec signal. However, complex digital storage units would be required to store the samples until we are ready to encode and transmit them. This increase in system complexity may not justify the savings in transmitted signal power.

13.10 CODING: INTRODUCTION

In Secs. 13.8 and 13.9 we studied one method of improving the efficiency of operation of a communications channel by transmitting a *coded* signal which consisted of 1 of M orthogonal signals. That is, given a channel of some capacity C, we examined how best to transmit information at a rate as nearly equal to C as possible with a minimum likelihood of error. We consider now alternative methods of coding signals. The use of coding rather than binary transmission is effective because it permits us to detect errors in transmission and in some cases to correct errors.

Coding accomplishes its purpose through the deliberate introduction of *redundancy* into messages. For example, orthogonal signaling required the transmission of 1 of M orthogonal messages. If each of these orthogonal messages were transmitted using binary digits, it can be shown that M binary digits/message are required. If, however, the original M quantization levels were binary-encoded, then only $N = \log_2 M$ binary digits/message are required. The excess number of binary digits required for orthogonal encoded signals is $M - N$. Hence the coding will produce an effective redundancy of $M - N$. As an additional example consider that we are transmitting information by means of binary PCM. Then we transmit a stream of binary digits, 0 or 1, and our concern is not to confuse a 0 for a 1, or a 1 for a 0. Suppose that when a 0 is to be transmitted, we transmit instead a sequence of three 0s, that is, 000, and transmit 111 to represent the digit 1. These triplets of 0s or 1s are certainly *redundant*, for two of the 0s in 000 add *no information* to the message. Suppose, however, that the signal-to-noise ratio on the channel is such that we can be nearly certain that not more than one error will be made in a triplet. Then if we received 001, 010, or 100, we would be rather certain that the transmitted message was actually 000. Similarly, if we received 011, 101, or 110, we would be rather certain that the message was actually 111.

Thus the redundancy, *deliberately introduced*, has enabled us to detect the occurrence of an error and even to correct the error.

The introduction of redundancy, however, can not *guarantee* that an error will be either detectable or correctable, for errors are caused by the *unpredictable* random process called noise. Hence, while the noise level, as we assumed, may be low enough so that more than a single error is *unlikely*, there is always a finite possibility that two errors will occur. In this case we would know that an error had occurred but we would be inclined to read a 0 as a 1 and a 1 as a 0. Even more, there is always the possibility, however small, that three errors had been made. In this case, not only would we misread the digit, but we would not even suspect that an error had been made. We are thus led to the conclusion that while coding may allow a great deal of detection and correction, it ordinarily cannot detect or correct all errors.

An essential feature which results from the introduction of redundancy is that not all sequences of symbols constitute a bona fide message. For example, with a triplet of binary digits, eight combinations are possible. However, only two of these combinations, that is, 000 and 111, are recognized as words which convey a message. The remaining combinations are not in our "dictionary." It is this fact that certain words are not in our "dictionary" that allows us to detect errors. Corrections are made on the basis of a similarity between unacceptable and acceptable words.

There is a correspondence between the redundancy deliberately introduced in coding messages prior to transmission over a channel and the redundancy which is part of language. For example, suppose that on a page of printed text we encounter the word *ekpensive*. We would immediately recognize that the printer had made an error since we know that there is no such word. If we were inclined to judge it unlikely that the printer had made an error in more than one letter, we would easily recognize the word to be *expensive*. On the other hand there is always the possibility that more than one error was made and that the intended word was, say, *offensive*. And in this latter case, of course, we would not make the proper correction.

In language, redundancy extends beyond combinations of letters in a word. It extends to entire words, and beyond to phrases and even sentences. For example, if we had come across the sentence "the army launched an ekpensive," we would not have much difficulty in recognizing that the proper correction of the last word is *offensive*, in spite of the fact that corrections in 3 letters are called for. In this case we would be taking advantage of the redundancy in the sentence.

Now let us return briefly to the binary PCM transmission scheme we described above, in which 3 bits 000 or 111 were transmitted as a code to represent the bits 0 and 1, respectively. If the redundant message is to be transmitted at the same rate as the original binary signal, we shall have to transmit 3 bits in the time T otherwise allocated to a single bit. As we have seen on many occasions when the time allocated to a bit transmission decreases, the error rate increases. Hence, the required increased bit rate will undo some of the advantage that will accrue from redundancy coding. We shall however, of course, find that coding may yield a very worthwhile net advantage.

13.11 PARITY CHECK CODING

The simplest error-detecting technique consists in adding an extra binary digit at the end of each word. This extra bit is called a *parity-check bit* and is chosen to make the number of 1s in each word an *even* number. Thus consider a case in which there are only 16 ($= 2^4$) possible messages to be transmitted. These 16 messages could be encoded into 4 bits. In adding an extra bit, the fifth bit, we now have $2^5 = 32$ possible words, only 16 of which are to be recognized as messages conveying information. Thus the parity-check bit has introduced redundancy. If an odd number of the first 4 bits were 1s, we would add a 1. Thus 1011 would become 10111. If an even number of the first 4 bits were 1s, we would add a 0. Thus 1010 would become 10100.

The parity-check bit is rather universally used in digital computers for error detection. This check bit will be effective if the probability of error in a bit is low enough so that we may ignore the likelihood of more than a *single* bit error in a word. If such a single error occurs, it will change a 0 to a 1 or a 1 to a 0 and the word, when received, would have an odd number of 1s. Hence it will be known that an error has occurred. We shall not, however, be able to determine which bit is in error. An error in *two* bits in a word is not detectable, since such a double error will again yield an even number of 1s.

13.12 ALGEBRAIC CODES

Coding has the advantage that it allows us to increase the rate at which information may be transmitted over a channel while maintaining a fixed error rate. The price we pay for this advantage is the increased complexity of both the transmitter where the *encoding* is done and the receiver where the *decoding* is performed. In principle, with ingenious enough coding and unlimited complexity we would be able to reach the Shannon limit. That is, we would be able to transmit at channel capacity and without error. One measure of the efficiency of a code is precisely the extent to which it allows us to approach the Shannon limit. Another measure of the merit of a code is the extent to which it improves transmission without introducing excessive complexity.   Many different types of codes have been developed and are in use.   The design of codes is a rather sophisticated topic which we shall not discuss.   We shall, however, describe and analyze some of the more commonly employed codes.   We consider now a class of codes which is described by the designation *algebraic codes.*

We assume transmission by binary PCM where the signal is a variable which has the value 0 or 1.   It is convenient to introduce now an *algebra for coding* for use with the binary variable.   In this algebra the operations of addition and multiplication are defined by the following rules:

$$
\begin{array}{ll}
\textit{Addition} & \textit{Multiplication} \\
0 + 0 = 0 & 0 \cdot 0 = 0 \\
0 + 1 = 1 & 0 \cdot 1 = 0 \\
1 + 0 = 1 & 1 \cdot 0 = 0 \\
1 + 1 = 0 & 1 \cdot 1 = 1
\end{array}
\tag{13.12-1}
$$

This algebra is the same as Boolean algebra with the single exception that here $1 + 1 = 0$ while in Boolean algebra $1 + 1 = 1$.   (We are using the plus sign to represent modulo-2 addition.)

The method of algebraic coding is the following.   Suppose that our message source can generate M equally likely messages.   Then initially we encode these in M message words each of length k where $M = 2^k$.   Each of these message code words has no redundancy.   Each bit in each message code word conveys an amount of information $I = 1$ bit.   We now add to each such message r redundant bits.   The transmitted code word therefore has $k + r \equiv n$ bits.   The total number of words is 2^n, while the number of possible messages is 2^k.   A typical transmitted code word will have the form

$$a_1 a_2 a_3 \; \cdot \; \cdot \; \cdot \; a_k c_1 c_2 \; \cdot \; \cdot \; \cdot \; c_r$$

where a_i is the ith bit of the message code word and c_j is the jth redundant bit.   These added redundant bits, as before, are called parity-check bits. Both a_i and c_j may assume the values 0 or 1.

The parity bits are selected to satisfy the linear equations

$$
\begin{aligned}
0 &= h_{11}a_1 + h_{12}a_2 + \cdots + h_{1k}a_k + 1c_1 + 0c_2 + \cdots + 0c_r \\
0 &= h_{21}a_1 + h_{22}a_2 + \cdots + h_{2k}a_k + 0c_1 + 1c_2 + \cdots + 0c_r \\
&\cdots \cdots \cdots \cdots \cdots \cdots \cdots \cdots \cdots \cdots \cdots \\
0 &= h_{r1}a_1 + h_{r2}a_2 + \cdots + h_{rk}a_k + 0c_1 + 0c_2 + \cdots + 1c_r
\end{aligned}
\tag{13.12-2}
$$

The coefficients h_{ij} are either 0 or 1. These coefficients are not variables; their value is fixed. The rules of algebra in Eq. (13.12-2) are given in Eq. (13.12-1). What distinguishes one algebraic code from another is the number of parity bits used and the selection of the coefficients h_{ij}.

Equations (13.12-2) may be written more conveniently in matrix notation. We define the matrix $\bar{H}$ as a rectangular matrix with r rows and n columns and given by

$$\bar{H} \equiv \begin{vmatrix} h_{11} & h_{12} & \cdots & h_{1k} & 1 & 0 & 0 & \cdots & 0 \\ h_{21} & h_{22} & \cdots & h_{2k} & 0 & 1 & 0 & \cdots & 0 \\ \cdots & \cdots & \cdots & \cdots & \cdots & \cdots & \cdots & \cdots & \cdots \\ h_{r1} & h_{r2} & \cdots & h_{rk} & 0 & 0 & 0 & \cdots & 1 \end{vmatrix} \qquad (13.12\text{-}3)$$

We also define a column matrix $\bar{T}$ representing the transmitted code word (which in turn represents a possible message) as

$$\bar{T} \equiv \begin{vmatrix} a_1 \\ a_2 \\ \cdot \\ \cdot \\ \cdot \\ a_k \\ c_1 \\ c_2 \\ \cdot \\ \cdot \\ \cdot \\ c_r \end{vmatrix} \qquad (13.12\text{-}4)$$

Then, in terms of these matrices, Eqs. (13.12-2) may be written

$$\bar{H}\bar{T} = 0 \qquad (13.12\text{-}5)$$

The column matrix $\bar{T}$ is called a *vector*, with *components* $a_1, \ldots, a_k, c_1, \ldots, c_r$.

Now let the received message be $\bar{R}$ which may or may not be the transmitted code word. Suppose that at the receiver we have an appropriate apparatus for forming the product $\bar{H}\bar{R}$. If $\bar{H}\bar{R} \neq 0$, we know that $\bar{R}$ is not a possible message and 1 or more errors have been made. If, however, we find that $\bar{H}\bar{R} = 0$, then we know that $\bar{R}$ is a possible message, and very likely, although not necessarily, $\bar{R}$ is the message that was transmitted. Still, if the possible messages are sufficiently *different* from one another, we may well be able to discount the possibility that the received possible message was not the one transmitted.

13.13 ELEMENTARY EXAMPLES OF ALGEBRAIC CODES

SINGLE PARITY-CHECK BIT CODE

The single parity-check bit code is an example of an algebraic code. In this case the parity-check bit is selected to satisfy the equation

$$a_1 + a_2 + \cdots + a_k + c_1 = 0 \tag{13.13-1}$$

Equation (13.13-1) is the first of the set in Eqs. (13.12-2) with

$$h_{11} = h_{12} = \cdots h_{1k} = 1$$

All other equations of the set are identically zero. In this case, then, $r = 1$ and $n = k + 1$. Keeping in mind the definition of addition, given in Eq. (13.12-1), we may see that Eq. (13.13-1) requires that the number of 1s in a word be even, for only the sum of an even number of 1s will add up to zero. If there were an odd number of 1s in the sum of Eq. (13.13-1), the result would be 1 rather than 0.

REPEATED CODES

In a *repeated* code a binary 0 is encoded as a sequence of $(2t + 1)$ zeros, and a binary 1 as a similar number of 1s. Thus $k = 1$, $r = 2t$, and $n = 2t + 1$. We considered such encoding in Sec. 13.10 for the case $t = 1$, so that $r = 2t = 2$ and $n = 2t + 1 = 3$. A repeated code is an algebraic code since the redundant bits are determined by Eqs. (13.12-2). We have $2t$ equations with a_1 equal to either 0 or 1, and

$$h_{11} = h_{21} = \cdots = h_{r1} = 1$$

while all other h's are set to zero. The redundant bits are therefore given by

$$
\begin{aligned}
0 &= a_1 + c_1 \\
0 &= a_1 \qquad + c_2 \\
&\cdot \cdot \cdot \cdot \cdot \cdot \cdot \cdot \cdot \cdot \\
0 &= a_1 \qquad\qquad + c_{2t}
\end{aligned}
\tag{13.13-2}
$$

Using again the coding arithmetic of Eq. (13.12-1), we find that if $a_1 = 0$, all the c's are 0, and if $a_1 = 1$, all the c's are 1. The matrix $\bar{H}$ for the repeated code with $n = 2t + 1 = 3$ is given by [see Eq. (13.12-3)]

$$\bar{H} = \begin{vmatrix} h_{11} & 1 & 0 \\ h_{21} & 0 & 1 \end{vmatrix} = \begin{vmatrix} 1 & 1 & 0 \\ 1 & 0 & 1 \end{vmatrix} \tag{13.13-3}$$

13.14 ERROR CORRECTION: THE SYNDROME

We have seen that when we find $\bar{H}\bar{R} \neq 0$, we know that an error has occurred. When $\bar{H}\bar{R} = 0$, either no error has been made or sufficient errors have occurred so that there has been a complete substitution of one possible message for another.   Fortunately, this latter contingency is very unlikely, since coding provides no mechanism of detecting this possibility. On the other hand, as we shall now see, when the occurrence of an error is detected, coding may actually allow us to *correct* the error.

Suppose that a transmitted message is $\bar{T}$ and that because of errors the received sequence of binary digits is $\bar{R}$.   Then we may always write

$$\bar{R} = \bar{T} + \bar{E} \tag{13.14-1}$$

in which the vector $\bar{E}$ is the *error*.   We keep in mind that the addition in Eq. (13.14-1) follows the rules in Eq. (13.12-1).   We may then readily verify that $\bar{E}$ is a vector with the same number of components as $\bar{R}$ and as $\bar{T}$, and that $\bar{E}$ contains 0s in all positions where errors did not occur and 1s in all positions where errors did occur.   If, for example, the transmitted binary sequence is 11111 (from the repeated code with $n = 5$) and if we receive 01101, then

$$\bar{R} = \begin{vmatrix} 0 \\ 1 \\ 1 \\ 0 \\ 1 \end{vmatrix} \qquad \bar{T} = \begin{vmatrix} 1 \\ 1 \\ 1 \\ 1 \\ 1 \end{vmatrix} \qquad \bar{E} = \begin{vmatrix} 1 \\ 0 \\ 0 \\ 1 \\ 0 \end{vmatrix} \tag{13.14-2}$$

To *correct* errors we need to know which binary digits in a received binary sequence are in error.   This determination may be made by computing the vector $\bar{S}$ defined by

$$\bar{S} \equiv \bar{H}\bar{R} \tag{13.14-3}$$

We have

$$\bar{S} = \bar{H}\bar{R} = \bar{H}\bar{T} + \bar{H}\bar{E} = \bar{H}\bar{E} \tag{13.14-4}$$

since $\bar{H}\bar{T} = 0$.   If $\bar{R} = \bar{T}$, that is, no errors, then $\bar{H}\bar{E} = 0$ as well.   Suppose now that only a single bit is in error, say, the ith bit.   Then in $\bar{E}$ all bits are zero except the ith bit.   Thus

$$\bar{E} = \begin{vmatrix} 0 \\ 0 \\ \cdot \\ \cdot \\ \cdot \\ 0 \\ 1 \\ 0 \\ \cdot \\ \cdot \\ \cdot \\ 0 \end{vmatrix} \rightarrow i\text{th position} \tag{13.14-5}$$

and

$$
\bar{S} =
\begin{vmatrix}
h_{11} & \cdots & h_{1k} & 1 & 0 & \cdots & 0 \\
h_{21} & \cdots & h_{2k} & 0 & 1 & \cdots & 0 \\
\multicolumn{7}{c}{\cdots\cdots\cdots\cdots\cdots\cdots} \\
h_{r1} & \cdots & h_{rk} & 0 & 0 & \cdots & 1
\end{vmatrix}
\begin{vmatrix}
0 \\ 0 \\ \cdot \\ \cdot \\ 0 \\ 1 \\ 0 \\ \cdot \\ \cdot \\ \cdot \\ 0
\end{vmatrix}
=
\begin{vmatrix}
i\text{th} \\ \text{column} \\ \text{of} \\ \bar{H}
\end{vmatrix}
\qquad (13.14\text{-}6)
$$

Thus, we find that the location of a single error may be determined by noting the column of the matrix $\bar{H}$ which appears when we form the product $\bar{S} \equiv \bar{H}\bar{R}$. The vector $\bar{S}$ is called the *syndrome*, a medical term which refers to a collection of symptoms which characterize a disease.

It is particularly to be noted that, for the syndrome to give an unambiguous determination of an error location, each column of $\bar{H}$ must be distinct, and no column consists of all zeros. For if, say, the ith column were all zeros, we would not be able to distinguish between the case of an error in the ith bit and the case of no error at all, since both cases would yield $\bar{S} = 0$.

13.15 AN EXAMPLE OF ALGEBRAIC CODING

Suppose that we are to encode 16 equiprobable messages. We then have $k = 4$; that is, 4 binary digits are required before we add digits for the sake of redundancy. We should like to add enough redundant digits to allow correction of single errors. It is rather apparent that a single redundant digit is not enough. We may also easily establish that 2 additional redundant digits are still not enough. If we add 2 such digits, then we have $r = 2$ and $n = k + r = 6$. The $\bar{H}$ matrix in Eq. (13.12-3) then has 2 rows and 6 columns. Each entry in the matrix is a 0 or a 1. It is obviously impossible to have each of the 6 columns different from every other. Hence 2 redundant digits are not enough.

With 3 redundant digits, $r = 3$, $n = k + r = 7$. We have 3 rows and 7 columns. With 3 digits in each row we have 8 possible *different* columns. We disallow a column in which all elements are 0, leaving 7. The $\bar{H}$ matrix may now be formed by simply writing down these 7 different columns. The order in which the columns are listed is of no consequence. Still, in order to preserve the form of $\bar{H}$ indicated in Eq. (13.12-3) (where the columns in which the digit 1 appears only a single time are on the right, and the 1s in these columns are on a diagonal), we write

$$
\begin{array}{ccccccc}
a_1 & a_2 & a_3 & a_4 & c_1 & c_2 & c_3
\end{array}
$$

$$
\bar{H} = \begin{vmatrix}
1 & 1 & 1 & 0 & 1 & 0 & 0 \\
1 & 1 & 0 & 1 & 0 & 1 & 0 \\
1 & 0 & 1 & 1 & 0 & 0 & 1
\end{vmatrix}
\tag{13.15-1}
$$

Since $n = 7$, there are $2^7 = 128$ possible seven-digit sequences. We may use the matrix $\bar{H}$ of Eq. (13.15-1) to determine the $2^4 \, (= 16)$ sequences of 7 digits which will actually be recognized as messages at the receiver. Consider one of the 16 possible messages where the first 4 digits are 1000. What, then, is the complete message sequence $1000c_1c_2c_3$? We write

$$
\bar{H}\bar{T} = \bar{H} \begin{vmatrix}
1 \\
0 \\
0 \\
0 \\
c_1 \\
c_2 \\
c_3
\end{vmatrix} = 0
\tag{13.15-2}
$$

and find

$$
\begin{vmatrix}
1 + 0 + 0 + 0 + c_1 + 0 + 0 \\
1 + 0 + 0 + 0 + 0 + c_2 + 0 \\
1 + 0 + 0 + 0 + 0 + 0 + c_3
\end{vmatrix} = 0
\tag{13.15-3}
$$

so that $c_1 = c_2 = c_3 = 1$. The message is therefore represented by the code word 1000111. In a similar manner the code words for the remaining 15 messages may be determined (Prob. 13.15-1).

To see how the coding may be used to correct single errors, let us assume that such a single error has been made in the code word 1000111 so that the received sequence is 1010111. We evaluate the syndrome and find

$$
\bar{S} = \bar{H}\bar{R} = \bar{H} \begin{vmatrix}
1 \\
0 \\
1 \\
0 \\
1 \\
1 \\
1
\end{vmatrix} = \begin{vmatrix}
1 \\
0 \\
1
\end{vmatrix}
\tag{13.15-4}
$$

Since the syndrome is identical with the third column of the matrix $\bar{H}$, we know that the third digit a_3 is in error, and hence the corrected word is 1000111.

It is to be noted that the selection of the 16 recognized words (messages) out of the possible 128 sequences is not unique. Suppose, for example, we had written the matrix $\bar{H}$ in Eq. (13.12-3) in the form

$$\begin{array}{ccccccc} a_1 & c_2 & a_2 & a_3 & a_4 & c_1 & c_3 \end{array}$$

$$\bar{H} = \begin{vmatrix} 1 & 0 & 1 & 1 & 0 & 1 & 0 \\ 1 & 1 & 1 & 0 & 1 & 0 & 0 \\ 1 & 0 & 0 & 1 & 1 & 0 & 1 \end{vmatrix} \tag{13.15-5}$$

In this case we have not maintained the order specified by Eqs. (13.12-2). However, it should be clear that the order is not mandatory. Let us again set $a_1 = 1$, $a_2 = a_3 = a_4 = 0$ corresponding to message number 8. Then it may be verified (Prob. 13.15-2) that the code word corresponding to this message would read 1100011. This code word is not only different from the previous code word found for this message but is also different from any of the code words corresponding to the matrix $\bar{H}$ in Eq. (13.15-1).

The acceptable code words depend on the manner in which the matrix columns are arranged. However, for any arrangement we would

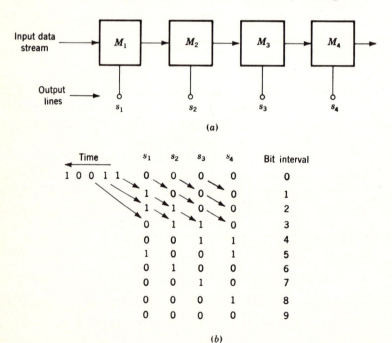

(a)

(b)

Fig. 13.16-1 (a) A four-stage shift register. (b) Successive states of stages for input train 11001.

always find that the 16 code words differ from each other in at least *three* bits. It is this feature that allows us to distinguish *single* errors. For if a single error is made in a word, it will not be confused with another message nor will it be confused with any other message which might have only *one* error. That is, two messages, both with a single error, will still differ by at least *one* bit. The minimum number of bits by which code words differ from one another is called the *distance* of the code, or the *Hamming distance*.

BLOCK CODES

The codes discussed in this section are examples of *block* codes. In block encoding the data stream to be encoded is separated into equal-length message blocks of k binary digits each. The encoder adds r redundant bits so that the code word has $n = k + r$ digits. Such a code is called an (n,k) block code and the number n is called the *block length* of the code word.

Each n-bit data stream block is encoded *independently* of preceding or succeeding n-bit blocks. We describe in the next section a different type of coding called *convolutional* coding, in which the message bit stream is encoded in a continuous fashion, rather than *piecemeal* as in block coding.

13.16 CONVOLUTIONAL CODES†

SHIFT REGISTERS

Convolutional codes are easily generated with a *shift register*. We therefore begin our discussion by describing the operation performed by a shift register on a binary encoded message. A shift register is a cascade of storage (memory) devices, each device being capable of storing 1 binary digit. Typically, the 1-digit storage device may be a flip-flop which stores the digit 1 or 0, depending on which of its two allowable states it occupies. A shift register employing four 1-digit memories, M_1 through M_4, is shown in Fig. 13.16-1a. A stream of binary encoded data is applied to M_1. Since the mechanism of operation of the shift register is not relevant to the present discussion, we shall therefore simply describe the response of the shift register to the input data stream.

The first memory device M_1 stores the most recent input bit and indicates its state on its output line. The output s_1 of M_1 is therefore a bit stream which duplicates the input bit stream. At the end of each bit interval the bit stored in *each* of the memory devices *shifts* one stage to the right. Hence the output s_2 of M_2 is the input bit stream except delayed by a 1-bit interval. Similarly the input bit stream also appears at

† The presentation given here follows the development given in Wozencraft and Jacobs (Ref. 2, pp. 405–431).

the output of M_3 and M_4 except delayed by 2- and 3-bit intervals, respectively. The input data stream enters the shift register at the left, is shifted through the register, and leaves the register at the right. At any particular time, the shift register remembers (i.e., serves as a temporary register) of the 4 successive most recent bits of the input data stream.

The operation of the shift register is illustrated in Fig. 13.16-1b. Here we have assumed that initially the shift register is *clear*, i.e., all memories are storing zeros. A data train of 5 bits is now applied to the input and this figure traces the passage of the data train through the register. The input data stream is 11001, but for convenience we have represented this train against a reversed time scale. Note that with each succeeding bit the contents of each memory device are shifted into the next device. Observe also that at the ninth interval the input data have passed completely through the register and the register returns to its "clear" state.

CODE GENERATION

A *convolutional code* is generated by combining the outputs of a K-stage shift register through the employment of v modulo-2 adders. Such a coder is illustrated in Fig. 13.16-2 for the case $K = 4$ and $v = 3$. The outputs v_1, v_2, and v_3 of the adders in Fig. 13.16-2 are

$$v_1 = s_1 \tag{13.16-1a}$$

$$v_2 = s_1 + s_2 + s_3 + s_4 \tag{13.16-1b}$$

$$v_3 = s_1 + s_3 + s_4 \tag{13.16-1c}$$

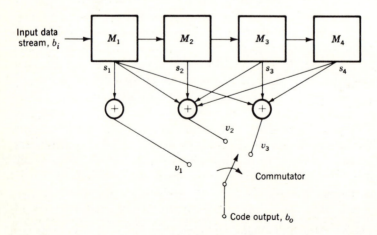

Fig. 13.16-2 An example of a convolutional coder.

The operation of the encoder proceeds as follows: We assume that initially the shift register is clear.   The first bit-of the input data stream is entered into M_1.   During this message bit interval the commutator samples, in turn, the modulo-2 adder outputs v_1, v_2, and v_3.   Thus a single message bit yields, in the present case, three coded output bits.   The next message bit then enters M_1, while the bit initially in M_1 transfers to M_2, and the commutator again samples the v adder outputs.   This process continues until eventually the last bit of the message has been entered into M_1. Thereafter, in order that every message bit may proceed entirely through the shift register, and hence be involved in the complete coding process, enough 0s are added to the message to transfer the last message bit through M_4 and, hence, out of the shift register.   The shift register then finds itself in its initial "clear" condition.

It may be verified, by way of example, that, for the encoder of Fig. 13.16-2, if the input bit stream to the encoder is given by the 5-bit sequence

$$m = 1 \quad 0 \quad 1 \quad 1 \quad 0 \qquad\qquad\qquad (13.16\text{-}2)$$

then the coded output bit stream is

$$c = 111 \quad 010 \quad 100 \quad 110 \quad 001 \quad 000 \quad 011 \quad 000 \quad 000 \qquad (13.16\text{-}3)$$

If the number of bits in the message stream is L, the number of bits in the output code is $v(L + K)$.   As a matter of practice however, L is ordinarily a very large number while K is a relatively small number.   Hence $v(L + K) \simeq vL$.   Thus the number of code bits is v times the number of message bits, v being the number of commutator segments.

Note the continuity of operation of the convolutional encoder.   Even if the input message bit stream were to consist of millions of bits, the stream would be run continuously through the encoder.   Each bit remains in the shift register for as many message bit intervals as there are stages in the shift register.   Hence each input bit has an influence on K groups of v bits.

13.17   DECODING A CONVOLUTIONAL CODE

THE CODE TREE

With a view toward exploring procedures for decoding a convolutional code, we consider the *code tree* of Fig. 13.17-1.   This code tree applies to the convolutional encoder of Fig. 13.16-2, for which $K = 4$ and $v = 3$, and which is constructed for the case $L = 5$ corresponding to a 5-bit message sequence.   We now discuss the interpretation of this code tree.

The starting point on the code tree is at the left and corresponds to the situation before the occurrence of the first message bit.   The first message bit may be either a 1 or a 0.   We adopt the convention that,

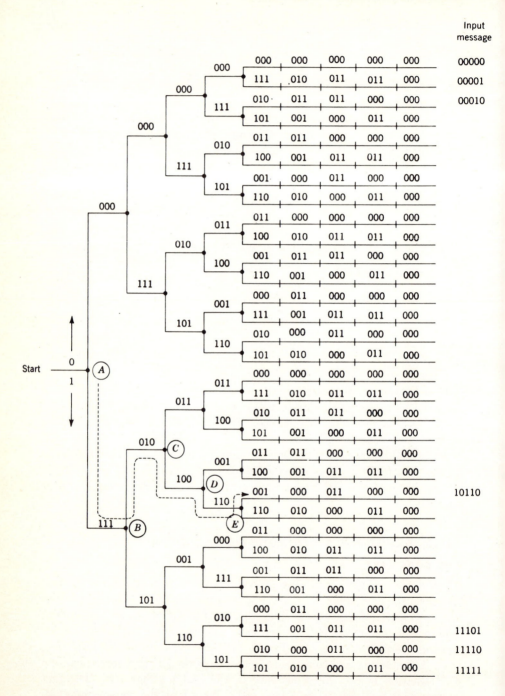

Fig. 13.17-1   Code tree for encoder of Fig. 13.6-2; $K = 4$, $L = 5$, $v = 3$. (*From Ref. 2.*)

when an input bit is a 0, we shall diverge upward from a node of the tree, and when the input bit is a 1 we shall diverge downward. Suppose then that the first input bit is 1. Then entering the tree at node A, we move downward to the lower branch and to node B. Now from Eqs. (13.16-2) and (13.16-3) we note that when the first input bit is 1, the output of the coder is 111. Hence the lower branch associated with node A in Fig. 13.17-1 has correspondingly been marked 111. Suppose, to continue, that the second input bit is 0. Then we now diverge upward from node B. We find, again, as to be noted from Eqs. (13.6-2) and (13.16-3), that, when the first two input bits to the encoder are 10, the encoder output during this second input message bit interval is 010. Hence the upper branch diverging from node B is correspondingly marked 010. Continuing in this fashion we find that the message $m = 10110$ of Eq. (13.16-2) indicates a downward divergence from node C to node D, a downward divergence from D to E, an upward divergence from node E, and from there out to the end of one of the branches of the tree. The path through the tree is shown by the dashed line. Reading, in order, the bits encountered from entrance to exit of the tree, we find precisely the code given in Eq. (13.16-3) for the message of Eq. (13.16-2).

Note that any path through the tree passes through only as many nodes (L) as there are bits in the input message. The node corresponds to a point where alternate paths are possible depending on whether the next message bit is 1 or 0. The extension of the terminal branches of the tree corresponds to the process of clearing the last message bit through the shift register.

DECODING IN THE PRESENCE OF NOISE

In the absence of noise, the code word will be received as transmitted. In this case it is a simple matter to reconstruct the original message. We simply follow the code word through the code tree v bits at a time. The message is then reconstructed from the path taken through the tree or, equivalently, from the terminal branch of the tree at which the path is completed. But suppose that, on account of noise, the word received is not as transmitted. How shall we undertake to reconstruct the transmitted code word in a manner which, hopefully, will correct errors? A recommended procedure is the following:

Consider the first message bit. This first message bit has an effect only on the first Kv bits in the code word. Thus with $K = 4$ and $v = 3$, as for the tree of Fig. 13.17-1, the first digit has an effect only on the first 4 groups of 3 digits. Hence to deduce this first bit, there is no point in examining the code word beyond its first 12 digits. On the other hand, we would not be taking full advantage of the redundancy of the code if we undertook to decide about the first message bit on the basis of anything less than an examination of the first 12 bits of the code. We see from the code tree of Fig. 13.17-1 that there are 16 possible combinations of the first 12 digits which are acceptable code words. These combinations correspond to the 16 possible (incomplete) paths through the code tree which penetrate into the tree only to the extent of passing the first 4 nodes. Let us then compare the first 12 bits of the received code with the 12 bits of each of the 16 acceptable possible paths. We next take a count, for each of the 16 paths, of the number of discrepancies between bits of the received code word and the acceptable code word corresponding to each path. If now the path that yields the minimum number of discrepancies is a path which diverges upward from the first node (node A in Fig. 13.17-1), we make the decision that the first message bit is 0. And, of course, if this path diverges downward from node A, we decide that the first bit is 1. Thus we see that a decision about the first message bit is not made until after a complete exploration of all possible paths which penetrate into the tree to the extent of including K nodes or equivalently Kv digits. It will be recognized that the operation of convolutional decoding described here is a digital operation which corresponds to the analog correlation detection process described in Sec. 11.6.

The second message bit is decided upon in the same manner. Thus, referring to Fig. 13.17-1, suppose it turns out that, using the above procedure, we decide that the first message bit is 1. Then we start at node B and again examine the 16 paths starting at node B and penetrating into the code tree to the extent of $Kv = 12$ code bits. We compare the 12 code bits corresponding to each of these 16 paths with the 12 received code bits after *discarding the first 3 received code bits*. A decision about the second message bit is now made on the basis of the same criterion used to decide about the first bit. Successive message bits are decoded in the same manner.

In summary, the decoding procedure is the following. The ith message bit m_i is decoded on the basis of previous decisions about the message bits $m_1, m_2, \ldots, m_{i-1}$ and on the basis of an examination of the Kv digit span of the received code word that is influenced by m_i. The previous decisions determine the starting node, the ith starting node, for the K node section of the code tree to be examined. Thereafter m_i itself is decoded by determining which one of the 2^K branch sections of the code tree diverging from the ith starting node exhibits the fewest discrepancies when compared with the Kv digit span of the received code word that is influenced by m_i. If this section of the code tree diverges upward from the ith starting node, the ith message digit is 0; otherwise it is 1.

It turns out that, when a message is decoded in the manner we have just described, the probability that a bit in the decoded message is in error decreases exponentially with K. Hence there is an interest in making K as large as possible. On the other hand, as noted, the decoding of each bit requires an examination of the 2^K branch sections of the code tree. Hence for large K the decoding procedure may involve such a lengthy procedure that it is simply not feasible. We therefore consider next a *sequential* decoding scheme that remains manageable even when K is large.

SEQUENTIAL DECODING

The principal advantage of sequential decoding is that such decoding generally allows the decoder to avoid the lengthy process of examining *every* branch of the 2^K possible branches of the code tree in the course of decoding a single message bit. In sequential decoding, at the arrival of the first v code bits, the encoder compares these bits with the two branches which diverge from the starting node. If one of the branches agrees exactly with these v code bits, then the encoder follows this branch. If, because of noise, there are errors in the received bits, the encoder follows the branch, in comparison with which the received bits exhibit the fewer discrepancies. At the second node a similar comparison is made between the diverging branches and the second set of v code bits, and so on, at succeeding nodes.

Now suppose that in the transmission of any v bits, corresponding to some tree, branch errors have found their way into more than half the bits. Then at the node from which this branch diverges the decoder will make a mistake. In such a case, the entire continuation of the path taken by the encoder must be in error. Consider, then, that the decoder keeps a record, as it progresses, of the total number of discrepancies between the received code bits and the corresponding bits encountered along its path. Then, after having taken a wrong turn at some node, the likelihood is very great that this total number of discrepancies will grow much more rapidly than would be the case if the decoder were following the correct path. The decoder may be programmed to respond to such a situation by retracing its path to the node at which an apparent error has been made and then taking the alternate branch out of that node. In this way the decoder will eventually find a path through K nodes. When such a path is found, the decoder decides about the first message digit on the basis of the direction of divergence of this path from the first node. Similarly, as before, the second message is determined on the basis of a path searched out by the decoder, again K branches long, but starting at the second starting node and on the basis of the received code bit sequence with the first v bits discarded.

We consider now how the decoder may judge that it has made an error and taken a wrong turn. Let the probability be q that a received

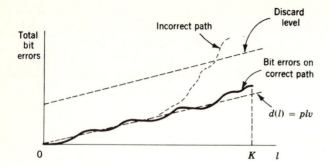

Fig. 13.17-2 Setting the threshold in sequential decoding.

code bit is in error, and let the encoder already have traced out a path through l nodes. We assume $l < K$, so that the decoder has not yet made a decision about the message bit in question. Since every branch is associated with v bits, then, on the average over a long path of many branches, we would expect the total number of bit differences between the decoder path and the corresponding received bit sequence to be $d(l) = qlv$ even when the correct path is being followed. In Fig. 13.17-2 we have plotted $d(l)$ in a coordinate system in which the abscissa is l, the number of nodes traversed, and the ordinate is the total bit error that has accumulated. We would expect that, if the encoder were following the correct path, the total bit errors accumulated would oscillate about $d(l)$. On the other hand, shortly after a wrong decoder decision has been made, we expect the accumulated bit error to diverge sharply as indicated. A *discard level* has also been indicated in the figure. When the plot of accumulated errors crosses the discard level, the decoder judges that a decoding error has been made. The decoder then returns to the nearest *unexplored* branch and starts moving forward again until, possibly, it is again reversed because the discard level is crossed. In this way, after some trial and error, an entire K-node section of the code tree is navigated. And at this point a decision is made about the message bit associated with this K-node section of the tree. In Fig. 13.17-2 the discard level line does not start at the origin in order to allow for the possibility that the initial bits of the received code sequence may be accompanied by a burst of noise.

The great advantage of sequential decoding of a convolutional code is that such decoding makes it unnecessary, generally, to explore every one of the 2^K paths in the code tree section. Thus, suppose it should happen that the decoder takes the correct turn at each node. Then, in this case, the decoder will be able to make a decision about the message bit in question on the basis of a *single* path. Let us assume that at some node the decoder errs and must return to this node to take an alternate branch. Then even the information that an error has been made is useful because thereafter the decoder may exclude from its searchings all paths which diverge in the original direction from this node. The end result is that sequential decoding may generally be accomplished with very much less computation than the direct decoding discussed earlier.

CONCLUSION

Sequential decoding of convolutional codes is one of the most powerful coding-decoding techniques in use today. Perhaps the major reason for employing this technique is, as can be shown, that while the error probability decreases exponentially with K, the complexity of the system increases linearly with K rather than exponentially. The error probability of block coding-decoding systems also decreases exponentially with increased block length. However, in such block coding systems the complexity also increases exponentially.

The convolutional encoder shown in Fig. 13.16-2 is not optimum. No optimum procedure currently exists for the selection of the number v or the equation of each v_i. In addition, there is no optimal way of tree searching to minimize simultaneously the time required to find the correct path and the probability of avoiding an incorrect path.

13.18 COMPARISON OF ERRORS IN CODED AND UNCODED TRANSMISSION (BLOCK CODING)

In this section we compare block-coded and uncoded systems to obtain some idea of their relative error rates. We do this under the condition that the rate of information transmission (messages or words per unit time) is the same in the two cases. Then, coded or uncoded, a word must have the same duration T_w. Since the coded word has more digits than does the uncoded word, the *bit* duration T in the coded case must be less than in the uncoded case. We have seen, however, that as the bit duration decreases, the probability of error in a bit increases. We have then to inquire whether coding yields a net advantage. For the sake of being specific, we shall compare k-bit uncoded transmission with transmission using the (n,k) block code discussed in Sec. 13.15. We shall, of course, assume that the signal power S and the thermal-noise power spectral density η are the same in both cases.

We shall use the symbols q and q_c to represent, respectively, the probability of error in a bit in the uncoded and coded cases, reserving the symbols P_e and $P_e^{(c)}$ for the probability of error of an uncoded or coded *word*. The energy in a bit in the uncoded case is $E_s = ST_w/k$ and

$E_s^{(c)} = ST_w/n$ in the coded case. Hence, from Eq. (11.5-8), assuming matched-filter reception,

$$q = \frac{1}{2} \operatorname{erfc} \sqrt{\frac{ST_w}{k\eta}} \tag{13.18-1}$$

and

$$q_c = \frac{1}{2} \operatorname{erfc} \sqrt{\frac{ST_w}{n\eta}} \tag{13.18-2}$$

In the uncoded case a word will be in error if 1 or more digits are in error. The probability that a digit is not in error is $1 - q$. The probability that all k digits are not in error is $(1 - q)^k$. The probability of at least 1 bit error, and hence a word error, is

$$P_e = 1 - (1 - q)^k \cong kq \tag{13.18-3}$$

since typically $kq \approx 10^{-3}$ or smaller.

In the coded case of Sec. 13.15 a word error occurs only if two or more errors occur. It may be verified (Prob. 13.18-1) that if $q_c \ll 1$ the likelihood of more than two errors is entirely negligible in comparison with the likelihood that one or two errors will occur. For $n = 7$ (7 digits in the coded word), the probability of just two errors, and hence the probability of a word error, is

$$P_e^{(c)} = \binom{7}{2} (1 - q_c)^5 q_c^2 \tag{13.18-4}$$

where $\binom{7}{2}$ is the number of combinations of seven things taken two at a time. Since $\binom{7}{2} = (7 \cdot 6)/2 = 21$ and assuming that $q_c \ll 1$, we have

$$P_e^{(c)} \cong 21 q_c^2 \tag{13.18-5}$$

From Eqs. (13.8-1) and (13.18-3) with $k = 4$ we have

$$P_e = 2 \operatorname{erfc} \sqrt{\frac{ST_w}{4\eta}} \tag{13.18-6}$$

while from Eqs. (13.18-2) and (13.18-5) we find

$$P_e^{(c)} = 5.25 \left(\operatorname{erfc} \sqrt{\frac{ST_w}{7\eta}} \right)^2 \tag{13.18-7}$$

For the purpose of comparing the performance of coded and uncoded transmission, Eqs. (13.18-6) and (13.18-7) are plotted in Fig. 13.18-1. Note that when $ST_w/\eta = E_s/\eta$ is greater than 9 dB, coded transmission results

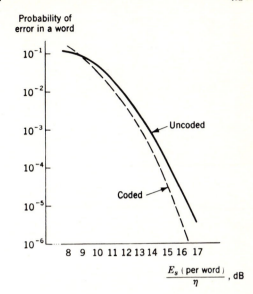

Probability of
error in a word

Fig. 13.18-1 Probability of error for an uncoded (4 bit) and a coded (7 bit) system.

$\dfrac{E_s \,(\text{per word})}{\eta}$, dB

in a lower probability of error than uncoded transmission. The difference, however, is not significant until E_s/η exceeds 16 dB. When E_s/η is less than 9 dB, the coded system has a higher probability of error than does the uncoded system.

Note that coding is not a cure-all. The complicated and expensive coding and decoding equipment needed does not yield significant improvement in this example until E_s/η is large. Here we see that at a probability of error in a word of 10^{-6}, the E_s/η difference between systems is only about 1 dB. Before deciding to employ coding, we should therefore carefully analyze the system response, with and without coding, to determine if the improvement obtained is worth the complication and expense.

The reader may well inquire at this point about why one studies coding if our improvement is only 1 dB. The answer is that the improvement may be considerably greater than 1 dB depending on the coding technique employed and the length of the code. In 1970 the National Aeronautics and Space Administration was employing a sequential decoder for a convolutional code which yielded an SNR improvement of more than 5 dB. It is possible to have greater improvement than this.

Today coding is used in computers, in data transmission over telephone cables, by the military for secure communications, and in some space communication systems. As the cost of the coding equipment decreases and the improvement obtained increases, the use of coding in communications will surely increase.

13.19 AN APPLICATION OF INFORMATION THEORY: AN OPTIMUM MODULATION SYSTEM

A generalized communication system is shown in Fig. 13.19-1.   The baseband signal $m(t)$ of bandwidth f_M is encoded or modulated, or both, onto a carrier and is then transmitted over a channel of bandwidth B.   Noise is added on the channel, and when the modulated/encoded signal arrives at the receiver it has a signal-to-noise ratio S_i/N_i at the receiver input. Here N_i is the noise power, at the receiver input, in the bandwidth B, that is $N_i = \eta B$.   The output signal obtained after decoding or demodulation, or both, is $\hat{m}(t) = m(t) + n_o(t)$, where $n_o(t)$ is the noise that accompanies the *output* signal.   The output signal-to-noise ratio is S_o/N_o, and the output waveform $m(t)$ is bandlimited to f_M.

From the Shannon-Hartley theorem, Eq. (13.6-1), the maximum rate at which information may be arriving at the receiver is

$$C_i = B \log_2 \left(1 + \frac{S_i}{N_i} \right) \tag{13.19-1}$$

while the maximum rate at which information may be issuing from the receiver is

$$C_o = f_M \log_2 \left(1 + \frac{S_o}{N_o} \right) \tag{13.19-2}$$

We now introduce the concept of an *optimum* or *ideal* modulation or encoding system as one in which $C_o = C_i$.   Such a system is defined as one in which the rate of information output of the receiver is equal to the rate of information input to the receiver.   That is, no information is lost as the information passes through the receiver, nor is any information accumulated in the receiver.   We assume that this feature of the

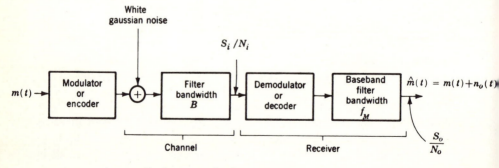

Fig. 13.19-1 A generalized communications system.

receiver persists at all rates of information flow up to the maximum as determined by the Shannon-Hartley theorem.

Setting $C_i = C_o$ and solving for S_o/N_o, we find

$$\frac{S_o}{N_o} = \left(1 + \frac{S_i}{N_i}\right)^{B/f_M} - 1 \tag{13.19-3}$$

An essential feature of the communication system represented in Fig. 13.19-1 is that while the baseband signal is bandlimited to f_M, we deliberately arranged that the encoded signal should occupy a larger bandwidth B. The *bandwidth expansion factor* B/f_M is a most important parameter of the system, and it is useful to rewrite Eq. (13.19-3) in a manner to emphasize this point. The input noise $N_i = \eta B$, and if we call the input-signal power S_i, we may write

$$\frac{S_i}{N_i} = \frac{S_i}{\eta B} = \frac{f_M}{B}\frac{S_i}{\eta f_M} \tag{13.19-4}$$

and from Eq. (13.19-3)

$$\frac{S_o}{N_o} = \left(1 + \frac{f_M}{B}\frac{S_i}{\eta f_M}\right)^{B/f_M} - 1 \tag{13.19-5}$$

Equation (13.19-5) characterizes the *optimum* communication system. For a given ratio $S_i/\eta f_M$ and a given bandwidth expansion factor B/f_M, any *physical* communication system will yield a smaller S_o/N_o. Plots of S_o/N_o as a function of $S_i/\eta f_M$ for various values of B/f_M are given in Fig. 13.19-2.

Equation (13.19-5) allows us to determine the extent to which we may improve the performance of an ideal system by making a sacrifice in bandwidth.   It is of interest to compare the ideal system with frequency modulation, which is a practical system in which such bandwidth sacrifice is made.   From Eq. (13.19-3), if $S_i/N_i \gg 1$, then

$$\frac{S_o}{N_o} \approx \left(\frac{S_i}{N_i}\right)^{B/f_M} \tag{13.19-6}$$

On the other hand, it may be verified (Prob. 13.19-2) that in a wideband FM system

$$\frac{S_o}{N_o} = \frac{3}{4}\left(\frac{B}{f_M}\right)^3 \frac{S_i}{N_i} \tag{13.19-7}$$

Thus, while the performance of the ideal system increases exponentially with bandwidth sacrifice, the performance of an FM system increases only with the cube of the bandwidth expansion factor.   Further, for $B/f_M > 1$ the performance of the ideal system increases much more rapidly with increase in S_i/N_i than does the FM system.

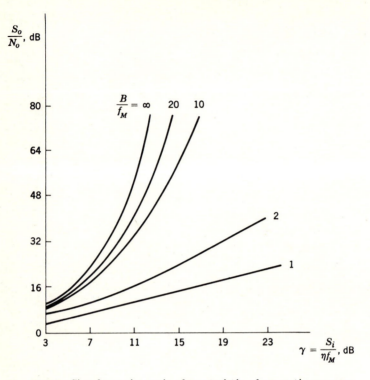

Fig. 13.19-2 Signal-to-noise ratio characteristic of an optimum communication system.

13.20 A COMPARISON OF AMPLITUDE-MODULATION SYSTEMS WITH THE OPTIMUM SYSTEM

In Chap. 8 we arrived at the following results for the amplitude-modulation systems studied there.

Single sideband, $B/f_M = 1$

$$\frac{S_o}{N_o} = \frac{S_i}{\eta f_M} \tag{13.20-1}$$

Double sideband, no carrier, synchronous detection, $B/f_M = 2$

$$\frac{S_o}{N_o} = \frac{S_i}{\eta f_M} \tag{13.20-2}$$

Square-law detection, $\overline{m^2(t)} \ll 1$, $B/f_M = 2$

$$\frac{S_o}{N_o} = \overline{m^2(t)} \frac{S_i}{\eta f_M} \frac{1}{1 + 3/4 S_i/\eta f_M} \tag{13.20-3}$$

Linear envelope detection, $\overline{m^2(t)} \ll 1$, $B/f_M = 2$

Above threshold $\dfrac{S_o}{N_o} = \overline{m^2(t)} \dfrac{S_i}{\eta f_M}$ (13.20-4)

Below threshold $\dfrac{S_o}{N_o} = \dfrac{\overline{m^2(t)}}{1.1} \left(\dfrac{S_i}{\eta f_M} \right)^2$ (13.20-5)

It is of interest to compare the performance of these systems with the performance of an ideal system with the same bandwidth expansion factor. In SSB, $B/f_M = 1$. If we set $B/f_M = 1$ in Eq. (13.19-5), which applies to the ideal system, we find that Eq. (13.19-5) reduces precisely to Eq. (13.20-1), which applies to SSB. Hence SSB is an *ideal* system. The plots of S_o/N_o vs. S_i/N_i, for $B/f_M = 1$ and for $B/f_M = 2$, shown in Fig. 13.19-2, are reproduced in Fig. 13.20-1. The plot for $B/f_M = 1$ applies both to the ideal system for $B/f_M = 1$ and also to SSB.

Equation (13.20-2) for the DSB system is identical with Eq. (13.20-1) for SSB. Hence again the plot in Fig. 13.20-1 for $B/f_M = 1$ applies to DSB. However, in DSB, $B/f_M = 2$. Hence if DSB were an optimum system, the plot corresponding to $B/f_M = 2$ would apply. DSB is therefore not an optimum system. By way of example, if $S_i/\eta f_M$ is chosen to be 20 dB, then the output S_o/N_o is also 20 dB. If, however, DSB were optimum, the output signal-to-noise ratio would be larger by 14 dB.

The performance curves for the asynchronous square-law and linear envelope detector are also given in Fig. 13.20-1 for $\overline{m^2(t)} = 0.1$

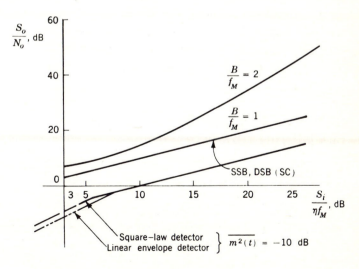

Fig. 13.20-1 A comparison of AM systems.

($= -10$ dB). The performance of these systems is seen to be poor in comparison with the optimum system for $B/f_M = 2$.

13.21 A COMPARISON OF FM SYSTEMS

The output SNR of each of the FM systems considered in Chaps. 9 and 10 are the same above threshold. The result, for sinusoidal modulation, is [see Eq. (9.2-21)]

$$\frac{S_o}{N_o} = \frac{3}{2}\beta^2\frac{S_i}{\eta f_M} \tag{13.21-1}$$

where $\beta \equiv \Delta f/f_M$, and Δf is the frequency deviation. To relate β and the bandwidth expansion factor, we note that $B = 2(\beta + 1)f_M$. Thus

$$\frac{B}{f_M} = 2(\beta + 1) \tag{13.21-2}$$

The expression for the output signal-to-noise ratio of the discriminator, including the effect of sinusoidal modulation and valid both above and below threshold, is found by combining Eqs. (10.6-6) and (13.21-2). We find

$$\frac{S_o}{N_o} = \frac{(3/2)\beta^2(S_i/\eta f_M)}{1 + (12\beta/\pi)(S_i/\eta f_M)\exp\left[-(f_M/B)(S_i/\eta f_M)\right]} \tag{13.21-3}$$

This equation is plotted in Fig. 13.21-1 for $\beta = 3$ and 12. The output SNR characteristics of the optimal demodulator and the second-order phase-locked loop are sketched on the same set of axes. The threshold extension presented for the PLL was obtained experimentally and theoretically verified.[6] We note that the PLL results in a 3-dB threshold extension when $\beta = 12$, and a 2-dB threshold extension when $\beta = 3$. We note that the performance of FM systems falls substantially short of the performance of ideal systems.

13.22 COMPARISON OF PCM AND FM COMMUNICATION SYSTEMS

The transmission of analog signals by modulating a binary PCM signal onto a PSK or FSK carrier was studied in Sec. 12.5. In Fig. 12.5-1 we saw that the threshold of FSK was 2.2 dB greater than the threshold using PSK. In Fig. 13.22-1 we compare PCM-PSK, the FM discriminator, and the optimal demodulator. The PCM characteristics are obtained from Eq. (12.5-1).

Consider the two FM discriminator characteristics first. When $\beta = 12$, the threshold occurs at 25 dB, and when $\beta = 3$, threshold occurs

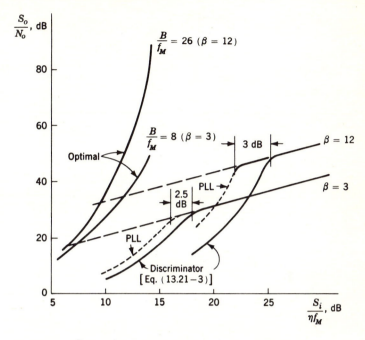

Fig. 13.21-1 Comparison of FM demodulators including the effect of sinusoidal modulation.

at 18 dB.   To compare these results with PCM, we consider both threshold and the bandwidths necessary for transmission, i.e., the bandwidth expansion factor.   We note that a bandwidth expansion factor $B/f_M = 8$ in PCM (which corresponds to $\beta = 3$) results in an output SNR which is approximately equal to that of the discriminator operating with $\beta = 12$, $B/f_M = 2(12 + 1) = 26$.   Thus, to obtain an output SNR of 48 dB requires a $B/f_M = 26$ when using a discriminator, and a $B/f_M = 8$ when employing PCM-PSK.   Hence 3.25 times more bandwidth is required of the discriminator.

If an output SNR of 28 dB is all that is required, the FM discriminator with $\beta = 3$, $B/f_M = 8$, can be employed.   However, the results can be obtained using PCM-PSK with $B/f_M = 5$.   In this case the FM discriminator requires only 1.6 times more bandwidth.   Thus, we see that the improvement of PCM over the discriminator increases with increased bandwidth expansion.

Comparing the PCM system with the optimal system for $B/f_M = 8$ indicates that the PCM system operating at threshold requires a signal-to-noise ratio which is 11 dB greater than required by the optimal demodulator threshold.

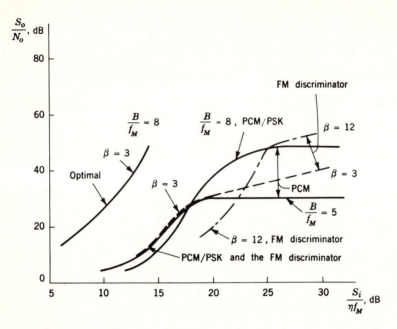

Fig. 13.22-1 Comparison of PCM-PSK, FM using discriminator demodulation, and the optimal system.

13.23 FEEDBACK COMMUNICATION

Many communication systems provide a two-way communication link. That is, not only is station A able to transmit and station B to receive, but station A is able also to receive and station B to transmit. In such a two-way link it is possible to incorporate *feedback* and thereby improve the performance of the communication system. We shall consider a type of feedback communication system, referred to as an *information-feedback* system, which is especially effective when transmission in one direction is much more reliable than transmission in the other direction. Such a situation might well result if one station were a fixed ground-based station while the other station was located in an airplane or a satellite. The ground-based station, having no essential limitations on weight, might be able to transmit hundreds of kilowatts of power while the station in space might be limited to tens of watts. We discuss now how feedback may be used to reduce the average energy required to transmit a bit of information from the low-powered station to the high-powered station, while maintaining a low probability of error.

SYSTEM DESCRIPTION

An information-feedback communication system is shown in Fig. 13.23-1. The two transceivers (transmitter-receiver combinations) are coupled

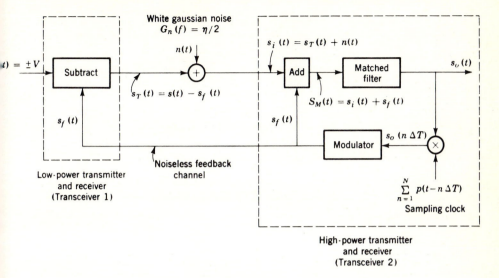

Fig. 13.23-1   An information-feedback communication system.

through two communication channels.   Over one channel, transceiver 1 (with low-power transmitter) transmits information to transceiver 2.   We have included a noise source in this channel.   Over the second channel (the feedback channel) transceiver 2 transmits back to transceiver 1. We have assumed that the transmitter in transceiver 2 has so much transmitter power available that we may ignore the effect of noise on the feedback channel.   Thus we allow the possibility that receiver 2 may make an error in determining the transmission of transmitter 1.   However, we assume that receiver 1 will receive the transmission of transmitter 2 with perfect reliability, that is, with negligible probability of error.

We assume transmission by binary pulse-code modulation.   The baseband signal $s(t)$ is a sequence of voltage levels $+V$ or $-V$ held for intervals T and representing binary 1s and 0s.   For simplicity we assume transmission at baseband, although in a physical system the baseband signal may be used to modulate a carrier, as in PSK or FSK.   We now describe the operation of the system.

Since transmission is at baseband, the matched filter in the receiver is the *integrate and dump* filter described in Sec. 11.1.   At the beginning of the interval T allocated to a bit, the output of the matched filter is zero. The interval T is divided into N subintervals, each of duration $\Delta T = T/N$, by the sampling clock pulses which occur at intervals of ΔT.   The modulation provides a feedback voltage $s_f(t)$ which, like the baseband signal $s(t)$, is $s_f(t) = +V$ or $s_f(t) = -V$.   The feedback signal is $s_f(t) = +V$ when the modulator input $s_0(n\,\Delta T)$ is positive and $s_f(t) = -V$ when $s_o(n\,\Delta T)$ is *negative*.

Now suppose that, during some bit interval T, the signal $s(t) = +V$ is transmitted. At the beginning of this interval, i.e., at $t = 0$, $s_o(t) = 0$. This output is sampled and the sample $s_o(t = 0) = 0$. The modulator output $s_f(t) = 0$. Thus, during the first subinterval, from $t = 0$ to $t = \Delta T$, there is no feedback signal. During this subinterval the input to the matched filter is (refer to Fig. 13.23-1)

$$s_M(t) = s_i(t) = s_T(t) + n(t) = s(t) + n(t) \tag{13.23-1}$$

At the end of this subinterval, the output of the matched filter is again sampled. Let us suppose that it turns out that $s_0(\Delta T)$ is positive. That is, let us suppose that at $t = \Delta T$ the matched filter *correctly* decides that the transmission is $s(t) = +V$. Then $s_f(t) = +V$. The transmitted signal in the second subinterval, from $t = \Delta T$ to $t = 2\Delta T$, would then be

$$s_T(t) = s(t) - s_f(t) = V - V = 0 \tag{13.23-2}$$

Accordingly, we see that during this second subinterval the transmitter will be *turned off*. On the other hand, the input to the matched filter will be

$$s_M(t) = s_i(t) + s_f(t) = s_T(t) + n(t) + s_f(t) \tag{13.23-3a}$$
$$= s(t) - s_f(t) + n(t) + s_f(t) \tag{13.23-3b}$$
$$= s(t) + n(t) \tag{13.23-3c}$$

precisely as in Eq. (13.23-1). Thus the matched filter will not "know" that the transmitter has been turned off. The signal plus noise present at the input to the filter will always be independent of the feedback signal. This result is apparent from Eq. (13.23-3) as well as from Fig. 13.23-1. For we see that the feedback signal is both added and subtracted from the filter input. Thus the matched filter will continue to integrate the signal plus noise $s(t) + n(t)$ in a manner which is entirely independent of the feedback signal.

Next, let us suppose that it turns out that $s_0(\Delta T)$ is negative. That is, let us suppose that at $t = \Delta T$ the matched filter *incorrectly* decides that the transmission is $s(t) = -V$. In this case we find $s_f(t) = -V$, and the transmitted signal during the subinterval $t = \Delta T$ to $t = 2\Delta T$ will be

$$s_T(t) = s(t) - s_f(t) = V - (-V) = 2V \tag{13.23-4}$$

In this case the transmitted signal voltage is doubled, and the transmitted signal power is increased by a factor of four. But again, as in the previous case, the input to the matched filter will remain as $s_M(t) = s(t) + n(t)$, independent of the feedback.

The process we have described over the first two subintervals continues over the remaining intervals up to the time T. Interval by interval the transmitter is turned off or turned on to a high power, depending on the preliminary and tentative estimates made by the matched filter at intervals of ΔT. However, a final determination of the transmitted bit is made only by examining the matched filter at the end of the bit interval, at $t = T$. As noted, the input to the matched filter is independent of the feedback. Hence the probability of an error in the determination of a bit is *exactly the same* as if no feedback were employed. This error probability is given by [see Eq. (11.2-3)]

$$P_e = \frac{1}{2} \operatorname{erfc} \sqrt{\frac{V^2 T}{\eta}} \qquad (13.23\text{-}5)$$

The merit of the information feedback system, then, lies not in that it reduces the probability of error, but rather that, for a fixed probability of error, the feedback system may use less average energy per bit than a system without feedback. For if the preliminary estimates of the matched filter, made at intervals ΔT, are rather more frequently right than wrong, the transmitter may be turned off often enough to produce a saving in energy.

CALCULATION OF AVERAGE TRANSMITTED SIGNAL ENERGY PER BIT

We now calculate the average signal energy per bit, that is, the energy transmitted in the interval T. We assume arbitrarily that $s(t) = +V$. Since $+V$ and $-V$ are transmitted with equal probability, this assumption does not affect the generality of the result.

During the first interval ΔT, the signal $+V$ is transmitted and the corresponding normalized energy is

$$E_1 = V^2 \Delta T \qquad (13.23\text{-}6)$$

During the second interval ΔT, $s_T(t) = 0$ if the decision made by the matched filter at $t = \Delta T$ was correct. In this case no energy is transmitted. If however, the decision was incorrect, $s_T(t) = 2V$ and the energy $(2V)^2 \Delta T$ is transmitted. Since the probability of an incorrect decision after processing the signal and noise for time ΔT is q_1, where

$$q_1 = \frac{1}{2} \operatorname{erfc} \sqrt{\frac{V^2 \Delta T}{\eta}} \qquad (13.23\text{-}7)$$

the average energy transmitted during the second ΔT second interval is

$$E_2 = (2V)^2 \Delta T q_1 \qquad (13.23\text{-}8)$$

Similarly, the average energy transmitted during the nth ΔT second interval is

$$E_n = (2V)^2 \Delta T q_{n-1} \qquad (13.23\text{-}9)$$

where

$$q_{n-1} = \frac{1}{2} \text{erfc} \sqrt{\frac{V^2(n-1)\,\Delta T}{\eta}} \qquad (13.23\text{-}10)$$

The average energy transmitted in the interval T is E_T, given by

$$E_T = E_1 + E_2 + \cdots + E_N \qquad (13.23\text{-}11a)$$
$$= V^2\,\Delta T + (2V)^2\,\Delta T q_1 + \cdots + (2V)^2\,\Delta T q_{N-1} \qquad (13.23\text{-}11b)$$
$$= V^2\,\Delta T + 2V^2\,\Delta T \sum_{n=2}^{N} \text{erfc} \sqrt{\frac{V^2(n-1)\,\Delta T}{\eta}} \qquad (13.23\text{-}11c)$$

For any given N, the total energy per bit E_T may be calculated from Eq. (13.23-11c). We find, for example (Prob. 13.23-1), that for $N = 2$, and if $V^2T/2\eta \gg 1$, that the feedback provides a 3-dB improvement. That is, the energy E_T is only one-half the energy that would be required, without feedback, for the same error probability.

It is of interest to explore the improvement possible in the limiting case as N becomes very large. For this purpose we introduce the variable

$$x_n \equiv \frac{V^2(n-1)\,\Delta T}{\eta} \qquad (13.23\text{-}12)$$

so that

$$\Delta x \equiv x_{n+1} - x_n = \frac{V^2\,\Delta T}{\eta} \qquad (13.23\text{-}13)$$

We note also that

$$\text{erfc} \sqrt{x_n} \equiv \frac{2}{\sqrt{\pi}} \int_{\sqrt{x_n}}^{\infty} e^{-u^2}\,du \qquad (13.23\text{-}14)$$

Substituting Eqs. (13.23-13) and (13.23-14) into Eq. (13.23-11c), we find

$$E_T = \eta \left(\Delta x + \sum_{n=2}^{N} \frac{4}{\sqrt{\pi}}\,\Delta x \int_{\sqrt{x_n}}^{\infty} e^{-u^2}\,du \right) \qquad (13.23\text{-}15)$$

In the limit as $N \to \infty$, Δx and ΔT may be replaced by the differentials dx and dt and the summation in Eq. (13.23-15) replaced by an integral. We then find (Prob. 13.23-2) that

$$E_T = \frac{4\eta}{\sqrt{\pi}} \int_0^{V^2T/\eta} dx \int_{\sqrt{x}}^{\infty} e^{-u^2}\,du \qquad (13.23\text{-}16)$$

This integral may be evaluated (Prob. 13.23-3) with the result

$$E_T = \eta \left(\frac{-2}{\sqrt{\pi}} \sqrt{\frac{V^2T}{\eta}}\, e^{-V^2T/\eta} + \text{erf} \sqrt{\frac{V^2T}{\eta}} + \frac{2V^2T}{\eta} \text{erfc} \sqrt{\frac{V^2T}{\eta}} \right)$$

$$(13.23\text{-}7)$$

For $V^2 T/\eta \gg 1$ (or, more specifically, for $V^2 T/\eta > 3$) the first and third terms in Eq. (13.23-17) may be neglected in comparison with the second term. Furthermore, erf $\sqrt{V^2 T/\eta} \simeq 1$. Hence we then find

$$F_m \simeq \eta \tag{13.23-18}$$

From Eqs. (13.23-5) and (13.23-18) we may now deduce the following interesting result. The error probability may be made arbitrarily small by increasing V. However, provided $V^2 T/\eta \gg 1$, the required average energy per bit is constant at $E_T = \eta$ and does not depend on the error probability P_e. Thus the characteristic of the transmitter which determines the extent to which the error probability may be reduced is not the power it can transmit but rather its *peak power*, i.e., the peak value V^2 which the transmitter can attain.

COMPARISON OF INFORMATION RATE WITH CHANNEL CAPACITY

It is of interest to compare the performance of an information-feedback communication system with the performance of an ideal system. The system of Fig. 13.23-1 is a binary PCM system. For such a system, the rate of information transmission R as given by Eq. (13.4-1) with $M = 2$ is

$$R = \frac{1}{T} \log_2 M = \frac{1}{T} \tag{13.23-19}$$

the channel capacity for a gaussian channel is [Eq. (13.7-2)]

$$C = 1.44 \frac{S_i}{\eta} \tag{13.23-20}$$

in which S_i is the input power at the receiver. From Eqs. (13.23-19) and (13.23-20), setting $R = C$, we find that in a system with ideal coding an arbitrarily small error probability is attainable with a power S_i that satisfies the condition

$$S_i T = \frac{1}{1.44} \eta = 0.69 \eta \tag{13.23-21}$$

The quantity $S_i T$ is the energy associated with a bit of duration T. We have found that the information-feedback system is capable of attaining arbitrarily small error with an energy per bit given by $E_T = \eta$. Hence we find that the feedback system requires an energy per bit which is only 1.44 times greater ($\simeq 1.5$ dB) than required by an optimal system.

PROBLEMS

13.3-1. One of four possible messages Q_1, Q_2, Q_3, and Q_4, having probabilities 1/8, 3/8, 3/8, and 1/8, respectively, is transmitted. Calculate the average information per message.

13.3-2. One of five possible messages Q_1 to Q_5 having probabilities $1/2$, $1/4$, $1/8$, $1/16$, $1/16$, respectively, is transmitted. Calculate the average information.

13.3-3. Messages $Q_1, \ldots, Q_M$ have probabilities $p_1, \ldots, p_M$ of occurring.

    (a) Write an expression for H.

    (b) If $M = 3$, write H in terms of p_1 and p_2, by using the result that $p_1 + p_2 + p_3 = 1$.

    (c) Find p_1 and p_2 for $H = H_{\max}$ by setting $\partial H/\partial p_1 = 0$ and $\partial H/\partial p_2 = 0$.

    (d) Extend the result of (c) to the case of M messages.

13.3-4. A code is composed of dots and dashes. Assume that the dash is 3 times as long as the dot, and has one-third the probability of occurrence.

    (a) Calculate the information in a dot and that in a dash.

    (b) Calculate the average information in the dot-dash code.

    (c) Assume that a dot lasts for 10 msec and that this same time interval is allowed between symbols. Calculate the average rate of information transmission.

13.4-1. In Example 13.4-1 we saw that the four messages have different probabilities and that as a result we transmit $4B$ binary digits per second and convey only $3.6B$ bits of information. Consider transmitting Q_3, Q_2, Q_1, and Q_4 by the symbols 0, 10, 110, 111.

    (a) Is this *code* uniquely decipherable? That is, for every possible sequence is there only one way of interpreting the message?

    (b) Calculate the average number of code bits per message. How does it compare with $H = 1.8$ bits per message?

13.4-2. Consider the four messages of Example 13.4-1. Let Q_1, Q_2, Q_3, Q_4 have probabilities $1/2$, $1/4$, $1/8$, $1/8$.

    (a) Calculate H.

    (b) Find R if $r = 1$ message per second.

    (c) What is the rate at which binary digits are transmitted if the signal is sent after encoding $Q_1, \ldots, Q_4$ as 00, 01, 10, 11?

    (d) What is the rate, if the code employed is 0, 10, 110, 111?

13.4-3. Consider five messages given by the probabilities $1/2$, $1/4$, $1/8$, $1/16$, $1/16$.

    (a) Calculate H.

    (b) One technique used in constructing an optimum code is to list the messages in order of decreasing probability and divide the list into two equally or almost equally probable groups. The messages in the top group are given the value 0, and the messages in the bottom group are given the value 1. The same procedure is now employed for each group, separately. The procedure is continued until no further division is possible. This is called the *Shannon-Fano* code. Find the code for each message.

    (c) Calculate the average number of bits per message. Compare with H.

13.6-1. A gaussian channel has a 1-MHz bandwidth. If the signal-power-to-noise spectral density $S/\eta = 10^5$ Hz, calculate the channel capacity C and the maximum information transfer rate R.

13.6-2. Suppose 100 voltage levels are employed to transmit 100 equally likely messages. Assume $\lambda = 3.5$ and the system bandwidth $B = 10^4$ Hz.

(a) Calculate S/η, using Eq. (13.6-3).

(b) If an integrate-and-dump filter is employed to determine which level is sent, calculate the probability of an error when sending the kth level. Assume that the only errors possible are in choosing the $k - 1$ or the $k + 1$ levels.

13.7-1. (a) Plot channel capacity C versus B, with $S/\eta =$ constant, for the gaussian channel.

(b) If the channel bandwidth $B = 5\,\text{kHz}$ and a message is being transmitted with $R = 10^6$ bits per sec, find S/η for $R \leq C$.

13.8-1. Show that the set of sinusoidal signals $\cos 2\pi nt/T$, $\sin 2\pi nt/T$ are orthogonal over the time interval T.

13.8-2. (a) Why is n_l, in Eq. (13.8-3), gaussian?

(b) Find $E(n_l)$ by interchanging integration and ensemble averaging. Why is this permitted?

(c) Show that $E(n_l^2) = \displaystyle\int_0^T dt \int_0^T d\lambda \; E[n(t)n(\lambda)]s_l(t)s_l(\lambda)$

(d) Using the result that $E[n(t)n(\lambda)] = R_n(t - \lambda) = (\eta/2)\delta(t - \lambda)$, show that $E(n_l^2) = \eta E_s/2$, and that $E(n_l n_k) = 0$ if $l \neq k$.

13.8-3. Show the following by changing variables:

(a) That as in Eq. (13.8-9)

$$P_L = \left(\frac{1}{\sqrt{\pi}} \int_{-\infty}^{\sqrt{E_s/\eta}+n_i/\sqrt{2}\,\sigma} e^{-x^2}\, dx \right)^{M-1}$$

(b) That as in Eq. (13.8-11)

$$P_e = \left(\frac{1}{\sqrt{\pi}} \right)^M \int_{-\infty}^{\infty} dy\; e^{-y^2} \left(\int_{-\infty}^{\sqrt{E_s/\eta}+y} e^{-x^2}\, dx \right)^{M-1}$$

(c) $\dfrac{E_s^2}{2\sigma^2} = \dfrac{E_s}{\eta} = \dfrac{S_i T}{\eta} = \dfrac{S_i}{\eta R} \log_2 M$

13.9-1. If $S_i/\eta =$ constant, plot T versus M for $P_e = 10^{-5}$ from Fig. 13.8-2.

13.9-2. An analog signal has a 4-kHz bandwidth. The signal is sampled at 3 times the Nyquist rate and quantized using 256 quantization levels. The S_i/η ratio at the receiver is $S_i/\eta = 1\,\text{MHz}$.

(a) Calculate the time T between samples.

(b) The quantized and sampled signal is encoded into a binary PCM waveform. Calculate the number of bits N and find the P_e.

(c) The quantized and sampled signal is encoded into 1 of 256 orthogonal signals. Find P_e.

13.9-3. Show that $T/\log_2 M =$ duration of a bit of a binary encoded signal.

13.9-4. (a) A scheme for constructing orthogonal binary PCM baseband waveforms is the following. We start with two signals each two bits in length. The first signal $s_1(t)$ holds the level $+1$ volt for both intervals and the second, $s_2(t)$, is $+1$ volt in the first interval and -1 volt in the second interval. Show that these two signals $s_1(t) = 1,1$ and $s_2(t) = 1,-1$ are orthogonal.

(b) Show that from the basis of $s_1(t)$ and $s_2(t)$ four orthogonal signals may be generated as

$$s_1'(t) = s_1(t), s_1(t) \quad = 1,1, \qquad\qquad 1,1$$
$$s_2'(t) = s_2(t), s_2(t) \quad = 1,-1, \qquad\quad 1,-1$$

$$s_3'(t) = s_1(t), -s_1(t) = 1,1, \qquad\quad -1,-1$$
$$s_4'(t) = s_2(t), -s_2(t) = 1,-1, \qquad\quad -1,1$$

(c) In part (b), dashed lines have been included to suggest a pattern. Use the pattern to form eight orthogonal signals. (Codes so generated are called *Haddamard* codes.)

(d) Note that the first bit of each message is always a 1 and hence this bit transmits no information and may be deleted. The code which results from this deletion is called a *Simplex* code. Is the Simplex code orthogonal? A Simplex code has better performance than an orthogonal code. Why?

13.10-1. We transmit either a 1 or a 0, and add redundancy by repeating the bit.

(a) Show that if we transmit 11111 or 00000, then 2 errors can be corrected.

(b) Show that in general if we transmit the same bit $2t + 1$ times we can correct t errors.

13.11-1. Show that a parity-check bit can be inserted which makes the number of 1s *odd* and that this detects errors.

13.11-2. An 8-bit binary code has a parity-check bit added. How many code words are now available that are not used.

13.13-1. The numbers 0 to 7 are binary-encoded.

(a) Write the 3 binary digits for each decimal number.

(b) Add a single parity-check bit to each code word.

(c) Each 4-bit code word forms a $\bar{T}$ matrix. Show that if $\bar{H} = [1111]$, $\bar{H}\bar{T} = 0$ for each $\bar{T}$. Also show that if a single error occurs, $\bar{H}\bar{T} = 1$.

13.13-2. The repeated code symbols 111 and 000 are the two values of $\bar{T}$. Show that if $\bar{H}$ is given by Eq. (13.13-3), $\bar{H}\bar{T} = 0$.

13.13-3. The repeated code symbols are 11111 and 00000. Find the $\bar{H}$ matrix.

13.14-1. The code word transmitted is either $T_1 = 111$ or $T_2 = 000$. The $\bar{H}$ matrix is given by Eq. (13.13-3). The signal 101 is received. Show that $\bar{H}\bar{R} = \bar{S}$ shows where the error is located.

13.15-1. For the $\bar{H}$ matrix given in Eq. (13.15-1) find the $2^4 = 16$ different code words.

13.15-2. If $a_1 = 1$, and $a_2 = a_3 = a_4 = 0$, show that with the $\bar{H}$ matrix given by Eq. (13.15-5) that the code word 1100011 results.

13.15-3. If the $\bar{H}$ matrix is

$$\bar{H} = \begin{bmatrix} 1 & 1 & 0 & 1 & 0 & 0 & 1 \\ 1 & 1 & 0 & 0 & 1 & 1 & 0 \\ 1 & 0 & 1 & 1 & 1 & 0 & 0 \end{bmatrix}$$

which is simply a shift of the columns of Eq. (13.15-1), find the 16 code words.

13.15-4. (a) Find a (15,11) block code.   Find all the code words.

(b) Find the $\bar{H}$ matrix.

(c) Show by example that single errors can be corrected.

13.16-1. Find the output sequence of the shift register connected as in Eq. (13.16-1) for the input given in Eq. (13.16-2).   Thereby verify Eq. (13.16-3).

13.16-2. Three shift registers are interconnected to provide 3 outputs $v_1v_2v_3$ for each input bit.   If $v_1 = s_1$, $v_2 = f_2(s_1,s_2,s_3)$, and $v_3 = f_3(s_1,s_2,s_3)$, how many different functions of f_2 and f_3 are possible?

13.17-1. Verify that the tree shown in Fig. 13.17-1 yields the same output as the encoder of Fig. 13.16-2.

13.17-2. Prove, using Eq. (13.16-1), that the output of any branch is 2 branches and that each bit of each branch differs.

13.17-3. The signal 001 101 001 000 100 is received by the tree shown in Fig. 13.17-1.   How many discrepancies are found by the time the output is reached? How many discrepancies if the correct path is always taken?

13.17-4. We note that a double error coming first leads us along an incorrect path. What is the effect of a double error coming last?   Consider, for example, that the signal 111 101 001 000 100 is transmitted and the signal 111 101 001 000 111 is received.

13.18-1. A single-error correcting coded word contains n bits.   If the probability of error in each bit is q_c, show that the probability of error in a word is

$$P_e^{(c)} = \binom{n}{2} (1 - q_c)^{n-2}q_c^2 + \binom{n}{3} (1 - q_c)^{n-3}q_c^3 + \cdots$$

$$= \sum_{i=2}^{n} \binom{n}{i} (1 - q_c)^{n-i}q_c^i$$

where $\binom{n}{i} = \dfrac{n!}{i!(n - i)!}$.

13.18-2. The binary digits 1 and 0 are transmitted by repeating the digits three times: 111 and 000.

(a) Calculate the probability of error in receiving the uncoded 1, in terms of ST/η, where T is the duration of the word.

(b) Calculate the probability of error in receiving the coded 1(111), in terms of ST/η.

(c) Plot the probability of error of (a) and (b) as a function of ST/η, and compare.

13.19-1. Using Eq. (13.19-5), find $\lim\limits_{B/f_M \to \infty} S_o/N_o$.

13.19-2. Verify Eq. (13.19-7).   *Hint:* Use Eq. (9.2-17) and recognize that $k^2\overline{m^2(t)} = 4\pi^2 \Delta f_{\text{rms}}^2$.   Assume that $B = 2 \Delta f_{\text{rms}}$.

13.19-3. Plot S_o/N_o versus S_i/N_i for the ideal and FM systems.   Assume that $B/f_M = 2\beta$.   Choose $\beta = 10$ and 100.

13.21-1. (a) Plot S_o/N_o and B/f_M versus $S_i/\eta f_M$ at the threshold of an FM discriminator. The abscissa of the graph should be $S_i/\eta f_M$ at threshold, and the two ordinate axes should be S_o/N_o and B/f_M.

(b) For each value of S_o/N_o (and its corresponding value of B/f_M), calculate and plot the value of $S_i/\eta f_M$ required by the ideal demodulator.

(c) Compare the results obtained in (a) and (b).

13.21-2. A signal $m(t)$ is gaussian and bandlimited to 100 kHz. The sensitivity of the FM modulator $k = 10^{-6}$ rad/sec/volt. An output SNR of 40 db is required.

(a) Calculate the rms frequency deviation, if a minimum $S_i/\eta f_M$ is to be employed.

(b) Calculate $S_i/\eta f_M$.

(c) Calculate B, the IF bandwidth.

(d) Calculate the $S_i/\eta f_M$ required by the ideal demodulator to give $S_o/N_o = 40$ dB for the same value of B. Compare results.

13.22-1. (a) Repeat Prob. 13.21-2 assuming that $m(t)$ is PCM-encoded and then transmitted using PSK.

(b) Compare results for Probs. 13.21-2 and 13.22-1.

13.22-2. (a) Plot S_o/N_o and N versus $S_i/\eta f_M$ at threshold for PCM.

(b) Compare your results with those obtained in Prob. 13.21-1.

(c) When would you use PCM rather than FM?

13.23-1. Show that, for $N = 2$ and when $V^2T/\eta \gg 1$, the information-feedback communication system allows a 3-dB reduction in energy per bit for a fixed error probability.

13.23-2. Show that in the limit as $N \to \infty$, Eq. (13.23-15) may be replaced by Eq. (13.23-16).

13.23-3. Evaluate Eq. (13.23-16):

$$E_T = \frac{4\eta}{\sqrt{\pi}} \int_0^{V^2T/\eta} dx \int_{\sqrt{x}}^{\infty} du e^{-u^2}$$

This evaluation is most easily accomplished by interchanging the order of integration. Refer to Fig. P13.23-3a. Here we see that the shaded area represents the region of integration of Eq. (13.23-16), i.e., we integrate over x for $0 \leq x \leq V^2T/\eta$, and over u for $\sqrt{x} \leq u < \infty$. The solid line at $x = x_1$, of thickness dx, illustrates that in Eq. (13.23-16) we integrate first over u and then over x. Now refer to Fig. P13.23-3b. Here the shaded area is the same as in Fig. P13.23-3a. However, we are now integrating first over x, from $0 \leq x \leq u^2 \leq V^2T/\eta$ and then over u, from 0 to infinity.

Show that using this new order of integration yields

$$E_T = \frac{4\eta}{\sqrt{\pi}} \left(\int_0^{\sqrt{V^2T/\eta}} du_1 e^{-u_1^2} \int_0^{u_1^2} dx + \int_{\sqrt{V^2T/\eta}}^{\infty} du_2 e^{-u_2^2} \int_0^{V^2T/\eta} dx \right)$$

$$= \frac{4\eta}{\sqrt{\pi}} \left(\int_0^{\sqrt{V^2T/\eta}} du_1 u_1^2 e^{-u_1^2} + \frac{V^2T}{\eta} \int_{\sqrt{V^2T/\eta}}^{\infty} du_2 e^{-u_2^2} \right)$$

Complete your evaluation of E_T by integrating both integrals. To do this change

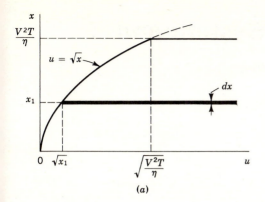
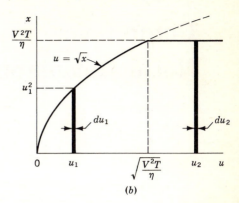

(a) (b)

Fig. P13.23-3

variables, so that

$$x = u_1 \quad \text{and} \quad dy = 2u_1 e^{-u_1^2} du_1$$

Show that

$$E_T = \eta \left(\frac{-2}{\sqrt{\pi}} \sqrt{\frac{V^2 T}{\eta}} e^{-V^2 T/\eta} + \operatorname{erf} \sqrt{\frac{V^2 T}{\eta}} + \frac{2V^2 T}{\eta} \operatorname{erfc} \sqrt{\frac{V^2 T}{\eta}} \right)$$

REFERENCES

1. Sakrison, D.: "Communication Theory," John Wiley & Sons, Inc., New York, 1968.
2. Wozencraft, J., and I. Jacobs: "Principles of Communication Engineering," John Wiley & Sons, Inc., New York, 1965.
3. Shannon, C. E.: A Mathematical Theory of Communication, *BSTJ*, vol. 27, pp. 379–623, 1948.
   Shannon, C. E.: Communication in the Presence of Noise, *Proc. IRE*, vol. 37, p. 10, 1949.
4. Viterbi, A., "Principles of Coherent Communications," McGraw-Hill Book Company, New York, 1966.
5. Osborne, P., and D. L. Schilling: Threshold Response of a Phase Locked Loop, *Proc. Intl. Conf. Commun.*, 1968.

14

Communication System and Noise Calculations

A received signal, whether it arrives over a wire communication channel or is received by an antenna, is accompanied by noise. As the signal is processed by the various stages of the receiver, each stage superimposes additional noise on the signal. We shall discuss in this chapter the parameters (noise temperature, noise figure, etc.) which are used to describe the extent to which the various component stages of a receiving system degrade the signal-to-noise ratio. We shall discuss, as well, how these parameters, relating to individual component stages, are combined to determine the overall signal-to-noise performance of a communication system. We shall then be able to determine, among other things, the minimum power level of the signal at the point of transmission which will ensure a final receiver output signal with an acceptable signal-to-noise ratio.

14.1 RESISTOR NOISE

As we have already noted in Sec. 7.1, resistors are a source of noise. The conductivity of a resistor results from the availability, within the resistor, of electrons which are free to move, and the resistor noise is due precisely to the random motion of these electrons. This random agitation at the atomic level is a universal characteristic of matter. It accounts for the ability of matter to store the energy which is supplied through the flow of heat into the matter. This energy, stored in the random agitation, is made manifest generally by an increase in temperature. Thus resistor noise as well as other noise of similar origin is called *thermal* noise.

It has been determined experimentally that the noise voltage $v_n(t)$ that appears across the terminals of a resistor R is gaussian and has a mean-square voltage, in a narrow frequency band df, equal to

$$\overline{v_n^2} = 4kTR \, df \tag{14.1-1}$$

where T is the temperature in degrees Kelvin, and k is the Boltzmann constant $k = 1.37 \times 10^{-23}$ joule/°K. At room temperature, taken as $T_0 = 290°$K, we have $kT_0 \approx 4 \times 10^{-21}$ watt-sec. Experiments indicate further that $\overline{v_n^2}$ is independent of the center frequency f_0 of the filter having the bandwidth df, for values of f_0 between 0 Hz and 4 GHz [see Sec. 14.5 for a discussion of the limitations to Eq. (14.1-1)]. Thus, we conclude that for the communication systems employed today (excluding optical communication systems), the normalized power supplied by $v_n(t)$ is also independent of f_0, and hence the power spectral density of the noise source is approximately white and equal to

$$G_v(f) = \frac{\overline{v_n^2}}{2 \, df} = 2kTR \tag{14.1-2}$$

The *noisy* resistor R can be represented by the equivalent circuit shown in Fig. 14.1-1. Here the physical resistor shown in Fig. 14.1-1a has been replaced in Fig. 14.1-1b by a Thévenin circuit representation consisting of a *noiseless* resistor in series with a noise voltage source $v_n(t)$ having the mean-square voltage $\overline{v_n^2} = 4kTR \, df$. If the terminals in Fig. 14.1-1b were short-circuited, a noise current $i_n(t)$ would flow having a

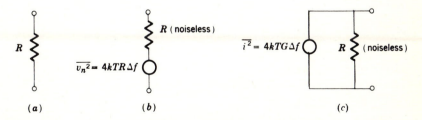

Fig. 14.1-1 A physical resistor in (*a*) of resistance R and at a temperature T is replaced in (*b*) by a Thévenin equivalent representation consisting of the noiseless resistor in series with a noise source. In (*c*) a Norton equivalent representation is shown.

mean-square value

$$\overline{i_n^2} = \frac{\overline{v_n^2}}{R^2} = \frac{4kTR\,df}{R^2} = 4kTG\,df \qquad (14.1\text{-}3)$$

where $G = 1/R$ is the conductance of the resistor. The Thévenin equivalent in Fig. 14.1-1b therefore corresponds to the Norton equivalent representation shown in Fig. 14.1-1c.

14.2 MULTIPLE-RESISTOR NOISE SOURCES

We saw in Sec. 2.15 that the sum of two (or more) independent gaussian random processes is itself a gaussian random process and that, further, the variance of the sum is equal to the sum of the variances of the individual processes. This result may be applied to the calculation of the noise power generated by combinations of resistors. Consider, for example, the case of two resistors, of resistances R_1 and R_2 both at the temperature T, connected in series. The mean-square values of the noise generated by each of the resistors and measured in the frequency band df are $\overline{v_{n_1}^2} = 4kTR_1\,df$ and $\overline{v_{n_2}^2} = 4kTR_2\,df$. Then, the mean-square voltage measured across the series combination of R_1 and R_2 is $\overline{v_n^2} = \overline{v_{n_1}^2} + \overline{v_{n_2}^2}$, where

$$\begin{aligned}
\overline{v_n^2} = \overline{v_{n_1}^2} + \overline{v_{n_2}^2} &= 4kTR_1\,df + 4kTR_2\,df \\
&= 4kT(R_1 + R_2)\,df \qquad (14.2\text{-}1)
\end{aligned}$$

This result is to have been anticipated since it indicates that two noisy resistors R_1 and R_2 can be replaced by a single noisy resistor $R = R_1 + R_2$. The mean-square noise voltage due to R is given by Eq. (14.2-1).

Rather obviously, this discussion can be extended to an arbitrary network of resistors, all at the same temperature T. If the resistance seen looking into some set of terminals is $R = 1/G$, then the open-circuit mean-square noise voltage at those terminals is $\overline{v_n^2} = 4kTR\,df$, and the mean-square short-circuit current is $\overline{i_n^2} = 4kTG\,df$.

14.3 NETWORKS WITH REACTIVE ELEMENTS

Consider a network composed of resistors, inductors, and capacitors as indicated in Fig. 14.3-1a. We arbitrarily select a set of terminals a-b and inquire now about an appropriate equivalent circuit to represent these terminals as a noise source.

The impedance $Z(f)$ seen looking back into these terminals a-b is generally a function of frequency and has a real and an imaginary part. That is,

$$Z(f) = R(f) + jX(f) \qquad (14.3\text{-}1)$$

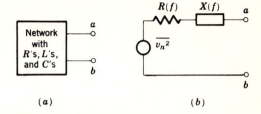

Fig. 14.3-1 (a) A network containing resistors, capacitors, and inductors with terminals a and b. (b) An equivalent circuit representing the terminals as a noise source.

(a)   (b)

where $R(f)$ is the resistive (real) component of the impedance, and $X(f)$ is the imaginary (reactive) component. We therefore replace the network by the equivalent circuit shown in Fig. 14.3-1b in which the resistance $R(f)$ and the reactance $X(f)$ are noiseless and the voltage source has a mean-squared value $\overline{v_n^2}$ which is to be determined.

In Fig. 14.3-2 we have bridged a load resistor R_1 across the terminals a-b of the network of Fig. 14.3-1. This load resistor is represented by a noiseless resistor R_1 in series with a noise generator of mean-squared voltage $\overline{v_{n1}^2}$. We assume that all the parts of the circuit of Fig. 14.3-2 are at the same temperature in thermal equilibrium. In this case the power delivered by $Z(f)$ to R_1 must equal the power delivered by R_1 to $Z(f)$. To show that this is indeed the case, consider what occurs if there is a net power flow from $Z(f)$ to R_1. Then the temperature of R_1 would increase, thereby exceeding the temperature of $Z(f)$. However, we have assumed that $Z(f)$ and R_1 are at the same temperature in thermal equilibrium. Hence the power flow to R_1 from $Z(f)$ must equal the power flow to $Z(f)$ from R_1 to maintain this equilibrium condition.

The mean-squared current due to the source $\overline{v_{n1}^2}$ is

$$\overline{i_{n1}^2} = \frac{\overline{v_{n1}^2}}{[R_1 + R(f)]^2 + [X(f)]^2} \tag{14.3-2}$$

and that due to the source $\overline{v_n^2}$ is

$$\overline{i_n^2} = \frac{\overline{v_n^2}}{[R_1 + R(f)]^2 + [X(f)]^2} \tag{14.3-3}$$

From the condition that there be no net transfer of power we have

$$\overline{i_{n1}^2}\,R(f) = \overline{i_n^2}R_1 \tag{14.3-4}$$

From Eqs. (14.3-2), (14.3-3), and (14.3-4), and using $\overline{v_{n1}^2} = 4kTR_1\,df$, we find

$$\overline{v_n^2} = 4kTR(f)\,df \tag{14.3-5}$$

Correspondingly, the two-sided power spectral density of the open-circuit voltage in Fig. 14.3-1 is

$$G_v(f) = 2kTR(f) \tag{14.3-6}$$

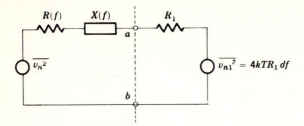

Fig. 14.3-2 A load resistor R_1 is bridged across the terminals $a - b$ of Fig. 14.3-1.

14.4 AN EXAMPLE

As an example of a noise calculation and also because the result is of interest in itself, we calculate the mean-squared noise voltage $\overline{v_n^2}$ at the terminals of the RC circuit shown in Fig. 14.4-1a.

Looking back into the terminals of the RC circuit, we calculate that the resistive component of the impedance seen is

$$R(f) = \frac{R}{1 + 4\pi^2 f^2 R^2 C^2} \tag{14.4-1}$$

Applying Eq. (14.3-6) to the entire frequency spectrum, we find

$$\overline{v_n^2} = 2kT \int_{-\infty}^{\infty} \frac{R\,df}{1 + 4\pi^2 f^2 R^2 C^2} = \frac{kT}{C} \tag{14.4-2}$$

The calculation of $\overline{v_n^2}$ may be performed in an alternate manner. In Fig. 14.4-1b we have replaced the noisy resistor by a noise generator of

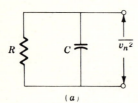

(a)

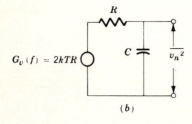

(b)

Fig. 14.4-1　(a) An RC circuit.　(b) The equivalent representation in which the resistor is replaced by a noise generator and a noiseless resistor.

power spectral density $G_v(f) = 2kTR$ and a noiseless resistor. The RC combination has a transfer function $H(f)$ from noise generator to output terminals given by

$$H(f) = \frac{1}{1 + j2\pi fRC} \tag{14.4-3}$$

The power spectral density of the output noise is $G_v(f)|H(f)|^2$, and $\overline{v_n^2}$ is given, as before, by

$$\overline{v_n^2} = 2kTR \int_{-\infty}^{\infty} \frac{df}{1 + 4\pi^2 f^2 R^2 C^2} = \frac{kT}{C} \tag{14.4-4}$$

Equation (14.4-4) may be written

$$\tfrac{1}{2}C\overline{v_n^2} = \tfrac{1}{2}kT \tag{14.4-5}$$

in which the left-hand member is the average energy stored on the capacitor. This result is an example of the famous equipartition theorem of classical statistical mechanics. The equipartition theorem states that a system in equilibrium with its surroundings, all at a temperature T, shares in the general molecular agitation and has an average energy which is $\tfrac{1}{2}kT$ for each degree of freedom of the system. Thus, an atom of a gas, which is free to move in three directions, has three degrees of freedom and correspondingly has an average kinetic energy $3 \times \tfrac{1}{2}kT = \tfrac{3}{2}kT$. At the other extreme, a macroscopic system such as a speck of dust suspended in a gas similarly flits about erratically and has an average energy associated with this random motion of $\tfrac{3}{2}kT$. Since the dust speck is much more massive than an atom, the average velocity of the dust speck will be correspondingly much smaller. As another example, consider a wall galvanometer, which, being free only to rotate, has a single degree of freedom. The kinetic energy associated with such rotation is $\tfrac{1}{2}I\dot{\theta}^2$ where I is the moment of inertia and $\dot{\theta}$ is the angular velocity. Such a galvanometer shares in the thermal agitation of the air in which it is suspended, and $\tfrac{1}{2}I\overline{\dot{\theta}^2} = \tfrac{1}{2}kT$. If the beam of light reflected from the galvanometer mirror is brought to focus on a scale sufficiently far removed, the slight random rotation of the galvanometer may be observed with the naked eye. Altogether, it is interesting to note that the noise generated by a resistor is not a phenomenon restricted to electrical systems alone, but is a manifestation of, and obeys, the same physical laws that characterize the general thermal agitation of the entire universe.

Returning now to the RC circuit of Fig. 14.4-1, we observe that it has one degree of freedom, i.e., the circuit has one mesh, and a single current is adequate to describe the behavior of the system. On this basis, then, Eq. (14.4-5) is seen to be an example of the equipartition theorem.

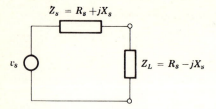

$Z_s = R_s + jX_s$

v_s

$Z_L = R_s - jX_s$

Fig. 14.5-1 A source of impedance Z_s is loaded by a complex conjugate impedance $Z_L = Z_s^*$ in order to draw maximum power.

14.5 AVAILABLE POWER

The *available power* of a source is defined as the maximum power which may be drawn from the source. If, as in Fig. 14.5-1, the source consists of a generator v_s in series with a source impedance $Z_s = R + jX$, then maximum power is drawn when the load is $Z_L = R - jX$, that is, $Z_L = Z_s^*$, the complex conjugate of Z_s. The available power is, therefore,

$$P_a = \frac{v_s^2}{4R_s} \tag{14.5-1}$$

Note that the available power depends only on the resistive component of the source impedance.

Using Eq. (14.5-1), we have that the available thermal-noise power (actual power, not normalized power) of a resistor R in the frequency range df is

$$P_a = \frac{4kTR\,df}{4R} = kT\,df \tag{14.5-2}$$

The two-sided available thermal-noise power spectral density is

$$G_a = \frac{P_a}{2\,df} = \frac{kT}{2} \tag{14.5-3}$$

Observe that G_a does not depend on the resistance of the resistor but only on the physical constant k and on the temperature. If the source consists of a combination of resistors (all at temperature T) together with inductors and capacitors, then in Eq. (14.5-2) the R in the numerator and the R in the denominator are both replaced by $R(f)$, where $R(f)$ is the (usually frequency-dependent) resistive component of the impedance seen looking back into the network. These $R(f)$'s will cancel, as do the R's. Hence, whether the network is a single resistor or a complicated RLC network, the available noise-power spectral density is $G_a = kT/2$ quite independently of its component values and circuit configuration.

Equation (14.5-3) expresses the available noise-power spectral density as predicted by the principles of classical physics, which also predict that this value of G_a applies at all frequencies; i.e., the noise is white. This result is manifestly untenable, since it predicts that the total

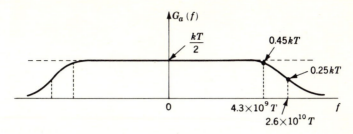

Fig. 14.5-2 Available power spectral density of thermal noise as given by Eq. (14.5-5).

available power

$$P_a = \int_{-\infty}^{\infty} G_a(f) \, df \tag{14.5-4}$$

is infinite. This prediction was one of a series of similar inconsistencies which were, in part, responsible for the development of the branch of physics called quantum mechanics. The quantum mechanical expression for $G_a(f)$ is

$$G_a(f) = \frac{hf/2}{e^{hf/kT} - 1} \tag{14.5-5}$$

in which $h = 6.62 \times 10^{-34}$ joule-sec is Planck's constant. Equation (14.5-5) yields a finite value for P_a and reduces to Eq. (14.5-3) when $hf \ll kT$.

The power spectral density of Eq. (14.5-5) is plotted in Fig. 14.5-2. Note that the density is lower than $kT/2$ by 1 dB or more only when $f \geq 4.3 \times 10^9 T$, which at room temperature $T_0 \cong 290°\mathrm{K}$ corresponds to $f \geq 1.3 \times 10^{12} = 1.3 \times 10^3 \, \mathrm{GHz}$. Hence we may certainly use $G_a = kT/2$ at radio and even microwave frequencies ($\approx 4 \, \mathrm{GHz}$). Note that a microwave receiver may employ a maser amplifier operating at a temperature as low as 4°K in order to minimize the noise due to the amplifier. Even at these low temperatures, it is still appropriate to assume that the noise is white. At optical frequencies this assumption is no longer valid, and Eq. (14.5-5) must be employed.

14.6 NOISE TEMPERATURE

Solving Eq. (14.5-2) for T, we have

$$T = \frac{P_a}{k \, df} \tag{14.6-1}$$

When we apply Eq. (14.6-1) to a passive RLC circuit in which the noise is due entirely to the resistors, then T is the actual common temperature of

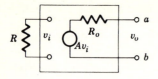

Fig. 14.6-1 Illustrating that the noise temperature seen looking back into a set of terminals $a - b$ may assume any value.

the resistors. Consider, however, the noise which may appear across a set of terminals connected to a more general type of circuit, including possibly active devices. Suppose that we measure the available power at the terminals and find that the noise is white, i.e., the available power P_a increases in proportion to the bandwidth, so that P_a/df is the same at all frequencies. We may then take Eq. (14.6-1) to be the definition of the noise temperature of the network. The noise temperature of the network need not be the temperature of any part of the network.

Consider, for example, the simple idealized situation represented in Fig. 14.6-1. Here a resistor R, which is a thermal-noise source at a temperature T, is connected to the input terminals of an amplifier of gain A. We assume that the input impedance of the amplifier is infinite and assume further, for simplicity, that the amplifier output resistance is a *noiseless* resistor R_o. Then the noise power, in a frequency range df, available at the amplifier output terminals is

$$P_a = \frac{\overline{v_o^2}}{R_o} = \frac{kTRA^2\,df}{R_o} \tag{14.6-2}$$

The noise temperature seen looking back into these terminals is, using Eq. (14.6-1),

$$T_n = \frac{P_a}{k\,df} = A^2\left(\frac{R}{R_o}\right)T \tag{14.6-3}$$

Thus, depending on the gain A and the ratio R/R_o, this noise temperature may assume any value including even $T_n = 0$.

As a further example to illustrate the concept of noise temperature, consider an antenna which may consist of nothing more than a loop of wire. If we assume that the wire has zero resistance, then the antenna by itself will generate no noise. Noise may, however, be induced in the antenna from a number of sources, some atmospheric and some man-made, including lightning, automobile ignition systems, fluorescent lights, etc. The spectral density of such noise falls off above about 50 MHz. Thus, while an AM radio may be affected by such noise, commercial FM, operating at higher carrier frequencies, is not appreciably affected. A second source of noise is the thermal radiation of any physical body which is at a temperature other than 0°K. Thus the earth, the atmosphere, the sun, the stars, and other cosmic bodies are all sources of noise. If it were possible to shield an antenna completely from all noise sources, then the antenna noise temperature would be zero. (It should also be noted that in this case the antenna would receive no signals either.) Otherwise, if, in the frequency range of interest, the available power spectral density of the antenna noise is constant, then Eq. (14.6-1) may be used to determine the antenna noise temperature. That is, if in a frequency band B the available noise power is P_a, then the antenna noise temperature is T (antenna) $= P_a/kB$.

14.7 TWO-PORTS

Received signals may be processed in a variety of ways. For example, the signal may need to be amplified through a number of amplifier stages or the signal may need to be subjected to frequency conversion to an intermediate frequency, as in a superheterodyne receiver. Each of these processing stages has input and output terminals, i.e., each is a two-port network. Each such two-port may, in general, contain resistors and active devices which are sources of noise. The signal at the input to a two-port will be accompanied by noise and will be characterized by some signal-to-noise ratio. Because of the noise sources within the two-port, the signal-to-noise ratio at the output will be lower than at the input. It is a great convenience to have available a means of describing the extent to which a signal is degraded in passing through a two-port. There are two methods which are commonly employed for this purpose. In one method, the two-port is characterized in terms of an *effective input noise temperature;* in a second method the two-port is characterized by a *noise figure.*

In preceding sections we have described noise sources in terms of their available noise power. In order to continue conveniently to use this concept when two-ports are involved, it is convenient to introduce, in connection with two-ports, the idea of the *available gain* of a two-port. The available gain $g_a(f)$ is generally frequency-dependent and is defined as

$$g_a(f) = \frac{\text{available power spectral density at the two-port output}}{\text{available power spectral density at the source output}}$$

$$(14.7\text{-}1)$$

Thus $g_a(f)$ is the ratio of available output power in a small frequency range df to the available source power in this same range.

A point which is immediately apparent from the definition of Eq. (14.7-1) is that the available gain is not a characteristic of a two-port alone, but depends on the driving source as well. The available gain does not depend on the two-port load. To consider the matter further, consider the situation represented in Fig. 14.7-1. Here 2 two-port networks,

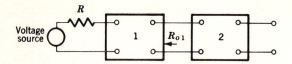

Fig. 14.7-1   A cascade of two-ports driven by a source of impedance R.

say two amplifiers, are in cascade, the first being driven by a source of impedance R.   The available gain of the first two-port depends on R as well as on the two-port itself.   The available gain of the second depends on the output impedance R_{o1} of the first two-port, which may in turn depend on R.   In the general case of a cascade of N two-ports, the available gain of the last two-port depends, in principle, on all the preceding $N - 1$ two-ports as well as on the source.   Of course, in a two-port cascade encountered in practice, the situation may not be as complicated as suggested here.   It may turn out, to a good approximation, that the output impedance of a stage is influenced only slightly by its driving source.   In spite of the possible complexity associated with calculating the available gain, particularly of a later stage in a cascade, fortunately the overall available gain of a cascade is related in a very simple manner to the available gains of the individual stages.   It may be verified (Prob. 14.7-1) that if the available gains of an N-stage cascade are $g_{a1}, g_{a2}, \ldots , g_{aN}$, then the overall available gain is the product

$$g_a = g_{a1}, g_{a2}, \ldots , g_{aN} \tag{14.7-2}$$

The concept of available gain is especially useful because it permits us to write in very simple form generalized results for the noise characteristics of two-ports.

For example, we note that if a source of available-power spectral density $G_a(f)$ is connected to the input of a two-port of available gain $g_a(f)$, the available-power spectral density at the output is $G_a(f)g_a(f)$, and the total available output power is

$$P_{ao} = \int_{-\infty}^{\infty} G_a(f)g_a(f) \, df \tag{14.7-3}$$

If the source has a noise temperature T, then $G_a(f) = kT/2$, and

$$P_{ao} = \frac{kT}{2} \int_{-\infty}^{\infty} g_a(f) \, df \tag{14.7-4}$$

14.8   NOISE BANDWIDTH

Bandpass amplifiers, as well as other two-ports of restricted bandpass, will typically have available gains with a frequency-dependence such as illustrated by the solid-line plot of Fig. 14.8-1.   We have indicated a

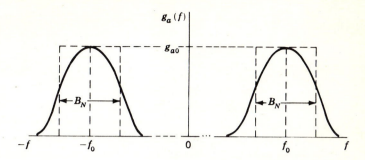

Fig. 14.8-1 Illustrating the concept of noise bandwidth.

passband characteristic which is symmetrical about some frequency f_0, since this is the type of characteristic most frequently encountered. However, this symmetry is not essential to the present discussion.

If this two-port is driven by a thermal-noise source of temperature T, then the two-port output available power will be

$$P_{ao} = \frac{kT}{2} \int_{-\infty}^{\infty} g_a(f) \, df \qquad (14.8\text{-}1)$$

It is frequently convenient to replace the actual available-gain characteristic by a rectangular characteristic, as shown by the dashed plot, which is equivalent for the purpose of computing available output-noise power.   Such a rectangular characteristic would have to have a bandwidth B_N determined by the condition of equal available noise output for the two cases, that is, B_N would be determined by

$$P_{ao} = g_{a0} kT B_N = \frac{kT}{2} \int_{-\infty}^{\infty} g_a(f) \, df \qquad (14.8\text{-}2)$$

where g_{a0} is the constant value of $g(f)$ over the passband for the rectangular characteristic.   The bandwidth B_N is called the *noise bandwidth* and, from Eq. (14.8-2), is given by

$$B_N = \frac{1}{2g_{a0}} \int_{-\infty}^{\infty} g_a(f) \, df \qquad (14.8\text{-}3)$$

In Fig. 14.8-1 we have selected g_{a0} to equal the value of g_a at $f = f_0$. Hence the noise bandwidth which results is the noise bandwidth with respect to the frequency f_0.   Customarily, as indicated, f_0 is selected to be the frequency at which $g_a(f)$ is a maximum.

14.9   EFFECTIVE INPUT-NOISE TEMPERATURE

Consider that a two-port is driven by a noise source that has a noise temperature T.   Then the source behaves like a resistor at temperature T, and from Eq. (14.5-3) the available noise-power spectral density of the source will be $kT/2$.   If the available gain of the two-port is g_a, and if the two-port itself is entirely noise-free, then the available two-sided power spectral density at the two-port output would be

$$G'_{ao} = g_a(f)\,\frac{kT}{2}$$
(14.9-1)

However, because the two-port itself will contribute noise, the available output noise power will be G_{ao}, which is larger than G'_{ao}. We may choose to make it appear that the two-port itself is noise-free, and to account for the increased noise by assigning to the source a new noise temperature higher than T by an amount T_e.   We would then have

$$G_{ao} = g_a(f)\,\frac{k}{2}\,(T + T_e)$$
(14.9-2)

The temperature T_e is called the *effective input-noise temperature* of the two-port.   It is to be kept in mind, however, that T_e, like the available gain, depends on the source as well as on the two-port itself.

Example 14.9-1   An antenna has a noise temperature $T_{\text{ant}} = 10°\text{K}$.   It is connected to a receiver which has an equivalent noise temperature $T_e = 140°\text{K}$.   The midband available gain of the receiver is $g_{ao} = 10^{10}$ and, with respect to its midband frequency, the noise bandwidth is $B_N = 1.5 \times 10^5$ Hz.   Find the available output noise power.

Solution   The available power spectral density of the antenna is $kT_{\text{ant}}/2$. By the definition of equivalent noise temperature T_e, the noise of the receiver may be taken into account by increasing the source temperature by T_e.   Hence the effective source temperature is $T = T_{\text{ant}} + T_e$, and the effective available noise-power spectral density at the input to the receiver is

$$G_a(f) = \frac{k}{2}\,(T_{\text{ant}} + T_e)$$
(14.9-3)

The available noise power at the output is, using Eq. (14.8-2),

$$
\begin{aligned}
P_{ao} &= g_{ao}k(T_{\text{ant}} + T_e)B_N \\
&= 10^{10} \times 1.38 \times 10^{-23} \times 150 \times 1.5 \times 10^5 = 3.1\ \mu\text{W}
\end{aligned}
$$
(14.9-4)

14.10   NOISE FIGURE

Let us assume that the noise present at the input to a two-port may be represented as being due to a resistor at the two-port input, the resistor being at room temperature T_0 (usually taken to be $T_0 = 290°\text{K}$).   If the two-port itself were entirely noiseless, the output available noise-power spectral density would be $G'_{ao} = g_a(f)(kT_0/2)$. However, the actual output noise-power spectral density is G_{ao}, which is greater than G'_{ao}. The ratio $G_{ao}/G'_{ao} \equiv F$ is the *noise figure* of the two-port, that is,

$$F(f) \equiv \frac{G_{ao}}{G'_{ao}} = \frac{G_{ao}}{g_a(f)(kT_0/2)} \tag{14.10-1}$$

If the two-port were noiseless, we would have $F = 1$ (0 dB). Otherwise $F > 1$. Using Eq. (14.9-2) with $T = T_0$, and Eq. (14.10-1), we find that the noise figure F and the effective temperature T_e are related by

$$T_e = T_0(F - 1) \tag{14.10-2}$$

or

$$F = 1 + \frac{T_e}{T_0} = \frac{T_e + T_0}{T_0} \tag{14.10-3}$$

The noise figure as defined by Eq. (14.10-1) is referred to as the *spot noise figure*, since it refers to the noise figure at a particular "spot" in the frequency spectrum. If we should be interested in the *average noise figure* over a frequency range from f_1 to f_2, then, as may be verified (Prob. 14.10-3), this average noise figure $\bar{F}$ is related to $F(f)$ by

$$\bar{F} = \frac{\displaystyle\int_{f_1}^{f_2} g_a(f)F(f)\,df}{\displaystyle\int_{f_1}^{f_2} g_a(f)\,df} \tag{14.10-4}$$

Two-ports are most commonly characterized in terms of noise figure when the driving noise source is at or near T_0, while the concept of effective noise temperature T_e is generally more convenient when the noise temperature is not near T_0.

When following a signal through a two-port, we are not so much interested in the noise level as in the signal-to-noise ratio. Consider, then, the situation indicated in Fig. 14.10-1. Here the noise at the two-port input is represented as being due to a resistor R so that the available input-noise-power spectral density is $G_{ai}^{(n)} = kT/2$. A signal is also present at the input with available power spectral density $G_{ai}^{(s)}$. The

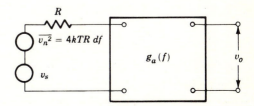

Fig. 14.10-1 A signal v_s and a noise source are superimposed and applied at the input of a two-port of available gain $g_a(f)$.

$$\overline{v_n^2} = 4kTR\,df$$

$$g_a(f)$$

$$v_s$$

$$v_o$$

available output-signal-power spectral density is

$$G_{ao}^{(s)} = g_a G_{ai}^{(s)} \qquad (14.10\text{-}5)$$

However, because of the noise added by the two-port itself, the available output-noise spectral density is

$$G_{ao}^{(n)} = g_a F G_{ai}^{(n)} \qquad (14.10\text{-}6)$$

Combining Eqs. (14.10-5) and (14.10-6), we have an alternative interpretation of the spot noise figure, that is,

$$F(f) = \frac{G_{ai}^{(s)}/G_{ai}^{(n)}}{G_{ao}^{(s)}/G_{ao}^{(n)}} \qquad (14.10\text{-}7)$$

Thus F is a ratio of ratios. The numerator in Eq. (14.10-7) is the input-signal-to-noise power spectral density ratio, while the denominator is the output-signal-to-noise power spectral density ratio.

Let us assume that in a frequency range from f_1 to f_2 the power spectral densities of signal and noise are uniform. In this case it may be verified (Prob. 14.10-5) that the average noise figure $\bar{F}$ defined by Eq. (14.10-4) has the significance

$$\bar{F} = \frac{S_i/N_i}{S_o/N_o} \qquad (14.10\text{-}8)$$

where S_i and N_i are, respectively, the total input available signal and noise powers in the frequency range f_1 to f_2, and similarly S_o and N_o are the total output available signal and noise powers.

The noise figure F (or $\bar{F}$) may be expressed in a number of alternative forms which are of interest. If the available gain g_a is constant over the frequency range of interest, so that $F = \bar{F}$, then $S_o = g_a S_i$. In this case Eq. (14.10-8) may be written

$$F = \frac{1}{g_a} \frac{N_o}{N_i} \qquad (14.10\text{-}9)$$

Further, the output noise N_o is

$$N_o = g_a N_i + N_{tp} \qquad (14.10\text{-}10)$$

where $g_a N_i$ is the output noise due to the noise present at the input, and N_{tp} is the additional noise due to the two-port itself. Combining Eqs. (14.10-9) and (14.10-10), we have

$$F = 1 + \frac{N_{tp}}{g_a N_i} \qquad (14.10\text{-}11)$$

or, the noise due to the two-port itself may be written, from Eq. (14.10-11), as

$$N_{tp} = g_a(F - 1)N_i \qquad (14.10\text{-}12)$$

14.11 NOISE FIGURE AND EQUIVALENT NOISE TEMPERATURE OF A CASCADE

In Fig. 14.11-1 is shown a cascade of 2 two-ports with a noise source at the input of noise temperature T_0. The individual two-ports have available gains g_{a1} and g_{a2} and noise figures F_1 and F_2. If the input-source noise power is N_i, the output noise due to this source is $g_{a1}g_{a2}N_i$. The noise output of the first stage due to the noise generated within this first two-port is $g_{a1}(F_1 - 1)N_i$ from Eq. (14.10-12). The corresponding noise at the output of the second stage is $g_{a1}g_{a2}(F_1 - 1)N_i$. Again, using Eq. (14.10-12), we find that the noise output due to the noise generated within the second two-port is $g_{a2}(F_2 - 1)N_i$. The total noise output is therefore

$$N_o = g_{a1}g_{a2}N_i + g_{a1}g_{a2}(F_1 - 1)N_i + g_{a2}(F_2 - 1)N_i \qquad (14.11\text{-}1)$$

If we use Eq. (14.10-9), the overall noise figure of the cascade is

$$F = \frac{1}{g_a}\frac{N_o}{N_i} = \frac{1}{g_{a1}g_{a2}}\frac{N_o}{N_i} \qquad (14.11\text{-}2)$$

$$= F_1 + \frac{F_2 - 1}{g_{a1}} \qquad (14.11\text{-}3)$$

from Eq. (14.11-1). If the calculation leading to Eq. (14.11-3) is extended to a cascade of k stages, the result is

$$F = F_1 + \frac{F_2 - 1}{g_{a1}} + \frac{F_3 - 1}{g_{a1}g_{a2}} + \cdots + \frac{F_k - 1}{g_{a1}g_{a2}\,\cdots\,g_{a(k-1)}} \qquad (14.11\text{-}4)$$

If the two-ports are characterized by equivalent temperatures rather than noise figures, then the equivalent temperature of the cascade T_e is related to the equivalent temperatures and available gains of the individual stages by

$$T_e = T_{e1} + \frac{T_{e2}}{g_{a1}} + \frac{T_{e3}}{g_{a1}g_{a2}} + \cdots + \frac{T_{ek}}{g_{a1}g_{a2}\,\cdots\,g_{a(k-1)}} \qquad (14.11\text{-}5)$$

Equation (14.11-5) may be established by combining Eq. (14.11-4) with Eq. (14.10-3) (see Prob. 14.11-2).

Suppose that the individual two-ports have comparable noise figures or equivalent temperatures. Then, especially if the gains are large, the

Fig. 14.11-1 A noise source at temperature T_0 drives a cascade of two-ports.

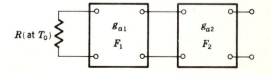

contribution to the net output noise of succeeding stages in the cascade becomes progressively smaller.   A very effective practice, for the purpose of securing a low-noise receiving system, is to design the first stage of the cascade with a low equivalent temperature and a high gain.   A gain of 30 dB is not uncommon.   Similarly, equivalent temperatures as low as $T_e = 4°K$ are obtained by cooling the amplifier with liquid nitrogen.

14.12 AN EXAMPLE OF A RECEIVING SYSTEM

The receiver shown in block diagram form in Fig. 14.12-1 is rather typical of microwave receivers such as are used for satellite communication.   In such cases, it is certainly justifiable to take considerable pains to keep the noise figure of the receiver as low as possible.   For variety, and also to be consistent with practice, we have characterized the noisiness of the first amplifier in terms of a noise temperature, while the other stages have been characterized by a noise figure.   We calculate now the overall noise figure of the receiver.   The antenna does not enter the calculation, since it is considered the driving source and not part of the receiver.   Using Eqs. (14.10-3) and (14.11-4), we have

$$F(\text{receiver}) = \left(1 + \frac{4}{290}\right) + \frac{4-1}{1000} + \frac{16-1}{100,000} = 1.017(=0.05 \text{ dB})$$

$$(14.12\text{-}1)$$

From Eq. (14.10-2) the equivalent temperature of the receiver is

$$T_e(\text{receiver}) = T_0[F(\text{receiver}) - 1]$$
$$= 290(0.017) = 4.93°K \qquad (14.12\text{-}2)$$

Note that, because of the high gain of the first amplifier stage, the travelling wave tube amplifier, the mixer, and the IF amplifier increase the effective receiver temperature only 0.93°K above the temperature of the maser amplifier.

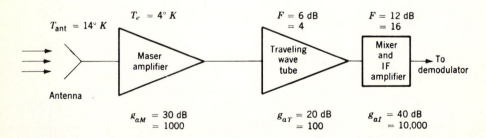

Fig. 14.12-1   A typical microwave receiver.

The available noise power present at the demodulator input in the bandwidth B is

$$P_a = \frac{k}{2}[T_{\text{ant}} + T_e(\text{receiver})](2B)g_a(\text{receiver}) \qquad (14.12\text{-}3)$$

The available gain of the receiver is $30 + 20 + 40 = 90$ dB $= 10^9$; hence

$$P_a = 1.38 \times 10^{-23}(14 + 4.93) \times 10^9 B$$

$$\approx 2.6 \times 10^{-13}B \text{ watts} \qquad (14.12\text{-}4)$$

Because of the high gain which precedes the demodulator, the noise which may be introduced by the demodulator or any succeeding processing will not further degrade the signal-to-noise ratio of the signal.

In the receiver of Fig. 14.12-1 each of the stages provides some band-limiting, and the IF amplifier may well consist of a number of stages, each one of which is bandlimited. Conceptually, however, it is very convenient to consider that all the stages of the system are of unlimited bandwidth and that bandlimiting is done in a single IF filter at the output end of the IF amplifier. Under these circumstances the noise input to this IF filter would have a white (uniform) spectral density. We have, as a matter of fact, throughout this text assumed that such was the case when we assumed that our communication channel was characterized by a two-sided noise-power spectral density $G_n(f) \equiv \eta/2$ [see Eq. (7.9-1)]. It is now of interest, in connection with the receiver of Fig. 14.12-1 to inquire into the magnitude of $\eta/2$. From the result given in Eq. (14.12-4) we have

$$G_n(f) \equiv \frac{\eta}{2} = \frac{P_a}{2B} = \frac{2.6 \times 10^{13}B}{2B} = 1.3 \times 10^{-13} \text{ watt/Hz}$$

$$(14.12\text{-}5)$$

As a further example of the performance of the system of Fig. 14.12-1 consider that the receiver is being used to receive a frequency-modulated signal with a baseband frequency range $f_M = 4$ MHz. The received signal power at the demodulator is S_i. Let us assume that, in order to keep the discriminator operating above threshold, we require $S_i/\eta f_M = 20$ dB. What then must be the value of available signal power at the output of the antenna? Since

$$\frac{S_i}{\eta f_M} = \frac{S_i}{2.6 \times 10^{-13} \times 4 \times 10^6} \geq 100 \ (= 20 \text{ dB}) \qquad (14.12\text{-}6)$$

we find

$$S_i \geq 10.4 \times 10^{-5} \text{ watt} \qquad (14.12\text{-}7)$$

Since the receiver gain is 90 dB$(= 10^9)$, the required minimum available signal power at the antenna must be

$$S_i(\text{antenna}) = \frac{10.4 \times 10^{-5}}{10^9} = 10.4 \times 10^{-14} \text{ watt} \qquad (14.12\text{-}8)$$

14.13 ANTENNAS

An antenna, as a noise source, is characterized by a noise temperature. The two-ports of a receiving system, so far as their noise generation is concerned, are characterized by equivalent input-noise temperatures or by noise figures. We have seen how, in terms of these characterizations, we may determine the signal power required to be available from a receiving antenna to ensure an acceptable signal-to-noise ratio. We refer briefly now to the manner in which the transmission performance of an antenna system is characterized, so that, given the required available power at the receiving antenna, we may determine the power required to be radiated by the transmitting antenna.

Consider a transmitting antenna which radiates a power P_T, and assume that the power is radiated uniformly in all directions, that is, isotropically. The power incident on an area A oriented perpendicularly to the direction of power flow, and at a distance d from the transmitting antenna, is

$$P_R = \frac{P_T}{4\pi d^2} A \qquad (14.13\text{-}1)$$

Equation (14.13-1) may be used to define an effective area A_e of an antenna. Thus, if the available power from a receiving antenna is P_R when the antenna is a distance d from an isotropic antenna transmitting a power P_T, then the effective area of the receiving antenna is

$$A_e \equiv 4\pi d^2 \frac{P_R}{P_T} \qquad (14.13\text{-}2)$$

The effective area of an antenna is related principally to the physical shape and dimensions of the antenna. Thus, for example, for a parabolic disk antenna the effective area is generally in the range 0.5 to 0.6 of the physical area of the disk.

Real antennas do not radiate isotropically but are rather directional. This directivity is of advantage when we are interested in transmitting from one antenna to a *particular* receiving antenna. In such a case we would be interested in making the transmitting antenna as directional as possible, and we would orient the transmitting antenna to radiate with maximum intensity toward the receiving antenna. A typical antenna-radiation pattern of a directional antenna is shown in Fig. 14.13-1. If we draw a line, in an arbitrary direction, from the antenna to the antenna

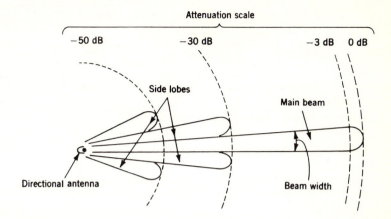

Fig. 14.13-1 Transmission pattern of a highly directional antenna.

pattern, the length of the line is proportional to the radiant power density in that direction. As indicated, the radiated power is principally confined to a *main beam*, while some power is also radiated in the direction of the *side lobes*. The *beam width* of the antenna is defined as the angle, at the antenna, between directions in which the radiated-power density is down 3 dB from the maximum. The direction of maximum radiation is referred to as the 0 dB direction. In a highly directional antenna, beam widths of 1°, with side lobes down 30 dB and 50 dB as indicated, are feasible.

There is a reciprocity relationship between an antenna used for radiation and the same antenna used for reception. An antenna has the same directivity pattern in the two cases. Thus a highly directional antenna radiates principally in one direction, and when used for reception it similarly absorbs most of the radiant energy from this same direction. Hence, for communication between two particular antennas, and to minimize interference and spurious signals, it is advantageous that both antennas be directional and, of course, with main lobes oriented toward one another.

The extent to which the principal direction of an antenna is favored is measured by the *antenna gain*. Thus suppose that an isotropic radiator would radiate a power per unit solid angle of p_i when furnished with a power P (that is, $p_i = P/4\pi$). If a directional antenna radiates a power per unit solid angle p_m in the direction of most intense radiation, then the antenna gain K is defined by

$$K \equiv \frac{p_m}{p_i} \qquad\qquad (14.13\text{-}3)$$

It turns out that the gain of an antenna and its effective area are related by

$$K = \frac{4\pi A_e}{\lambda^2} \qquad (14.13\text{-}4)$$

where λ is the wavelength of the radiation. Note that, for fixed λ, the gain increases with effective area and hence with the physical dimensions of the antenna. A large antenna, therefore, absorbs more power and can also be made more directional than a small antenna.

Consider now a receiving antenna of effective area A_{eR} facing a transmitting antenna of gain K_T. Then from Eq. (14.13-1) and from the definition of antenna gain we find that the received power is

$$P_R = \frac{P_T}{4\pi d^2} A_{eR} K_T \qquad (14.13\text{-}5)$$

where K_T is the gain of the transmitting antenna. Applying Eq. (14.13-4) to the receiving antenna, i.e., replacing K by K_R and A_e by A_{eR} and combining this result with Eq. (14.13-5), we have

$$\frac{P_R}{P_T} = \frac{K_T K_R}{(4\pi d/\lambda)^2} \qquad (14.13\text{-}6)$$

Thus we find that the received-to-transmitted power ratio P_R/P_T depends on the ratio d/λ called the *effective distance* and on the gains K_T and K_R of the two antennas.

Example 14.13-1   The available power required at a receiving antenna is 10^{-6} watt (that is, -60 dB with respect to 1 watt). Transmitting and receiving antennas have gains of 40 dB each. The carrier frequency used is 4 GHz, and the distance between antennas is 30 miles. Find the required transmitter power.

Solution   Using Eq. (14.13-6), we have $(1.6 \times 10^3 \text{ m} = 1 \text{ mile})$

$$10 \log P_R = 10 \log P_T + 10 \log K_T + 10 \log K_R - 20 \log \left(4\pi \frac{d}{\lambda}\right)$$
$$(14.13\text{-}7)$$

$$-60 = 10 \log P_T + 40 + 40 - 20 \log \left[4\pi \frac{30 \times 1.6 \times 10^3}{(3 \times 10^8)/(4 \times 10^9)} \right]$$
$$= 10 \log P_T + 40 + 40 - 138$$

so that

$$10 \log P_T = -2$$

or $P_T = -2$ dB $(14.13\text{-}8)$

That is, P_T is 2 dB below 1 watt or $P_T = 0.64$ watt.

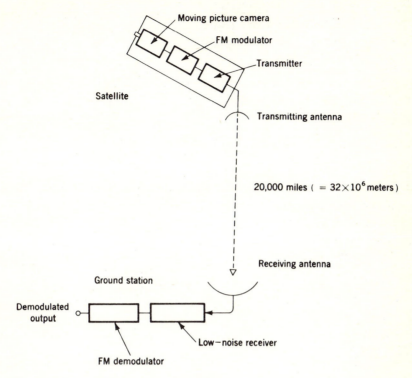

Fig. 14.14-1   A satellite communication system.

14.14   A SYSTEM CALCULATION

For the sake of tying together a number of the ideas developed in this chapter as well as several of the concepts encountered in connection with frequency modulation, let us undertake some calculations on a proposed satellite-to-earth communication system as represented in Fig. 14.14-1. We propose a system with the following specifications:

1. Frequency modulation is to be used to transmit from the satellite a TV signal with $f_M = 4$ MHz on a carrier of frequency $f_c = 3$ GHz.
2. The satellite antenna gain is to be $K_T = 20$ dB, while the receiving antenna on the ground is to have $K_R = 50$ dB.   (Note that the earth-bound antenna may be much larger than the satellite antenna and hence may have a larger gain.)   The satellite is assumed to be at a distance of 32×10^6 m ($= 20,000$ miles).
3. The noise temperature of the receiving antenna is to be $T_{ant} = 14°$K.
4. The receiver (a cooled maser amplifier is employed) is to have a total

noise figure $F = 0.2$ dB(1.047), and the overall available gain of the receiver up to the demodulator is to be $g_a = 70$ dB.
5. The demodulated signal-to-noise ratio is to be $S_o/N_o = 40$ dB.

We are to find:

1. The rms frequency deviation (Δf_{rms})
2. The IF bandwidth
3. The minimum required transmitter power

For simplicity we shall assume that demodulation is performed by an FM discriminator. To minimize the required transmitted power, we assume operation at or just above threshold. We shall further assume that the TV signal can be approximated by a gaussian process. Thus, the received FM signal can be written as

$$v(t) = A \cos\left[\omega_c t + k \int_{-\infty}^{t} m(\lambda)\, d\lambda\right] + n(t) \tag{14.14-1}$$

where $m(t)$ is the TV signal. Hence the rms frequency deviation produced is

$$(\Delta f)_{rms} = \frac{\sqrt{k^2 \overline{m^2(t)}}}{2\pi} \tag{14.14-2}$$

From Eq. (10.6-16) we have

$$\frac{S_o}{N_o} = \frac{3(\Delta f_{rms}/f_M)^2 (S_i/\eta f_M)}{1 + 6\sqrt{2/\pi}\,(\Delta f_{rms}/f_M)(S_i/\eta f_M)e^{-(f_M/B)(S_i/\eta f_M)}} \tag{14.14-3}$$

Equation (14.14-3) is a function of the input SNR $S_i/\eta f_M$ and the rms modulation index $\Delta f_{rms}/f_M$ [note that using Eq. (4.13-5) $B/f_M = 4.6\,\Delta f_{rms}/f_M$]. Thus, for a given output SNR, an infinite number of combinations of these ratios are possible. However, we have a constraint in our problem; that is, we are to operate at or above threshold. Let us solve Eq. (14.14-3) subject to the constraint that we are operating *at* threshold. The 1-dB dropoff associated with threshold, as noted in Fig. 10.1-1, occurs when the denominator in Eq. (14.14-3) has the value 1.26. Hence, at threshold, since an output SNR of 40 dB $(= 10^4)$ is required, we have, from Eq. (14.14-3), the following two results:

$$0.26 = 6\sqrt{\frac{2}{\pi}}\,\frac{\Delta f_{rms}}{f_M}\,\frac{S_i}{\eta f_M}\,e^{-(f_M/4.6\Delta f_{rms})(S_i/\eta f_M)} \tag{14.14-4}$$

$$3\left(\frac{\Delta f_{rms}}{f_M}\right)^2 \frac{S_i}{\eta f_M} = 1.26 \times 10^4 \tag{14.14-5}$$

Equations (14.14-4) and (14.14-5) can be solved by solving Eq.

(14.14-5) for $S_i/\eta f_M$ and substituting this result in Eq. (14.14-4), thereby eliminating $S_i/\eta f_M$ (Prob. 14.14-2). The resulting equation is

$$13 \times 10^{-6} \frac{\Delta f_{\mathrm{rms}}}{f_M} \simeq e^{-913(f_M/\Delta f_{\mathrm{rms}})^3} \tag{14.14-6}$$

It may be shown (Prob. 14.14-3) that this equation has the solution

$$\frac{\Delta f_{\mathrm{rms}}}{f_M} \approx 4.5 \tag{14.14-7}$$

The corresponding input SNR is

$$\frac{S_i}{\eta f_M} = 23 \text{ dB} \tag{14.14-8}$$

Good design requires that a margin of safety be employed to ensure operation at or above but not below threshold. We therefore decrease the ratio $\Delta f_{\mathrm{rms}}/f_M$, making it

$$\frac{\Delta f_{\mathrm{rms}}}{f_M} = 4 \tag{14.14-9}$$

The input SNR required to obtain an $S_o/N_o = 40$ dB is still

$$\frac{S_i}{\eta f_M} = 23 \text{ dB} \tag{14.14-10}$$

since we are now operating above threshold.

The IF bandwidth can now be calculated by letting

$$B = 4.6 \, \Delta f_{\mathrm{rms}} = 18.4 f_M = 73.6 \text{ MHz} \tag{14.14-11}$$

We need now determine only the required transmitter power, to complete our design. We begin this calculation with Eq. (14.14-10). Thus

$$\frac{S_i}{\eta} = 23 \text{ dB} + 10 \log f_M = 89 \text{ dB} \tag{14.14-12}$$

The noise-power spectral density $\eta/2$ at the input to the IF filter was shown in Sec. 14.7 to be a function of the noise figure (equivalent temperature) of the receiver and the receiver gain. Using Eqs. (14.9-3) and (14.12-5), we have

$$\eta = k(T_{\mathrm{ant}} + T_e)g_a(f) \tag{14.14-13}$$

where, from the problem specifications,

$$T_{\mathrm{ant}} = 14°\text{K}$$

$$T_e = T_0(F - 1) = 290(1.047 - 1) \approx 13.6°\text{K}$$

$$g_a(f) = 70 \text{ dB}$$

Thus

$$\eta = 1.38 \times 10^{-23}(27.6)g_a(f) = 38 \times 10^{-23}g_a(f) \tag{14.14-14}$$

We can now compute the signal power measured at the IF filter. We find [Eq. (14.14-12)]

$$S_i = 89 \text{ dB} + 10 \log \eta \approx (89 - 224) \text{ dB} + 10 \log g_a(f) \tag{14.14-15}$$

The input-signal power measured at the *antenna* is $S_i/g_a(f)$. This result, in decibels, is

$$(S_i)_{\text{antenna}} = (S_i)_{\text{IF filter}} - 10 \log g_a(f) = -135 \text{ dB} \tag{14.14-16}$$

The transmitter power required to deliver a signal power of -135 dB and meeting the problem specifications is found from Eq. (14.13-6).

$$P_T = P_R \frac{(4\pi \, d/\lambda)^2}{K_T K_R} \tag{14.14-17}$$

Performing the calculations in decibels yields

$$10 \log P_T = 10 \log P_R + 20 \log \left(\frac{4\pi \, d}{\lambda}\right) - 10 \log K_T - 10 \log K_R$$

$$= -135 + 20 \log \left[\frac{4\pi \times 32 \times 10^6}{(3 \times 10^8)/(3 \times 10^9)}\right] - 20 - 50$$

Thus

$$P_T = -13 \text{ dB} \tag{14.14-18}$$

Hence $P_T \geq 50$ mW must be transmitted.

PROBLEMS

14.2-1. The three resistors are at a temperature T. A bandlimited rms voltmeter is placed across ab, bc, and ac in succession.

    (a) Is $V_{ac}(\text{rms}) = V_{ab}(\text{rms}) + V_{bc}(\text{rms})$? Why, or why not?

    (b) Is $\overline{V_{ac}^2} = \overline{V_{ab}^2} + \overline{V_{bc}^2}$? Why or why not?

    (c) Calculate $V_{ac}(\text{rms})$ read by a meter having a bandwidth B, if R_a and R_c are at temperature T, and R_b is at temperature T_b. Give your answer in terms of symbols.

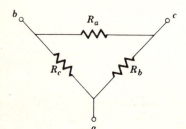

Fig. P14.2-1

14.3-1. A parallel RLC circuit centered at 3 GHz has a bandwidth of 10 MHz. If the resistance R is 10 kilohms, calculate $R(f)$ and the power spectral density $G_v(f)$ of the noisy circuit.

14.3-2. (a) Develop an expression for the power spectral density of the noise voltage e_n.

(b) The noise voltage e_n is passed through a low-pass filter with cutoff frequency at ω_c and then through an amplifier of gain $A = 9$. Develop an expression for the total noise output power of the amplifier.

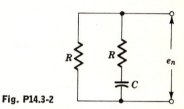

Fig. P14.3-2

14.4-1. Refer to Fig. 14.4-1a. Assume that the resistor has an inductance L in series with it. Find $R(f)$, $G_v(f)$, and $\overline{v_n^2}$ when the integral is over all frequencies.

14.5-1. Comment on the difference between $G_a(f) = kT/2$ in Eq. (14.5-3) and $G_v(f) = 2kTR(f)$ in Eq. (14.3-6).

14.5-2. If $T = 4°$K, find $G_a(f)$ from Eq. (14.5-5) when the wavelength $\lambda \approx 1$ mm, 1 μm. Is the noise white in these regions?

14.7-1. Verify Eq. (14.7-2).

14.8-1. Calculate the noise bandwidth of a parallel RLC filter having a 3-dB bandwidth B.

14.8-2. Calculate the noise bandwidth of an RC low-pass filter having a 3-dB bandwidth f_c.

14.8-3. A gaussian filter has the characteristic

$$H(f) = e^{-k^2 f^2} \qquad -\infty \le f \le \infty$$

(a) Calculate the 3-dB bandwidth.

(b) Calculate the noise bandwidth.

14.9-1. An antenna is connected to a receiver having an equivalent noise temperature $T_e = 100°$K. The available gain of the receiver is $g_a(f) = 10^8$ and the noise bandwidth is $B_N = 10$ MHz. If the available output-noise power is 10 μw, find the antenna temperature.

14.9-2. An antenna has a noise temperature $T_{\text{ant}} = 4°$K. It is connected to a receiver which has an equivalent noise temperature $T_e = 100°$K; the midband available gain of the receiver is $g_{a0} = 10^{10}$, and can be represented by a parallel RLC filter having a 3-dB bandwidth of 10 MHz. Find:

(a) B_N.

(b) The available output-noise power.

14.10-1(a). Explain why F cannot be less than 1.

(b) Verify Eqs. (14.10-2) and (14.10-3).

14.10-2. The noise figure of an amplifier is 0.2 dB.   Find the equivalent temperature T_e.

14.10-3. Show that the average value of the noise figure is given by Eq. (14.10-4).

14.10-4. The noise present at the input to a two-port is 1 μw.   The noise figure F is 0.5 dB.   The two-port gain $g_a = 10^{10}$.   Calculate:

(a) The available noise power contributed by the two-port.

(b) The output available noise power.

14.10-5. Show that Eq. (14.10-8) applies under the conditions specified in the text.

14.10-6. With the entire setup operating at 290°K the measured noise figure of the amplifier circuit is $F = 3$ and the voltmeter reads 12 volts.   When the resistor R_s alone is cooled to 25°K, what will be the reading of the voltmeter?

Fig. P14.10-6

14.10-7. The entire setup is at $T = 290$°K.   Boltzmann's constant $k = 1.38 \times 10^{-23}$.   When $e = 0$V, the voltmeter reads 3 volts.   When $e = 10\mu$V rms, the voltmeter reads 5 volts.   Find the noise figure F.

Fig. P14.10-7

14.10-8. A noisy amplifier has flat gain from 0 to B Hz and zero gain above B Hz, source impedance R_s, and load impedance R_L.   When the input voltage is zero, it is found that P_o watts are dissipated in the load impedance.   When the input voltage is white noise with (one-sided) spectral density η (volts)^2/Hz, the power dissipated in the load is $2P_o$.   Find an expression for the noise figure of the amplifier.   Assume $R_s = R_L$.

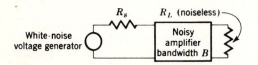

Fig. P14.10-8

14.11-1. Derive Eqs. (14.11-3) and (14.11-4).

14.11-2. Derive Eq. (14.11-5).

14.12-1. Refer to Fig. 14.12-1. Let $T_{ant} = 10°$K, T_e(maser) $= 4°$K with a gain of 20 dB, the Travelling Wave Tube (TWT) has a noise figure $F = 3$ dB and a gain $g_a = 40$ dB, and a mixer and IF amplifier with $F = 10$ dB and a gain $g_{aT} = 60$ dB. Calculate the available noise power at the receiver output.

14.12-2. Repeat Prob. 14.12-1. Now, however, consider that the available noise power at the receiver output, measured in a noise bandwidth of 10 MHz, is 1 mw. Find the value of T_{ant}.

14.14-1. Verify Eqs. (14.14-4.) and (14.14-5).

14.14-2. Verify Eq. (14.14-6).

14.14-3. Verify Eq. (14.14-7).

The Complementary Error Function

$$\text{erfc}\,(x) \equiv \frac{2}{\sqrt{\pi}} \int_x^\infty e^{-u^2}\,du$$

x†	erfc (x)	x	erfc (x)
0.0	1.000	2.2	1.86×10^{-3}
0.2	0.777	2.4	$6.9 \ \times 10^{-4}$
0.4	0.572	2.6	$2.4 \ \times 10^{-4}$
0.6	0.396	2.8	$7.9 \ \times 10^{-5}$
0.8	0.258	3.0	$2.3 \ \times 10^{-5}$
1.0	0.157	3.3	$3.2 \ \times 10^{-6}$
1.2	$8.97 \ \times 10^{-2}$	3.7	$1.7 \ \times 10^{-7}$
1.4	$4.87 \ \times 10^{-2}$	4.0	$1.5 \ \times 10^{-8}$
1.6	$2.37 \ \times 10^{-2}$	5.0	$1.5 \ \times 10^{-12}$
1.8	$1.09 \ \times 10^{-2}$		
2.0	$7.21 \ \times 10^{-3}$		

† For large values of x, erfc $(x) \simeq e^{-x^2}/(x \sqrt{\pi})$

Index

Index